Ambient Ionization Mass Spectrometry

New Developments in Mass Spectrometry

Editor-in-Chief:
Professor Simon J Gaskell, *Queen Mary University of London, UK*

Series Editors:
Professor Ron M A Heeren, *FOM Institute AMOLF, The Netherlands*
Professor Robert C Murphy, *University of Colorado Denver, USA*
Professor Mitsutoshi Setou, *Hamamatsu University School of Medicine, Japan*

Titles in the Series:
1: Quantitative Proteomics
2: Ambient Ionization Mass Spectrometry

How to obtain future titles on publication:
A standing order plan is available for this series. A standing order will bring delivery of each new volume immediately on publication.

For further information please contact:
Book Sales Department, Royal Society of Chemistry, Thomas Graham House, Science Park, Milton Road, Cambridge, CB4 0WF, UK
Telephone: +44 (0)1223 420066, Fax: +44 (0)1223 420247
Email: booksales@rsc.org
Visit our website at www.rsc.org/books

Printed and bound by CPI Group (UK) Ltd, Croydon, CR0 4YY

Ambient Ionization Mass Spectrometry

Edited by

Marek Domin
Boston College, Chestnut Hill, MA, USA
Email: marek.domin@bc.edu

and

Robert Cody
JEOL USA Inc, Peabody, MA, USA
Email: cody@jeol.com

THE QUEEN'S AWARDS
FOR ENTERPRISE:
INTERNATIONAL TRADE
2013

New Developments in Mass Spectrometry No. 2

Print ISBN: 978-1-84973-926-9
PDF eISBN: 978-1-78262-802-6
ISSN: 2044-253X

A catalogue record for this book is available from the British Library

Published by The Royal Society of Chemistry,
Thomas Graham House, Science Park, Milton Road,
Cambridge CB4 0WF, UK

Registered Charity Number 207890

For further information see our web site at www.rsc.org

Preface

A decade ago, it would have been unthinkable to hold an object in front of a mass spectrometer in open air and record mass spectra in real time. The DART was originally conceived as an atmospheric-pressure electron capture source for the analysis of volatile components with an ion-mobility spectrometer. In late 2002 and early 2003 when we pointed the newly constructed DART source toward the atmospheric pressure interface of a time-of-flight mass spectrometer, we realized that the DART was something new. We now had a way to measure mass spectra in real time for liquids, solids, and gases sampled directly.

After decades of dissolving and injecting samples into GC/MS and LC/MS systems or introducing materials into vacuum chambers, this was truly liberating! Now we could analyze almost anything instantaneously. We could not analyze *everything* directly, but the probability of getting useful information was high, and the amount of effort required to attempt the analysis was minimal. We began examining everything we could find, from foods and beverages to clothing and rubber tires.

Graham Cooks and I gave the first public presentations on DESI and DART in successive talks at the ASMS Sanibel Conference in Clearwater, FL in 2005. The topic that year was "MS in Forensic Science and Counterterrorism". This was an exciting session. Not just one, but *two* new "ambient ionization" techniques appeared almost simultaneously!

The response to the commercial introduction of the JEOL AccuTOF-DART mass spectrometer at the Pittsburgh Conference two months later was enthusiastic. The DART won the PittCon Editors' Gold Award for best new product. An R&D 100 Award followed and DART was featured on the television series "CSI: NY". Within a short time, early adopters were finding a variety of innovative uses for the new technology.

New Developments in Mass Spectrometry No. 2
Ambient Ionization Mass Spectrometry
Edited by Marek Domin and Robert Cody
© The Royal Society of Chemistry 2015
Published by the Royal Society of Chemistry, www.rsc.org

Following the introduction of DESI and DART, a large number of new ambient ionization sources have been developed. The contributions to this book represent cutting-edge developments in ion-source design and applications.

A preface is an appropriate place to comment about the terminology used in this book. The term "ambient ionization" first appeared in the original 2004 DESI publication in *Science*. The novel aspect of both DESI and DART was that samples could be analyzed directly under *laboratory-ambient* conditions; that is, in open air. Therefore, we used the same term to describe DART, the other open-air ionization source at that time.

Terminology can be confusing. Being a stickler for the correct use of terminology, David Sparkman pointed out to me that the term "ambient" means "the surrounding environment" and that the ambient condition of a sample in vacuum is *vacuum*. By that definition, any ion source could be considered "ambient"! Nevertheless, it is understood that ambient ionization sources are all atmospheric-pressure ion sources.

"Direct analysis" refers to analysis of samples in their native state with little or no sample preparation. Direct analysis does not necessarily require ambient ionization. The term could just as easily refer to a direct insertion probe sample introduced into an electron ionization (EI) source in vacuum as long as the sample was analyzed in its native state.

Despite these concerns, the general understanding is that we are referring to "ambient ionization" as occurring under *laboratory-ambient* conditions. In the broadest sense, ambient ionization has come to mean *ionization at atmospheric pressure with little or no sample preparation*.

The information that we can obtain from an ambient ionization method depends on the information content that the mass spectrometer (and/or ion mobility spectrometer) can provide. High-resolution accurate-mass data and tandem mass spectrometry are valuable tools for extracting the maximum information from data obtained without the benefit of chromatography. Although it stretches the meaning of the term "ambient", you will find that sample preparation methods and chromatography can be combined with many of the ion sources described in this book.

Ambient ionization is a rapidly growing topic within mass spectrometry. Even as this book nears completion, publications describing new ambient ionization methods have appeared in the literature. Given the "moving target" nature of an evolving technology, it is necessary to limit our goals and present a snapshot of the field at the time of publication. We have attempted to offer a wide variety of innovative approaches to instrumentation and chemical analysis presented by the leaders in the field. I have no doubt that we will continue to see exciting new developments in the years to come.

Robert B. Cody

I would like to dedicate this book to my loving wife, Michelle and daughter Hayley, who sacrificed to lot in order to allow me to pursue my dreams.

Most importantly I would like to acknowledge the support of my loving parents, my mother Wanda and late father Kazimierz Domin.

Contents

New Developments in Mass Spectrometry No. 2
Ambient Ionization Mass Spectrometry
Edited by Marek Domin and Robert Cody
© The Royal Society of Chemistry 2015
Published by the Royal Society of Chemistry, www.rsc.org

Chapter 8   Spray Desorption Collection and DESI Mechanisms   196
Andre R. Venter, Kevin A. Douglass and
Gregg Hasman, Jr.

Chapter 9   Easy Ambient Sonic-Spray Ionization   220
Carlos H. V. Fidelis and Marcos N. Eberlin

Chapter 12   Desorption Electrospray Mass Spectrometry   307
Joshua S. Wiley, Zoltan Takats, Zheng Ouyang and
R. Graham Cooks

Chapter 13   Surface Acoustic Wave Nebulization   334
Yue Huang, Scott Heron, Sung Hwan Yoon and
David R. Goodlett

Chapter 14   Laser Ablation Electrospray Ionization Mass
Spectrometry: Mechanisms, Configurations and Imaging
Applications   348
Peter Nemes and Akos Vertes

Chapter 15 Electrospray Laser Desorption Ionization Mass Spectrometry

Min-Zong Huang, Siou-Sian Jhang, Ya-Ting Chan, Sy-Chi Cheng, Chun-Nian Cheng and Jentaie Shiea

Chapter 16 Paper Spray

Jiangjiang Liu, Nicholas E. Manicke, Xiaoyu Zhou, R. Graham Cooks and Zheng Ouyang

CHAPTER 1

An Introduction to Ambient Ionization Mass Spectrometry

MARÍA EUGENIA MONGE AND FACUNDO M. FERNÁNDEZ*

School of Chemistry and Biochemistry, Georgia Institute of Technology, Atlanta, GA 30332, USA
*Email: facundo.fernandez@chemistry.gatech.edu

1.1   Introduction

The introduction of desorption electrospray ionization mass spectrometry (DESI MS) by Cooks and coworkers in 2004 brought, for the first time, widespread attention to the concept of open-air surface analysis under ambient conditions.[1] Contemporary with the disclosure of DESI, work carried in parallel by other research teams explored a similar philosophy in chemical analysis. Examples include the patent on the ion source named direct analysis in real time (DART) filed in December 2003,[2] Shiea's work on open-air laser-based ion sources,[3] and work by the Van Berkel group at Oak Ridge National Laboratory on surface sampling probes (SSPs) for direct sampling of thin-layer chromatography plates first published in 2002.[4] DESI, DART, and other ambient MS techniques enabled an exciting new perspective on ways to perform both qualitative and quantitative chemical investigations on samples not typically amenable to direct MS analysis. As a bonus, direct analysis on native surfaces could be done, in most cases, without sample preparation.

Considering the pressures on modern analytical laboratories in terms of workload, turnaround time, and cost per sample, it is not surprising, in perspective, that ambient MS would rise so quickly to the forefront of

New Developments in Mass Spectrometry No. 2
Ambient Ionization Mass Spectrometry
Edited by Marek Domin and Robert Cody
© The Royal Society of Chemistry 2015
Published by the Royal Society of Chemistry, www.rsc.org

analytical science. Our particular interest, as a research group, stemmed from our involvement in large surveys to study the quality of anti-infective medicines used to treat malaria. These surveys required that we rapidly screen large numbers (thousands) of drug tablets for the presence of falsified and other poor-quality medicines. Being able to simply hold a tablet in front of the atmospheric-pressure inlet of a high-resolution mass spectrometer while exposing it to an ionizing plasma and obtaining a pass/fail result in seconds almost seemed like magic. It goes without saying that initial experiments that our group did with DART before the technique was officially released, got us interested almost instantly. Typical chromatographic analysis for testing drug quality requires hours of sample preparation and at least tens of minutes for chromatographic analysis. The time savings with ambient MS instantly made this type of techniques a central step in our multitiered approach to detect and source falsified and other poor-quality pharmaceuticals. After more than 8 years using various ambient MS methods for falsified drug analysis, we can confidently say that this family of desorption/ionization techniques have definitely enabled unique analytical workflows that were not possible before 2004.

As skeptical scientists, we should still ask ourselves what is truly new with respect to ambient MS approaches. Will the excitement about ambient MS withstand the challenge of time? Will we see ambient MS approaches being routinely used in laboratories worldwide 20 years from now? The answers to these questions lie in the true usefulness of ambient MS. For analytical technologies to become widely adopted they have to offer capabilities that are sufficiently different from existing approaches. The key advantages of ambient MS approaches are in the *format* in which well-established ionization mechanisms are implemented to enable surface analysis. DART, for example, makes use of ionization mechanisms that predominantly follow atmospheric-pressure chemical ionization (APCI) pathways, but in an open-air format. This technique, however, has enabled experiments that are not easily performed by APCI. APCI, in its most common format, requires a liquid sample. This is not the case with DART, by which one can also readily examine solids and gases directly. Direct infusion APCI, which overcomes the chromatographic step, still requires sample dissolution and pre- and postanalysis rinsing of the tubing connecting the pump propeling the liquid into the ion source. The lack of any plumbing makes DART much more impervious to memory effects that arise from rapidly injecting samples with analytes in a widely varying concentration range. Therefore, the advantages that make DART MS attractive compared to direct infusion APCI MS are related to its minimum carryover, as all parts in contact with the sample are disposable, and its shorter analysis time, as there is no need for cleaning parts. More importantly, DART has also been shown to have advantages over APCI when compared side-by-side as ion sources in LC MS. For example, LC DART MS has shown less that 11% ionization suppression in the analysis of parabens in sewage-plant effluents in comparison to APCI, which showed ionization suppression that ranged between 20 and >90%.[5] DART MS has enabled applications such as rapid forensic screening,[6–8] rapid metabolomic profiling,[9–14] rapid bacterial typing,[15] chemical profiling of live animals to study pheromone-mediated behavior,[16] fast screening of counterfeit drugs,[17–19] low cost authentication of food products,[20] and rapid detection of warfare agents,[21,22] among others.

DESI, on the other hand, is in many ways complementary to DART (Table 1.1) and makes use of desorption mechanisms that involve a continuous solid–liquid extraction process, while capitalizing on the known ionization mechanisms of ESI. As with DART, DESI has enabled applications that are not possible by ESI. Examples include imaging of tissues in reactive mode,[23] *in vivo* imaging of secondary metabolites,[24] intraoperative lipid profiling of brain tumor tissue sections,[25] direct detection of chemical warfare agents,[26] imaging of counterfeit pharmaceuticals from developing-world countries,[27] clustering based on sample composition,[28] and imaging products of heterogeneous model prebiotic reactions on the surface of minerals,[29] to name a few examples. We have chosen the following two case studies to showcase in more detail unique advantages brought by DART and DESI, the two flagships of ambient MS approaches:

(i) The work by Musah *et al.*[30] illustrates the effectiveness of DART MS in forensic drug chemistry, demonstrated by the detection of synthetic cannabinoids in herbal blends. Detection and identification of these compounds is challenging given the wide range of active ingredients, and the variety of botanical matrices in which they are found. In addition, these substances are not part of routine drug screens, and metabolites in urine would not show positive for marijuana use. To make this application even more challenging, none of the synthetic cannabinoids trigger a positive drug test using standard immunological screening procedures, and they are particularly problematic for screening methods that rely on a library search for identification because these substances are rarely included in standard databases. Additionally, the matrix in which synthetic cannabinoids are found can be comprised of several types of plants, making their detection even more arduous. This work shows that DART MS is capable of overcoming these difficulties and identifying this type of compound with a high degree of confidence without using a database. Plant leaves doped with the AM-251 and JWH-015 synthetic cannabinoids can be analyzed by simply holding the leaves with tweezers between the ionization source and the mass-spectrometer inlet. Despite the complex mass-spectral profiles given by the plant matrices, DART MS was capable of identifying the target compounds without the need for extraction or other sample preparation steps. Ionization suppression did not significantly affect analysis since 300 µg of cannabinoid were easily detected within an excess of background matrix. Unique advantages of DART MS illustrated in this case study, and that can be extended to other applications include: (i) no need for solvents,

Table 1.1 Comparison between DART and DESI.

	DART	DESI
Detection of large M_w analytes	Requires extra heating, may induce unwanted fragmentation. Derivatization increases mass range by increasing volatility.[12,127]	Readily achievable, even for large proteins.[128]
Spectral background	Simpler. Depends on ion-source construction materials and laboratory air contaminants.	More complex. Depends on solvent type and purity.
Species observed in positive-ion mode	H^+, NH_4^+ (inducible by doping with NH_3). Suitable for low-polarity analytes.[113]	H^+, NH_4^+, Na^+, K^+ (similar to ESI). Generally suitable for highly polar analytes.
Ease of implementation of "reactive modes"	Somewhat limited. Different discharge gases (He, Ar, N_2) can be combined with volatile species to induce desired gas-phase ion chemistries.[113,129,130]	Virtually unlimited. Reagents can be added to spray to increase ionization efficiency, decrease fragmentation, or increase selectivity.[131,132]
Robustness towards ion-source geometric configuration	Simpler, standardized geometry. Suitable for robust open-access operation.	Sensitivity and spatial resolution depend on a number (5–10) of geometric and experimental variables.
Sample throughput	Very high (up to hundreds h^{-1})	Very high (up to hundreds h^{-1})
Sensitivity	Depends on analyte volatility, basicity/acidity, fluid dynamic ion transfer effects and temperature gradients within ionization region.[133,134]	Depends on variables affecting ESI response, such as ion fugacity. Additionally, ion-source geometry and spray parameters affect dynamics of splashing mechanism resulting in changes in droplet size, charge and analyte dissolution extent.
Portability/fieldability	No solvent and simple geometry can aid in field applications.	Solvent requirements and source geometry limit fieldability in standard configuration. Geometry-independent and transmission modes solve these issues.
Degree of sample preservation	Depends on gas temperature used. Sample damage increases with increasing temperature and exposure time.	Generally high. High-velocity nebulizing gas can mechanically ablate delicate samples/powders.

Table 1.1 (*Continued*)

	DART	DESI
Specificity	Given by type of discharge gas used, temperature of drift gas and detector resolution. First coupled to TOF MS, now available on most MS platforms.	High in reactive mode.[131] Specificity given by wet chemistry (extraction and reaction) between solvent solution and surface-bound analyte. First coupled to ion trap MS, now available on most MS platforms.
Ability to produce spatially resolved information	Limited. Nozzle modifications can give mm-level lateral resolution.[135] Good for obtaining average response from inhomogeneous samples. Coupling to laser desorption/ablation increases resolution.[93,136]	High. Depends on spray focusing. Typically 50–200 µm range. Spray can be defocused for obtaining averaged measurement. (large-area DESI)[137,138]

extractions, sample processing, or preparation steps; (ii) resulting spectrum produced in seconds; (iii) high-throughput analysis with no carryover between samples, and (iv) no plant matrices interference in the detection of target compounds. In contrast, comparative analysis performed by gas chromatography/mass spectrometry (GC/MS) required approximately 3 days to be completed. Analyte solvent extraction is usually undesirable given the drawbacks associated with additional sample preparation steps such as higher blanks, low recovery, and decreased sample throughput. Given that illicit drug manufacturers have demonstrated the capability to rapidly modify the components and formulations that they market, instrumentation and methodologies that can readily identify the presence of illicit substances are highly desirable.

(ii) The study by Wu *et al.*[31] is another example highlighting how ambient techniques can tackle challenging applications. This work illustrates the capability of reactive DESI to detect cholesterol in human serum and in rat brain tissues with high sensitivity and selectivity by incorporating betaine aldehyde into the spray solvent. The experiment combines a chemical derivatization *in situ* that takes place in the short timescale of the solvent extraction–ionization–detection process to efficiently detect a nonpolar compound. A rapid and selective nucleophilic addition occurs at the spot being sampled, generating a positively charged hemiacetal, which allows the detection of a low proton affinity analyte that is hardly ionized by common soft ionization techniques. The capability of quantitative analysis of free cholesterol in human serum was demonstrated using the standard addition method. Serum calibration solutions, spiked with isotopically labeled cholesterol-d_7 as internal standard, were manually spotted on a glass slide and analyzed with reactive DESI MS. Matrix interferences were mitigated due to the high selectivity of the nucleophilic addition reaction. The performance of this quantitation method was comparable to GC/MS and ESI MS, but accomplished in a shorter timescale. A detection limit of 1 ng was achieved with reactive DESI when a solution of 1 µg mL^{-1} was spotted onto the surface. In addition, using 65 ppm betaine aldehyde in acetonitrile:water: dimethylformamide (8:3:1), cholesterol in rat brain tissue was imaged under ambient conditions giving a full 2D image at a 200 µm pixel size resolution within an hour. This ambient MS technique provides unique advantages for cholesterol detection in comparison to colorimetric extraction assays, and traditional hyphenated techniques in terms of high-throughput analysis, as there is no need for sample pretreatment such as derivatization, extraction or other time-consuming steps. If the sample amount is not a critical limitation, this approach can enable successful quantitation of cholesterol in biological fluids. Similarly, when high lateral resolution is not needed, reactive DESI provides imaging capabilities for mapping low-polar compounds in biological tissues under atmospheric pressure and using a soft ionization technique with no need for matrix addition.

With so many new ionization techniques being reported since 2004, distinguishing ambient ionization techniques from more conventional atmospheric-pressure ionization techniques can help delineate the different applications that are best paired with each approach. To this purpose, we propose a set of basic characteristics that should be present in techniques to be part of the "ambient ionization/sampling" MS field. First, ambient MS techniques should be able to carry ionization in the open air. This is a critical feature when examining objects of unusual shape or size such as plants, solid phase extraction fibers/bars, tablets, fabrics, *etc.* in direct analysis applications. Direct *surface* analysis capability is another key attribute of ambient MS techniques. This is particularly useful for surface analysis of solids, avoiding many, if not all sample preparation steps typically required in MS-based chemical analysis. Ambient MS ion sources are easily swappable in most types of mass spectrometers fitted with atmospheric-pressure interfaces. No modification to the ion transfer optics or the vacuum interface are generally needed for ambient MS operation, with the exception, in some cases, of suction interfaces to reduce the gas load and prevent damage to the vacuum system. It goes without saying that ambient MS techniques should generate ions without significant fragmentations, as is the case with their atmospheric-pressure counterparts.

As in ambient MS, many newly reported ionization approaches also strive to incorporate sample-preparation steps into the ionization process or analyze samples in its native form. Examples include paperspray ionization,[32,33] extractive electrospray ionization (EESI),[34,35] and fused droplet ESI (FD-ESI).[36] Sometimes these approaches are bundled into the ambient MS field, but not being surface-analysis techniques, we argue that this may not be correct. Paperspray ionization incorporates simple chromatographic separation and/or solid-phase extraction processes so they occur simultaneously with the ionization process, allowing direct analysis of dried biofluid samples.[33] EESI and FD-ESI incorporate a continuous liquid–liquid extraction step into the ionization process, leading to increased salt tolerance than when compared with ESI. This feature is useful for the simplified extraction of trace analytes, such as melamine, from complex samples such as milk.[35] Extensions of the paperspray concept can be found in the recently described "tissue-spray",[37] and leaf-spray techniques.[38] In these cases it is possible to perform electrospray ionization of tissue components directly by wetting the sample with solvent. The sample is usually cut to have a sharp end from which to initiate the electrospray process. Paperspray, EESI/FD-ESI, tissue-spray, leaf-spray and similar techniques are best classified as direct ionization techniques more closely related to ESI than to DESI and DART.

A number of review articles and tutorials are available on the topic of ambient ionization and ambient imaging.[6,7,39–54] The classification of the various ambient MS techniques in subclasses varies in these reviews, with a certain degree of overlap. Our group[39,42,51] and others[49,50] have classified ambient MS techniques based on their intrinsic desorption/ionization mechanisms, but these divisions are sometimes debatable. This is the case when several concurrent desorption/ionization mechanisms occur. The subdivisions that we propose are as follows:

- One-step techniques where desorption occurs by solid–liquid extraction followed by ESI, APPI, sonic spray, or CI ion production mechanisms.
- One-step plasma-based techniques involving thermal or chemical sputtering neutral desorption followed by gas-phase chemical ionization.
- Two-step techniques involving thermal desorption or mechanical ablation in the first step followed by a second, separate step where secondary ionization occurs.
- Two-step techniques involving laser desorption/ablation followed by an independent secondary ionization step.
- Two-step methods involving acoustic desorption approaches.
- Multimode techniques combining two or more ambient MS techniques.
- One-of-a-kind techniques that make use of other principles for desorption or ionization that do not belong to any of the previous categories.

Table 1.2 describes the techniques that fall into the aforementioned division. It is clear from this table that not all reported techniques are truly different from each other, sometimes the differences simply being a specific

Table 1.2 List of ambient MS techniques, their abbreviations and relevant references to first reports. Techniques in italics are available commercially.

Abbreviation	Name	First Report
Group 1: Solid–liquid extraction-based		
DESI	*Desorption electrospray ionization*	1
EASI	Easy ambient sonic spray ionization	60
DAPPI	Desorption atmospheric-pressure photoionization	68
DICE	Desorption ionization by charge exchange	58
LMJ-SSP	Liquid microjunction-surface sampling probe (*flowprobe*™)	4
LESA	*Liquid-extraction surface analysis*	64
Group 2: Plasma-based		
DART	*Direct analysis in real-time*	83
FAPA	Flowing atmospheric-pressure afterglow	72
ASAP	Atmospheric solids analysis probe	82
LTP	Low-temperature plasma probe	73
DAPCI	Desorption atmospheric-pressure chemical ionization	75
DBDI	Dielectric barrier discharge ionization	74
DCBI	Desorption corona beam ionization	76
PADI	Plasma-assisted desorption ionization	77
APTDI	Atmospheric-pressure thermal desorption/ ionization	79
HAPGDI	Helium atmospheric-pressure glow-discharge ionization	80
PPAMS LTP	Plasma pencil atmospheric mass spectrometry LTP	139
Ambient MHCD	Ambient microhollow cathode discharge ionization	81
Group 3: Two-step thermal/mechanical desorption/ablation (nonlaser)		
ND-EESI	Neutral desorption extractive electrospray ionization	86
BADCI	Beta electron-assisted direct chemical ionization	87
AP-TD/SI	Atmospheric-pressure thermal desorption- secondary ionization	88
PESI	Probe electrospray ionization	89
Group 4: Two-step laser-based desorption ablation		
ELDI	Electrospray-assisted laser desorption ionization	3
MALDESI	Matrix-assisted laser desorption electrospray ionization	100
LAESI	*Laser-ablation electrospray ionization mass spectrometry*	92
LADESI	Laser-assisted desorption electrospray ionization	94
LDESI	Laser-desorption electrospray ionization	95
LEMS	Laser electrospray mass spectrometry	98
LD-APCI	Laser-desorption atmospheric-pressure chemical ionization	140

Table 1.2 (*Continued*)

Abbreviation	Name	First Report
IR-LAMICI	Infrared laser-ablation metastable-induced chemical ionization	93
PAMLDI	Plasma-assisted multiwavelength laser desorption ionization	136
LAAPPI	Laser-ablation atmospheric-pressure photoionization	104
Group 5: Acoustic desorption		
RADIO	Radio-frequency acoustic desorption and ionization	141
LIAD/ESI	Laser-induced acoustic desorption-electrospray ionization	109
LIAD/APCI	Laser-induced acoustic desorption/atmospheric-pressure chemical ionization	110
SAWN	Surface acoustic wave nebulization	112
Group 6: Multimode		
DEMI	Desorption electrospray/metastable-induced ionization	113
Group 7: Other-techniques		
REIMS	Rapid evaporative ionization mass spectrometry	114
LDI	Laser-desorption ionization	115
SwiFerr	Switched ferroelectric plasma ionizer	116
LSI	Laserspray ionization	117

set of experimental conditions (such as type of laser used, flow regime, *etc.*). Distinguishing between true innovation and the simple rebranding of already-reported techniques continues to be a challenge. For these reasons, authors are strongly discouraged to give existing techniques new names. In order to provide a general overview of the most common ambient MS approaches, a brief description of them is given here.

1.1.1 One-step Techniques where Desorption occurs by Solid–Liquid Extraction followed by ESI, APPI, Sonic Spray, or CI Ion Production Mechanisms

Two groups of techniques, and desorption atmospheric-pressure photo-ionization (DAPPI) fall in this category. The first group includes DESI and its variants such as reactive DESI, transmission mode-DESI (TM-DESI), desorption ionization by charge exchange (DICE). It can be tempting to include easy ambient sonic-spray ionization (EASI) in this group, but it must be taken into account that the ionization mechanisms in DESI and EASI are different. The second group is based on the formation of liquid micro-junctions (LMJs), and comprises LMJ-surface sampling probe (LMJ-SSP), liquid-extraction surface analysis (LESA), and nanospray desorption electrospray ionization (nano-DESI).

DESI is a one-step technique where desorption occurs by solid–liquid extraction followed by ESI-like ion-production mechanisms. Solid-phase analytes are extracted into a thin liquid film that is formed on the sample surface. Small, charged solvent microdroplets continuously impact this thin film, producing secondary droplets during the collision, which are driven upwards while transporting the extracted analyte. These droplets are suctioned into an extended atmospheric-pressure ion-transfer capillary that transports ions and droplets towards the vacuum regions. A pneumatically assisted ESI probe[55] is attached to an XYZ adjustable mount to aim the spray nozzle at a surface/sample,[1] and the coaxial high flow nebulizing gas is supplied from a pressurized cylinder (100–200 psi) to induce the formation of small, charged solvent microdroplets. The five main geometrical variables to consider when fine tuning a DESI source are the incident angle, the collection angle, the sample spot-to-MS inlet distance, the tip-to-surface height, and the MS orifice-to-surface height. As in ESI, a high-voltage connection is required to produce the primary electrically charged droplets that are emitted from the spray nozzle. When sampling is performed through a transmissive/porous material the operation mode is known as TM-DESI,[56] and if specific reagents are incorporated in the solvent system to enhance the selectivity and sensitivity for certain analytes, the technique is named reactive DESI.[57] A similar technique to reactive DESI has been named DICE,[58] which targets low-polarity analytes by utilizing nonpolar spray solvents and electrochemical oxidation at the metal spray needle/solvent interface.[59] The process proceeds via charge exchange between the charged spray solvent (commonly toluene) and the low polarity (neutrally desorbed) analyte that are not efficiently ionized with conventional DESI spray solvents. When the same experimental setup as DESI (the nebulizing gas flow and polar solvent system) is utilized, but no high voltage or heat is applied, the technique is known as EASI. This technique was initially referred to as desorption sonic spray ionization (DeSSI)[60] given the prevailing ionization mechanisms involving bipolar-charged droplets formed at atmospheric pressure that lead to an ion current dependent on the supersonic gas velocity.[61]

The second family of one-step solid–liquid extraction-based techniques is based on the formation of a liquid junction.[62] When surface sampling is performed by a semistatic liquid junction,[63] the technique is known as LMJ-SSP,[4] now commercially available under the name of flowprobe™. The LMJ probe assembly consists of two concentric tubes, with the outer (larger internal diameter) tube supplying the fresh extraction solvent to the surface, and the inner tube applying suction to pull the solution to an ESI probe for direct analysis, or for offline collection prior to a chemical separation. The variables that influence performance are the liquid junction height and the inner capillary retraction. Similarly, when the microliquid extraction from a solid surface is integrated with nanoelectrospray mass spectrometry, the technique is known as LESA, which is an adaptation of the Nanomate® robotic pipette chip-based infusion nano-ESI system.[64,65] LESA is fully automated by means of a robotic arm and uses disposable pipette tips creating LMJs with spatial resolutions of ∼1 mm, and single-use nano-ESI nozzles eliminating spot-to-spot sample carryover. Nanoelectrospray is initiated by applying the appropriate high voltage to the pipette tip and gas pressure on the liquid. The smallest scale LMJ approach to ambient surface analysis has been called nano-DESI.[66,67] This approach uses two fused-silica capillaries connected by a solvent bridge formed on the sample surface. It involves a solid–liquid extraction mechanism as part of the desorption process, and does not employ a nebulizing gas.

The remaining technique belonging to this classification group is desorption atmospheric-pressure photoionization (DAPPI),[68] which is based on a combination of thermo/chemical desorption processes and atmospheric-pressure photoionization mechanisms.[69] A heated mix of atomized gas and solvent vapor produced by a microchip nebulizer is aimed at the sample. In addition, the desorption region is exposed to ultraviolet (UV) radiation. In positive mode, ions are produced by photoionization of desorbed neutral analytes, charge-transfer reactions with solvent species or dopant molecules, and/or ion–molecule reactions involving protonated solvent/dopants.[70] In negative-ion mode similar pathways lead to ion production, with the additional electron-transfer mechanisms.

1.1.2 One-step Plasma-based Techniques Involving Thermal or Chemical Sputtering Neutral Desorption followed by Gas-phase Chemical Ionization

The techniques that fall into this category involve metastables and reactive ions. These species react with the analyte directly or indirectly through proton- and charge-transfer reactions. Optional heating can be used to enhance desorption, and samples can be placed in a glancing or transmission geometry. Plasma-based ambient MS techniques can be grouped based on the following: (a) design features for removal of plasma species from the flowing gas stream (nitrogen, helium, argon *etc.*) previous to interaction with the sample, (b) presence or absence of heating used for enhancing desorption, and type of heating employed, (c) discharge operation mode (DC, AC pulsed mode), (d) current–voltage regime chosen for operation to distinguish between glow, corona, spark, *etc.*, (e) discharge configuration (annular, point to plane), and (f) applicability to miniaturized instrumentation based on the discharge gas consumption (mL min^{-1}–L min^{-1}). Based on the number of publications, DART, flowing atmospheric-pressure afterglow (FAPA),[71,72] and low-temperature plasma (LTP)[73] are the main techniques, followed by others such as dielectric barrier discharge ionization (DBDI),[74] desorption atmospheric-pressure chemical ionization (DAPCI),[75] and desorption corona beam ionization (DCBI).[76] Other plasma-based ambient techniques with a smaller number of examples include PADI,[77] a technique that employs a radiofrequency-driven glow discharge plasma in direct contact with the sample;[78] atmospheric-pressure thermal desorption/ionization[79]

(APTDI), a technique suitable for the analysis of organic salts; helium atmospheric-pressure glow-discharge ionization[80] (HAPGDI), an early name given to FAPA; and microhollow cathode discharge (MHCD) microplasmas.[81] The "atmospheric solids analysis probe" or ASAP ion source,[82] which is commercially available, is based on the introduction of a probe directly into the plasma of a modified APCI ion source, and probably shares ionization mechanisms with standard APCI ion sources, with desorption occurring by a stream of heated gas as in DART and FAPA.

DART uses a negatively biased point-to-plane atmospheric-pressure glow discharge physically separated from the ionization region by one or several electrodes. The metastable species formed within the discharge supporting gas, typically He or N_2, generate protonated water clusters by Penning ionization of atmospheric water molecules from naturally present moisture.[83] These hydronium ion–water clusters are the reactive reagents involved in proton-transfer reactions with analyte molecules thermally desorbed by the heated gas stream.[39,83] The thermal conductivity of the gas used for desorption is a critical parameter affecting the sensitivity due to the thermal desorption step, with He being approximately one order of magnitude more conducting than N_2. In the majority of applications, the utilization of DART has mainly been focused on analytes with masses below 1 kDa. In contrast to DART, in FAPA there is no filtering of plasma species by any electrodes before interaction with the sample, leading to a higher reactivity, and is operated in the current-controlled glow-to-arc regime (~ 25 mA), whereas DART typically operates at lower currents. A third difference with DART is that FAPA achieves heating of the gas stream through Joule heating within the electrical discharge and not by an external heater. DAPCI[84] and DCBI[76] also use APCI-like ionization mechanisms, but the desorption process is based on chemical sputtering by charged solvent gaseous species formed by the plasma. In DAPCI, gas-phase solvent vapors are ionized by corona discharge ionization. DCBI uses a helium plasma sustained in the corona regime (10–40 µA discharge current under a 3 kV potential difference) and can be operated in modes that seem equivalent to temperature-ramped DART,[85] low-current FAPA and DAPCI. Thermal mechanisms and chemical sputtering mechanisms are most likely responsible for desorption in DCBI. Solvent vapors can be selectively added to the DCBI probe, achieving DAPCI-like desorption and ionization.

1.1.3   Two-step Techniques Involving Thermal Desorption or Mechanical Ablation in the First Step followed by a Second, Separate Step where Secondary Ionization Occurs

Neutral desorption extractive electrospray ionization (ND-EESI),[86] beta electron-assisted direct chemical ionization (BADCI),[87] atmospheric-pressure thermal desorption–secondary ionization (AP-TD/SI),[88] and probe electrospray ionization (PESI)[89] belong to this family of two-step desorption–ionization techniques. In ND-EESI[90] a plume of neutrals is generated from the surface of a sample by the impact of a gas stream. This plume is subsequently ionized in an ion cloud created by ESI of solvents.[34] Uncharged analytes in the sample spray beam are ionized through charge-transfer reactions taking place during collisions between neutral aerosols and ESI ions in the gas phase. In this way desorption and ionization events are separated in time and space, enhancing ionization efficiency in complex samples. Probe electrospray ionization (PESI) is a two-step technique that uses disposable acupuncture needles as solid-sampling electrospray probes.[89] When the probe picks material up from a biological sample, a very small amount of water ($\sim$ pL)[91] carried on the needle surface is sufficient to induce an electrospray when a high voltage is applied to the probe. Automation can be achieved if the needle is attached to a motorized controller to control the depth and rate of sampling. Samples with complex matrices such as biological tissues can be analyzed intact with a single needle with no clogging, and no carryover if the needle surface is clean after analysis in repeated measurements.

1.1.4 Two-step Techniques Involving Laser Desorption/ Ablation followed by an Independent Secondary Ionization Step

In this group of techniques the analyte is desorbed or ablated from a surface by an IR or UV laser with or without a matrix. The generated sample plume is subsequently merged with an electrospray droplet cloud or a plasma stream, depending on the ionization source utilized for the second step. Analytes are ionized through ESI mechanisms, or charge/proton-transfer reactions. As desorption and ionization processes are separated in space and time, samples are not directly in contact with the ionizing plume, and desorption and ionization can be optimized independently.

The most used name for IR laser-sampling/ESI ionization hybrid techniques has been laser-ablation electrospray ionization (LAESI),[92] which is commercially available. However, the first technique that coupled laser sampling to an ESI source was electrospray-assisted laser desorption ionization (ELDI),[3] which used a 337 nm nanosecond pulsed nitrogen laser. In contrast, when the IR-ablated sample is picked up in the open air by a plasma stream operated in the glow regime, the technique has been named infrared laser ablation metastable-induced chemical ionization (IR-LAMICI).[93] Additional techniques that belong to this family and share some instrumentation aspects with LAESI are: infrared laser-assisted desorption electrospray ionization (IR LADESI),[94] laser desorption electrospray ionization (LDESI),[95] laser ablation mass spectrometry (LAMS),[96] laser desorption spray postionization (LDSPI),[97] and laser electrospray mass spectrometry (LEMS).[98] The latter uses a high-intensity nonresonant femtosecond laser (laser pulse $\sim 10^{13}$ W cm^{-2}) allowing analysis of anhydrous samples.

An interesting LEMS capability is the preservation, at atmospheric pressure, of condensed-phase protein conformation upon transfer into the gas phase for capture and ionization in the electrospray plume.[99]

When the IR or UV laser excites an exogenous matrix that cocrystallizes with the analyte, and a voltage ($\sim$500 V) is applied to the stainless steel target plate, the technique is called matrix-assisted laser desorption electrospray ionization (MALDESI).[100] The parameter settings that influence performance in this group of techniques comprise the source geometry, *i.e.* the laser incidence angle (90° or 45°) and the ionization source; the laser wavelength; the duration of the laser pulse; the pulse energy; the repetition frequency, and the use (or not) of a matrix (endogenous or exogenous). IR lasers are usually tuned for 2940 nm with pulses of 5 ns duration at 2–20 Hz and pulse energy between 100 µJ and 2.5 mJ. As the IR laser resonantly couples with the O–H water stretch, endogenous water inherently present in biological samples can act as ionization matrix and facilitate desorption.[101–103] Infrared laser ablation can also be combined with photoionization, improving the analysis of compounds with different polarities in a technique named laser ablation atmospheric-pressure photoionization (LAAPPI).[104]

An additional group of two-step techniques are those that involve laser desorption/ablation in a transmission or reflection geometry to produce a sample plume that is subsequently transferred to a liquid phase, such as a droplet, for analysis.[105] The droplet-capture approach can also be replaced by a continuous-flow LMJ-SSP, operated in a noncontact, surface sampling mode, providing additional means for mass-spectrometry imaging (MSI).[106] The laser-ablated analyte can also be captured on solvent droplets for transferring to a MALDI target,[107] or can be directly captured on a slide to be used as the target for vacuum MALDI MSI, allowing additional imaging capabilities such as high spatial resolution.[108]

1.1.5 Two-step Methods Involving Acoustic Desorption Approaches

In this group of techniques the analyte is desorbed through a laser-induced acoustic wave or by a piezoelement generating an aerosol plume. The neutral plume is subsequently entrained by reactive ion species or charged solvent droplets from an external ionization source (ESI or APCI).[109,110] As desorption and ionization are two separate steps, the original sample is not in contact with the ionization plume directly. Laser-induced acoustic desorption (LIAD) is a nonresonant laser-based matrix-free desorption approach in which the sample is deposited onto a thin ($\sim$10–15 µm thickness) metal foil (*e.g.* titanium or aluminum), which is irradiated from the backside with a series of high-energy laser pulses. The acoustic waves created by the laser propagate through the metal foil, causing desorption of nonvolatile, and thermally labile compounds on the other side of the foil. In particular, photosensitive compounds can be analyzed because there is no direct sample exposure to the laser. Metal foils used as substrates should have low reflectivity at the laser wavelength used, low thermal conductivity, and high thermal expansion coefficient to minimize the amount of thermal energy that reaches the analyte.[111]

Surface acoustic wave nebulization (SAWN) is also a matrix-free method, which generates low internal energy ions from a planar piezoelectric surface using acoustic waves of 400 μm wavelength. This method produces an aerosol from the liquid sample drop that is deposited on a chip,[112] by applying a radiofrequency signal ($\sim$ 9.56 MHz) in a pulsed or continuous mode to an interdigitated transducer patterned onto a piezoelectric $LiNbO_3$ wafer. The liquid surface tension in the droplet is disrupted, atomizing the sample, with additional desolvation of the generated aerosols occurring in the mass-spectrometer inlet.

1.1.6 Multimode Techniques Combining Two or More Ambient MS Techniques

Experiments combining DESI and DART-type ionization processes can be performed with desorption electrospray/metastable-induced ionization (DEMI).[113] This ion source can be operated in three modes (plasma mode, spray mode or combined mode) affording a versatile platform for ambient MS.

1.1.7 One-of-a-kind Techniques that Make Use of Other Principles for Desorption or Ionization that do not Belong to any of the Previous Categories

Rapid evaporative ionization mass spectrometry (REIMS),[114] laser desorption ionization (LDI),[115] switched ferroelectric plasma ionizer (SwiFerr),[116] and laserspray ionization (LSI),[117] are one-of-a-kind techniques that do not belong to any of the previous categories. REIMS and LDI are of particular interest since they exemplify the contribution that ambient MS approaches may have in a surgical scenario. When a high-frequency electric current is applied to surgical blades as in REIMS,[114] or when utilizing surgical CO_2 lasers as in LDI,[115] ablated tissues produce aerosols, and charged species by the heat dissipated during the process. These species created during surgery can be removed by suction from the surgical site and transported to the mass spectrometer for analysis.

As with any rapidly moving technology field, predicting the future directions where we will see important developments in ambient MS is a risky game. Several research themes that we suggest will become increasingly important are:

- Hybrid multimode ion sources: Only a handful of examples on the combination of two ambient ion sources exist. One such example DEMI, which allows for experiments that combine DESI and DART-type ionization processes.[113] Another example is the combination of desorption ionization by charge exchange (DICE) with DESI (DICE/DESI).[118] In this approach, a tee union creates a zone where immiscible solvents can intermix before being directed to the spray needle. These solvents allow different ionization chemistries to take place. In both of these cases, the ability of obtaining additional chemical information in a single experiment from a broader spectrum of chemical species remains the key advantage.

- Robotization and full automation of ambient MS: Interfacing of ambient MS approaches to robotics and digital microfluidics is an exciting, yet unexplored area of research that could potentially enable automation and integration of many of the existing approaches into platforms that can respond to the demands of the modern analytical laboratory. Ion transfer in ambient MS techniques occurs under atmospheric-pressure conditions where there is a complex interaction between fluid dynamic forces, thermal gradients and electric fields. This results in a scenario where small changes in sample position and orientation result in large changes in sensitivity. Although this can be viewed as a disadvantage, it can also be considered an opportunity. By reproducibly placing the sample in different positions in space one could imagine that different ionization conditions could be achieved in a dynamic, continuous fashion.

- Quantitative measurements: Quantitation by ambient MS, although completely feasible from the perspective of the governing basic principles, is generally quite challenging due to the variability in the sample introduction conditions. If proper care is taken in introducing the sample reproducibly, it has been shown in numerous times that quantitation is indeed possible.[119–121] The remaining challenges involve a better basic understanding of how analytes can be differentially enriched during ablation/desorption, the extent of ion suppression that exists due to charge competition, and further insights into controlling the prevailing ionization mechanisms for a more rational development of quantitative applications.

- Nanoscale imaging: Nanoimaging by ambient MS will enable further reduction in the dimensions of the systems that we are able to study, *i.e.* from tissues to cells to subcellular compartments. Ambient MS experiments where nanoscale resolution is obtained have been recently reported.[122] In another example, a multimodal imaging platform that allows coregistering bright-field fluorescence and mass-spectral chemical images has been recently developed by coupling a laser capture microdissection instrument with transmission geometry for laser ablation to a mass spectrometer with an APCI ion source.[123]

- Machine learning and expert systems: Machine learning of ambient MS fingerprints will allow expert-independent decision making based on highly complex MS data. Combination of high-throughput ambient MS approaches with advanced classification and pattern recognition algorithms will be useful in developing automated systems used in diagnostics and food safety applications, among others.

- Inclusion of ambient MS technology in the operating room: Medical questions can be translated to chemical questions that ambient MS approaches can address to help physicians make decisions.[124] Ions and aerosols created during electrosurgery are removed by suction from the surgical site and transported to the mass spectrometer for analysis by remote sampling,[115,125,126] providing metabolic information and chemical histology. Identifying the margins of pathological tissue during operation is one example that can improve surgical decision making in real time.

References

1. Z. Takats, J. M. Wiseman, B. Gologan and R. G. Cooks, *Science*, 2004, **306**, 471.
2. R. B. Cody and J. A. Laramee, *USA Pat.*, 6949741, 2005.
3. J. Shiea, M. Z. Huang, H. J. HSu, C. Y. Lee, C. H. Yuan, I. Beech and J. Sunner, *Rapid Commun. Mass Spectrom.*, 2005, **19**, 3701.
4. G. J. Van Berkel, A. D. Sanchez and J. M. E. Quirke, *Anal. Chem.*, 2002, **74**, 6216.
5. S. Beissmann, W. Buchberger, R. Hertsens and C. W. Klampfl, *J. Chromatogr. A*, 2011, **1218**, 5180.
6. F. M. Green, T. L. Salter, P. Stokes, I. S. Gilmore and G. O'Connor, *Surf. Interface Anal.*, 2010, **42**, 347.
7. D. R. Ifa, A. U. Jackson, G. Paglia and R. G. Cooks, *Anal. Bioanal. Chem.*, 2009, **394**, 1995.
8. R. R. Steiner and R. L. Larson, *J. Forensic Sci.*, 2009, **54**, 617.
9. T. Cajka, K. Riddellova, M. Tomaniova and J. Hajslova, *Metabolomics*, 2011, **7**, 500.
10. A. D. M. Dove, J. Leisen, M. S. Zhou, J. J. Byrne, K. Lim-Hing, H. D. Webb, L. Gelbaum, M. R. Viant, J. Kubanek and F. M. Fernández, *PLoS One*, 2012, **7**, e49379.
11. H. Gu, Z. Pan, B. Xi, V. Asiago, B. Musselman and D. Raftery, *Anal. Chim. Acta*, 2011, **686**, 57.
12. C. M. Jones and F. M. Fernández, *Rapid Commun. Mass Spectrom.*, 2013, **27**, 1311.
13. S. W. Kim, H. J. Kim, J. H. Kim, Y. K. Kwon, M. S. Ahn, Y. P. Jang and J. R. Liu, *Plant Methods*, 2011, **7**, 14.
14. M. Zhou, W. Guan, L. D. Walker, R. Mezencev, B. B. Benigno, A. Gray, F. M. Fernández and J. F. McDonald, *Cancer Epidemiol., Biomarkers Prev.*, 2010, **19**, 2262.
15. S. Singh and S. K. Verma, *Anal. Lett.*, 2012, **45**, 2562.
16. J. Y. Yew, R. B. Cody and E. A. Kravitz, *Proc. Natl. Acad. Sci. U. S. A.*, 2008, **105**, 7135.
17. F. M. Fernández, R. B. Cody, M. D. Green, C. Y. Hampton, R. McGready, S. Sengaloundeth, N. J. White and P. N. Newton, *ChemMedChem*, 2006, **1**, 702.

18. W. C. Samms, Y. J. Jiang, M. D. Dixon, S. S. Houck and A. Mozayani, *J. Forensic Sci.*, 2011, **56**, 993.
19. E. S. Chernetsova, P. O. Bochkov, M. V. Ovcharov, S. S. Zhokhov and R. A. Abramovich, *Drug Test. Anal.*, 2010, **2**, 292.
20. J. Hajslova, T. Cajka and L. Vaclavik, *TrAC, Trends Anal. Chem.*, 2011, **30**, 204.
21. J. M. Nilles, T. R. Connell and H. D. Durst, *Anal. Chem.*, 2009, **81**, 6744.
22. J. M. Nilles, T. R. Connell, S. T. Stokes and H. Dupont Durst, *Propellants, Explos., Pyrotech.*, 2010, **35**, 446.
23. C. Wu, D. R. Ifa, N. E. Manicke and R. G. Cooks, *Anal. Chem.*, 2009, **81**, 7618.
24. A. L. Lane, L. Nyadong, A. S. Galhena, T. L. Shearer, E. P. Stout, R. M. Parry, M. Kwasnik, M. D. Wang, M. E. Hay, F. M. Fernández and J. Kubanek, *Proc. Natl. Acad. Sci. U. S. A.*, 2009, **106**, 7314.
25. L. S. Eberlin, I. Norton, D. Orringer, I. F. Dunn, X. H. Liu, J. L. Ide, A. K. Jarmusch, K. L. Ligon, F. A. Jolesz, A. J. Golby, S. Santagata, N. Y. R. Agar and R. G. Cooks, *Proc. Natl. Acad. Sci. U. S. A.*, 2013, **110**, 1611.
26. P. A. D'Agostino and C. L. Chenier, *Rapid Commun. Mass Spectrom.*, 2010, **24**, 1617.
27. L. Nyadong, A. M. McKenna, C. L. Hendrickson, R. P. Rodgers and A. G. Marshall, *Anal. Chem.*, 2011, **84**, 7131.
28. P. N. Newton, F. M. Fernández, A. Plancon, D. C. Mildenhall, M. D. Green, L. Ziyong, E. M. Christophel, S. Phanouvong, S. Howells, E. McIntosh, P. Laurin, N. Blum, C. Y. Hampton, K. Faure, L. Nyadong, C. W. Soong, B. Santoso, W. Zhiguang, J. Newton and K. Palmer, *PLoS Med.*, 2008, **5**, e32.
29. R. V. Bennett, H. J. Cleaves, J. M. Davis, D. A. Sokolov, T. M. Orlando, J. L. Bada and F. M. Fernández, *Anal. Chem.*, 2013, **85**, 1276.
30. R. A. Musah, M. A. Domin, M. A. Walling and J. R. Shepard, *Rapid Commun. Mass Spectrom.*, 2012, **26**, 1109.
31. C. P. Wu, D. R. Ifa, N. E. Manicke and R. G. Cooks, *Anal. Chem.*, 2009, **81**, 7618.
32. H. Wang, J. Liu, R. G. Cooks and Z. Ouyang, *Angew. Chem., Int. Ed.*, 2010, **49**, 877.
33. J. Liu, H. Wang, N. E. Manicke, J.-M. Lin, R. G. Cooks and Z. Ouyang, *Anal. Chem.*, 2010, **82**, 2463.
34. H. W. Chen, A. Venter and R. G. Cooks, *Chem. Commun.*, 2006, **19**, 2042.
35. L. Zhu, G. Gamez, H. W. Chen, K. Chingin and R. Zenobi, *Chem. Commun.*, 2009, 559.
36. D. Y. Chang, C. C. Lee and J. Shiea, *Anal. Chem.*, 2002, **74**, 2465.
37. S. L. Chan, M. Y. Wong, H. W. Tang, C. M. Che and K. M. Ng, *Rapid Commun. Mass Spectrom.*, 2011, **25**, 2837.
38. J. Liu, H. Wang, R. G. Cooks and Z. Ouyang, *Anal. Chem.*, 2011, **83**, 7608.
39. M. E. Monge, G. A. Harris, P. Dwivedi and F. M. Fernández, *Chem. Rev.*, 2013, **113**, 2269.

40. X. X. Ma, S. C. Zhang and X. R. Zhang, *TrAC, Trends Anal. Chem.*, 2012, **35**, 50.

41. M. Z. Huang, S. C. Cheng, Y. T. Cho and J. Shiea, *Anal. Chim. Acta*, 2011, **702**, 1.

42. G. A. Harris, A. S. Galhena and F. M. Fernández, *Anal. Chem.*, 2011, **83**, 4508.

43. A. L. Dill, L. S. Eberlin, D. R. Ifa and R. G. Cooks, *Chem. Commun.*, 2011, **47**, 2741.

44. D. J. Weston, *Analyst*, 2010, **135**, 661.

45. D. R. Ifa, C. Wu, Z. Ouyang and R. G. Cooks, *Analyst*, 2010, **135**, 669.

46. M.-Z. Huang, C.-H. Yuan, S.-C. Cheng, Y.-T. Cho and J. Shiea, *Annu. Rev. Anal. Chem.*, 2010, **3**, 43.

47. R. M. Alberici, R. C. Simas, G. B. Sanvido, W. Romao, P. M. Lalli, M. Benassi, I. B. S. Cunha and M. N. Eberlin, *Anal. Bioanal. Chem.*, 2010, **398**, 265.

48. H. Chen, G. Gamez and R. Zenobi, *J. Am. Soc. Mass Spectrom.*, 2009, **20**, 1947.

49. A. Venter, M. Nefliu and R. G. Cooks, *TrAC, Trends Anal. Chem.*, 2008, **27**, 284.

50. G. J. Van Berkel, S. P. Pasilis and O. Ovchinnikova, *J. Mass Spectrom.*, 2008, **43**, 1161.

51. G. A. Harris, L. Nyadong and F. M. Fernández, *Analyst*, 2008, **133**, 1297.

52. R. G. Cooks, Z. Ouyang, Z. Takats and J. M. Wiseman, *Science*, 2006, **311**, 1566.

53. Z.-P. Yao, *Mass Spectrom. Rev.*, 2012, **31**, 437.

54. C. P. Wu, A. L. Dill, L. S. Eberlin, R. G. Cooks and D. R. Ifa, *Mass Spectrom. Rev.*, 2013, **32**, 218.

55. A. P. Bruins, T. R. Covey and J. D. Henion, *Anal. Chem.*, 1987, **59**, 2642.

56. J. Chipuk and J. Brodbelt, *J. Am. Soc. Mass Spectrom.*, 2008, **19**, 1612.

57. Z.-X. Zhao, H.-Y. Wang and Y.-L. Guo, *Curr. Org. Chem.*, 2011, **12**, 3734.

58. C.-C. Chan, M. S. Bolgar, S. A. Miller and A. B. Attygalle, *J. Am. Soc. Mass Spectrom.*, 2010, **21**, 1554.

59. G. J. Van Berkel, S. A. McLuckey and G. L. Glish, *Anal. Chem.*, 1992, **64**, 1586.

60. R. Haddad, R. Sparrapan and M. N. Eberlin, *Rapid Commun. Mass Spectrom.*, 2006, **20**, 2901.

61. A. Hirabayashi, M. Sakairi and H. Koizumi, *Anal. Chem.*, 1995, **67**, 2878.

62. E. D. Lee, W. Mück, J. D. Henion and T. R. Covey, *Biol. Mass Spectrom.*, 1989, **18**, 844.

63. T. Wachs and J. Henion, *Anal. Chem.*, 2001, **73**, 632.

64. V. Kertesz and G. J. Van Berkel, *J. Mass Spectrom.*, 2010, **45**, 252.

65. P. Marshall, V. Toteu-Djomte, P. Bareille, H. Perry, G. Brown, M. Baumert and K. Biggadike, *Anal. Chem.*, 2010, **82**, 7787.

66. P. J. Roach, J. Laskin and A. Laskin, *Analyst*, 2010, **135**, 2233.

67. P. J. Roach, J. Laskin and A. Laskin, *Anal. Chem.*, 2010, **82**, 7979.

68. M. Haapala, J. Pol, V. Saarela, V. Arvola, T. Kotiaho, R. A. Ketola, S. Franssila, T. J. Kauppila and R. Kostiainen, *Anal. Chem.*, 2007, **79**, 7867.

69. D. B. Robb, T. R. Covey and A. P. Bruins, *Anal. Chem.*, 2000, **72**, 3653.

70. L. Luosujärvi, V. Arvola, M. Haapala, J. Pól, V. Saarela, S. Franssila, T. Kotiaho, R. Kostiainen and T. J. Kauppila, *Anal. Chem.*, 2008, **80**, 7460.

71. F. J. Andrade, J. T. Shelley, W. C. Wetzel, M. R. Webb, G. Gamez, S. J. Ray and G. M. Hieftje, *Anal. Chem.*, 2008, **80**, 2654.

72. F. J. Andrade, J. T. Shelley, W. C. Wetzel, M. R. Webb, G. Gamez, S. J. Ray and G. M. Hieftje, *Anal. Chem.*, 2008, **80**, 2646.

73. J. D. Harper, N. A. Charipar, C. C. Mulligan, X. Zhang, R. G. Cooks and Z. Ouyang, *Anal. Chem.*, 2008, **80**, 9097.

74. N. Na, M. Zhao, S. Zhang, C. Yang and X. Zhang, *J. Am. Soc. Mass Spectrom.*, 2007, **18**, 1859.

75. Z. Takats, I. Cotte-Rodriguez, N. Talaty, H. Chen and R. G. Cooks, *Chem. Commun.*, 2005, **15**, 1950.

76. H. Wang, W. Sun, J. Zhang, X. Yang, T. Lin and L. Ding, *Analyst*, 2010, **135**, 688.

77. L. V. Ratcliffe, F. J. M. Rutten, D. A. Barrett, T. Whitmore, D. Seymour, C. Greenwood, Y. Aranda-Gonzalvo, S. Robinson and M. McCoustra, *Anal. Chem.*, 2007, **79**, 6094.

78. J. Kratzer, Z. Mester and R. E. Sturgeon, *Spectrochim. Acta, Part B*, 2011, **66**, 594.

79. H. Chen, O. Y. Zheng and R. G. Cooks, *Angew. Chem., Int. Ed.*, 2006, **45**, 3656.

80. F. J. Andrade, W. C. Wetzel, G. C. Y. Chan, M. R. Webb, G. Gamez, S. J. Ray and G. M. Hieftje, *J. Anal. At. Spectrom.*, 2006, **21**, 1175.

81. J. M. Symonds, A. S. Galhena, F. M. Fernández and T. M. Orlando, *Anal. Chem.*, 2010, **82**, 621.

82. C. N. McEwen, R. G. McKay and B. S. Larsen, *Anal. Chem.*, 2005, **77**, 7826.

83. R. B. Cody, J. A. Laramee and H. D. Durst, *Anal. Chem.*, 2005, **77**, 2297.

84. T. Ast, D. E. Riederer, S. A. Miller, M. Morris and R. G. Cooks, *Org. Mass Spectrom.*, 1993, **28**, 1021.

85. J. M. Nilles, T. R. Connell and H. D. Durst, *Analyst*, 2010, **135**, 883.

86. H. Chen, A. Wortmann and R. Zenobi, *J. Mass Spectrom.*, 2007, **42**, 1123.

87. J. Steeb, A. S. Galhena, L. Nyadong, J. Janata and F. M. Fernández, *Chem. Commun.*, 2009, **31**, 4699.

88. F. Basile, S. F. Zhang, Y. S. Shin and B. Drolet, *Analyst*, 2010, **135**, 797.

89. K. Hiraoka, K. Nishidate, K. Mori, D. Asakawa and S. Suzuki, *Rapid Commun. Mass Spectrom.*, 2007, **21**, 3139.

90. H. Chen, S. Yang, A. Wortmann and R. Zenobi, *Angew. Chem., Int. Ed.*, 2007, **46**, 7591.

91. L. C. Chen, Z. Yu, H. Nonami, Y. Hashimoto and K. Hiraoka, *Environ. Control Biol.*, 2009, **47**, 73.

92. P. Nemes and A. Vertes, *Anal. Chem.*, 2007, **79**, 8098.

93. A. S. Galhena, G. A. Harris, L. Nyadong, K. K. Murray and F. M. Fernández, *Anal. Chem.*, 2010, **82**, 2178.
94. Y. H. Rezenom, J. Dong and K. K. Murray, *Analyst*, 2008, **133**, 226.
95. J. S. Sampson and D. C. Muddiman, *Rapid Commun. Mass Spectrom.*, 2009, **23**, 1989.
96. K. Jorabchi and L. M. Smith, *Anal. Chem.*, 2009, **81**, 9682.
97. J. Liu, B. Qiu and H. Luo, *Rapid Commun. Mass Spectrom.*, 2010, **24**, 1365.
98. J. J. Brady, E. J. Judge and R. J. Levis, *Rapid Commun. Mass Spectrom.*, 2009, **23**, 3151.
99. J. J. Brady, E. J. Judge and R. J. Levis, *Proc. Natl. Acad. Sci. U. S. A.*, 2011, **108**, 12217.
100. J. S. Sampson, A. M. Hawkridge and D. C. Muddiman, *J. Am. Soc. Mass Spectrom.*, 2006, **17**, 1712.
101. S. Berkenkamp, M. Karas and F. Hillenkamp, *Proc. Natl. Acad. Sci. U. S. A.*, 1996, **93**, 7003.
102. V. V. Laiko, N. I. Taranenko, V. D. Berkout, M. A. Yakshin, C. R. Prasad, H. S. Lee and V. M. Doroshenko, *J. Am. Soc. Mass Spectrom.*, 2002, **13**, 354.
103. P. Nemes and A. Vertes, *TrAC, Trends Anal. Chem.*, 2012, **34**, 22.
104. A. Vaikkinen, B. Shrestha, T. J. Kauppila, A. Vertes and R. Kostiainen, *Anal. Chem.*, 2012, **84**, 1630.
105. O. S. Ovchinnikova, V. Kertesz and G. J. Van Berkel, *Anal. Chem.*, 2011, **83**, 1874.
106. O. S. Ovchinnikova, V. Kertesz and G. J. Van Berkel, *Rapid Commun. Mass Spectrom.*, 2011, **25**, 3735.
107. S. G. Park and K. K. Murray, *J. Am. Soc. Mass Spectrom.*, 2011, **22**, 1352.
108. S.-G. Park and K. K. Murray, *Anal. Chem.*, 2012, **84**, 3240.
109. S.-C. Cheng, T.-L. Cheng, H.-C. Chang and J. Shiea, *Anal. Chem.*, 2009, **81**, 868.
110. J. Gao, D. J. Borton, II, B. C. Owen, Z. Jin, M. Hurt, L. M. Amundson, J. T. Madden, K. Qian and H. I. Kenttamaa, *J. Am. Soc. Mass Spectrom.*, 2011, **22**, 531.
111. R. C. Shea, C. J. Petzold, J. L. Campbell, S. Li, D. J. Aaserud and H. I. Kenttamaa, *Anal. Chem.*, 2006, **78**, 6133.
112. S. R. Heron, R. Wilson, S. A. Shaffer, D. R. Goodlett and J. M. Cooper, *Anal. Chem.*, 2010, **82**, 3985.
113. L. Nyadong, A. S. Galhena and F. M. Fernández, *Anal. Chem.*, 2009, **81**, 7788.
114. K. C. Schafer, J. Denes, K. Albrecht, T. Szaniszlo, J. Balog, R. Skoumal, M. Katona, M. Toth, L. Balogh and Z. Takats, *Angew. Chem., Int. Ed.*, 2009, **48**, 8240.
115. K. C. Schafer, T. Szaniszlo, S. Gunther, J. Balog, J. Denes, M. Keseru, B. Dezso, M. Toth, B. Spengler and Z. Takats, *Anal. Chem.*, 2011, **83**, 1632.
116. E. L. Neidholdt and J. L. Beauchamp, *Anal. Chem.*, 2011, **83**, 38.
117. E. D. Inutan and S. Trimpin, *J. Am. Soc. Mass Spectrom.*, 2010, **21**, 1260.

118. C.-C. Chan, M. Bolgar, S. Miller and A. Attygalle, *J. Am. Soc. Mass Spectrom.*, 2011, **22**, 173.
119. R. Vismeh, D. J. Waldon, Y. Teffera and Z. Y. Zhao, *Anal. Chem.*, 2012, **84**, 5439.
120. P. D'Aloise and H. Chen, *Sci. Justice*, 2012, **52**, 2.
121. D. S. Saang'onyo and D. L. Smith, *Rapid Commun. Mass Spectrom.*, 2012, **26**, 385.
122. O. S. Ovchinnikova, M. P. Nikiforov, J. A. Bradshaw, S. Jesse and G. J. Van Berkel, *Acs Nano*, 2011, **5**, 5526.
123. M. Lorenz, O. S. Ovchinnikova, V. Kertesz and G. J. Van Berkel, *Rapid Commun. Mass Spectrom.*, 2013, **27**, 1429.
124. J. K. Nicholson, E. Holmes, J. M. Kinross, A. W. Darzi, Z. Takats and J. C. Lindon, *Nature*, 2012, **491**, 384.
125. J. Balog, T. Szaniszlo, K. C. Schaefer, J. Denes, A. Lopata, L. Godorhazy, D. Szalay, L. Balogh, L. Sasi-Szabo, M. Toth and Z. Takats, *Anal. Chem.*, 2010, **82**, 7343.
126. K.-C. Schäfer, J. Balog, T. Szaniszló, D. Szalay, G. Mezey, J. Dénes, L. Bognár, M. Oertel and Z. Takáts, *Anal. Chem.*, 2011, **83**, 7729.
127. M. Zhou, J. F. McDonald and F. M. Fernández, *J. Am. Soc. Mass Spectrom.*, 2010, **21**, 68.
128. C. N. Ferguson, S. A. Benchaar, Z. Miao, J. A. Loo and H. Chen, *Anal. Chem.*, 2011, **83**, 6468.
129. J. Wright, M. Heywood, G. Thurston and P. Farnsworth, *J. Am. Soc. Mass Spectrom.*, 2013, **24**, 335.
130. H. M. Yang, D. B. Wan, F. R. Song, Z. Q. Liu and S. Y. Liu, *Anal. Chem.*, 2013, **85**, 1305.
131. I. Cotte-Rodriguez, Z. Takats, N. Talaty, H. W. Chen and R. G. Cooks, *Anal. Chem.*, 2005, 77, 6755.
132. L. Nyadong, M. D. Green, V. R. De Jesus, P. N. Newton and F. M. Fernández, *Anal. Chem.*, 2007, **79**, 2150.
133. G. A. Harris, C. E. Falcone and F. M. Fernández, *J. Am. Soc. Mass Spectrom.*, 2012, **23**, 153.
134. G. A. Harris and F. M. Fernández, *Anal. Chem.*, 2009, **81**, 322.
135. G. Morlock and E. S. Chernetsova, *Cent. Eur. J. Chem.*, 2012, **10**, 703.
136. J. Zhang, Z. Zhou, J. Yang, W. Zhang, Y. Bai and H. Liu, *Anal. Chem.*, 2012, **84**, 1496.
137. S. Soparawalla, G. A. Salazar, R. H. Perry, M. Nicholas and R. G. Cooks, *Rapid Commun. Mass Spectrom.*, 2009, **23**, 131.
138. S. Soparawalla, G. A. Salazar, E. Sokol, R. H. Perry and R. G. Cooks, *Analyst*, 2010, **135**, 1953.
139. M. J. Stein, E. Lo, D. G. Castner and B. D. Ratner, *Anal. Chem.*, 2012, **84**, 1572.
140. J. J. Coon, H. A. Steele, P. J. Laipis and W. W. Harrison, *J. Mass Spectrom.*, 2002, **37**, 1163.
141. R. B. Dixon, J. S. Sampson and D. C. Muddiman, *J. Am. Soc. Mass Spectrom.*, 2009, **20**, 597.

CHAPTER 2

Direct Analysis in Real Time (DART®)

ROBERT B. CODY* AND A. JOHN DANE

JEOL USA, Inc., 11 Dearborn Rd., Peabody, MA 01960, USA
*Email: cody@jeol.com

2.1 Introduction

The direct analysis in real time (DART®) ion source[1-4] is a form of atmospheric-pressure chemical ionization (APCI) where the initial ion formation step involves Penning ionization.[5] DART can be considered a type of chemi-ionization, which is defined as ionization that occurs as the result of chemical reactions between two neutral atoms or molecules at collision energies below the threshold energy for ionization. Reactions of long-lived electronically excited atoms or vibronically excited molecules ("meta-stables") with atmospheric gases and sample molecules ultimately result in the formation of characteristic ions from the analytes.

DART grew out of discussions between coinventors Laramée and Cody about developing an alternative to the radioactive ion sources in hand-held chemical agent monitors. The original concept was to develop an electron-capture source for ionizing volatile compounds. When the first DART spectra were observed in late 2002 and early 2003, it became apparent that DART was a much more versatile source that was capable of ionizing gases, liquid and solids in the open air under laboratory-ambient conditions. This was quite surprising, because nothing like it had ever been seen before (the initial DESI publication[6] appeared almost 2 years later). A patent was filed in April 2003, but DART was not discussed publicly until both DESI and DART were

New Developments in Mass Spectrometry No. 2
Ambient Ionization Mass Spectrometry
Edited by Marek Domin and Robert Cody
© The Royal Society of Chemistry 2015
Published by the Royal Society of Chemistry, www.rsc.org

presented in successive talks at the ASMS Sanibel Conference in Clearwater, FL in January 2005.

Exclusive rights to the DART technology were granted to IonSense LLC,[7] the commercial supplier of the DART source and related technology. The DART source was introduced by JEOL USA, Inc. at the 2005 Pittsburgh Conference on Analytical Chemistry and Applied Spectroscopy where it won the Pittsburgh Editor's Gold Award for Best New Product. The initial DART publication appeared in *Analytical Chemistry* shortly afterward[1] and JEOL USA, Inc. received an R&D 100 Award for DART that same year. The AccuTOF-DART was featured on the popular USA television series "CSI:NY" within a few months after its introduction.

2.2 The DART Ion Source

A schematic diagram of the DART ion source is shown in Figure 2.1 and photographs of DART sources are shown in Figure 2.2. The source typically consists of two chambers through which the DART gas flows. In the first region, a glow discharge is initiated when the gas (typically helium or nitrogen) is exposed to a high electric field. The commercial DART source uses a DC point-to-plane glow discharge. A potential of approximately −3500 V is applied to the needle electrode and the perforated counter-electrode is grounded. Although the commercial source applies a negative potential to the needle electrode, the choice of polarity is not critical. Various kinds of electrical discharges were considered for producing metastable atoms and molecules in the DART prototypes, but a DC glow discharge was chosen for the commercial source.

A number of factors influence the nature of the DART electrical discharge including geometry, voltage, gas flow and gas composition. Although a homebuilt DART-like device was reported to operate in the corona-to-glow discharge transition regime,[8] the commercial DART source operates with a geometry and conditions that produce a robust point-to-plane DC atmospheric-pressure glow discharge.[9]

The plasma in the glow discharge contains a variety of transient highly energetic species, including ions, electrons, and excimers. Ion–electron recombination within the glow discharge also results in the formation of

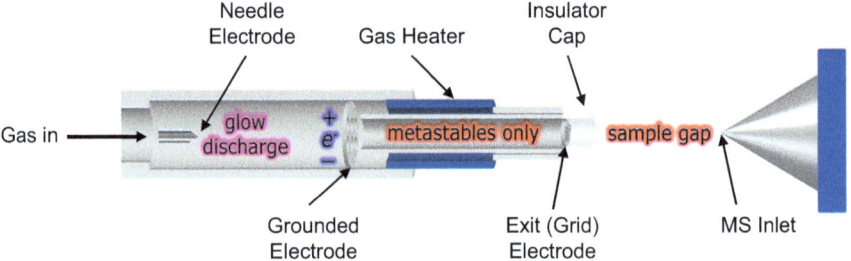

Figure 2.1   Schematic of the DART ion source.

(a)

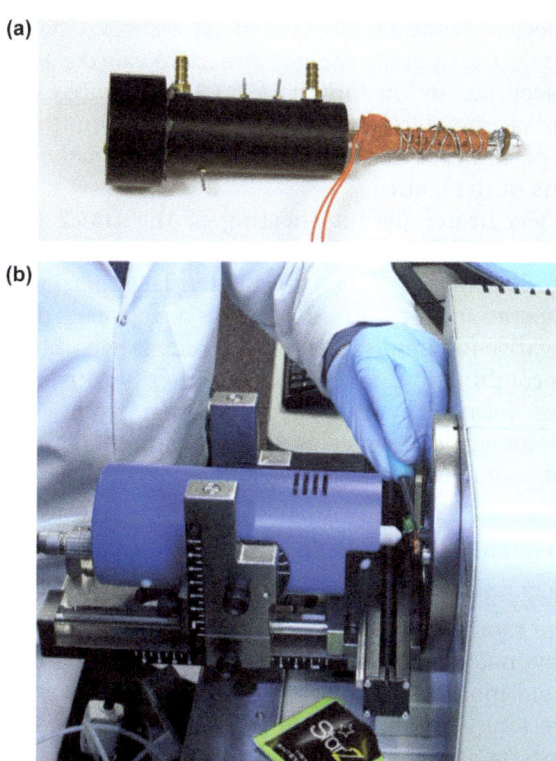

(b)

(c)

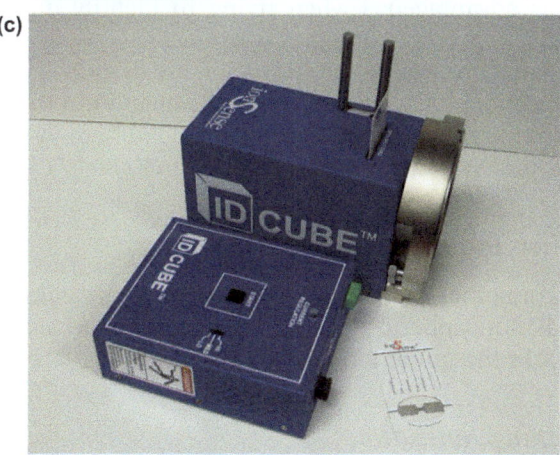

Figure 2.2 (a) Original DART prototype constructed in 2002. The gas heater was made from a GC injector liner and the exit electrode was made from aluminum foil. (b) IonSense DART-SVP source used to analyze drugs on plant material. (c) IonSense ID-Cube source. Samples are spotted onto the sample card at the center of the mesh strip. The sample is heated by applying an electric current to the mesh strip; the narrow section of mesh acts as a resistor.

excited-state species. Some of the excited-state species have long-enough lifetimes to survive transport in the gas stream to exit the DART source. An intermediate electrode in the original DART-100 source was thought to remove ions and electrons from the DART gas stream, but this electrode was determined to be unnecessary and was removed from the DART-SVP and later generations of the source.

An optional gas heater permits heating of the DART gas to facilitate desorption of less-volatile compounds from the target surface. The exit or "grid" electrode is biased to a positive potential (typically in the range of 50–500 V) for operation in positive-ion mode or to a negative potential for operation in negative-ion mode. The exit electrode is hypothesized to prevent ion–electron recombination by removing either positively or negatively charged species formed by Penning ionization from the gas stream. A ceramic insulator cap is provided as a safety feature to protect the operator from accidental contact with the exit electrode.

2.3   Ion Formation

DART is primarily a small-molecule technique. It is not suitable for the analysis of large biomolecules such as proteins. To a first approximation, DART forms protonated and/or ammoniated molecules in positive-ion mode and deprotonated molecules in negative-ion mode. Unlike spray methods, multiple charge ions and alkali metal cation adducts are not observed in DART mass spectra. Ions formed by DART have relatively low internal energies, although it does not appear to be quite as "soft" an ionization technique as electrospray ionization (ESI).[10] The relative simplicity of DART mass spectra is advantageous when trying to identify unknowns and to interpret the mass spectra of complex mixtures.

The mechanisms involved in ion formation are described in greater detail in several publications[1,11–13] and in a separate chapter in this book. For the sake of completeness, an overview will be included here. A summary of ion types observed in DART mass spectra is given in Table 2.1.

2.3.1   Positive Ions

DART differs from other forms of atmospheric-pressure chemical ionization (APCI), such as atmospheric-pressure photoionization (APPI) and corona-discharge APCI in the initial ion-formation step.

2.3.1.1   *Penning Ionization*

The initial step in DART ionization is Penning ionization, where an excited-state neutral atom or molecule E* interacts with a neutral analyte N to produce a positive ion $N^{+\bullet}$ and an electron e^-:

$$E^* + N \rightarrow N^{+\bullet} + e^- \tag{2.1}$$

Table 2.1 Typical ions observed in dart mass spectra.

Ions observed	Reaction or method	Compound properties	Examples
$[M+H]^+$	Protonation	Polar or basic, proton affinity (PA) higher than PA of water	Amines, unsaturated hydrocarbons, carbonyls and esters, many organometallics
$[M+NH_4]^+$	Ammoniation (favored with Vapur interface)	Polar	Ethers, esters, carbonyls, esters, peroxides
$[M-H]^-$	Deprotonation	Acidic	Carboxylic acids, phenols, nitroaromatic explosives
M^+	Penning ionization	Low ionization energies (IE), or IE competitive with proton affinity	Phenylenediamine antioxidants, PAHs
M^-	Electron capture	High electron affinities	Nitroaromatic explosives
Cation$^+$, Anion$^-$	Chemical sputtering or thermal desorption (?)	Salts	Organic salts, ionic liquids, organometallics
$[M+Cl]^-$, $[M+acetate]^-$, $[M+NO_3]^-$	Anion attachment	High anion affinities	High explosives (*e.g.* PETN, tetryl, HMX, RDX)
$[M+O_2]^-$	Anion attachment with supersonic expansion and rapid cooling	Polarizable, hydrogen bonding. Cannot be observed with Vapur interface installed	Large alkanes, alcohols, mono-, di- and triglycerides
Compound dependent	Derivatization	Thermally labile group	Alcohols, carbohydrates, labile phosphate or sugar groups
None or excessive fragmentation	Lack of desorption or thermal decomposition	Completely nonvolatile or too large for intact desorption	Metals, proteins, large biomolecules

For Penning ionization to occur, N must have an ionization energy that is lower than the internal energy of E*. In contrast to vacuum Penning ionization sources, such as the metastable atom bombardment (MAB) source,[14] DART ionization is carried out at atmospheric pressure and atmospheric gases play a key role in the ionization mechanism. Related techniques are Tsuchiya's liquid surface ionization technique[15] and Hiraoka's atmospheric-pressure penning ionization (APPeI) source.[16]

The majority of DART applications have used helium as the DART gas. The long-lived helium triplet state 2^3S_1 has an internal energy of 19.8 eV, and a predicted lifetime of ~ 8000 s.[17] Consequently, its internal energy is sufficiently high to ionize virtually all organic compounds. Although the role of the metastable singlet state 2^1S_0 with an internal energy of 20.6 eV cannot be ruled out, its lifetime is significantly shorter (on the order of 19 ms).[17]

Analytes with very low ionization energies can undergo direct Penning ionization with DART. Under low-humidity conditions when the DART is positioned close to the mass spectrometer sampling orifice with no additional gas-transfer interface present, Penning ionization can even be observed for saturated alkanes.[18]

2.3.1.2   Proton Transfer

The DART positive-ion background mass spectrum (Figure 2.3(a)) is dominated by protonated water and the corresponding proton-bound dimer. The most commonly cited mechanism for the formation of protonated water clusters with DART involves Penning ionization of atmospheric water by helium metastables He* followed by proton transfer and clustering:

$$He(2^3S) + H_2O \rightarrow H_2O^{+\bullet} + He(1^1S) + e^- \tag{2.2}$$

$$H_2O^{+\bullet} + H_2O \rightarrow H_3O^+ + OH^\bullet \tag{2.3}$$

$$H_3O^+ + nH_2O \rightarrow [(H_2O)_nH]^+ \tag{2.4}$$

The positive-ion DART background mass spectrum (Figure 2.3(a)) is dominated by H_3O^+ and $[(H_2O)_2 + H]^+$, but ammonium NH_4^+ is also observed. The origin of the NH_4^+ is probably environmental. It is well known that human breath contains trace amounts of ammonia.[19] Ammonia's relatively high proton affinity (853.6 kJ mol^{-1})[20] ensures that trace levels will be efficiently ionized by proton transfer. Evidence for the presence of radical formation (reaction (2.3)) in DART was reported by Curtis[13,21] using radical traps to capture the hydroxyl radical in reaction (3) and amidogen radicals formed when ammonium is present in the DART gas stream.[13,21]

Water has an ionization energy of 12.62 eV.[20] The reaction cross section for the ionization of water by the He 2^3S_1 state is extremely large, estimated to be 100 Å^2.[22] Because of the efficiency of the reaction and the fact that the DART glow discharge is carried out in an enclosed chamber isolated from

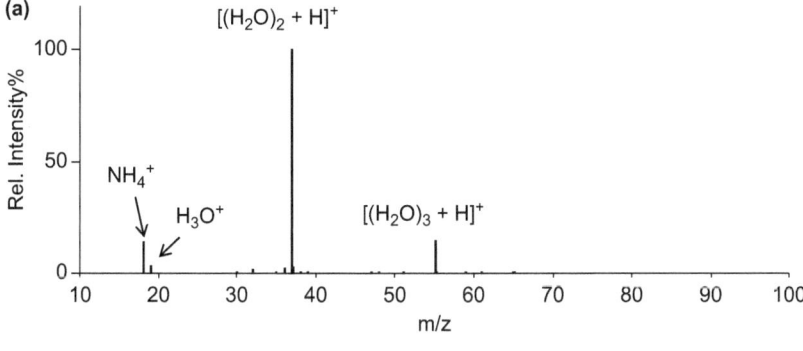

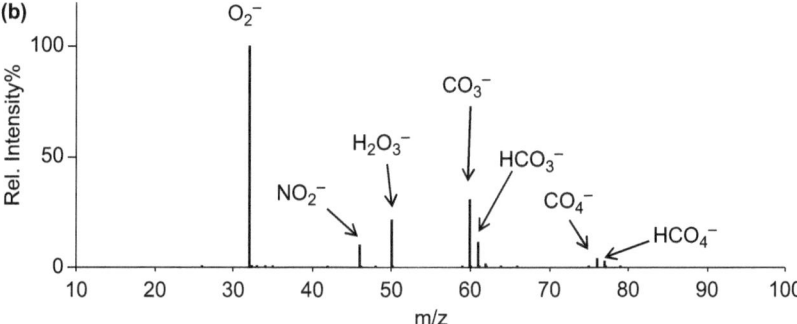

Figure 2.3 (a) Positive-ion background mass spectrum with helium DART gas. (b) Negative-ion background mass spectrum with helium DART gas.

the laboratory atmosphere, changes in ambient humidity have little effect on DART ionization. The only known exception is an increase in the observation of molecular ions $M^{+\bullet}$ under low-humidity conditions.[18]

The protonated water and water clusters formed by this series of reactions can undergo proton-transfer reactions with analytes M having proton affinities higher than that of water ($691\ kJ\ mol^{-1}$)[20] or water dimer ($808 \pm 6\ kJ\ mol^{-1}$).[23]

$$[(H_2O)_nH]^+ + M \rightarrow [M+H]^+ + nH_2O \qquad (2.5)$$

Analyte molecules can also be ionized directly by Penning ionization and then participate in proton-transfer reactions with other analyte molecules. This process is referred to as "self-CI".[24]

Corona-discharge APCI is generally recognized to proceed by a series of reactions involving $N_2^{+\bullet}$ and $N_4^{+\bullet}$.[25] The role of these species in the DART ionization mechanism cannot be ruled out. $N_2^{+\bullet}$ and $N_4^{+\bullet}$ are not detected in the DART background mass spectrum with either nitrogen or helium DART gas, even if the DART exit is positioned within less than 0.5 mm from the mass spectrometer sampling orifice (designated "orifice 1" on the JEOL

AccuTOF mass spectrometer) as described by Dzidic and coworkers.[25] Other species besides He* may play a role in the initial ion-formation step. Shelley *et al.* investigated a flowing atmospheric-pressure afterglow by optical means and speculated that helium excimer He_2* may also participate in DART ionization.[26]

$H_2O^{+\bullet}$ is not be detected when helium is use as a DART gas and the DART is positioned close to orifice 1, although protonated water clusters are readily observed. The one exception is observed when the DART exit voltage is set to its maximum value (+530 V) and an arc is observed between the DART exit and orifice 1. In that case, we can observe $H_2O^{+\bullet}$, $N_2^{+\bullet}$, and $O_2^{+\bullet}$ formed in the high-voltage discharge external to the DART source.

2.3.2   Negative Ions

The DART negative-ion background mass spectrum (Figure 2.3(b)) shows abundant $O_2^{-\bullet}$. Penning ionization (reaction (2.1)) results in the formation of a positive ion and an electron. The kinetic energy of the electron depends on the energy difference between E* and N. When the exit grid is biased to a negative potential, the dominant reactions observed result from an initial electron capture step. At atmospheric pressure, the Penning electron e^{-}* experiences rapid collisional cooling with gas molecules G (reaction (2.6)).

$$e^{-}* + G \rightarrow G* + e^{-} \tag{2.6}$$

In addition to Penning ionization of atmospheric gases, electrons can also be produced from interactions of metastables E* with the negatively biased exit grid (S) by surface Penning ionization (reaction (2.7)).

$$E* + S \rightarrow E + S + e^{-} \tag{2.7}$$

Thermalized electrons e^{-} are readily captured by atmospheric oxygen to form $O_2^{-\bullet}$ (reaction (2.8))

$$e^{-} + O_2 \rightarrow O_2^{-\bullet} \tag{2.8}$$

$O_2^{-\bullet}$ can react with the analyte by proton abstraction (reaction (2.9)), charge exchange (reaction (2.10)), or attachment (reaction (2.11)).

$$O_2^{-\bullet} + M \rightarrow [M-H]^{-} + OOH^{\bullet} \tag{2.9}$$

$$O_2^{-\bullet} + M \rightarrow M^{-\bullet} + O_2 \tag{2.10}$$

$$O_2^{-\bullet} + M \rightarrow [M+O_2]^{-\bullet}* + G \rightarrow [M+O_2]^{-\bullet} + G* \tag{2.11}$$

Proton abstraction is the most common reaction observed for compounds having an acidic proton. The use of $O_2^{-\bullet}$ attachment for the DART analysis of nonpolar compounds will be discussed later in this chapter.

Sample reactions with $O_2^{-\bullet}$ are most commonly observed in negative-ion DART, but other species observed in the negative-ion DART background mass spectrum (Figure 2.3(b)) can play a role. NO_2^- and NO_3^-, the dominant species in atmospheric-pressure corona and glow discharges, have relatively low abundances in the DART background spectrum because the DART glow discharge is carried out in an enclosed chamber, purged with DART gas and isolated from atmosphere. Analytes can undergo electron capture directly in DART, although this appears to be less common than ionization by reaction with $O_2^{-\bullet}$. Song and coworkers observed similarities between the mechanisms of negative-ion formation by DART and atmospheric-pressure photoionization (APPI).[27]

2.3.3 Matrix Effects

Pure compounds are ionized by the reaction mechanisms described in the preceding sections. Matrix effects in mixtures can enhance or diminish the analyte signal. Steiner and Larson reported that fragmentation to form $[M + H - H_2O]^+$ was dramatically reduced relative to the abundance of $[M + H]^+$ for codeine in the presence of acetaminophen.[28] Stout and Ropero-Miller[73] showed a loss of signal for trace levels of oxazepam in urine because of the presence of excess creatinine.

These observations can be explained by the transient microenvironment mechanism (TMEM) proposed by Song and coworkers.[12] In brief, the DART gas stream ionizes the matrix molecules that can either react further to ionize the sample or shield the sample from ionization. In the codeine example, acetaminophen $[M + H]^+$ transfers a proton to codeine, with a smaller difference in proton affinities resulting in less fragmentation than direct proton transfer from protonated water to codeine. Atmospheric water and oxygen may be thought of as the matrix for positive-ion DART and negative-ion DART ionization of pure compounds, respectively.

In contrast, the excess creatinine in urine shields oxazepam from ionization. The number of protonated water molecules generated by the DART source is the limiting reagent. If there are enough matrix molecules with sufficiently high proton affinities, then the analyte molecules with lower proton affinities will not be ionized. Evidence for this mechanism was found in an experiment where DART ionization of 50 nL droplets showed significantly less suppression of oxazepam in the presence of excess creatinine than 2 μL droplets.[29] A tentative hypothesis is that there are excess protonated reagent ions in the case of the 50 nL droplets, so there is no competition between matrix and analyte for the charge.

2.3.4 Desorption

Thermal desorption clearly plays a major role in DART analysis. The DART gas or the sample support must be heated to observe analytes with low volatility. However, there is some evidence that other desorption processes may be occurring. DART was used to observe self-assembled monolayers on gold[30] and covalently bound substrates on silicon nitride surfaces.[31] The results suggest a chemical sputtering mechanism. Possible evidence that DART desorption is not strictly thermal is that DART can be used to detect ionic liquids with very low vapor pressures.[32,33] Increasing the gas temperature can result in increased fragmentation,[10,18,34] so the optimum gas temperature may vary for a given analysis.

2.3.5 DART Gases

As mentioned previously, most DART applications use helium DART gas. Nitrogen is generally used to keep the DART source purged while in standby mode, but it can also be used as a DART gas. Nitrogen has a higher breakdown voltage than helium and does not heat the sample as effectively. Therefore, when using nitrogen, the DART gas-heater temperature setting must be increased to desorb samples with low volatility. However, if the sample is heated directly as in the IonSense ID Cube transmission-mode source, then the DART gas does not have to be heated and nitrogen can be effective for less-volatile analytes. The mechanisms for DART ionization with nitrogen are not well understood, although NO^+ may play a role.

Neon DART gas gives almost exactly the same results as helium. The neon $^3P_2$ state has an internal energy of 17 eV and a lifetime of 24 s.[35] This is sufficient to ionize atmospheric water in positive-ion mode and to produce electrons in negative-ion mode to ionize atmospheric oxygen.

The internal energy of the argon $^3P_2$ state is 11.55 eV and the $^3P_0$ state has an internal energy of 11.72 eV. Therefore, argon does not ionize atmospheric water and it can be used for highly selective Penning ionization of compounds with low ionization energies. Argon was used as a DART gas for the selective ionization of melamine in contaminated milk powder by choosing a multistep reaction sequence that involved Penning ionization of acetyl acetone, followed by proton transfer to pyridine and then a subsequent transfer from protonated pyridine to melamine.[36]

2.3.6 Dopants

Dopants are volatile additives that are introduced into the DART gas stream to influence ion formation. Dopants are added to the gas stream after it exits the DART source; introduction methods include liquid on a cotton swab, liquid introduction *via* syringe pump and tubing, aerosol generators[37] and headspace vapor.

Ammonium adduct formation can be enhanced by introducing headspace vapor from dilute aqueous ammonium hydroxide into the DART gas stream. Ammonium adducts $[M + NH_4]^+$ are commonly observed for polar compounds with relatively low proton affinities, such as peroxides[38–40] and esters.[41–44] Enhanced ammonium adduct formation is observed when the

DART exit grid voltage is reduced to around 50 V.[45] The IonSense Vapur™ interface (described in the following section) also enhances ammonium adduct formation because of its longer reaction path.

Anion adducts can be observed in negative-ion mode for many compounds such as the high explosives RDX, HMX, tetryl, PETN, and nitroglycerine.[1,46–48] Dichloromethane is a convenient dopant to use to form chloride adducts $[M + Cl]^-$ for these compounds.

Hydrogen/deuterium exchange studies provide structural information by determining the number of exchangeable protons. H/D exchange can be easily carried out by introducing a deuterated dopant such as D_2O into the DART gas stream. An interesting application of H/D exchange for structure confirmation by DART was the detection of melamine in contaminated pet food.[37] H/D exchange combined with collisional activation and accurate mass measurements was also used to examine ion structures for fragments of caffeine, theophylline, and theobromine.[49]

2.4 Sample Introduction

2.4.1 Interfacing DART to Mass Spectrometers

The DART was initially developed on the JEOL AccuTOF time-of-flight mass spectrometer. The accessibility, rugged design and helium pumping ability of the AccuTOF atmospheric-pressure interface make it a convenient platform for atmospheric-pressure ion-source development. The exit from the DART source is typically positioned at a distance of 1 cm or less from the mass spectrometer sampling orifice (termed "orifice 1"). Ions formed in the gap between the DART and the mass spectrometer orifice are transported into the mass spectrometer by entrainment in the gas stream. The effects of sample position and gas turbulence on ion transport in DART have been modeled[50] and visualized by Schlieren imaging.[13]

Turbulence can be reduced by depositing samples on a porous mesh and operating DART in transmission mode.[44,51,52] This provides a more reproducible signal with a concomitant improvement in quantitative response.

Many mass spectrometer pumping systems cannot handle helium. The DART can be interfaced to any mass spectrometer with an atmospheric-pressure interface (trademarked as Vapur™ by IonSense) by using a splitter and gas-transport system consisting of a gas-transfer tube and vacuum pump (Figure 2.4). The Vapur interface can be configured to mount the DART to various atmospheric-pressure interfaces and can significantly improve quantitative reproducibility and peak shape for extracted ion current profiles. The Vapur interface also provides flexibility for fitting automated sample handling devices onto different mass spectrometers. However, increasing the reaction path increases the likelihood of ion suppression. Ion–molecule reactions occurring during transport through the interface result in the loss of signal for nonpolar compounds and atmospheric reagent ions.

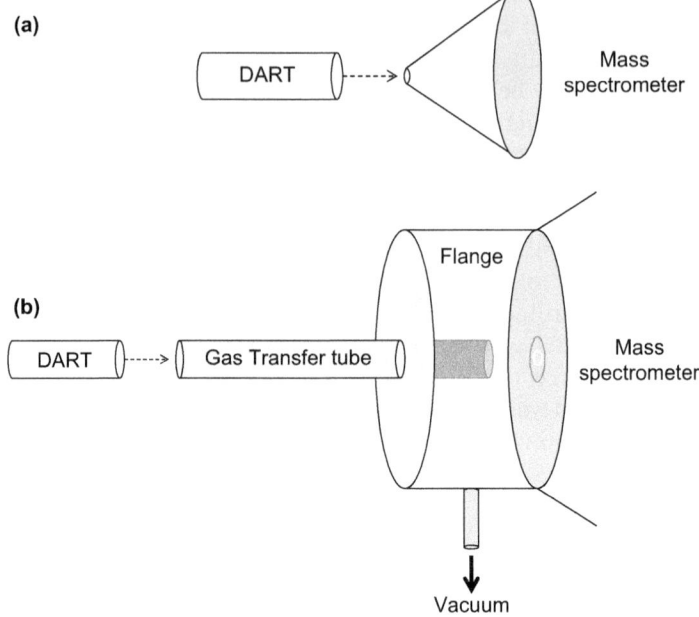

Figure 2.4   Interfacing DART to mass spectrometers (a) Direct mounting (JEOL
AccuTOF) (b) IonSense Vapur™ interface.

2.4.2   Sampling Devices

Glass melting-point tubes are convenient, inexpensive, and disposable
sampling devices. The sealed end of the tube can be used to introduce
liquids or some powdered samples into the DART gas stream. Automated
sampling with sealed glass capillaries has been carried out with a modified
commercial autosampler[53] and a 12-tube sample holder mounted on a linear
rail.[7] Samples can also be introduced with forceps[54–56] or manually. Swabs
are useful for detecting materials on surfaces.[48,52,57–61]

The open-air nature of the DART source has led to the development of
innovative sample-introduction methods. Jones and coworkers used a custom stage to distinguish writing inks on paper.[62] Grange reported a novel
autosampler constructed from an N scale model railroad track.[59]

Commercially available transmission-mode sampling devices include
multiple sample holders, a two-dimensional transmission-mode stage in a
96-well format, and a business-card sized holder that provides a sampling
mesh that can be indirectly or directly (resistively) heated.[7]

2.4.3   Sample Handling, Extraction, and Derivatization

Sample preparation methods that are rapid and convenient can greatly
increase the range of analyses that can be carried out with an ambient

ionization method. Solid-phase microextraction (SPME)[63] is effective in combination with DART for sampling volatile components and trace components in solution. SPME was used to monitor banana ripening[64] and flavor compounds in liquors.[65,66] Microextraction with a packed sorbent (MEPS) improved the detection limits for cocaine and its metabolites in urine.[67] A different approach using a transmissive thin film coated with adsorbent SPME material also improved detection limits of cocaine and methadone in urine. Stir-bar sorptive extraction (SBSE) permitted part-per-trillion detection limits for contaminants in water in two independent reports.[68,69] The SBSE approach was also applied to the detection of pesticides and spoilage indicators in wine.[70] In an alternative approach, Li reported a confined DART source for the identification of volatile components from plant materials.[71] Corns and coworkers reported a purge-and-trap method for detecting solvent vapors[72] by DART.

Other sample cleanup procedures have proven effective for DART analysis. Disposable-pipette extraction (DPX) tips were effective in reducing creatinine and reducing suppression[73] for oxazepam in urine.[74] In our laboratory, we have used liquid–liquid extraction with ethyl acetate to extract flavonoid antioxidants from wine and reduce background from sugars and other interferences. The QuEChERS procedure was used to clean up samples for the detection of pesticides,[52] fungicides[75] and fungal toxins[76] in agricultural products.

Many compound classes are difficult to detect by DART directly because they are difficult to desorb, they fragment easily, or they are difficult to ionize by atmospheric-pressure chemical ionization reactions. Derivatization offers a solution for many of the problem compound classes.

Trimethylsilylimidazole (TMSI) in pyridine was used for *in vitro* silylation of hydroxyl groups on various compounds including compounds of forensic interest,[77] cyclodextrins, sulfated aminoglycoside antibiotics, and polyphenolic antioxidants in tea.[65] *In vitro* silylation with *N*-methyl-*N*-(trimethylsilyl) trifluoroacetamide (MSTFA) increase the number of metabolites that could be detected for serum metabolite studies.[78,79] The hallucinogenic drug psilocybin was silylated *in situ* by applying MSTFA on the sealed end of a melting-point tube together with the sample and placing the tip of the tube in the heated DART gas stream. Silylation protected the labile phosphate group and permitted detection of the intact molecule.[3]

Tetramethylammonium hydroxide (TMAH) is a very effective method for thermal hydrolysis and methylation (THM)[80] with DART. The reagent is deposited on the sampling device together with the analyte and exposed to the heated DART gas stream. A DART-SVP heater temperature of 350 °C is sufficient to achieve complete permethylation of compounds with multiple hydroxyls. A good example is γ-cyclodextrin (cyclo-octaamylose), which contains 24 hydroxyl groups.

Unless other easily ionized functional groups are present, alcohols tend to cluster, dehydrate and undergo complex ion–molecule reactions leading to complex DART mass spectra. Laramée *et al.* reported that derivatization of alcohols with phenyl isocyanate to form *O*-alkyl carbamates was effective for the analysis of alcohols[81] and subsequently reported an improved protocol using modified reagents.[82] An alternative approach to the analysis of alcohols based upon O_2^- adduct formation does not require derivatization. That method is described in a subsequent section of this chapter.

2.5   Quantitative Analysis

Most people are familiar with DART as a qualitative analysis method. However, DART can also be applied to quantitative analysis.[61,67,69,70,75,76,83–121]

The ion current produced by DART for a pure standard compound is essentially proportional to the amount of material present. However, gas turbulence affects the response for an object placed in the gas stream.[13,122] The simplest solution is to add an internal standard to the sample being measured. The response of the analyte and the standard are affected by gas turbulence in the same way. The internal standard should have similar desorption and ionization characteristics as the target analytes. Homologous compounds or isotopically labeled standards are ideal. We first tested the feasibility of quantitative analysis by DART by using promazine as an internal standard to create a linear working curve for standard solutions of chlorpromazine.[123] The first validated method for qualitative analysis by DART was the determination of levels of gamma hydroxybutyrate (GHB) in urine[83,124] by using a deuterated internal standard.

The effects of gas turbulence can be reduced by using automation to improve the reproducibility of sample introduction. The IonSense Vapur™ interface and transmission-mode DART[97] described in an earlier section significantly improve quantitative reproducibility. In the absence of an internal standard, reasonably good quantitative results can be obtained by summing or averaging three replicate measurements.[92] DART was also used to assay the reactivity of the toxin ricin by monitoring the release of adenosine.[125]

As with any mass spectrometric method for quantitative analysis, method development is critical for obtaining good results. Detection limits are determined by the presence of background interferences that cannot be distinguished by mass resolving power and/or tandem mass spectrometry. Matrix effects must be evaluated for complex mixtures where sample signal suppression or enhancement may occur.

2.6   Chemometrics and Pyrolysis DART

Pattern-recognition algorithms such as principal component analysis (PCA) and linear discriminant analysis (LDA) are well suited for classification of samples based upon DART data. DART and chemometrics were first proposed for classification of bacterial fatty acid profiles[126] and then used to classify different brands and formulations of sinus medications[127] and to assess olive oil quality.[43] Later applications included the differentiation of *Piper betel* cultivars,[128] authentication of animal fats,[129] fish metabolomics,[130] beer profiling,[66,131] and investigating heat-related degradation of vegetable oils.[44] The use of chemometrics with oxygen anion attachment chemical ionization with the DART ion source will be described in a subsequent section.

An exciting application of chemometrics with DART data involves plasma metabolomics for ovarian cancer screening with very high accuracy.[78,79] Nuclear magnetic resonance (NMR) spectroscopy combined with DART mass spectrometry has shown promise for breast cancer screening.[132]

Gas-heater temperatures of 350 °C and above can be used to induce pyrolysis of samples that cannot otherwise be analyzed by DART. Pyrolysis DART was first applied to the qualitative analysis of industrial polymers,[2,3,133] to study the release of volatiles from eucalypts,[54] and to analyze printing and writing papers.[55] An enclosed pyrolysis chamber was constructed by researchers at the Ames Laboratory to permit DART and FTIR analysis of the products of stepwise pyrolysis of poplar biomass.[134]

Chemometric analysis of the compounds produced by pyrolysis is well suited for classification problems. This approach was first used to distinguish red oak and white oak[135] and later applied by the US Fish and Wildlife Forensic Laboratory to identify products from endangered species including wood from Dalbergia species[136] and agarwood.[137]

2.7 Combination with Chromatography

TLC plates can be sampled by grazing the spot with the DART gas stream, scanning the plate with an angled gas stream, or by scraping material off the plate and doing a solvent extraction. Morlock and coworkers reported extensively report on the combination of thin-layer chromatography with DART detection.[84,86,87,100,138,139] Combined TLC/DART has been applied to both qualitative and quantitative analyses in a variety of laboratories for applications ranging from natural product identification to support for organic synthesis and forensic drug identification.[91,96,140–145]

The effluent from a gas chromatograph (GC) can be readily sampled by using a heated transfer line and positioning the exit of GC column in the DART gas stream. The results closely resemble GC/MS analysis carried out with a chemical ionization source, although molecular ions and some peaks observed in electron ionization (EI) mass spectra can be observed under dry conditions if the DART source is positioned very close to the mass spectrometer sampling orifice with no Vapur interface.[18] The principal benefits of this approach are that the GC effluent does not have to be introduced into a vacuum chamber and that no reagent gas is required.

An early observation that analyte response in DART is not significantly affected by the presence of salts or buffers[2] suggested that DART could be used as an LC/MS interface. Initial results were presented at the 2005 Montreux LC/MS Symposium for flow injections of 10 ppb to 10 ppm reserpine in methanol, LC/MS analysis of opiates and of LSD and its metabolites and isomers, and flow injections in normal-phase solvents.[146]

The first published HPLC/MS results using a DART interface were reported by researchers at Johannes Kepler University who observed that no ion suppression was observed for HPLC/DART analysis, even in the presence of 120 mM phosphate buffer.[147] The method was refined and applied to environmental and waste water extracts with excellent results.[148] The salt and buffer tolerance of HPLC/DART was confirmed in a poster presentation by Cheng at the 2012 ASMS Conference.[149] Recently, DART was combined with capillary electrophoresis[150] and chiral separations with normal-phase chromatography.[108]

2.8 Nonpolar Compounds: O_2^--Adduct Chemical Ionization

Atmospheric-pressure chemical ionization reactions that play the major role in DART ionization are effective for polar, basic and acidic compounds and compounds with sites of unsaturation that will accept a proton. Saturated hydrocarbons and alcohols are problematic because of their low proton affinities. Under normal positive-ion DART conditions, alkanes produce complex mass spectra with low relative abundance characterized by hydride abstraction, oxidation and fragmentation. Trace levels of NO^+ in the DART background are the most likely reagent ions responsible for these species. Molecular ions can be formed from alkanes under low-humidity conditions[18] but it is difficult to accomplish this under routine conditions. Unsubstituted alcohols can produce complex and confusing mass spectra. Although derivatization is effective,[81,82] it is often useful to be able to detect underivatized samples.

A surprising observation is that compounds that lack strongly acidic or basic functional groups such as saturated hydrocarbons, alcohols and esters can form O_2^- adducts $[M + O_2^-]^-$ [151] by combining DART and a form of inlet ionization. Formation of O_2^- in the DART source was described in the section about negative-ion mechanisms. Samples are dissolved in hexane or a similar volatile nonpolar solvent and aspirated directly into the mass spectrometer sampling orifice where rapid expansion cools the weakly bound adducts (Figure 2.5). Volatile alcohols can be aspirated directly into the orifice without requiring a solvent. A low orifice 1 or "cone" voltage is used to minimize collisional activation. Adduct formation is believed to occur because of polarizability and/or hydrogen bonding. This is an extremely "soft" ionization method because little or no fragmentation is observed.

The formation of O_2^- adducts is only observed for large, polarizable hydrocarbons (roughly C20 or larger), but alcohols as small as methanol and as large as cholesterol readily form O_2^- adducts. Sensitivity is relatively poor, with detection limits in the high picogram to low nanogram range

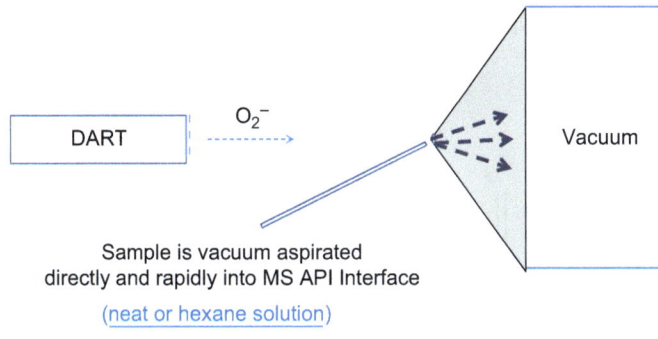

Figure 2.5 Schematic diagram of the O_2^- adduct formation experiment with an inset mass spectrum showing alkanes from Parafilm®.

under optimized conditions. Nevertheless, the method has been applied to the detection of species-specific hydrocarbons in blowfly puparial casings,[121] forensic identification of biodiesel feedstocks,[152] and typing of crude oils.[49] The information obtained by this method is complementary to the information obtained by normal DART methods. This is illustrated by comparing the positive-ion DART mass spectrum (Figure 2.6(a)) with the negative-ion O_2^- adduct mass spectrum for a synthetic motor oil (Figure 2.6(b)). The positive-ion mass spectrum is dominated by polar additives, whereas the negative-ion mass spectrum shows primarily non-polar hydrocarbons.

2.9 Applications

2.9.1 Forensics

The forensic community was an early adopter of DART technology, with some of the first commercial units being delivered to the US Army Edgewood Chemical and Biological Center and the FBI Lab. A large number of DART publications have described methodology, validation, and application of DART to paper[55] and ink analysis,[62] screening for drugs of abuse and pharmaceuticals,[1,28,46,56,57,67,73,74,102,143,153–166] clandestine laboratories,[167] counterfeit drugs,[28,40,157,168–175] sexual assault investigations,[176] explosives and arson accelerants,[1,38,46,48,177–183] chemical warfare agents,[1,46,81,92,184–188] illegal importation of endangered species,[136,137,189] and trace evidence.[190]

The forensic analysis of licit and illicit drugs by DART is now well established. Compounds are identified by combining information about elemental compositions from exact masses and isotopic abundances with fragment-ion mass spectra obtained by collisional activation. An example of the synthetic cannabinoid AM-2201 is shown in Figure 2.7; the in-source collision energy is rapidly switched under computer control to measure the

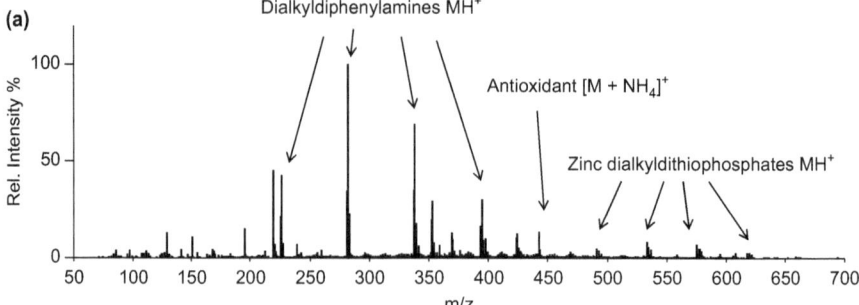

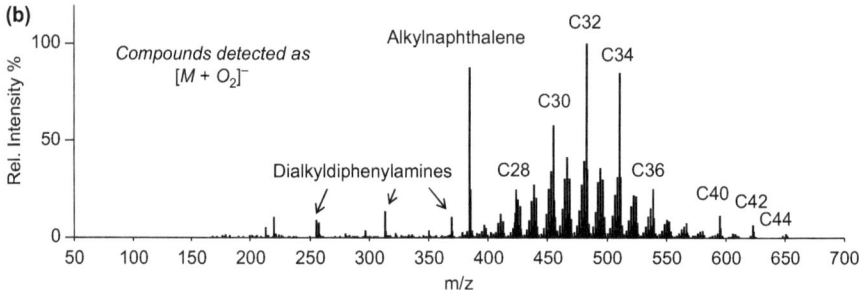

Figure 2.6   Synthetic motor oil. (a) Positive-ion DART mass spectrum of neat oil
sampled on a glass rod. (b) Mass spectrum of a hexane solution of the oil
aspirated into the mass spectrometer atmospheric-pressure interface
with DART and mass spectrometer operated in negative-ion mode to
produce O_2^- adducts.

protonated molecule (Figure 2.7(a)) and fragments (Figure 2.7(b)) in a single
acquisition. DART analysis has been validated for drug screening,[28,160] and
for confirmation of alprazolam.[157,191] The state of Virginia has approved the
use of DART for confirmation of pharmaceuticals by combining the DART
data with tablet markings.[192] A searchable database compiled by the Virginia
Department of Forensic Science consisting of 3217 positive ion spectra of
828 substances measured on the JEOL AccuTOF-DART mass spectrometer is
available from NIST[193] and ForensicDB.org also contains a set of DART mass
spectra in electronic form.[194]

   The appearance of a large number of designer drugs being sold as "bath
salts" or "herbal incense" has created problems for analysts and legislators
alike. DART has proven to be a powerful tool for the detection of new
synthetic cannabinoids, cathonines and other threats. Drugs can be detected
directly in or on plant material held with forceps in the DART gas stream. In
addition to rapid screening for known designer drugs,[56,115,195,196] the exact
mass and isotopic information for protonated molecules and fragment ions
obtained from DART measurements provides structural information that
can be used to identify new and unknown drugs.

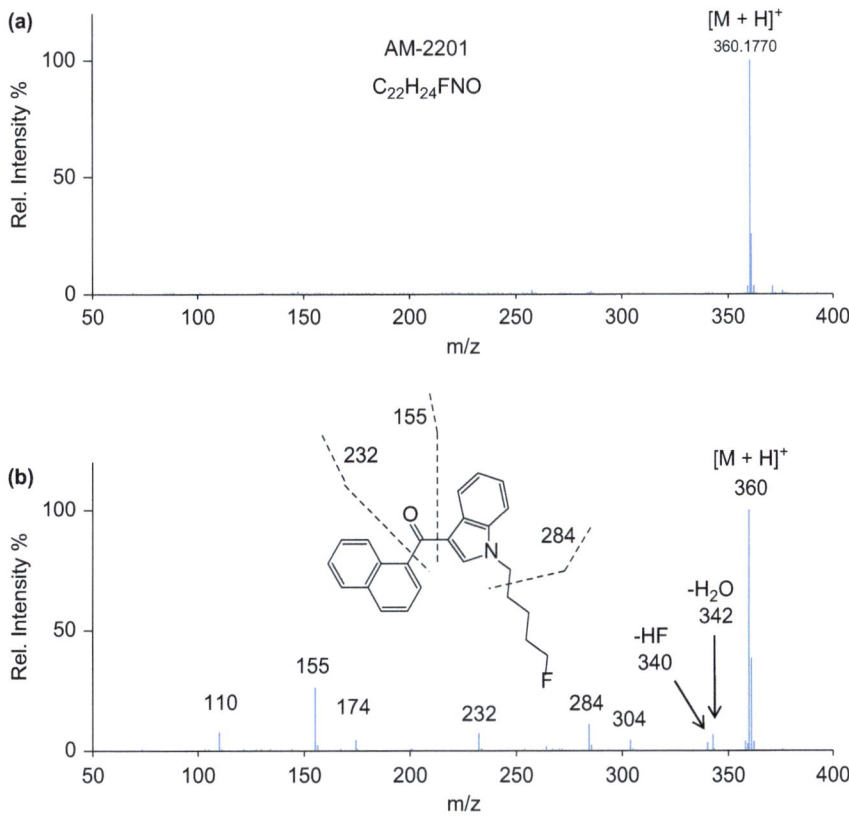

Figure 2.7 Positive-ion DART mass spectra of the synthetic cannabinoid AM-2201 measured with an in-source CID voltage of (a) 20 V and (b) 90 V. (a) shows only the protonated molecule, whereas (b) shows fragmentation related to structure. Although exact masses were measured, only nominal masses are shown in (b) for readability.

2.9.2 Food and Beverage

An important area of DART application is the characterization of foods and beverages for both quality control and safety. A number of early application notes from JEOL dealt with food analysis.[41,197–206] The Hajslova research group at the Institute of Chemical Technology in Prague has been particularly prolific in this area[43,44,66,75,76,98,104,109,121,131,207–221] with publications on such diverse topics as characterization of vegetable oils,[43,44,121] beer,[66,131,210,212] detection of fungal toxins in wheat grains[75,76,210,212,216] and caffeine content in soft drinks.[208] Chernetsova and Morlock[110] applied DART to the quantitative analysis of 5-hydroxymethylfurfural in honey as an indicator of overheating, age, or poor storage. DART was used to support a fluorescence/planar chromatography method for detecting trace levels of acrylamide in water[87] Herbal and "health food" formulations have been analyzed for the purpose of standardization and detection of adulterants by DART.[88,96,112–114,141,156,158,222–230]

An example of quantitative analysis of caffeine in coffee is shown in Figure 2.8. A coffee vendor claimed that his coffee beans were lower in caffeine than competing brands. Standard dilutions of caffeine spiked with theophylline internal standard were used to create a quantitative

(a)

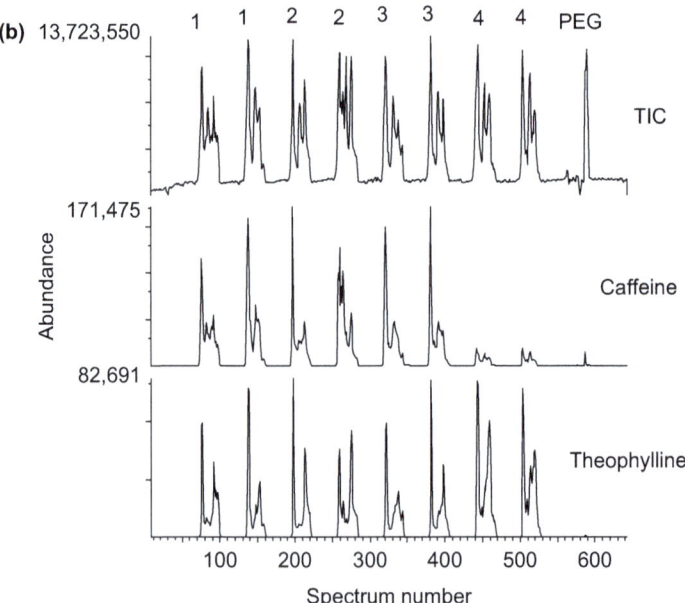

Figure 2.8   Determination of caffeine levels on coffee. (a) Coffee samples with added internal standard are spotted onto the transmission sampling module. (b) Response for protonated caffeine for two replicates each of samples 1 and 2 (from the vendor in question), sample 3 (a similar varietal from a high-end grocery store) and sample 4 (decaffeinated coffee). Polyethylene glycol (PEG 600) was measured as a mass reference standard. (c) Corresponding response for the protonated theophylline internal standard.

Table 2.2 Quantitation of caffeine in coffee.

Coffee[a]	Caffeine ($\mu g\ ml^{-1}$)[b]	Caffeine per cup (mg)[c]
1. Ethiopian	37	87.7
2. Sumatra dark A	34.6	82.0
3. Sumatra dark B	38.3	90.8
4. Decaffeinated	1.77	4.2

[a]Samples 1 and 2 are from the vendor claiming low caffeine levels. Samples 3 and 4 are a high-end grocery store brand.
[b]Coffee was diluted 1/10 with high-purity water before DART measurement.
[c]These values are obtained by correcting for dilution and taking into account that the volume of a standard cup is 237 ml.

working curve. Four identically prepared coffee samples were spiked with internal standard and measured by DART using the transmission sampling module (Figure 2.8(a)). The response for samples 1–4 is shown in Figure 2.8(b). There was no significant difference between the samples from the vendor in question and a standard coffee from a high-end grocery store. However, the decaffeinated coffee showed low levels of caffeine, as expected (Table 2.2).

2.9.3 Environmental and Consumer Products

Grange developed an integrated wipe/sample transport/autosampler method for collecting field samples to map the accidental or deliberate dispersion of toxins in the environment.[59,60,94,231] He also reported the use of DART for semiquantitative analysis of contaminants in soils[61] and for certifying decontamination of clandestine laboratories.[57] Detection of trace environmental contaminants in water was reported by Loftin *et al.*[68,232] and Haunschmidt *et al.*[69,103] DART has also been used to detect regulated phthalate plasticizers in toys[233] and additives in packaging.[93,234] A study by Maleknia *et al.*[54] examined the temperature dependence for the release of environmentally significant volatile organic compounds from eucalypts.

2.9.4 Art Conservation

Adams at the US Library of Congress reported on the use of DART to characterize printing and writing papers.[55] Langlois and coworkers at CNRS (France) used DART to analyze resins, waxes, and lipids in art materials.[235] Armitage and coworkers at Eastern Michigan University applied DART to the characterization of heme as a biomarker for blood in an encrustation on an African mask[236] and for the determination of dyes in fibers.[237]

2.9.5 Natural Products

Many active ingredients and components in herbal and natural products can be analyzed directly by DART.[54–56,71,91,96,112,114–116,128,135–137,141,154–156,158,189,196,218,226,228–230,238–248]

A novel application of DART is to the detection of unstable and reactive compounds in *Allium* species and related plants.[242–244,249–251] The intermediate 2-propenesulfinic acid has long been postulated to form when garlic (*Allium sativum*) is crushed but had never been detected because of its short lifetime and high reactivity. The first experimental evidence for the existence of this intermediate was the detection of a transient peak with an exact mass corresponding to the elemental composition $C_3H_5SO^-$ in the negative-ion DART mass spectrum of garlic crushed at room temperature. This species was found to have a half-life of less than one second.[242,243,251]

2.9.6   Pharmaceutical and Chemical Synthesis

One of the benefits of ambient ionization mass spectrometry is the capability to rapidly characterize synthetic products and monitor chemical reactions. A note in the Spring 2008 Newsletter from the Chemistry Department of the University of Tennessee Knoxville reported that DART reduced the time it took to characterize the intermediates for a 5-step synthesis from a month and a half to just four days.[252]

DART reaction monitoring in drug discovery was first reported by Gomez *et al.*[253] and Petucci *et al.*[254] DART reaction monitoring for acetylation of 1,6-hexanediol was described in 2007 in a JEOL application note.[255] Smith and coworkers reported detection of compounds related to organic synthesis analyzed directly from TLC plates with exact mass measurements.[140] Applications to other types of chemical reactions have appeared in the literature.[44,105,149] In addition, DART was found to be useful for the analysis of organometallic compounds.[256–259]

DART can be applied to the detection of counterfeit drugs, a growing problem for the pharmaceutical industry. Counterfeit drugs can be recognized by the presence or absence of active ingredients, excipients, and impurities. The DART mass spectra for genuine and counterfeit Cialis tablets are shown in Figure 2.9(a) and (b), respectively. Instead of the expected active ingredient (tadalafil), the counterfeit tablet was found to contain sildenafil, the active ingredient in Viagra®.

2.9.7   Surface and Material

DART can detect additives directly in polymers and adhesives[260–263] and lubricants.[264] If the DART gas-heater temperature is set to approximately 350 °C or higher, pyrolysis can occur, permitting the analyst to identify polymers and industrial materials from their "fingerprint" mass spectra.[133,232,265,266] DART can also analyze ionic liquids.[32,33]

Evidence that DART does not rely solely on thermal desorption is found in the report by Nesnas and coworkers of the detection of self-assembled monolayers.[30] Manova and coworkers reported that DART can detect organic monolayers covalently bound to a silicon nitride surface.[267]

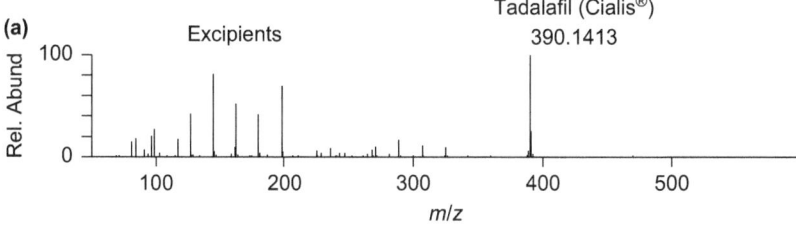

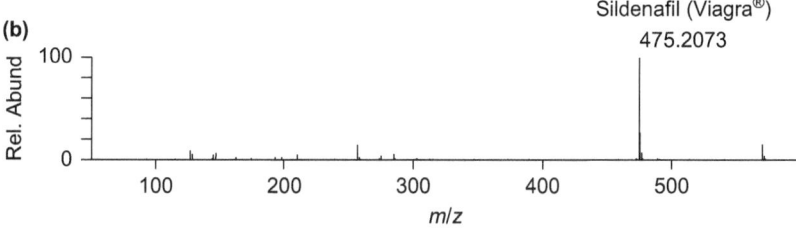

Figure 2.9 (a) Positive-ion DART mass spectra of (a) a genuine Cialis® tablet, showing the active ingredient (tadalafil) and excipients and (b) a counterfeit Cialis® tablet, showing the presence of sildenafil, the active ingredient in Viagra®.

DART is generally not useful for the analysis of metals and other purely inorganic compounds. Only a few exceptions have been reported. Thermal ionization with a handheld butane torch was used to detect lead in solder and metals in a pipette plunger.[268] Although the measurement was carried out with a DART source installed, this was not really DART ionization. Ligand-exchange reactions with ammonium acetyl acetonate were used to volatilize a surprisingly large number of elements, including many elements that are not known to form stable complexes with acetyl acetone.[269] This approach was strictly qualitative, and it was not particularly sensitive. A more promising approach to ambient ionization of metals and inorganics is the report by He and coworkers on mass spectrometry of solid samples in open air by laser ionization and laser ionization combined with DART.[270]

2.10 Conclusion

DART can be applied to the analysis of small molecules with a pyrolysis mode available for the characterization and classification of some large molecules including polymers and natural products. Although no single analytical method is universal, DART has found application in a wide range of analytical applications ranging from simple target compound identification (*e.g.* drugs and explosives) to the analysis of more complex mixtures such as natural products, formulations and materials. The open-source concept has led to the development of innovative sampling schemes. DART will undoubtedly continue to play a role in the development of new analytical techniques in coming years, either as a standalone ion source or in combination with other ambient ionization methods.

References

1. R. B. Cody, J. A. Laramée and H. D. Durst, *Anal. Chem.*, 2005, 77, 2297–2302.
2. J. A. Laramée and R. B. Cody, in *The Encyclopedia of Mass Spectrometry. Volume 6: Ionization Methods*, ed. M. L. Gross and R. M. Caprioli, Elsevier, Amsterdam, 2007, vol. 6.
3. R. B. Cody and A. John Dane, *Direct Analysis in Real-Time Ion Source*, John Wiley & Sons, Ltd, 2006.
4. J. Gross, *Anal. Bioanal. Chem.*, 2014, **406**(1), 63–80.
5. F. M. Penning, *Naturwissenschaften*, 1927, **15**, 818.
6. Z. Takats, J. M. Wiseman, B. Gologan and R. G. Cooks, *Science*, 2004, **306**, 471–473.
7. Ionsense LLC (Saugus MA), Saugus, MA.
8. J. Shelley, J. Wiley, G. Y. Chan, G. Schilling, S. Ray and G. Hieftje, *J. Am. Soc. Mass Spectrom.*, 2009, **20**, 837–844.
9. R. B. Cody, in *Pittsburgh Conference on Analytical Chemistry and Applied Spectroscopy*, Chicago, IL, 2009.
10. G. A. Harris, D. M. Hostetler, C. Y. Hampton and F. M. Fernández, *J. Am. Soc. Mass Spectrom.*, 2010, **21**, 855–863.
11. L. Song, A. B. Dykstra, H. Yao and J. E. Bartmess, *J. Am. Soc. Mass Spectrom.*, 2009, **20**(1), 42–50.
12. L. Song, S. C. Gibson, D. Bhandari, K. D. Cook and J. E. Bartmess, *Anal. Chem.*, 2009, **81**, 10080–10088.
13. M. Curtis, PhD thesis, University of the Pacific, 2012.
14. C. K. Fagerquist, M. K. Hellerstein, D. Faubert and M. J. Bertrand, *J. Am. Soc. Mass Spectrom.*, 2001, **12**, 754–761.
15. M. Tsuchiya, *Mass Spectrom. Rev.*, 1998, **17**, 51–69.
16. K. Hiraoka, S. Fujimaki, S. Kambara, H. Furuya and S. Okazaki, *Rapid Commun. Mass Spectrom.*, 2004, **18**, 2323–2330.
17. K. Baldwin, *Contemp. Phys.*, 2005, **46**, 105–120.
18. R. B. Cody, *Anal. Chem.*, 2009, **81**, 1101–1107.
19. T. Hibbard and A. J. Killard, *J. Breath Res.*, 2011, **5**, 037101.
20. NIST Chemistry Web Book. http://webbook.nist.gov/chemistry/.
21. M. E. Curtis, P. R. Jones and O. D. Sparkman, in *International Mass Spectrometry Conference*, Bremen, Germany, 2009.
22. H. C. Mastwijk, PhD Thesis, University of Utrecht, Netherlands, 1997.
23. D. J. Goebbert and P. G. Wentold, *Eur. J. Mass Spectrom.*, 2004, **10**, 837–846.
24. S. Ghaderi, P. S. Kulkarni, E. B. J. Ledford, C. L. Wilkins and M. L. Gross, *Anal. Chem.*, 1981, **53**, 428–437.
25. I. Dzidic, D. I. Carroll, R. N. Stillwell and E. C. Horning, *Anal. Chem.*, 1976, **48**, 1763–1768.

26. J. Shelley, G. Y. Chan and G. Hieftje, *J. Am. Soc. Mass Spectrom.*, 2012, **23**, 407–417.
27. L. Song, A. B. Dykstra, H. Yao and J. E. Bartmess, *J. Am. Soc. Mass Spectrom.*, 2009, **20**, 42–50.
28. R. R. Steiner and R. L. Larson, *J. Forensic Sci.*, 2009, **54**, 617–622.
29. R. B. Cody, A. J. Dane, A. D. Sauter Jr. and A. D. Sauter III, in *58th ASMS Annual Conference on Mass Spectrometry and Allied Topics*, Salt Lake City, UT, 2010.
30. K. Kpegba, T. Spadaro, R. B. Cody, N. Nesnas and J. A. Olson, *Anal. Chem.*, 2007, **79**, 5479–5483.
31. L. M. Sanchez, M. E. Curtis, B. E. Bracamonte, K. L. Kurita, G. Navarro, O. D. Sparkman and R. G. Linington, *Org. Lett.*, 2011, **13**, 3770–3773.
32. Mass Spectrometry Application Group, *MS Tips*, JEOL Ltd., 2008, vol. D031.
33. D. Smith, M. Mazzotta, R. Pace and S. Morton III, Proceedings of the 60th Annual Conference on Mass Spectrometry and Allied Topics, Vancouver, Canada, 2012.
34. C. Lapthorn and F. Pullen, *Eur. J. Mass Spectrom.*, 2009, **15**, 587–593.
35. J. G. C. Tempelaars, PhD thesis, Eindhoven University of Technology, 2001.
36. A. J. Dane and R. B. Cody, *Analyst*, 2009, **135**, 696–699.
37. T. M. Vail, P. R. Jones and O. D. Sparkman, *J. Anal. Toxicol.*, 2007, **31**, 304–312.
38. JEOL USA Inc, in *Detection of the Peroxide Explosives TATP and HMTD*, JEOL Application Note, 2007.
39. J. A. Laramée, H. D. Durst, T. R. Connell and J. M. Nilles, *Am. Lab.*, 2009, **2**(2), 1–5.
40. F. M. Fernández, R. B. Cody, M. D. Green, C. Y. Hampton, R. McGready, S. Sengaloundeth, N. J. White and P. N. Newton, *ChemMedChem*, 2006, **1**, 702–705.
41. JEOL USA Inc, in *"No-prep" Analysis of Lipids in Cooking Oils and Detection of Adulterated Olive Oil,* JEOL Application Note MS-0510A, 2005.
42. K. P. Madhusudanan, 12th ISMAS Symposium cum Workshop on Mass Spectrometry Cidade de Goa, Dona Paula, Goa, India, 2007.
43. L. Vaclavik, T. Cajka, V. Hrbek and J. Hajslova, *Anal. Chim. Acta*, 2009, **645**, 56–63.
44. L. Vaclavik, B. Belkova, Z. Reblova, K. Riddellova and J. Hajslova, *Food Chem.*, 2013, **138**, 2312–2320.
45. JEOL USA Inc., in *Rapid Detection of Melamine in Dry Milk Using Accu-TOF-DART*, Application Note, 2009.
46. R. B. Cody and J. A. Laramée, in *17th ASMS Sanibel Conference on Mass Spectrometry: MS in Forensic Science and Counterterrorism. 2005:* Clearwater, FL., 2005.
47. J. M. Nilles, T. R. Connell, S. T. Stokes and H. D. Durst, *Propellants, Explos., Pyrotech.*, 2010, **35**(5), 446–451.

48. C. Hubert, X. Machuron-Mandard and J.-C. Tabet, *Spectroscopy*, 2013, Special Issue: Current Trends in Mass Spectrometry, 8–13.

49. R. B. Cody, *Mass Spectrom.*, 2013, **2**, S0007.

50. G. A. Harris and F. M. Fernández, *Analytical Chemistry*, 2009, **81**(1), 322–329.

51. J. J. Pérez, G. A. Harris, J. E. Chipuk, J. S. Brodbelt, M. D. Green, C. Y. Hampton and F. M. Fernández, *Analyst*, 2010, **135**, 712–719.

52. S. E. Edison, L. A. Lin, B. M. Gamble, J. Wong and K. Zhang, *Rapid Commun. Mass Spectrom.*, 2011, **25**, 127–139.

53. Leap technologies Inc.

54. S. D. Maleknia, T. M. Vail, R. B. Cody, D. O. Sparkman, T. L. Bell and M. A. Adams, *Rapid Commun. Mass Spectrom.*, 2009, **23**, 2241–2246.

55. J. Adams, *Int. J. Mass Spectrom.*, 2011, **301**, 109–126.

56. R. A. Musah, M. A. Domin, M. A. Walling and J. R. E. Shepard, *Rapid Commun. Mass Spectrom.*, 2012, **26**, 1109–1114.

57. A. H. Grange and G. W. Sovocool, *Rapid Commun. Mass Spectrom.*, 2011, **25**, 1271–1281.

58. C. M. Coates, S. Coticone, P. D. Barreto, A. E. Cobb, R. B. Cody and J. Barreto, *J. Forensic Ident.*, 2008, **58**(6), 624–631.

59. A. H. Grange, *Environ. Forensics*, 2008, **9**, 127–136.

60. A. H. Grange, *Environ. Forensics*, 2008, **9**, 137–143.

61. A. H. Grange, *Rapid Commun. Mass Spectrom.*, 2013, **27**, 305–318.

62. R. W. Jones, R. B. Cody and J. F. McClelland, *J. Forensic Sci.*, 2006, **51**, 915–918.

63. J. Pawliszyn, *Solid Phase Microextraction: Theory and Practice*, Wiley-VCH, New York, 1997.

64. A. J. Dane and R. B. Cody, *LC/GC: The Application Notebook*, 2009, http://www.chromatographyonline.com/lcgc/Articles/Using-Solid-Phase-Microextraction-with-AccuTOF-DAR/ArticleStandard/Article/detail/581399.

65. R. B. Cody, 55th Annual Conference on Mass Spectrometry and Allied Topics, Indianapolis, IN, 2007.

66. T. Cajka, K. Riddellova, M. Tomaniova and J. Hajslova, *J. Chromatogr. A*, 2010, **1217**, 4195–4203.

67. E. Jagerdeo and M. Abdel-Rehim, *J. Am. Soc. Mass Spectrom.*, 2009, **20**, 891–899.

68. K. B. Loftin, T. P. Griffin, C. A. Clausen III, R. B. Cody and A. J. Dane, in *57th ASMS Conference on Mass Spectrometry and Allied Topics*, Philadelphia, PA, 2009.

69. M. Haunschmidt, C. Klampfl, W. Buchberger and R. Hertsens, *Anal. Bioanal. Chem.*, 2010, **397**(1), 269–275.

70. E. Crawford, P. Domizio, B. J. Musselman, C. M. Lucy, L. F. Bisson, B. C. Weimer and R. Jeannotte, 61st Annual Conference on Mass Spectrometry and Allied Topics, Minneapolis, MN, 2013.

71. Y. Li, *Rapid Commun. Mass Spectrom.*, 2012, **26**, 1194–1202.

72. H. Corns, C. Hornwood and G. Risbridger, in *5th European Academy of Forensic Science Conference*, Glasgow, Scotland UK, 2009.
73. P. R. Stout and J. D. Ropero-Miller, in *NCJRS Abstract*, 2008, http://www.ncjrs.gov/App/Publications/abstract.aspx?ID=246488.
74. R. B. Cody, A. J. Dane and W. E. Brewer, in *59th Annual Conference on Mass Spectrometry and Allied Topics*, Denver, CO, 2011.
75. J. Schurek, L. Vaclavik, H. Hooijerink, O. Lacina, J. Poustka, M. Sharman, M. Caldow, M. W. F. Nielen and J. Hajslova, *Anal. Chem.*, 2008, **80**, 9567–9575.
76. L. Vaclavik, M. Zachariasova, V. Hrbek and J. Hajslova, *Talanta*, 2010, **82**, 1950–1957.
77. J. N. Leibowitz, J. A. Clark, E. Jagerdeo, L. G. Schumacher and M. A. LeBeau, in *Pittsburgh Conference on Analytical Chemistry and Applied Spectroscopy*, 2006.
78. M. Zhou, W. Guan, L. D. Walker, R. Mezencev, B. B. Benigno, A. Gray, F. M. Fernandez and J. F. McDonald, *Cancer Epidemiol., Biomarkers Prev.*, 2010, **19**(9), 2262–2271.
79. M. Zhou, J. F. McDonald and F. M. Fernández, *J. Am. Soc. Mass Spectrom.*, 2010, **21**, 68–75.
80. J. M. Challinor, *J. Anal. Appl. Pyrolysis*, 2001, **61**, 3–34.
81. J. A. Laramée, H. D. N. Durst, J. Michael and T. R. Connell, *Am. Lab.*, 2009, **41**, 24–27.
82. J. A. Laramée, H. D. Durst, J. M. Nilles and T. R. Connell, *Am. Lab.*, 2009, **41**, 25–27.
83. R. P. Karas, E. Jagerdeo, A. L. Deakin, M. A. LeBeau and R. B. Cody, in *SOFT Conference*, Nashville, TN, 2005.
84. G. Morlock and W. Schwack, *Anal. Bioanal. Chem.*, 2006, **385**, 586–595.
85. O. P. Haefliger and N. Jeckelmann, *Rapid Commun. Mass Spectrom.*, 2007, **21**, 1361–1366.
86. G. Morlock and Y. Ueda, *J. Chromatogr. A*, 2007, **1143**, 243–251.
87. A. Alpmann and G. Morlock, *J. Sep. Sci.*, 2008, **31**, 71–77.
88. S. Ayers, B. J. Roschek, J. M. Williams and R. S. Alberte, *Online Journal of Pharmacology and PharmacoKinetics*, 2008, **5**, 6–21.
89. Y. Zhao, M. Lam, D. Wu and R. Mak, *Rapid Commun. Mass Spectrom.*, 2008, **22**, 3217–3224.
90. S. Yu, E. Crawford, J. Tice, B. Musselman and J.-T. Wu, *Anal. Chem.*, 2009, **81**, 193–202.
91. H. J. Kim and Y. P. Jang, *Phytochem. Anal.*, 2009, **20**, 372–377.
92. J. M. Nilles, T. R. Connell and H. D. Durst, *Anal. Chem.*, 2009, **81**, 6744–6749.
93. L. K. Ackerman, G. O. Noonan and T. H. Begley, *Food Addit. Contam., Part A*, 2009, **26**(12), 1611–1618.
94. A. H. Grange, *Environ. Forensics*, 2009, **10**, 183–195.
95. N. Jeckelmann and O. P. Haefliger, *Rapid Commun. Mass Spectrom.*, 2010, **24**, 1165–1171.
96. H. J. Kim, M. S. Oh, J. Hong and Y. P. Jang, *Phytochem. Anal.*, 2010, **22**, 258–262.

97. J. J. Pérez, G. A. Harris, J. E. Chipuk, J. S. Brodbelt, M. D. Green, C. Y. Hampton and F. M. Fernández, *Analyst*, 2010, **135**, 712–719.
98. J. Hajslova, T. Cajka and L. Vaclavik, *TrAC, Trends Anal. Chem.*, 2011, **30**, 204–218.
99. E. Chernetsova and G. Morlock, *Bioanal. Rev.*, 2011, **3**, 1–9.
100. E. S. Chernetsova, A. I. Revelsky and G. E. Morlock, *Rapid Commun. Mass Spectrom.*, 2011, **25**, 2275–2282.
101. E. S. Chernetsova and G. E. Morlock, *Bioanal. Rev.*, 2011, **3**, 1–9.
102. E. S. Chernetsova and G. E. Morlock, *Mass Spectrom. Rev.*, 2011, **30**, 875–883.
103. M. Haunschmidt, W. Buchberger, C. W. Klampfl and R. Hertsens, *Anal. Methods*, 2011, **3**, 99–104.
104. T. Cajka, K. Riddellova, P. Zomer, H. Mol and J. Hajslova, *Food Addit. Contam., Part A*, 2011, **28**(10), 1372–1382.
105. D. S. Cho, S. C. Gibson, D. Bhandari, M. E. McNally, R. M. Hoffman, K. D. Cook and L. Song, *Rapid Commun. Mass Spectrom.*, 2011, **25**, 3575–3580.
106. D. Saang'onyo, G. Selby and D. L. Smith, *Anal. Methods*, 2012, **4**, 3460–3465.
107. D. S. Saang'onyo and D. L. Smith, *Rapid Commun. Mass Spectrom.*, 2012, **26**, 385–391.
108. C. Chang, Z. Zhou, Y. Yang, Y. Han, Y. Bai, M. Zhao and H. Liu, *Electrophoresis*, 2012, **33**, 3387–3393.
109. J. Lojza, T. Cajka, V. Schulzova, K. Riddellova and J. Hajslova, *J. Sep. Sci.*, 2012, **35**, 476–481.
110. E. S. Chernetsova and G. E. Morlock, *Int. J. Mass Spectrom.*, 2012, **314**, 22–32.
111. G. E. Morlock and E. S. Chernetsova, *Cent. Eur. J. Chem.*, 2012, **10**, 703–710.
112. T. A. van Beek, Y. Shen, T. Verweij, A. Villela and F. Claassen, *Planta Med.*, 2012, **78**, CL44.
113. Y.-J. Li, Z.-Z. Wang, Y.-A. Bi, G. Ding, L.-S. Sheng, J.-P. Qin, W. Xiao, J.-C. Li, Y.-X. Wang and X. Wang, *Rapid Commun. Mass Spectrom.*, 2012, **26**, 1377–1384.
114. Y. Shen, T. A. van Beek, F. W. Claassen, H. Zuilhof, B. Chen and M. W. F. Nielen, *J. Chromatogr. A*, 2012, **1259**, 179–186.
115. S. J. B. Dunham, P. D. Hooker and R. M. Hyde, *Forensic Sci. Int.*, 2012, **223**, 241–244.
116. G. Krishnakumar, K. B. Rameshkumar, P. Srinivas, K. Satheeshkumar and P. N. Krishnan, *Asian Pac. J. Trop. Biomed.*, 2012, **2**, S727–S731.
117. P. Nemes, W. J. Hoover and D. A. Keire, *Anal. Chem.*, 2013, **85**(13), 7405–7412.
118. M. N. Chan, T. Nah and K. R. Wilson, *Analyst*, 2013, **138**, 3749–3757.
119. E. Crawford, P. Domizio, B. Musselman, C. M. L. Joseph, L. F. Bisson, B. C. Weimer and R. Jeannotte, 61st Annual Conference on Mass Spectrometry and Allied Topics, Minneapolis, MN, 2013.

120. H. Chen, G. Gamez and R. Zenobi, *J. Am. Soc. Mass Spectrom.*, 2009, **20**, 1947–1963.
121. E. Moravcova, L. Vaclavik, O. Lacina, V. Hrbek, K. Riddellova and J. Hajslova, *Anal. Bioanal. Chem.*, 2012, **402**, 2871–2883.
122. G. A. Harris and F. M. Fernandez, *Anal. Chem.*, 2009, **81**, 322–329.
123. R. B. Cody, J. A. Laramée, J. M. Nilles and H. D. Durst, *JEOL News*, 2005, **40**, 8–12.
124. J. A. Laramée and R. B. Cody, in *The Encyclopedia of Mass Spectrometry. Volume 6: Ionization Methods*, ed. M. L. Gross and R. M. Caprioli, Elsevier, Amsterdam, 2007, vol. 6, pp. 377–387.
125. V. L. H. Bevilacqua, J. M. Nilles, J. S. Rice, T. R. Connell, A. M. Schenning, L. M. Reilly and H. D. Durst, *Anal. Chem.*, 2010, **82**, 798–800.
126. C. Y. Pierce, J. R. Barr, R. B. Cody, R. F. Massung, A. R. Woolfitt, H. Moura, H. A. Thompson and F. M. Fernandez, *Chem. Commun.*, 2007, 807–809.
127. R. B. Cody, in *Pittsburgh Conference on Analytical Chemistry and Applied Spectroscopy*, Chicago, IL, 2009.
128. V. Bajpai, D. Sharma, B. Kumar and K. P. Madhusudanan, *Biomed. Chromatogr.*, 2010, **24**, 1283–1286.
129. L. Vaclavik, V. Hrbek, T. Cajka, B.-A. Rohlik, P. Pipek and J. Hajslova, *J. Agric. Food Chem.*, 2011, **59**(11), 5919–5926.
130. T. Cajka, H. Danhelova, A. Vavrecka, K. Riddellova, V. Kocourek, F. Vacha and J. Hajslova, *Talanta*, 2013, **115**, 263–270.
131. T. Cajka, K. Riddellova, M. Tomaniova and J. Hajslova, *Metabolomics*, 2011, **7**, 1–9.
132. H. Gu, Z. Pan, B. Xi, V. Asiago, B. Musselman and D. Raftery, *Anal. Chim. Acta*, 2011, **686**, 57–63.
133. JEOL USA Inc., in *Identification of Polymers*, JEOL USA, Inc., Application Note MS-0510B-2, 2005.
134. R. W. Jones, T. Reinot and J. F. McClelland, *Energy Fuels*, 2010, **24**, 5199–5209.
135. R. B. Cody, A. J. Dane, B. Dawson-Andoh, E. O. Adedipe and K. Nkansah, *J. Anal. Appl. Pyrolysis*, 2012, **95**, 134–137.
136. C. Lancaster and E. Espinoza, *Rapid Commun. Mass Spectrom.*, 2012, **26**, 1147–1156.
137. C. Lancaster and E. Espinoza, *Rapid Commun. Mass Spectrom.*, 2012, **26**, 2649–2656.
138. G. Morlock and W. Schwack, *LC · GC Eur.*, 2008, 366–371.
139. G. Morlock and E. Chernetsova, *Cent. Eur. J. Chem.*, 2012, **10**, 703–710.
140. N. J. Smith, M. A. Domin and L. T. Scott, *Org. Lett.*, 2008, **10**, 3493–3496.
141. E. H. Jee, C. W. Jeong, S. D. Jeong, H. J. Kim and Y. P. Jang, *Planta Med.*, 2009, **75**, PG38.
142. H. Kim, E. Jee, K. Ahn, H. Choi and Y. Jang, *Arch. Pharmacal Res.*, 2010, **33**, 1355–1359.
143. S. E. Howlett and R. R. Steiner, *J. Forensic Sci.*, 2011, **56**, 1261–1267.

144. J. L. Wood and R. R. Steiner, *Drug Test. Anal.*, 2011, **3**, 345–351.
145. H. Djelal, C. Cornée, R. Tartivel, O. Lavastre and A. Abdeltif, *Arabian J. Chem.*, 2013, DOI: 10.1016/j.arabjc.2013.06.003.
146. R. B. Cody, J. A. Laramée, H. D. Durst, J. M. Nilles, D. Simmons and Z. Wu, in *Montreux LC/MS Symposium*, Ithaca, New York, 2005.
147. W. Eberherr, W. Buchberger, R. Hertsens and C. W. Klampfl, *Anal. Chem.*, 2010, **82**, 5792–5796.
148. S. Beißmann, W. Buchberger, R. Hertsens and C. W. Klampfl, *J. Chromatogr. A*, 2011, **1213**, 7.
149. G. Cheng, in *60th Annual Conference on Mass Spectrometry and Allied Topics*, Vancouver, Canada, 2012.
150. C. Chang, G. Xu, Y. Bai, C. Zhang, X. Li, M. Li, Y. Liu and H. Liu, *Anal. Chem.*, 2012, **85**, 170–176.
151. R. B. Cody and A. J. Dane, *J. Am. Soc. Mass Spectrom.*, 2013, **24**, 329–334.
152. R. B. Cody, A. J. Dane, M. Ubukata and E. Christenson, unpublished results.
153. A. H. Grange and G. W. Sovocool, *Rapid Commun. Mass Spectrom.*, 2011, **25**(9), 1271–1281.
154. N. Uchiyama, R. Kikura-Hanajiri, N. Kawahara and Y. Goda, *Forensic Toxicol.*, 2009, **27**, 61–66.
155. M. Kawamura, R. Kikura-Hanajiri and Y. Goda, *Yakugaku zasshi*, 2009, **129**, 719–725.
156. H. J. Kim, E. H. Jee, K. S. Ahn, H. S. Choi and Y. P. Jang, *Arch. Pharmacal Res.*, 2010, **33**, 1355–1359.
157. W. C. Samms, Y. J. Jiang, M. D. Dixon, S. S. Houck and A. Mozayani, *J. Forensic Sci.*, 2011, **56**, 993–998.
158. Z. Zhou, J. Zhang, W. Zhang, Y. Bai and H. Liu, *Analyst*, 2011, **136**, 2613–2618.
159. M. Kawamura, R. Kikura-Hanajiri and Y. Goda, *Yakugaku Zasshi*, 2011, **131**, 827–833.
160. S. E. Howlett and R. R. Steiner, *J. Forensic Sci.*, 2011, **56**, 1261–1267.
161. R. A. Musah, M. A. Domin, R. B. Cody, A. D. Lesiak, A. J. Dane and J. R. E. Shepard, *Rapid Commun. Mass Spectrom.*, 2012, **26**, 2335–2342.
162. R. A. Musah, M. A. Domin, R. B. Cody, A. D. Lesiak, A. John Dane and J. R. E. Shepard, *Rapid Commun. Mass Spectrom.*, 2012, **26**, 2335–2342.
163. R. P. Karas, E. Jagerdeo, A. L. Deakin, M. A. LeBeau and R. B. Cody, in *Society of Forensic Toxicologists (SOFT) Annual Meeting*, 2005.
164. M. J. Bennett and R. R. Steiner, *J. Forensic Sci.*, 2009, **54**, 370–375(376).
165. P. Stout, N. Bynum, E. Minden Jr. and J. Miller, in *Society of Forensic Toxicologists (SOFT) Annual Meeting*, Raleigh-Durham, NC., 2007.
166. E. J. Minden Jr., N. D. Bynum, J. D. Ropero-Miller and P. R. Stout, in *Society of Forensic Toxicologists (SOFT), Annual Meeting*, Raleigh-Durham, NC., 2007.

167. R. R. Steiner, *Microgram J.*, 2010, 7, 3–6.
168. T. Moffat, R. Cody, R. Jee and A. O'Neil, in *Federation of Analytical Chemistry and Spectroscopy Societies (FACSS), Fall 2007*, 2007.
169. P. N. Newton, F. M. Fernandez, A. Plancon, D. C. Mildenhall, M. D. Green, L. Ziyong, E. M. Christophel, S. Phanouvong, S. Howells, E. McIntosh, P. Laurin, N. Blum, C. Y. Hampton, K. Faure, L. Nyadong, C. W. R. Soong, B. Santoso, W. Zhiguang, J. Newton and K. Palmer, *PLoS Med.*, 2008, 5, e32.
170. F. M. Fernandez, M. D. Green and P. N. Newton, *Ind. Eng. Chem. Res.*, 2008, 47, 585–590.
171. L. Nyadong, A. S. Galhena and F. M. Fernaìndez, *Anal. Chem.*, 2009, 81(18), 7788–7794.
172. L. Nyadong, G. A. Harris, S. Balayssac, A. S. Galhena, M. Malet-Martino, R. Martino, R. M. Parry, M. D. Wang, F. M. Fernández and V. Gilard, *Anal. Chem.*, 2009, 81, 4803–4812.
173. A. S. Galhena, G. A. Harris, K. K. Murray and F. M. Fernandez, *Anal. Chem.*, 2010, 82, 2178–2181.
174. J. Zhang, F. Huo, Z. Zhou, Y. Bai and H. Liu, *Prog. Chem.*, 2012, 24(1), 101–109.
175. E. S. Chernetsova, R. A. Abramovich and I. A. Revel'skii, *Pharm. Chem. J.*, 2012, 45, 698–700.
176. R. A. Musah, R. B. Cody, A. J. Dane, A. L. Vuong and J. R. E. Shepard, *Rapid Commun. Mass Spectrom.*, 2012, 26, 1039–1046.
177. J. A. Laramée, H. D. Durst, T. R. Connell and J. M. Nilles, *Am. Lab.*, 2009, 2, 1–5.
178. J. M. Nilles, T. R. Connell, S. T. Stokes and H. D. Durst, *Propellants, Explos., Pyrotech.*, 2010, 35, 446–451.
179. F. Rowell, J. Seviour, A. Y. Lim, C. G. Elumbaring-Salazar, J. Loke and J. Ma, *Forensic Sci. Int.*, 2012, 221, 84–91.
180. E. Sisco, J. Dake and C. Bridge, *Forensic Sci. Int.*, 2013, 232, 160–168.
181. C. M. Coates, S. Coticone, P. D. Barreto, A. E. Cobb, R. B. Cody and J. C. Barreto, *J. Forensic Ident.*, 2008, 58, 624–631.
182. J. A. Meyers, in *Combined Regional Forensic Meeting*, Orlando, FL, 2009.
183. A. J. Peña-Quevedo, N. Mina-Calmide, N. Rodríguez, D. Nieves, R. B. Cody and S. P. Hernández-Rivera, in *Synthesis, characterization and differentiation of high energy amine peroxides by MS and vibrational microscopy*, 2006; pp. 62012E-62012E-10.
184. J. A. Laramée, H. D. Durst, T. R. Connell and J. M. Nilles, *Am. Lab.*, 2008, 40, 16–20.
185. G. A. Harris and F. M. Fernandez, in *58th ASMS Conference on Mass Spectrometry and Allied Topics*, Salt Lake City, UT, 2010.
186. J. L. Rummel, J. D. Steill, J. Oomens, C. S. Contreras, W. L. Pearson, J. Szczepanski, D. H. Powell and J. R. Eyler, *Anal. Chem.*, 2011, 83, 4045–4052.
187. G. A. Harris, C. E. Falcone and F. M. Fernández, *J. Am. Soc. Mass Spectrom.*, 2012, 23, 153–161.

188. C. E. Kolb, J. L. Beauchamp, R. A. Beaudet, J. B. Berkowitz, H. Chen, A. T. Cooper, F. M. Fernandez, R. T. Gibbons, J. A. Mclean, M. D. Morris, D. W. Murphy, C. S. Reese, L. R. Rhomberg, A. A. Vigginao, *Agent Monitoring Strategies for the Blue Grass and Pueblo Chemical Agent Destruction Pilot Plants*, The National Academies Press, 2012.

189. E. O. Espinoza, C. A. Lancaster, N. M. Kreitals, M. Hata, R. B. Cody and R. A. Blanchette, *Rapid Commun. Mass Spectrom.*, 2014, **28**, 1–9.

190. J. A. Laramée, R. B. Cody, J. M. Nilles and H. D. Durst, in *Forensic Analysis on the Cutting Edge: New Methods for Trace Evidence Analysis*, ed. R. D. Blackledge, Wiley-Interscience, Hoboken, NJ, 2007.

191. S. Houck, Y. J. Jiang, M. Dixon, A. Mozayani and L. A. Sanchez, in S*outhwestern Association of Toxicologists*, Galveston, TX, 2007.

192. State of Virginia Department of Forensic Science Controlled Substance Procedures Manual, http://www.dfs.virginia.gov/wp-content/uploads/2013/09/221-D100-Controlled-Substances-Procedures-Manual.pdf.

193. NIST DART Forensics Library, http://chemdata.nist.gov/mass-spc/ms-search/DART_Forensic.html.

194. RTI International.

195. A. D. Lesiak, R. A. Musah, R. B. Cody, M. A. Domin, A. J. Dane and J. R. E. Shepard, *Analyst*, 2013, **138**, 3424–3432.

196. A. D. Lesiak, R. A. Musah, M. A. Domin and J. R. E. Shepard, *J. Forensic Sci.*, 2014, **59**(2), 337–343.

197. JEOL USA Inc., in *Using Solid Phase Microextraction with AccuTOF-DART® for Fragrance Analysis (Application Note)*, JEOL Application Note.

198. JEOL USA Inc., in *Rapid Detection of Fungicide on Orange Peel*, JEOL Application Note.

199. JEOL USA Inc, in *Detection of Oleocanthal in Freshly Pressed Exra-Virgin Olive Oil*, JEOL Application Note, 2005.

200. JEOL USA Inc, in *Detection of Unstable Compound Released by Chopped Chives*, JEOL Application Note, 2005.

201. JEOL USA Inc, in *Distribution of Capsaicin in Chili Peppers*, JEOL Application Note, 2005.

202. JEOL USA Inc, in *Instantaneous Detection of Opiates in Single Poppy Seeds*, JEOL Application Note, 2005.

203. JEOL USA Inc, in *Rapid Detection of Fungicide in Orange Peel*, JEOL Application Note, 2005.

204. JEOL USA Inc, in *Detection of Lycopene in Tomato Skin*, JEOL Application Note, 2005.

205. JEOL USA Inc, in *Flavones and Flavor Components in Two Basil Leaf Chemotypes*, JEOL Application Note, 2006.

206. JEOL USA Inc, in *Rapid Detection of Melamine in Dry Milk Using AccuTOF-DART*, JEOL Application Note, 2009.

207. L. Vaclavik, J. Schurek, T. Cajka and J. Hajslova, *Chem. Listy*, 2008, **102**, s324–s327.

208. T. Cajka, L. Vaclavik, K. Riddellova and J. Hajslova, *LC · GC Eur.*, 2008, **21**, 250–256.

209. J. Hajslova, L. Vaclavik, T. Cajka, J. Poustka, O. Lacina and J. Schurek, *DART Workshop*, Institute Of Chemical Technology, Department of Food Chemistry and Analysis, Prague, Czech Republic, 2008.

210. J. Hajslova, L. Vaclavik, T. Cajka, J. Poustka and J. Schurek, *LCGC: Chromatography Online*, 2008, http://chromatographyonline. findanalytichem.com/lcgc/Articles/Analysis-of-Deoxynivalenol-in-Beer/ ArticleStandard/Article/detail/547873.

211. T. Cajka, J. Hajslova and K. Mastovska, in *Handbook of Food Analysis Instruments*, ed. S. Ötleş, CRC Press, Taylor & Francis Group, 2008, pp. 197–228.

212. M. Zachariasova, T. Cajka, M. Godula, A. Malachova, Z. Veprikova and J. Hajslova, *Rapid Commun. Mass Spectrom.*, 2010, **24**, 3357–3367.

213. L. Vaclavik, J. Rosmus, B. Popping and J. Hajslova, *J. Chromatogr. A*, 2010, **1217**, 4204–4211.

214. R. Kubec, P. Krejĉova, P Simek, L Vaclavik, J Hajslova and J Schraml, *J. Agric. Food Chem.*, 2011, **59**, 5763–5770.

215. L. Vaclavik, V. Hrbek, T. Cajka, B. A. Rohlik, P. Pipek and J. Hajslova, *J. Agric. Food Chem.*, 2011, **59**, 5919–5926.

216. J. Hajslova, M. Zachariasova and T. Cajka, in *Mass Spectrometry in Food Safety: Methods and Protocols*, ed. J. Zweigenbaum, Humana Press, 2011, pp. 233–258.

217. K. Kalachova, J. Pulkrabova, L. Drabova, T. Cajka, V. Kocourek and J. Hajslova, *Anal. Chim. Acta*, 2011, **707**, 84–91.

218. H. Novotná, O. Kmiecik, M. Gałązka, V. Krtková, A. Hurajová, V. Schulzová, E. Hallmann, E. Rembiałkowska and J. Hajšlová, *Food Addit. Contam., Part A*, 2012, **29**, 1335–1346.

219. T. Cajka, H. Danhelova, M. Zachariasova, K. Riddellova and J. Hajslova, *Metabolomics*, 2013, **9**, 545–557.

220. T. Cajka, H. Danhelova, A. Vavrecka, K. Riddellova, V. Kocourek, F. Vacha and J. Hajslova, *Talanta*, 2013, **115**, 263–270.

221. V. Hrbek, L. Vaclavik, O. Elich and J. Hajslova, *Food Control*, 2014, **36**, 138–146.

222. D. S. Wishart, C. Knox, A. C. Guo, R. Eisner, N. Young, B. Gautam, D. D. Hau, N. Psychogios, E. Dong, S. Bouatra, R. Mandal, I. Sinelnikov, J. Xia, L. Jia, J. A. Cruz, E. Lim, C. A. Sobsey, S. Shrivastava, P. Huang, P. Liu, L. Fang, J. Peng, R. Fradette, D. Cheng, D. Tzur, M. Clements, A. Lewis, A. De Souza, A. Zuniga, M. Dawe, Y. Xiong, D. Clive, R. Greiner, A. Nazyrova, R. Shaykhutdinov, L. Li, H. J. Vogel and I. Forsythe, *Nucleic Acids Res.*, 2009, **37**, D603–D610.

223. X.-l. Cheng, W.-j. Li, F. Wei, X.-y. Xiao and R.-c. Lin, *Chin. J. Pharm. Anal.*, 2011, **31**(3), 438–442.

224. M. Farre and D. Barcelo, *TrAC, Trends Anal. Chem.*, 2012, **43**, 240–253.

225. B. Roschek Jr. and R. S. Alberte, *Online Journal of Pharmacology and PharmacoKinetics*, 2008, **4**, 1–17.

226. B. Roschek Jr., R. C. Fink, D. Li, M. McMichael, C. M. Tower, R. D. Smith and R. S. Alberte, *J. Med. Food*, 2009, **12**, 615–623.

227. B. Roschek Jr., R. C. Fink, M. D. McMichael, D. Li and R. S. Alberte, *Phytochemistry*, 2009, **70**, 1255–1262.
228. J. Ra, S. Lee, H. J. Kim, Y. P. Jang, H. Ahn and J. Kim, *J. Ethnopharmacol.*, 2010, **128**, 241–247.
229. S. M. Lee, H.-J. Kim and Y. P. Jang, *Phytochem. Anal.*, 2012, **23**, 508–512.
230. H. Zhu, C. Wang, Y. Qi, F. Song, Z. Liu and S. Liu, *Anal. Chim. Acta*, 2012, **752**, 69–77.
231. A. H. Grange, *Am. Lab.*, 2008, **40**, 11–13.
232. K. B. Loftin, T. P. Griffin and C. A. Clausen, in *57th ASMS Conference on Mass Spectrometry and Allied Topics*, Philadelphia, PA, 2009.
233. T. Rothenbacher and W. Schwack, *Rapid Commun. Mass Spectrom.*, 2009, **23**, 2829–2835.
234. T. Rothenbacher and W. Schwack, *Rapid Commun. Mass Spectrom.*, 2010, **24**, 21–29.
235. J. Langlois, A. Kusai and E. Van Elslande, in *MaSC Group LDMS and MALDI Workshop and Meeting*, Cambridge, MA, USA, 2011.
236. D. Fraser, C. S. DeRoo, R. B. Cody and R. A. Armitage, *Analyst*, 2013, **138**, 4470–4474.
237. C. S. Deroo and R. A. Armitage, *Anal. Chem.*, 2011, **83**, 6924–6928.
238. S. Banerjee, K. P. Madhusudanan, S. K. Chattopadhyay, L. U. Rahman and S. P. S. Khanuja, *Biomed. Chromatogr.*, 2008, **22**, 830–834.
239. K. P. Madhusudanan, S. Banerjee, S. P. S. Khanuja and S. K. Chattopadhyay, *Biomed. Chromatogr.*, 2008, **22**, 596–600.
240. S. Banerjee, K. P. Madhusudanan, S. P. S. Khanuja and S. K. Chattopadhyay, *Biomed. Chromatogr.*, 2008, **22**, 250–253.
241. S. D. Maleknia, T. L. Bell and M. A. Adam, *Int. J. Mass Spectrom.*, 2009, **279**, 126–133.
242. E. Block, A. J. Dane, S. Thomas and R. B. Cody, *J. Agric. Food Chem.*, 2010, **58**, 4617–4625.
243. E. Block, R. B. Cody, A. J. Dane, R. Sheridan, A. Vattekkatte and K. Wang, *Pure Appl. Chem.*, 2010, **82**, 535–539.
244. E. Block, *Volatile Sulfur Compounds in Food*, American Chemical Society, 2011, vol. 1068, pp. 35–63.
245. S. W. Kim, H. J. Kim, J. H. Kim, Y. K. Kwon, M. S. Ahn, Y. P. Jang and J. R. Liu, *Plant Methods*, 2011, **7**, 1–10.
246. V. Singh, A. K. Gupta, S. P. Singh and A. Kumar, *Sci. World J.*, 2012, **2012**, 6.
247. S. Singh and S. K. Verma, *Nat. Prod. Bioprospect.*, 2012, **2**, 206–209.
248. E. O. Espinoza, C. A. Lancaster, N. M. Kreitals, M. Hata, R. B. Cody and R. A. Blanchette, *Rapid Commun. Mass Spectrom.*, 2014, **28**, 1–9.
249. E. Block, *Garlic and Other Alliums. The Lore and the Science*, RSC Publishing, Cambridge, UK, 2010.
250. R. Kubec, R. B. Cody, A. J. Dane, R. A. Musah, J. Schraml, A. Vattekkatte and E. Block, *J. Agric. Food Chem.*, 2010, **58**, 1121–1128.

251. E. Block, A. J. Dane and R. B. Cody, *Phosphorus, Sulfur Silicon Relat. Elem.*, 2011, **186**, 1085–1093.
252. S. Francis, in *J. Chem. Tenn*, UTK Dept. of Chemistry, Knoxville, TN, 2008, p. 4, http://www.chem.utk.edu/pdf/newsletter/Spring08Newsletter.pdf.
253. M. Gomez, R. L. Johnson and J. D. Williams, 54th Annual Conference on Mass Spectrometry and Allied Topics, Seattle, WA, 2006.
254. C. Petucci, J. Diffendal, D. Kaufman, B. Mekonnen, G. Terefenko and B. Musselman, *Anal. Chem.*, 2007, **79**, 5064–5070.
255. JEOL USA Inc, in *Chemical Reaction Monitoring*, JEOL Application Note, 2007.
256. JEOL USA Inc, in *Direct Analysis of Organometallic Compounds*, JEOL Application Note, 2006.
257. D. L. G. Borges, R. E. Sturgeon, B. Welz, A. J. Curtius and Z. Mester, *Anal. Chem.*, 2009, **81**, 9834–9839.
258. B. Gulbakan, J. R. Sommer, J. L. Rummel, K. S. Schanze, W. Tan and D. H. Powell, in *58th ASMS Conference on Mass Spectrometry and Allied Topics*, Salt Lake City, UT, 2010.
259. Q. Zhang, J. Bethke and M. Patek, in *58th ASMS Conference on Mass Spectrometry and Allied Topics*, Salt Lake City, UT, 2010.
260. JEOL USA Inc., in *Rapid Analysis of p-Phenylenediamine Antioxidants in Rubber*, JEOL Application Note MS-0105C, 2005.
261. M. Haunschmidt, C. W. Klampfl, W. Buchberger and R. Hertsens, *Analyst*, 2010, **135**, 80–85.
262. A. Mess, J.-P. Vietzke, C. Rapp and W. Francke, *Anal. Chem.*, 2011, **83**, 7323–7330.
263. N. Aminlashgari and M. Hakkarainen, *Advances in Polymer Science*, Springer, Berlin/Heidelberg, 2012, pp. 1–37.
264. E. Dytkiewitz and G. E. Morlock, *J. AOAC Int.*, 2008, **91**, 1237–1244.
265. JEOL USA Inc., in *Rapid Analysis of Glues, Cements, and Resins*, JEOL Application Note MS-0509C, 2005.
266. JEOL USA Inc., in *Direct Analysis of Adhesives*, Application Note MS-0509, 2005.
267. R. K. Manova, F. W. Claassen, M. W. F. Nielen, H. Zuilhof and T. A. van Beek, *Chem. Commun.*, 2013, **49**, 922–924.
268. R. B. Cody, in *55th ASMS Conference on Mass Spectrometry and Allied Topics*, Indianapolis, IN, 2007.
269. R. B. Cody and A. J. Dane, in *Analysis of Fragile, Reactive, Large or Inorganic Samples with the Direct Analysis in Real Time (DART) Ion Source (poster presented at the International Mass Spectrometry Conference)*, Bremen, Germany, 2009.
270. X. N. He, Z. Q. Xie, Y. Gao, W. Hu, L. B. Guo, L. Jiang and Y. F. Lu, *Spectrochim. Acta, Part B*, 2012, **67**, 64–73.

Ionization Mechanisms of Direct Analysis in Real Time (DART)

LIGUO SONG AND JOHN E. BARTMESS*

Department of Chemistry, University of Tennessee, Knoxville, Tennessee 37996-1600, USA
*Email: bartmess@utk.edu

3.1 Introduction

Over the last several years, the interest in direct analysis has led to the development of a myriad of ionization techniques that aim to directly sample and ionize samples in their ambient states with no or minimal effort for sample preparation. They are now more collectively reviewed as ambient ionization and often referenced as equivalent to atmospheric-pressure ionization (API).[1-7] Direct ionization in real time (DART) and desorption electrospray ionization (DESI), which were respectively reported in early 2005[8] and late 2004,[9] are the two pioneer ambient ionization techniques. Now more than 30 ambient ionization techniques have been developed; and they are often classified according to the method used for desorption and/or the process involved in ionization. APCI and ESI are two typical ionization processes used in the classification of ambient ionization techniques,[1-7] and DART and DESI are the respective primary representative. According to a recent review of ambient ionization by Fernandez and coworkers[7] for the January 2009 to January 2011 period, DART and DESI were the top two

New Developments in Mass Spectrometry No. 2
Ambient Ionization Mass Spectrometry
Edited by Marek Domin and Robert Cody

ambient ionization techniques with respectively 27% and 32% of the relevant publications.

The DART source consists of a few chambers aligned consecutively where a gas, usually helium, flows through before entering the ambient atmosphere to interact with the sample. First, an electric glow discharge of the gas[8] produces ions, electrons, and metastable species. Then, downstream perforated electrodes remove ions and allow metastable species to remain in the gas stream. A heated chamber further adjusts the temperature of the gas stream. At the end, a grid electrode is used to prevent ion–ion and ion–electron recombination; an insulation cap is used to protect the sample and the operator from any exposure to high voltage. With regard to the ionization process of DART, it seems to be agreed in the literature that it is APCI-like.[1–8,10–14] However, the DART ionization process can be greatly affected by the ionization environment and as such the nature of the ionization products can be affected.

3.2   Ionization Mechanism of Positive-ion DART

The ionization products of positive-ion DART depend on the nature of the DART gas, DART conditions, analyte polarity and reactivity, and ionization environment. Frequently obtained ionization products are $[M+H]^+$, $M^{+\bullet}$, $[M–H]^+$ and $[M+NH_4]^+$ ions, but ion–molecule adducts, *i.e.* $[(M_a+H)+M_b]^+$ where the ion and molecule can both be from the analytes, have been also reported. Unlike electrospray ionization (ESI), alkali-metal adducts, such as $[M+Na]^+$ ions, have never been observed.[8]

In order to understand why the various analyte ions form, the processes that lead to the background ions, in the absence of analyte, must be examined first.

3.2.1   Generation of Background Ions

The initial step of positive-ion DART is Penning ionization (Scheme 3.1)[15] in which a long-lived ("metastable") excited-state neutral atom or molecule A* transfers energy to some molecule M, resulting in the formation of a molecular ion $M^{+\bullet}$ and an electron e^-. This reaction will occur if the molecule M has an ionization energy (IE) that is lower than the internal energy of the metastable A*, *i.e.* ME(A).

Table 3.1 shows the IEs and proton affinities (PAs) of atmospheric gases and species used in DART, obtained from the NIST Chemistry WebBook (http://webbook.nist.gov).[16]

Scheme 3.2 shows the generation of the various "air ions" that are observed in the positive-ion helium spectrum, with no analytes present.

$$A^* + M \rightarrow A + M^{+\bullet} + e^- \qquad \text{if ME(A)>IE(M)}$$

Scheme 3.1   Generation of $M^{+\bullet}$ ions by Penning ionization.

Table 3.1   IEs and PAs of atmospheric gases and species used in DART.

Gases	Formula	Isotopic mass	IE^a (eV)	PA^a (kJ mol^{-1})
Helium	He	4.0026	24.58741	177.8
Argon	Ar	39.9624	17.759	369.2
Nitrogen	N_2	28.0062	15.581	493.8
Water	H_2O	18.0106	12.621	691
Oxygen	O_2	31.9898	12.0697	421
Ammonia	NH_3	17.0266	10.070	853.6
Nitric oxide	NO	29.9980	9.2642	531.8

^{a}Data obtained from NIST Chemistry WebBook (http://webbook.nist.gov).

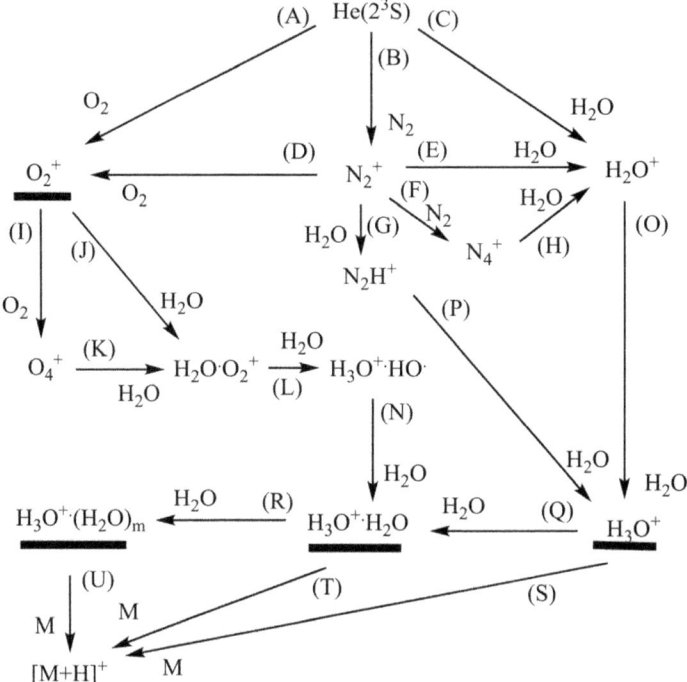

Scheme 3.2   Generation of background ions by positive-ion DART.

Underlined ions are those thought to be responsible for much of the ana-lytes' spectra. The electric glow discharge of helium primarily produces long-lived helium 2^3S electronic excited state atoms with an internal energy of 19.8 eV, which is higher than the ionization energies of common atmos-pheric gases (Table 3.1). These metastable helium atoms rapidly react with O_2, N_2, and atmospheric moisture with extremely high efficiency (reactions (A)–(C), Scheme 3.2). N_2^+, however, is known to rapidly react with O_2 and H_2O by charge exchange, as well as by hydrogen abstraction from water, and clustering with itself (reactions (D)–(G), Scheme 3.2).[17] Ionized water then yields protonated water and water cluster ions *via* reactions O–R, because water (PA = 691 kJ mol^{-1} [16]) and water clusters have PA values stronger than that of OH$^•$ (PA = 593 kJ mol^{-1} [16]). O_2^+ likewise can contribute to the $H_3O^+ \cdot (H_2O)_n$ ion pool, *via* clustering reactions (I)–(N) in Scheme 3.2. It will be seen, however, that under proper conditions, enough O_2^+ remains to react in its own right (see Section 3.2.2).

A typical DART background mass spectrum under standard DART conditions is shown in Figure 3.1(a). Protonated water and water clusters $[(H_2O)_m + H]^+$ were observed at *m/z* 19.0184 (*m* = 1) and 37.0299 (*m* = 2). Due to the ubiquitous existence of ammonia in ambient air,[18] protonated ammonia was also observed at *m/z* 18.0341. Trace solvent vapors in the laboratory environment produced peaks assigned as $[M + H]^+$ for methanol (*m/z* 33.042), acetonitrile (*m/z* 42.0342), ethanol (*m/z* 47.0493), and acetone (*m/z* 59.0503) (indicated by asterisks in Figure 3.1(a)). These protonated ions should be generated as shown in Scheme 3.2, reactions (S)–(U).

A NO$^+$ peak at *m/z* 29.9983 was also observed, which is significant despite its relatively low abundance, because NO$^+$ is highly reactive and can undergo a variety of ion–molecule reactions including addition, hydride abstraction,

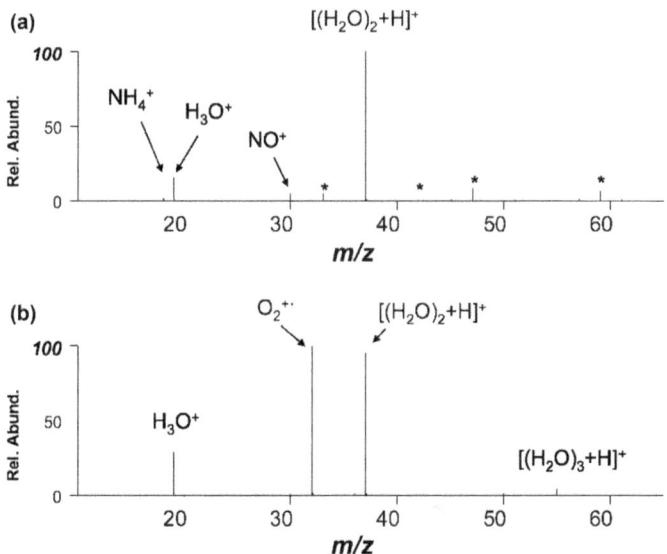

Figure 3.1 (a) A typical positive-ion DART background mass spectrum. The DART source potentials were set to needle = 3500 V, electrode 1 = +150 V, electrode 2 (grid) = +250 V. This spectrum was obtained with the exit of the DART source positioned approximately 10 mm from the mass spectrometer orifice. (b) An atypical DART background mass spectrum under atypical DART condition. The DART source potentials were set to needle = 3500 V, electrode 1 = +150 V, electrode 2 (grid) = 650 V. This spectrum was obtained with the exit of the DART source positioned approximately 3 mm from the mass spectrometer orifice.

Reprinted with permission from R. B. Cody, *Anal. Chem.*, 2009, **81**, 1101–1107. Copyright 2009 American Chemical Society.

$$He(2^3S) + N_2 \rightarrow N_2^* + He \qquad (A)$$

$$N_2^* \rightarrow N^* + N \qquad (B)$$

$$N^* + O_2 \rightarrow NO + O \qquad (C)$$

$$NO + O_2 \rightarrow NO_2 + O \qquad (D)$$

$$NO + O_2^+ \rightarrow NO^+ + O_2 \qquad (E)$$

Scheme 3.3   Generation of NO and NO^+ by neutral and charge exchange reactions.[22]

charge exchange, and oxidation.[19] It is known[17] that NO^+ is not produced in air from N_2^+ or O_2^+, but only N^+ or O^+. Those atomic ions are not thermochemically accessible from the He metastables. It is thus suspected that NO^+ may arise by ionization of neutral NO, produced in a neutral atom reaction. Scheme 3.3 gives a likely scenario.

It is known that rare-gas metastables can transfer that energy to N_2.[20] There is a wealth of N_2 metastable states accessible, from 6.18 to 11.88 eV in the ground vibrational state.[21] Many of those metastables in vibrationally excited states rapidly dissociate to ground and metastable nitrogen atoms, as per reaction (B) in Scheme 3.3. The nitrogen atoms so produced can react with O_2 to yield NO,[22] and in a second step NO_2. These species can be ionized by charge exchange with O_2^+, due to the IE of 9.26 eV for NO.[16]

3.2.2   Generation of $[M+H]^+$ Ions

The most frequently observed ions by positive-ion DART are $[M+H]^+$ ions. An ionization mechanism for the generation of $[M+H]^+$ ions by positive-ion DART was proposed by Cody *et al.*[8,10]

Table 3.2 shows the boiling points (BPs) and IEs of common solvents that were also obtained from the NIST Chemistry WebBook (http://webbook.nist.gov).[23] Table 3.3 shows the PAs of common solvents in different forms such as monomer, dimer, and $[S-H]^\bullet$. The PA values of their monomers were obtained from the NIST Chemistry WebBook (http://webbook.nist.gov).[23] The PA of an $[S-H]^\bullet$ radical was calculated as follows, from enthalpies of formation, IEs, and bond dissociation energies (BDEs):

$$PA(A^\bullet) = \Delta_f H^\circ(A^\bullet) + \Delta_f H(H^+) - \Delta_f H(AH^{+\bullet})$$

$$\Delta_f H^\circ(A^\bullet) = BDE(A-H) - \Delta_f H^\circ(H^\bullet) + \Delta_f H^\circ(AH)$$

$$\Delta_f H(AH^{+\bullet}) = \Delta_f H^\circ(AH) + IE(AH)$$

$$\text{thus } PA(A^\bullet) = BDE(A-H) - \Delta_f H^\circ(H^\bullet) + \Delta_f H^\circ(AH)$$

$$+ \Delta_f H(H^+) - \Delta_f H^\circ(AH) + IE(AH)$$

$$= BDE(A-H) + IE(H^\bullet) + IE(AH)$$

Table 3.2 BPs and IEs of common solvents.

Solvent	Formula	Isotopic mass	bp (°C)	IE^a (eV)
Water	H_2O	18.0110	100	12.62
MeCN	CH_3CN	41.0270	82	12.20
$CHCl_3$	$CHCl_3$	117.9140	61	11.37
CH_2Cl_2	CH_2Cl_2	83.9530	40	11.33
MeOH	CH_3OH	32.0260	65	10.84
EtOH	C_2H_5OH	46.0420	78	10.50
iPrOH	$(CH_3)_2CHOH$	60.0580	82	10.17
Hex	C_6H_{14}	86.1100	69	10.13
EtAcetate	$CH_3COOC_2H_5$	88.0520	77	10.01
Hep	C_7H_{16}	100.1250	98	9.93
C_6F_6	C_6F_6	185.9900	81	9.90
IsoOct	C_8H_{18}	114.1410	99	9.89
CyHex^b	C_6H_{12}	84.0940	81	9.88
Acetone	CH_3COCH_3	58.0420	56	9.70
THF	C_4H_8O	72.0580	66	9.40
Benzene	C_6H_6	78.0470	80	9.24
PhF	C_6H_5F	96.0380	85	9.20
DMF	$HCON(CH_3)_2$	73.0530	153	9.13
DMSO	CH_3SOCH_3	78.0140	189	9.10
PhCl	C_6H_5Cl	112.0080	132	9.07
$PhCH_3$	$C_6H_5CH_3$	92.0630	111	8.83
PhC_2H_5	$C_6H_5C_2H_5$	106.0780	136	8.77
o-Xylene	$C_6H_4(CH_3)_2$	106.0780	144	8.56
p-Xylene	$C_6H_4(CH_3)_2$	106.0780	138	8.44
$PhOCH_3$	$C_6H_4OCH_3$	107.0500	154	8.20

^{a}Data obtained from NIST Chemistry WebBook (http://webbook.nist.gov).
^{b}Cyclohexane.

This ignores the fact that IEs are 0 K quantities, while BDEs and PAs are 298 K quantities. There should be only a small temperature effect on IEs, however, because the integrated heat capacities of AH and $AH^{+\bullet}$ should approximately cancel, due to their similar structures.[24]

The PA corresponding to an S_2H^+ ion as show in Table 3.3 can be taken as the PA of the monomer plus the binding enthalpy of the proton-bound dimer, $\Delta H_{01}(S\cdots SH^+)$. This assumes that on proton loss, the conjugate base of S_2H^+ is two free bases, and not a neutral dimer. Neutral dimer binding enthalpies are typically 10–20 kJ mol^{-1}, but entropy favors separation into two free S species on proton loss. In addition, $\Delta H_{01}(S\cdots SH^+)$ is roughly constant at 130 ± 8 kJ mol^{-1} for the range of bases here. This is because for constant structure at the binding site of $OH^+\cdots O$, as the monomer S becomes a stronger base and thus stronger hydrogen-bond acceptor, its conjugate acid SH^+ becomes a weaker hydrogen-bond donor. These effects roughly cancel to give a near constant ΔH_{01}. For those bases where $\Delta H_{01}(S\cdots SH^+)$ is not known, we assign the PA of the dimer as that of the monomer plus 126 kJ mol^{-1}, the experimental value of $\Delta H_{01}(S\cdots SH^+)$ for acetone.[23]

Table 3.3 PAs (kJ mol^{-1}) of common solvents in different forms.

Solvent	Species	PA^a	Solvent	Species	PA^a	Solvent	Species	PA^a
MeCN	[S–H]$^\bullet$	538	Acetone	[S–H]$^\bullet$	778	PhOCH$_3$	S	840
Water	[S–H]$^\bullet$	593	MeCN	S	779	o-Xylene	[S–H]$^\bullet$	845
CH$_2$Cl$_2$	S	~629	PhCH$_3$	S	784	p-Xylene	[S–H]$^\bullet$	866
CHCl$_3$	S	~635	CyHex^b	Cyhexene^c	784	DMSO	[S–H]$^\bullet$	879
C$_6$F$_6$	S	648	PhC$_2$H$_5$	S	788	DMSO	S	884
MeOH	[S–H]$^\bullet$	660	THF	[S–H]$^\bullet$	789	DMF	S	888
Hexanes	S	~680	iPrOH	S	793	MeOH	S$_2$	891
Heptane	S	~680	p-Xylene	S	794	Benzene	[S–H]$^\bullet$	895
Isooctane	S	~680	o-Xylene	S	796	MeCN	S$_2$	909
CyHex^b	S	687	Heptane	Isobutene	802	EtOH	S$_2$	910
Water	S	691	Isooctane	Isobutene	802	PhCl	[S–H]$^\bullet$	913
EtOH	[S–H]$^\bullet$	689	Hexanes	Hexene	~805	iPrOH	S$_2$	927
iPrOH	[S–H]$^\bullet$	714	Acetone	S	812	Acetone	S$_2$	938
EtOAc	[S–H]$^\bullet$	723	THF	S	822	THF	S$_2$	948
Benzene	S	750	Water	S$_2$	825	EtAc	S$_2$	963
PhCl	S	753	DMF	[S–H]$^\bullet$	824	PhF	[S–H]$^\bullet$	900
MeOH	S	754	PhC$_2$H$_5$	[S–H]$^\bullet$	828	PhOCH$_3$	[S–H]$^\bullet$	995
PhF	S	756	EtAc	S	836	DMSO	S$_2$	1013
EtOH	S	776	PhCH$_3$	[S–H]$^\bullet$	838	DMF	S$_2$	1014

^{a}Data obtained from NIST Chemistry WebBook (http://webbook.nist.gov) or estimated as described in text.
^{b}Cyclohexane.
^{c}Cyclohexene.

Proton transfer to produce [M + H]$^+$ ions will occur if the analyte molecule M has a higher proton affinity than protonated water and water cluster ions (reactions (S)–(U), Scheme 3.2). Because water and water clusters have PA values weaker than most polar compounds, *e.g.* common solvents (Table 3.3), a large number of analytes have been reported to generate abundant [M + H]$^+$ ions. This is documented in the literature and reviews on DART application for example.[1–7,13,14] Likewise, the chemical ionization mass spectrometry (CIMS) literature can be used as a guide in this.[25]

3.2.3 Generation of M$^{+\bullet}$ and [M–H]$^+$ Ions

Figure 3.1(b) shows a DART background mass spectrum with abundant O$_2$$^{+\bullet}$ ions (*m/z* 31.9900), which was reported by Cody[10] when the DART ion source was positioned closer to the mass spectrometry sampling orifice and the grid potential was high. The abundance of O$_2$$^{+\bullet}$ was strongly dependent on the proximity of the DART source to the mass spectrometer orifice, the grid potential of the DART source, and the presence or absence of excess moisture around the DART source.

The distance between the exit of the DART ion source and the sampling orifice of the mass spectrometer used in the study had a major influence on the DART background mass spectrum. The relative abundance of O$_2$$^{+\bullet}$ diminished rapidly as the DART/orifice was increased from 5 to 15 mm, whereas the relative abundance of the water dimer $[(H_2O)_2 + H]^+$ was maximized at a distance of 15 mm.

When additional moisture was introduced between the DART source and the orifice of the mass spectrometer, the relative abundance of $O_2^{+\bullet}$ diminished and the relative abundance of the protonated water clusters increased. Conversely, the relative abundance of $O_2^{+\bullet}$ could be increased by bleeding oxygen gas into the gap between the DART source and the orifice of the mass spectrometer. Excess moisture could also be reduced by flaming the ceramic insulator cap on the DART ion source, by increasing the temperature of the DART gas heater, by minimizing the distance between the DART source and the orifice of the mass spectrometer, and/or introducing a small amount of dry air or oxygen into the DART sampling region.

Figure 3.2 shows positive-ion DART mass spectra of *n*-hexadecane measured under conditions that produce abundant $O_2^{+\bullet}$ in the background mass spectrum as shown in Figure 3.1(b). An abundant peak representing $M^{+\bullet}$ was observed at nominal m/z 226, together with a peak that represented the $[M-H]^+$ ion, and peaks that represented fragment ions dominated by $[M-(CH_2)_n-H]^+$ ($n = 1, 2, 3 \cdots$) ions. At a gas temperature of 200 °C (Figure 3.2(a)), the base peak (100% relative abundance) in the mass spectrum represented $M^{+\bullet}$ and the peaks representing the $[M-(CH_2)_n-H]^+$ ($n = 1, 2, 3 \cdots$) fragments had relative abundances of approximately 20% or less. As the gas temperature was increased to 300 °C (Figure 3.2(b)), the base peak was $C_6H_{13}^+$ and the peak representing $M^{+\bullet}$ was reduced to a relative

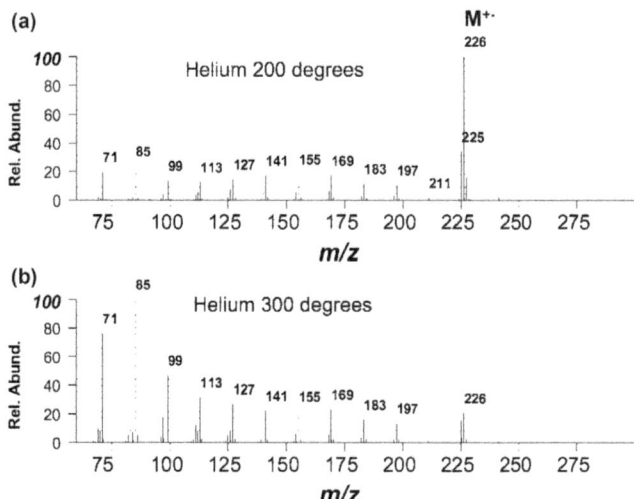

Figure 3.2   Positive-ion DART mass spectra of *n*-hexadecane measured under conditions that produced abundant $O_2^{+\bullet}$ in the background mass spectrum as shown in Figure 3.1(b): (a) gas heater set to 200 °C; (b) gas heater set to 300 °C.

Reprinted with permission from R. B. Cody, *Anal. Chem.*, 2009, **81**, 1101–1107. Copyright 2009 American Chemical Society.

abundance of approximately 20%. The generation of $M^{+\bullet}$ and $[M–H]^+$ ions by positive-ion DART through $O_2^{+\bullet}$, which is typical for alkanes, can be ascribed to Scheme 3.4.

Direct Penning ionization to generate $M^{+\bullet}$ ions is also possible since it is responsible for the presence of $O_2^{+\bullet}$ ions in the DART background mass spectrum. Indeed, Cody[10] has also observed $M^{+\bullet}$ ions for *n*-hexadecane when no significant $O_2^{+\bullet}$ was visible in the background mass spectrum (Figure 3.3). The generation of $M^{+\bullet}$ and $[M–H]^+$ ions by positive-ion DART through Penning ionization, which is typical for alkanes, can be ascribed to Scheme 3.5.

$$He(2^3S) + O_2 \rightarrow He(1^1S) + O_2^{+\bullet} + e^- \qquad \text{as ME(He)>IE(O}_2) \qquad (A)$$

$$O_2^{+\bullet} + M \rightarrow O_2 + M^{+\bullet} \qquad \text{If IE(O}_2)\text{>IE(M)} \qquad (B)$$

$$O_2^{+\bullet} + M \rightarrow O_2 + [M - R]^{+\bullet} + R^\bullet \qquad \text{If IE(O}_2)\text{>IE(M)} \qquad (C)$$

Scheme 3.4   Generation of $M^{+\bullet}$ and $[M–H]^+$ ions by positive-ion DART through $O_2^{+\bullet}$ charge exchange, which is typical of alkanes.

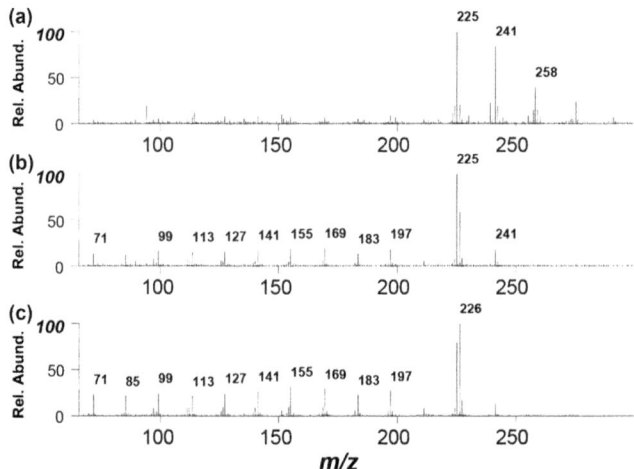

Figure 3.3   Positive-ion DART mass spectra of *n*-hexadecane measured under different grid potential: (a) 50 V; (b) 350 V; (c) 650 V. No significant $O_2^{+\bullet}$ is visible in the background mass spectra.
Reprinted with permission from R. B. Cody, *Anal. Chem.*, 2009, **81**, 1101–1107. Copyright 2009 American Chemical Society.

$$He(2^3S) + M \rightarrow He(1^1S) + M^{+\bullet} + e^- \qquad \text{as ME(He)>IE(M)} \qquad (A)$$

$$M^{+\bullet} + M \rightarrow [M - H]^\bullet + C_nH_{2(n + 1)} + [M - (CH_2)_n - H]^+ \qquad (n\text{=0, 1, 2, 3...}) \qquad (B)$$

Scheme 3.5   Generation of $M^{+\bullet}$ and $[M–H]^+$ ions by positive-ion DART through Penning ionization, which is typical of alkanes.

Hydride abstraction is commonly observed in the methane CIMS of alkanes.[19] In addition, the dominant reagent ion in propane or isobutane CIMS at high pressure is an $[M–H]^+$ ion, which reacts with propane or isobutane primarily by hydride abstraction.[19] Therefore, it may be reasonable to assume that the $M^{+\bullet}$ of alkanes by positive-ion DART is also self-reactive for hydride/alkide abstraction reactions under appropriate DART conditions (reaction (B), Scheme 3.5). However, this reaction is obviously not rapid enough to result in the total absence of $M^{+\bullet}$ ions under some DART conditions where $M^{+\bullet}$ ions are observed (Figure 3.3(c)).

As shown in Figure 3.3(a) and (b), mass spectra obtained for alkanes also contained peaks that represented numerous oxidation products including $[(M+O)–H]^+$, $[(M+O)–3H]^+$, $[(M+2O)–H]^+$, $[(M+2O)–3H]^+$, and $[(M+3O)–H]^+$ ions. Chemical ionization with NO^+ is known to produce both hydride abstraction and oxidation.[19] Although the relative abundances of the peaks representing the $[M–H]^+$ ion and the oxidized species in mass spectra obtained with DART correlated with the abundance of NO^+ in the background, the exact origin of the oxidized species was not well understood. It may be noted that the relative abundance of these species was dependent on ion-source parameters, humidity, and analyte concentration. Increasing the potential on the discharge needle resulted in an increase in total ion current but also increased the fractional abundance of the peak representing NO^+, which correlated with analyte oxidation. Analyte oxidation was diminished at reduced discharge potential, but sensitivity would also be decreased.

3.2.4 Generation of Both $M^{+\bullet}$ and $[M+H]^+$ Ions

The coexistence of $M^{+\bullet}$ and $[M+H]^+$ ions requires the coexistence of the Penning (Scheme 3.1) and proton-transfer ionization (reactions (S)–(U), Scheme 3.2). This frequently occurs in positive-ion DART of aromatic hydrocarbons, although some alkyl aromatic hydrocarbons can also generate $[M–H]^+$ ions.[12,26] For nonpolar compounds such as alkanes which have PA values weaker than water and water clusters, proton transfer ionization (reactions (S)–(U), Scheme 3.2) cannot be enabled to generate $[M+H]^+$ ions. On the other hand, while polar compounds may be ionized through Penning ionization to generate $M^{+\bullet}$ ions, fast self-protonation[25] can occur to transform $M^{+\bullet}$ to $[M+H]^+$ ions because they usually have PA values stronger than their $[M–H]^\bullet$, which is similar to the self-protonation of water radical ions as in reaction (O), Scheme 3.2.

Examples of the coexistence of $M^{+\bullet}$ and $[M+H]^+$ ions are positive-ion DART of benzene, toluene, ethyl benzene, *p*-xylene, *o*-xylene, 1,2,4,5-tetramethylbenzene (1,2,4,5-TMB), naphthalene, anthracene and 9-methylanthracene, which will be described in the following sections. Generally, as the number of conjugated benzene rings increases, the PA of an aromatic hydrocarbon becomes stronger, and more abundant $[M+H]^+$ relative to $M^{+\bullet}$ ions should be observed.[26]

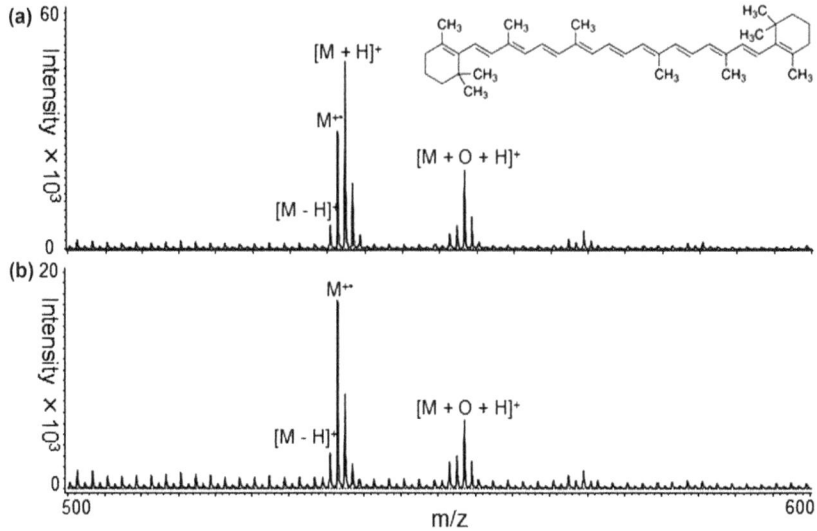

Figure 3.4 Positive-ion DART mass spectra of β-carotene at 300 °C DART gas
temperature: (a) sample was prepared as a 0.2 mg mL^{-1} solution
using heptane as solvent, then sampled with a melting-point capillary
and air dried for approximately 3 min before DART ionization;
(b) solid power of β-carotene was directly sampled with a melting-
point capillary. The DART source potentials were set to needle = 3500 V,
electrode 1 = +250 V, electrode 2 (grid) = +60 V. This spectrum
was obtained with the exit of the DART source positioned approximately 10 mm from the orifice of a JEOL AccuTOF mass spectrometer.

Figure 3.4 shows a positive-ion DART mass spectrum of β-carotene, a
highly conjugated nonaromatic hydrocarbon. In Figure 3.4(a), sample was
prepared as a 0.2 mg mL^{-1} solution using heptane as solvent, then sampled
with a melting-point capillary and air dried for approximately 3 min before
DART ionization. $[M–H]^+$ (m/z 535.4304), $M^{+\bullet}$ (m/z 536.4382) and $[M+H]^+$
(m/z 537.4460) together with their oxidized ions were observed. In addition,
numerous small fragment ions (not shown) were also present. In
Figure 3.4(b), DART ionization was accomplished by directly sampling solid
powder of β-carotene using a melting-point capillary. The relative abundance
of $M^{+\bullet}$ *versus* $[M+H]^+$ ions was obviously increased, which indicated that
more sample amount promoted the generation of $M^{+\bullet}$ ions. This can be
interpreted by the transient microenvironment mechanism (TMEM), which
will be described in detail later.

Figure 3.5 shows a positive-ion DART mass spectrum of C_{60} fullerene,
a highly conjugated carbon molecule. In Figure 3.5(a), the sample
was prepared as a 0.2 mg mL^{-1} solution using toluene as solvent, then
sampled with a melting-point capillary and air dried for approximately
3 min before DART ionization. Highly abundant $[M+H]^+$ (m/z 721.0078)
relative to $M^{+\bullet}$ (m/z 720.0000) ions were obtained. In addition, their

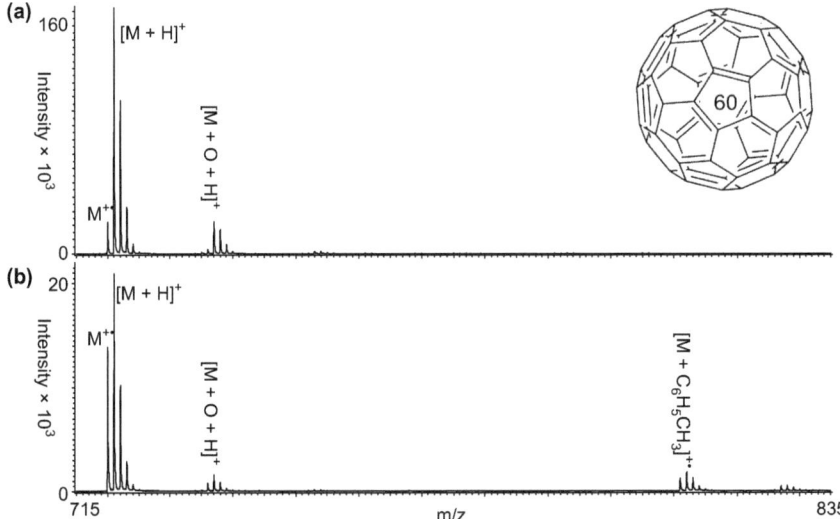

Figure 3.5 Positive-ion DART mass spectra of C_{60} fullerene at 500 °C DART gas temperature: (a) sample was prepared as a 0.2 mg mL^{-1} solution using toluene as solvent, then sampled with a melting-point capillary and air dried for approximately 3 min before DART ionization; (b) 0.2 mg mL^{-1} sample solution in toluene was directly sampled with a melting-point capillary. Other instrumental parameters were the same as described in Figure 3.4.

oxidized ions were also obvious. In Figure 3.5(b), DART ionization of C_{60} fullerene was accomplished by direct sampling of its toluene solution with a melting-point capillary. The relative abundance of $M^{+\bullet}$ *versus* $[M+H]^+$ ions was obviously increased, which indicated that toluene promoted the generation of $M^{+\bullet}$ ions. In addition, $[M+C_6H_5CH_3-H]^+$ and $[M+C_6H_5CH_3]^{+\bullet}$ ions together with their oxidized ions were also observed. These can also be interpreted by TMEM, which will be described in detail later.

Positive-ion DART of other less-polar compounds is also able to generate $M^{+\bullet}$ and $[M+H]^+$ ions. An example is positive-ion DART of dibenzosuberone.[10] The mass spectra obtained under the conditions associated with Figure 3.1(a) were dominated by $[M+H]^+$ ions; abundant $M^{+\bullet}$ ions are not observed. The same compound measured under the conditions that produce the background spectrum shown in Figure 3.1(b) gave a very different result, which contained both $M^{+\bullet}$ and $[M+H]^+$ ions and also the odd-electron fragment ion at *m/z* 180.0928, corresponding to $[M-CO]^{+\bullet}$. Another example is positive-ion DART of *p*-phenylenediamine (PPD) derivative compounds,[27] which are commonly used as antioxidants and antiozonants in black rubber. Both $M^{+\bullet}$ and $[M+H]^+$ ions were obtained for *N*-phenyl-*p*-phenylenediamine, *N*-isopropyl-*N'*-phenyl-*p*-phenylenediamine and *N*-(1,3-dimethylbutyl)-*N'*-phenyl-*p*-phenylenediamine.

3.2.5   Generation of $[M+NH_4]^+$ Ions

As an ambient ionization technique, the ambient environment has a great effect on DART ionization. Due to the ubiquitous existence of ammonia in ambient air,[18] NH_4^+ can often be observed in DART background mass spectrum.[8,10] Consequently, $[M+NH_4]^+$ ions can be frequently obtained for some polar compounds. The intensity of $[M+NH_4]^+$ ions can usually be enhanced by opening a bottle of dilute ammonium hydroxide or holding a cotton swab wetted with dilute ammonium hydroxide aqueous solution near the DART source[8] (Scheme 3.6).

Analytes for which ammonium attachment is commonly observed are compounds that do not contain strongly basic amine groups, but usually contain a few O atoms in the form of carbonyl, ketone, hydroxyl and/or ether functional groups.[28–40] These compounds largely have PA values somewhat less than that of ammonia, *i.e.* 853.6 kJ mol^{-1}, which make them unable to deprotonate NH_4^+, but their O functional groups are still able to form hydrogen bonds with NH_4^+.[25] This hypothesis is supported by positive-ion DART of triacetone triperoxide (TATP) and hexamethylenetriperoxide diamine (HMTD) where TATP was readily detected as $[M+NH_4]^+$ at *m/z* 240.1447 and HMTD was observed as $[M+H]^+$ at *m/z* 209.0776 (Figure 3.6).[37] This hypothesis is also supported by a publication[33] where positive-ion DART of neopentyl-2-nitro-4-(trifluoromethyl) phenyl carbamate generated a single $[M+NH_4]^+$ peak of high relative intensity, but positive-ion DART of neopentyl-*N,N*-dimethylamino phenyl carbamate generated $[M+H]^+$, $M^{+\bullet}$ and a series of oxygenated adducts ions. While we were unable to find the PA values of neopentyl-2-nitro-4-(trifluoromethyl) phenyl carbamate and neopentyl-*N,N*-dimethylamino phenyl carbamate in the literature, the PA values of *p*-nitroaniline and *N,N*-dimethyl-1,4-benzenediamine are 866.0 and 955.0 kJ mol^{-1}, respectively.[23] Therefore, it is reasonable to assume that neopentyl-2-nitro-4-(trifluoromethyl) phenyl carbamate has a PA value close to ammonia and neopentyl-*N,N*-dimethylamino phenyl carbamate has a PA value much stronger than ammonia.

The above hypothesis is further tested by positive-ion DART of D-(+)-glucose. A saturated solution of glucose in methanol (approximately 1–10 mg mL^{-1}) was sampled with a melting-point capillary and air dried for approximately 3 min before analysis. In a normal laboratorial environment only small-fragment ions of glucose, *i.e.* $[C_6H_4O_2+H]^+$ (*m/z* 109.0290), $[C_6H_6O_3+H]^+$ (*m/z* 127.0395) and $[C_6H_8O_4+H]^+$ (*m/z* 145.0501), were abundant (Figure 3.7(a)).[41] However, when a bottle of ammonium hydroxide was opened near the DART source, larger fragment ions, *i.e.* $[C_6H_{10}O_5+H]^+$ (*m/z* 163.0607), and $[M+NH_4-H_2O]^+$ (*m/z* 180.0872), as well as $[M+NH_4]^+$ (*m/z* 198.0978) ions were also abundant (Figure 3.7(b)).[41]

$$M + NH_4^+ \rightarrow [M+NH_4]^+ \qquad \text{If PA(M)} \approx \text{PA(NH}_3)$$

Scheme 3.6   Generation of $[M+NH_4]^+$ ions by positive-ion DART.

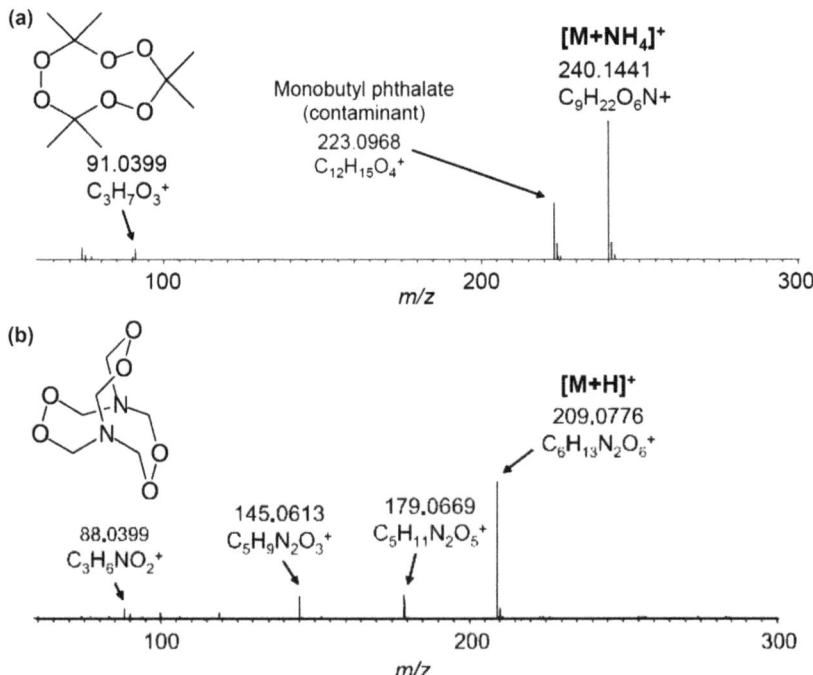

Figure 3.6 Positive-ion DART mass spectrum of the explosive compounds: (a) triacetone triperoxide (TATP) and (b) hexamethylenetriperoxide diamine (HMTD).) Reprinted with permission from ref. 30.

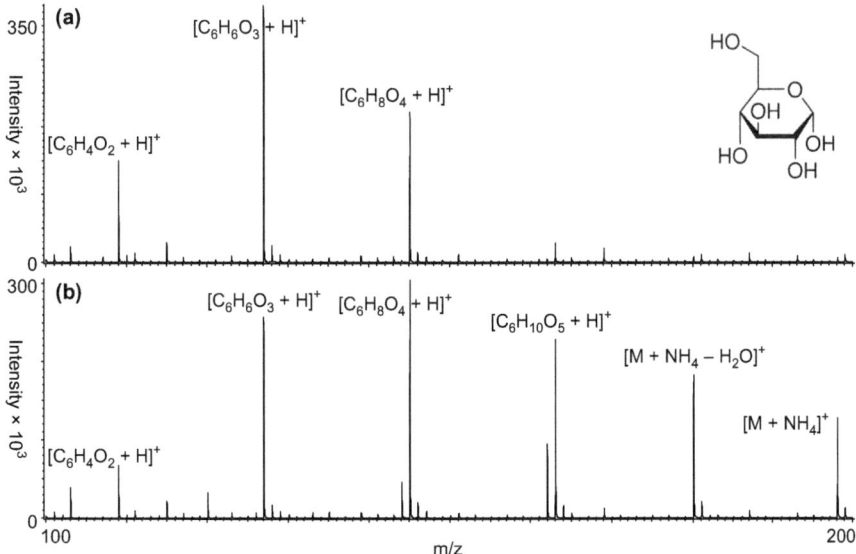

Figure 3.7 Positive-ion DART mass spectrum of D-(+)-glucose at 300 °C DART gas temperature: (a) at normal laboratorial environment and (b) when a bottle of ammonium hydroxide was opened near the DART source. Other instrumental parameters were the same as described in Figure 3.4.

3.2.6 Positive-ion DART with a Dopant

Theoretically, water can be considered as a dopant in positive-ion DART, although the water content is high enough in the ambient environment so that additional delivery of water into the DART source is unnecessary. In comparison with water, the ammonia content in the ambient environment is much lower so that additional delivery of ammonia into the DART source can be necessary and has been achieved by opening a bottle of dilute ammonium hydroxide or holding a cotton swab wetted with dilute ammonium hydroxide aqueous solution nearby the DART source.[8] Other reagents, which are not ubiquitous in the ambient environment, have also been used as DART dopants.

Fluorobenzene has been used as a dopant for atmospheric-pressure photoionization (APPI).[42] It has an ionization potential of 9.2 eV and a proton affinity of 755.9 kJ mol^{-1} (Tables 3.2 and 3.3) and acts as a charge-transfer reagent to ionize compounds that have ionization energies lower than 9.2 eV. Introduction of fluorobenzene vapor during DART ionization of cholesterol resulted in a dramatic change in the mass spectrum.[10] Under typical DART conditions where $O_2^{+\bullet}$ is not present in the DART background, cholesterol was ionized as low-abundant $[M-H]^+$ and high-abundant $[(M+H)-H_2O]^+$ ions (Figure 3.8(a)). However, if fluorobenzene vapor was

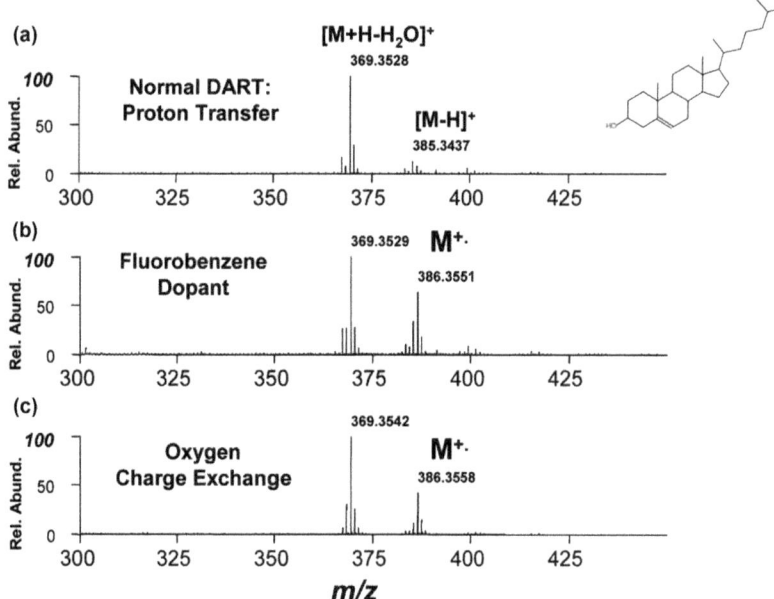

Figure 3.8   Mass spectra obtained with positive-ion DART for cholesterol with (a) no dopant, (b) with the addition of fluorobenzene vapor, and (c) decreased DART/orifice distance and increased grid voltage.
Reprinted with permission from R. B. Cody, *Anal. Chem.*, 2009, **81**, 1101–1107. Copyright 2209 American Chemical Society.

present, abundant $M^{+\bullet}$ ion was observed in the mass spectrum (Figure 3.8(b)). Abundant $M^{+\bullet}$ ion was also observed if the DART parameters are adjusted to maximize the abundance of $O_2^{+\bullet}$ (Figure 3.8(c)).

In fact, in positive-ion DART of solution samples, the volatile solvents can generate sufficient vapors to be a dopant for the analytes and the corresponding ionization mechanism will be described in detail next.

3.2.7 Transient Microenvironment Mechanism (TMEM)

The TMEM was originally proposed to address the matrix effect on positive-ion DART ionization.[12] It appears now that it can also address the effect of ionization environment, especially positive-ion DART with a dopant by considering the dopant as an artificial sample matrix, although the dopant can create a more permanent environment.

The TMEM states that when the DART gas stream, which contains a significant amount of both metastable helium atoms and $H_5O_2^+$, is in contact with a sample, a TME, which can shield the analytes from direct ionization by the DART gas stream, may be generated through desorption of the volatile matrix of the analyte. The DART gas stream will directly ionize the volatile matrix molecules in the TME, and then those matrix ions are the species that ionize the analytes *via* gas-phase ion/molecule reactions.

Scheme 3.7 lists a series of reactions that describe what can happen when a solution is analyzed. There are three steps:

(1) When the helium gas stream, containing metastable atoms, is in contact with the atmosphere, molecular ions of water are formed (reaction (A)) which in turn produce protonated water clusters (reaction (B));

$$He^* + H_2O \rightarrow He + H_2O^{+\bullet} + e^-, \qquad \text{as } ME(He) > IE(H_2O) \qquad (A)$$

$$H_2O^{+\bullet} + (H_2O)_m \rightarrow HO^\bullet + [(H_2O)_m + H]^+, \qquad \text{as } PA((H_2O)_m) > PA(HO^\bullet) \qquad (B)$$

$$He^* + S \rightarrow He + S^{+\bullet} + e^-, \qquad \text{as } ME(He) > IE(S) \qquad (C)$$

$$S^{+\bullet} + S_n \rightarrow [S - H]^\bullet + [S_n + H]^+, \qquad \text{if } PA(S_n) > PA([S-H]^\bullet) \qquad (D)$$

$$[(H_2O)_m + H]^+ + S_n \rightarrow (H_2O)_m + [S_n + H]^+, \qquad \text{if } PA(S_n) > PA((H_2O)_m) \qquad (E)$$

$$[S_n + H]^+ + M \rightarrow S_n + [M + H]^+, \qquad \text{if } PA(M) > PA(S_n) > PA([S-H]^\bullet) \qquad (F)$$

$$S^{+\bullet} + M \rightarrow [S - H]^\bullet + [M + H]^+, \qquad \text{if } PA(M) > PA([S-H]^\bullet) > PA(S_n) \qquad (G)$$

$$S^{+\bullet} + M \rightarrow S + M^{+\bullet}, \qquad \text{if } PA([S-H]^\bullet) > PA(S_n)$$
$$\text{and } IE(S) > IE(M) \qquad (H)$$

$$[(H_2O)_m + H]^+ + M \rightarrow (H_2O)_m + [M + H]^+, \qquad \text{if the TME is thin} \qquad (I)$$

Scheme 3.7 Main reactions in positive-ion DART. ME(He) is helium's metastable energy, 19.8 eV; $m = 1, 2,$ or 3; $n = 1$ or 2. Reaction (D) has a few variants for alkanes and chlorinated methanes as described in the text. Reprinted with permission from L. G. Song, S. C. Gibson, D. Bhandari, K. D. Cook and J. E. Bartmess, *Anal. Chem.*, 2009, **81**, 10080–10088. Copyright 2009 American Chemical Society.

(2) When the stream of He metastable atoms is in contact with the solvent molecules that constitute a TME, reaction (C) will take place resulting in solvent molecular ions that in turn react with other solvent molecules to produce protonated solvent molecules (reaction (D)). Protonated water clusters can also react with solvent molecules to produce protonated solvent molecules (reaction (E));

(3) The analyte molecules are ionized to form protonated molecules through gas-phase ion/molecule reactions with protonated solvent molecules (reaction (F)). Solvent molecular ions can react with analyte molecules to produce both protonated analyte molecules and analyte molecular ions (reactions (G) and (H)).

3.2.7.1 TME Generated by Common Solvents

In positive-ion DART of solution samples, the TME is generated by desorption of the solvent of a sample. The TME of a total 25 common solvents, *i.e.* methanol, ethanol, acetonitrile, 2-propanol, acetone, tetrahydrofuran (THF), ethyl acetate, *N,N*-dimethylformamide (DMF), dimethyl sulfoxide (DMSO), hexafluorobenzene, benzene, chlorobenzene, fluorobenzene, toluene, ethyl benzene, *p*-xylene, *o*-xylene, anisole, hexanes, heptane, iso-octane, cyclohexane, methylene chloride and chloroform, have been investigated by us, by analyzing approximately 1 μL each of individual solvents. The observed ions are listed in Table 3.4. Representative mass spectra of the solvents are shown in Figure 3.9.

In Table 3.4, the solvents are organized in four groups: proton acceptors, benzene derivatives, alkanes, and chlorinated methanes. Proton-acceptor solvents, listed in Table 3.4 according to the increasing order of their PA values, include methanol, ethanol, acetonitrile, 2-propanol, acetone, THF, ethyl acetate, DMF and DMSO. Because all of these solvents have IE values (Table 3.2) lower than helium's metastable energy (19.8 eV), they can be ionized through reaction (C) in Scheme 3.7 to generate $S^{+\bullet}$ ions. These should further undergo reaction (D) in Scheme 3.7 to become $[S+H]^+$ ions because they have PA values stronger than their (S–H) radicals (Table 3.3). It is noted that these solvents can form clusters, mostly dimers, possessing PA values stronger than the corresponding monomers. These clusters may be ionized through reactions (C) and (D) in Scheme 3.7. In addition, the PA values of these solvents (monomer and dimer) are also stronger than the PA values of water (monomer and dimer); these solvents may be ionized through reaction (E) in Scheme 3.7. For the alcoholic solvents, *i.e.* methanol, ethanol and 2-propanol, $[S_2+H-H_2O]^+$ ions were also observed. They are produced by the condensation reaction of a protonated alcohol with a neutral molecule to form a protonated ether ion.[43] However, we believe that these ions do not play a dominant role in the TME due to their low abundance. Overall, the mass spectra of the proton-acceptor solvents were dominated by $[S_2+H]^+$ and/or $[S+H]^+$ ions.

Table 3.4 Observed ion peaks with relative intensity over 5% in the positive-ion DART mass spectra of common solvents.[a]

Solvents		$[S-H]^+$	$S^{+\bullet}$	$[S+H]^+$	$[S_2+H]^+$	Other detected ions	
Proton acceptors	MeOH			72%	100%	$[S_2+H-H_2O]^+$	15%
	EtOH			74%	100%	$[S_2+H-H_2O]^+$	16%
	MeCN			100%	63%		
	iPrOH			14%	100%	$[S_2+H-H_2O]^+$	16%
	Acetone			90%	100%		
	THF			46%	100%		
	EtOAc			31%	100%		
	DMF				100%		
	DMSO				100%		
Benzene derivatives	C_6F_6		90%			$[S-F+OH]^+$	68%
	Benzene		100%	30%			
	PhCl		100%	18%			
	PhF		100%	39%		$[S-F+OH]^+$	63%
						$[2S-2F+H_2O]^+$	14%
						$[S-F+H_2O]^+$	11%
	PhCH$_3$		72%	100%			
	PhC$_2$H$_5$	18%	34%	100%			
	p-Xylene		26%	100%			
	o-Xylene		49%	100%			
	PhOCH$_3$		14%	100%			

Table 3.4 (Continued)

Solvents		$[S-H]^+$	$S^{+\bullet}$	$[S+H]^+$	$[S_2+H]^+$	Other detected ions	
Alkanes	Hexanes	100%				$C_4H_9^+$ 11% $[S-4H]^+$ 11% $[S-2H]^+$ 15%	$C_5H_{11}^+$ 10% $[S-3H]^+$ 62%
	Heptane	11%				$C_4H_9^+$ 100% $C_5H_{11}^+$ 52% $[S-2H]^+$ 4%	$C_5H_9^+$ 14% $[S-3H]^+$ 25%
	Isooctane					$C_4H_9^+$ 100%	
	CyHex[b]	100%					
CH_mCl_n	CH_2Cl_2[c]					$C_3H_7^+$ 7% $C_5H_{11}^+$ 60% $C_3H_7^+$ 11%	$C_4H_7^+$ 6% $CHCl_2^+$ 9%
	$CHCl_3$[d]					$C_5H_{11}^+$ 100% $C_6H_{11}^+$ 12% $C_6H_{11}Cl^+$ 11%	$C_5H_9^+$ 26% $CHCl_2^+$ 45% CCl_3^+ 2%
	$CHCl_3$[e]					$[C_2H_5OH+H]^+$ 100% $CHCl_2^+$ 5%	$[(C_2H_5OH)_2+H]^+$ 46%

[a]Contribution from the isotopic peak of $[m/z-1]$ was subtracted.
[b]Cyclohexane.
[c]15–100 ppm amylene or 40–100 ppm cyclohexene is present as preservative.
[d]50 ppm pentene is present as preservative.
[e]0.75% ethanol is present as preservative.
Adapted with permission from L. G. Song, S. C. Gibson, D. Bhandari, K. D. Cook and J. E. Bartmess, *Anal. Chem.*, 2009, **81**, 10080–10088. Copyright 2009 American Chemical Society.

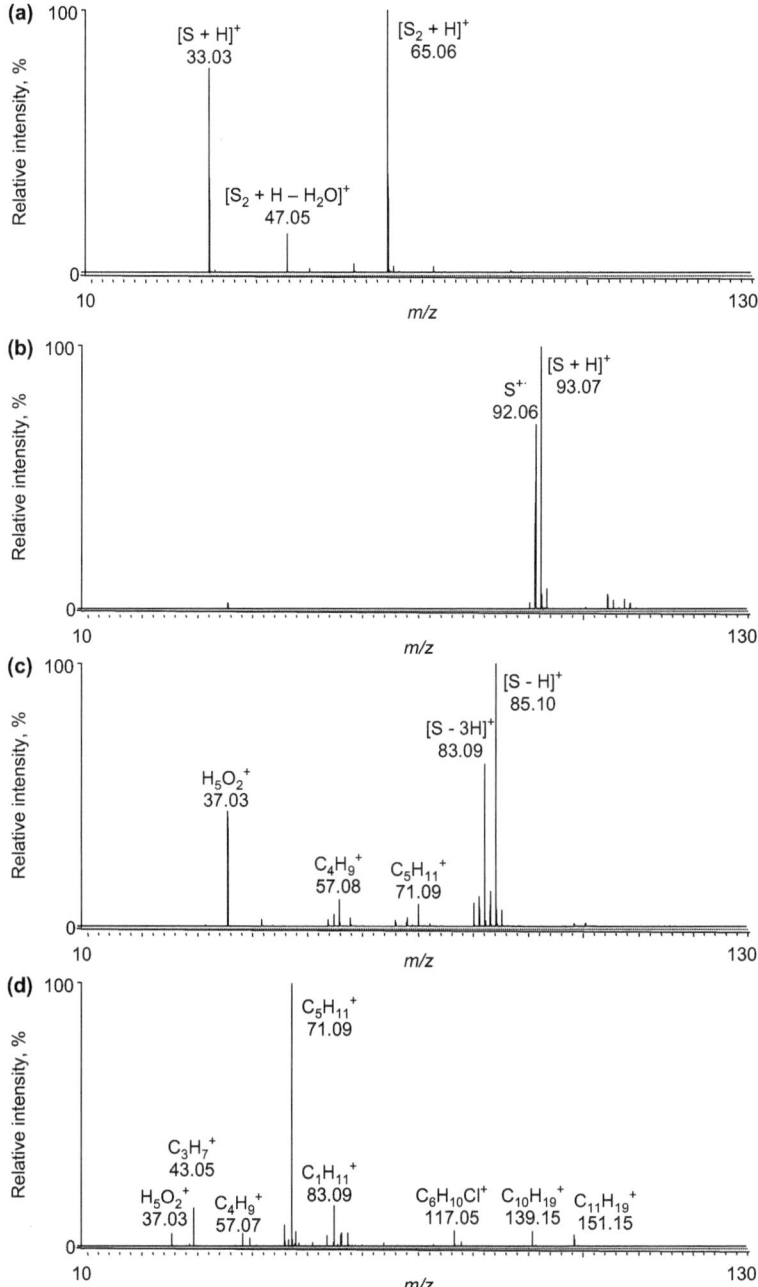

Figure 3.9 Positive-ion DART mass spectra of (a) methanol; (b) toluene; (c) hexanes; (d) chloroform with approximately 50 ppm pentene as preservative.
Adapted with permission from L. G. Song, S. C. Gibson, D. Bhandari, K. D. Cook and J. E. Bartmess, *Anal. Chem.*, 2009, **81**, 10080–10088.

Benzene-derivative solvents, listed in Table 3.4 according to the increasing order of their PA values, include hexafluorobenzene, benzene, chlorobenzene, fluorobenzene, toluene, ethyl benzene, *p*-xylene, *o*-xylene, and anisole. All of these solvents have IE values (Table 3.2) lower than helium's metastable energy and may be ionized through reaction (C) in Scheme 3.7 to generate $S^{+\bullet}$ ions. However, the $S^{+\bullet}$ ions cannot undergo reaction D in Scheme 3.7 to become $[S+H]^{+}$ ions due to weaker PA values than their (S–H) radicals (Table 3.3). Conversely, these solvents, with the sole exception of hexafluorobenzene, possess PA values stronger than the monomer of water and may also be ionized by reaction (E) in Scheme 3.7 to generate $[S+H]^{+}$ ions. As the PA values of these solvents increase, the abundance of the $[S+H]^{+}$ ions should increase, but the abundance of the $S^{+\bullet}$ ions may decrease. This is confirmed in Table 3.4. Additional ions were also observed for C_6F_6 and C_6H_5F. The $[S-F+OH]^{+}$ ions are thought to arise from a nucleophilic aromatic substitution reaction of $S^{+\bullet}$ with H_2O. This reaction is exothermic by *ca.* -84 kJ mol^{-1} for both C_6F_6 and C_6H_5F, which is driven by the extremely strong bond strength of the neutral HF product. The presence of $[S+H]^{+}$ ions for C_6H_5F but not C_6F_6, is consistent with the PA value for the former being 756 kJ mol^{-1}, 65 kJ mol^{-1} stronger than water, and 648 kJ mol^{-1} for the later, 43 kJ mol^{-1} weaker (Table 3.3). Similarly, the $[S-F+H_2O]^{+}$ ions, seen with C_6H_5F but not C_6F_6, probably arise from the reaction of the protonated C_6H_5F with water, losing HF. Likewise, the presence of the $[2S-2F+H_2O]^{+}$ ions with C_6H_5F but not C_6F_6, implies that this likely arises ultimately from the $[S+H]^{+}$ ion. For ethylbenzene, a $[S-H]^{+}$ ion was also observed, which could be interpreted similarly to the alkanes in the next paragraph. It should be noted that the $H_5O_2{}^{+}$ ion, *i.e. m/z* 37.1, was also observed for most of the benzene derivatives (data not shown in Table 3.4). Overall, the mass spectra of the benzene-derivative solvents were dominated by $S^{+\bullet}$ and $[S+H]^{+}$ ions.

Alkane solvents listed in Table 3.4 include hexanes, heptane, isooctane, and cyclohexane. Because all of these solvents have IE values (Table 3.2) lower than helium's metastable energy, they can be ionized through reaction (C) in Scheme 3.7 to generate $S^{+\bullet}$ ions. However, the $S^{+\bullet}$ ions of alkanes may be able to undergo a hydride/alkide abstraction reaction to form $[S-(CH_2)_n-H]^{+}$ ions:

$$S^{+\bullet} + S \rightarrow [S-H]^{\bullet} + C_nH_{2(n+1)} + [S-(CH_2)_n-H]^{+} \quad n = 0, 1, 2, \cdots$$

Such $[S-(CH_2)_n-H]^{+}$ ($n = 0, 1, 2, \cdots$) ions can be considered as $[S+H]^{+}$ ions of the corresponding alkenes. It should be noted that the $H_5O_2{}^{+}$ ion, *i.e. m/z* 37.1, also appeared in their mass spectra (data not show in Table 3.1). Overall, the mass spectra of the alkane solvents were dominated by $[S-(CH_2)_n-H]^{+}$ ($n = 0, 1, 2, \cdots$) ions.

Chlorinated-methane solvents in Table 3.4 include methylene chloride and chloroform. Because both methylene chloride and chloroform have IE values (Table 3.2) lower than helium's metastable energy, they can be ionized through reaction (C) in Scheme 3.7 to generate $S^{+\bullet}$ ions. It is possible that the $S^{+\bullet}$ ions of methylene chloride and chloroform undergo a hydride/HCl abstraction reaction to form $[S-H]^+$ and/or $[S-Cl]^+$ ions:

$$CH_2Cl_2^{+\bullet} + CH_2Cl_2 \rightarrow CHCl_2^{\bullet} + H_2 + CHCl_2^+$$
$$CHCl_3^{+\bullet} + CHCl_3 \rightarrow CCl_3^{\bullet} + HCl + CHCl_2^+$$
$$CHCl_3^{+\bullet} + CHCl_3 \rightarrow CCl_3^{\bullet} + H_2 + CCl_3^+$$

However, the appearance energy for $CHCl_2^+$ from $CHCl_3$ is only 0.1 eV above the IE,[23] and 0.8 eV above the IE for CH_2Cl_2 as the source. It may be that the $[S-H]^+$ and/or $[S-Cl]^+$ ions are simply a fragment ions formed on ionization of the halogenated matrices. It should be noted that preservatives such as 15–100 ppm "amylene" (pentenes) or 40–100 ppm cyclohexene are usually added to commercial methylene chloride for stabilization. Likewise, 50 ppm pentene or 0.75% ethanol is usually added to commercial chloroform. These preservatives can be ionized through the TMEs of methylene chloride and chloroform. Corresponding peaks representing the preservatives' ions are therefore observed in the mass spectra of methylene chloride and chloroform as shown in Table 3.4 (please refer to the next section for the interpretation of the observed ions).

3.2.7.2 Ionization in the TMEs of Solvents

The TME mechanism has been well elucidated in positive-ion DART ionization of ten representative analytes, *i.e.* naphthalene, 1,2,4,5-TMB, decanoic acid, 1-naphthol, anthracene, 1,3-dimethoxybenzene (1,3-DMOB), 9-methylanthracene, 12-crown-4, *N,N*-dimethylaniline (PhNMe$_2$) and tributylamine, each at a concentration of 1 µg mL^{-1} in 25 common solvents, *i.e.* methanol, acetonitrile, 2-propanol, acetone, THF, ethyl acetate, DMF, DMSO, anisole, *o*-xylene, toluene, chlorobenzene, fluorobenzene, hexanes, heptane, isooctane, methylene chloride and chloroform.

The boiling points, IEs and PAs of the ten representative analytes are listed in Table 3.5. Most of the IE and PA values were obtained from the NIST Chemistry WebBook.[23] The IE and PA of decanoic acid are estimated from that of smaller carboxylic acids *via* standard estimation schemes.[44] The PA of 1,2,4,5-tetramethylbenzene is taken as that of 1,2,3,5-tetramethylbenzene.[23] The PA of 1-naphthol is taken as that of phenol plus the difference between the PAs of naphthalene and benzene; this estimation agrees very well with computational values, for protonation para to the hydroxyl group.

The ions observed in positive-ion DART ionization of the 10 representative analytes, each at a concentration of 1 µg mL^{-1} in 25 common solvents, are listed in Table 3.6. Representative mass spectra of the analyte solutions are shown in Figure 3.10.

When analytes are dissolved in proton acceptor solvents, *i.e.* methanol, acetonitrile, 2-propanol, acetone, THF, ethyl acetate, DMF and DMSO, they should be ionized through reaction (F) in Scheme 3.7 as the TMEs of theses

Table 3.5 Boiling point (bp), ionization energy (IE) and proton affinity (PA) of ten representative analytes for the TMEM investigation.

Compound	Formula	$M^{+\bullet}$	$[M+H]^+$	Boiling point, °C	IE^a (eV)	PA^a (kJ mol^{-1})
Naphthalene	$C_{10}H_8$	128.063	129.070	218.0	8.14	803
1,2,4,5-TMB	$C_{10}H_{14}$	134.110	135.117	/	8.06	~846
Decanoic acid	$C_{10}H_{20}O_2$	172.146	173.154	268–270	~9.90	~848
1-Naphthol	$C_{10}H_8O$	144.058	145.065	278–280	7.76	~867
Anthracene	$C_{14}H_{10}$	178.078	179.086	340.0	7.44	877
1,3-DMOB	$C_8H_{10}O_2$	138.068	139.076	85–87 (7 mm Hg)	8.20	~892
9-Methylanthracene	$C_{15}H_{12}$	192.094	193.102	196–197 (12 mm Hg)	7.31	897
12-Crown-4	$C_8H_{16}O_4$	176.105	177.113	61–70 (0.5 mm Hg)	8.80	927
N,N-DMA	$C_8H_{11}N$	121.089	122.097	193–194	7.12	941
Tributylamine	$C_{12}H_{27}N$	185.214	186.222	216.0	7.40	999

^{a}Data obtained from NIST Chemistry WebBook (http://webbook.nist.gov) or estimated as described in the text.

Table 3.6 Observed ion peaks in the positive-ion DART mass spectra of the ten representative analytes in solutions of common solvents for the TMEM investigation.[a]

Solvents[b]	PhNMe$_2$ M$^+$	[M+H]$^+$	Naphthalene M$^+$	[M+H]$^+$	1,2,4-TMB M$^+$	[M+H]$^+$	1,3-DMOB M$^+$	[M+H]$^+$	1-Naphthol M$^+$	[M+H]$^+$
MeOH	/	49	/	/	/	/	/	18	/	/
MeCN	/	13	/	/	/	/	/	/	/	/
iPrOH	/	B	/	/	/	/	/	B	/	/
Acetone	/	12	/	/	/	/	/	/	/	/
THF	/	5	/	/	/	/	/	/	B	B
EtOAc	/	2	/	/	/	/	/	/	/	/
PhOCH$_3$	25	36	/	/	/	/	/	8	/	/
o-Xylene	B	B	/	/	5	4	11	9	6	18
PhCH$_3$	27	35	4	/	7	2	8	17	9	6
PhCl	34	27	B	B	15	6	16	17	12	4
PhF	51	63	3	/	5	3	10	28	8	5
Hexanes	31	78	9	1	11	8	13	46	12	16
Heptane	20	52	8	8	10	6	10	37	10	15
Isooctane	12	56	2	B	4	3	4	33	4	8
CH$_2$Cl$_2$	B	36	/	/	/	3	/	B	/	3
[d]CHCl$_3$	B	100	/	/	/	5	/	B	/	12

Solvents[b]	Decanoic acid M$^+$	[M+H]$^+$	12-Crown-4 M$^+$	[M+H]$^+$	Anthracene M$^+$	[M+H]$^+$	Tributyl-amine[c] M$^+$	[M+H]$^+$	9-Methyl-anthracene M$^+$	[M+H]$^+$
MeOH	/	/	/	31	/	/	/	34	/	7
MeCN	/	/	/	27	/	/	/	21	/	/
iPrOH	/	/	/	18	/	/	/	17	/	/
Acetone	/	/	/	18	/	/	/	19	/	/
THF	/	/	/	6	/	/	/	3	/	/
EtOAc	/	/	/	B	B	B	/	0	/	/
PhOCH$_3$	/	3	/	12	5	/	19	15	5	0
o-Xylene	/	0	/	18	11	1	17	20	13	3
PhCH$_3$	/	0	/	32	16	12	20	33	19	15
PhCl	/	5	/	17	15	8	20	21	15	9
PhF	B	5	/	36	13	8	19	63	13	12
Hexanes	/	7	/	72	13	22	20	80	12	19
Heptane	/	9	/	49	14	24	11	42	16	28
Isooctane	/	3	/	41	6	17	5	36	9	16
CH$_2$Cl$_2$	/	5	/	30	/	6	/	27	/	9
[d]CHCl$_3$	/	6	/	70	/	19	/	76	/	23

[a]Ion intensity was normalized to the most intensive one as a percentage. Contribution of ion intensity from the isotopic peak of [m/z − 1] was subtracted.
[b]No relevant peaks were observed when DMF and DMSO were used.
[c]A m/z 142.16 ion, which may be $(C_4H_9)_2NCH_2^{+\bullet}$, was observed. It may be a fragment from an unstable M$^{+\bullet}$ ion.
[d]0.75% ethanol is present as preservative. B: background ion.
Adapted with permission from L. G. Song, S. C. Gibson, D. Bhandari, K. D. Cook and J. E. Bartmess, *Anal. Chem.*, 2009, **81**, 10080–10088. Copyright 2009 American Chemical Society.

solvents are dominated by [S$_2$ + H]$^+$ ions. Therefore, only [M + H]$^+$ ions can be observed and their intensities should be dependent on PA(M)–PA(S$_2$) values (Tables 3.3 and 3.5). Methanol should be the best solvent for the ionization of all the analytes because it has the weakest PA(S$_2$) among the proton-acceptor solvents. DMF and DMSO should be the worst solvents

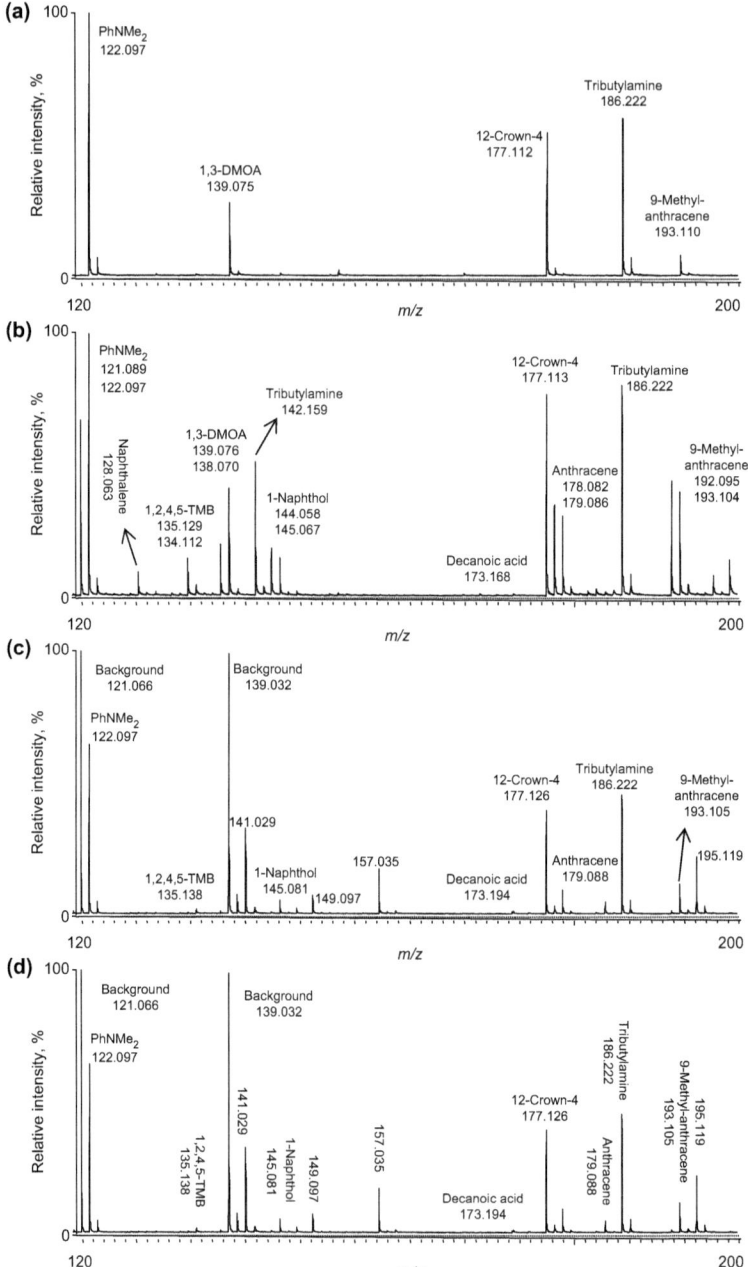

Figure 3.10 Positive-ion DART mass spectra of the 10 representative analytes, *i.e.*
naphthalene, 1,2,4,5-TMB, decanoic acid, 1-naphthol, anthracene, 1,3-
DMOB, 9-methylanthracene, 12-crown-4, PhNMe$_2$ and tributylamine, each
at a concentration of 1 µg mL^{-1} in (a) methanol; (b) toluene; (c) hexanes;
(d) chloroform with approximately 50 ppm pentene as preservative.
Adapted with permission from L. G. Song, S. C. Gibson, D. Bhandari, K. D.
Cook and J. E. Bartmess, *Anal. Chem.*, 2009, **81**, 10080–10088. Copyright
2009 American Chemical Society.

for the ionization of all the analytes because they have the strongest PA(S$_2$). Both conclusions are confirmed in Table 3.6. Even with methanol, half of the analytes including naphthalene, 1,2,4,5-TMB, 1-naphthnol, decanoic acid and anthracene were still not ionized as shown in Figure 3.10(a). This is because the corresponding PA(M)–PA(S$_2$) values are negative. For both DMF and DMSO, none of the analytes were ionized, also because the corresponding PA(M)–PA(S$_2$) values are negative.

When analytes are dissolved in benzene-derivative solvents, *i.e.* anisole, *o*-xylene, toluene, chlorobenzene, fluorobenzene, they should be ionized through reactions (F), (G) and/or (H) in Scheme 3.7 because the TMEs of these solvents are dominated by S$^{+\bullet}$ and [S + H]$^+$ ions. Therefore, [M + H]$^+$ ions can be observed when PA(M)–PA(S) values (Tables 3.3 and 3.5) are positive. M$^{+\bullet}$ ions can also be observed when IE(M)–IE(S) values (Tables 3.2 and 3.5) are negative. This is confirmed in Table 3.6. No significant M$^{+\bullet}$ ions were observed for decanoic acid due to its high IE; and anisole was the worst solvent to ionize the analytes *via* their M$^{+\bullet}$ ions due to its lowest IE among the solvents. The most favorable benzene-derivative solvent to ionize the 10 representative analytes is toluene; the corresponding mass spectrum is shown in Figure 3.10(b).

When analytes are dissolved in alkane solvents, *i.e.* hexanes, heptane and isooctane, they should be ionized first by reaction (F) in Scheme 3.7 because the TMEs of these solvents are dominated by [S–(CH$_2$)$_n$–H]$^+$ ($n = 0$, 1, 2, $\cdots$) ions which can be considered as [S + H]$^+$ ions of the corresponding alkenes. Therefore, [M + H]$^+$ ions can be observed for all the analytes because the PA(M)–PA(S) values (Tables 3.3 and 3.5) are positive. This is confirmed in Table 3.6. In addition, most of the analytes were also ionized as M$^{+\bullet}$ ions (please refer to the next section for the interpretation of the absence of M$^{+\bullet}$ ions from decanoic acid and 12-crown-4). This should take place through reaction (H) in Scheme 3.7 and requires both the existence of S$^{+\bullet}$ ions and the IE(M)–IE(S) values (Tables 3.2 and 3.5) to be negative. Although peaks representing S$^{+\bullet}$ ions of alkane solvents were not observed, they did exist as the precursors of [S–(CH$_2$)$_n$–H]$^+$ ($n = 0$, 1, 2, $\cdots$) ions. Such alkane radical cations are thermochemically higher in energy as reactants for reaction (D) than benzene-derivative radical cations, and thus may have shorter lifetime in the source so that they are not observed. There was no significant difference among the alkane solvents in the ionization of all the analytes. Figure 3.10(c) shows the corresponding mass spectrum when hexanes were used.

When methylene chloride and chloroform were used as solvent, the ionization of the analytes appeared to be similar as alkanes. However, no significant numbers of M$^{+\bullet}$ ions were observed, possibly implying a greater reactivity of S$^{+\bullet}$ ions from methylene chloride and chloroform than alkanes. This is consistent with the IE of methylene chloride and chloroform being higher than the alkanes (Table 3.2). In addition, the ionization of the analytes and the stabilizers in the solvents appeared similar. With 15–100 ppm amylene (presumably a pentene mixture) as stabilizer in methylene chloride, protonated pentene was observed. With 0.75% ethanol as stabilizer in chloroform, the protonated monomer and dimer of ethanol were observed. With 50 ppm pentene as stabilizer in chloroform, protonated pentene was observed along with (M–H) pentene positive ion (similar as $[S-(CH_2)_n-H]^+$ ($n = 0$, 1, 2, $\cdots$) ions of alkanes) and $C_6H_{10}Cl^+$ of unknown provenance. Figure 3.10(d) shows the corresponding mass spectrum when chloroform were used.

3.2.7.3   TMEs Generated by Volatile Solids

A TME can also consist of vapors from solids that can be desorbed by the DART gas stream and further ionized by DART. The TMEs generated by the ten representative analytes as listed in Table 3.5 (most of them are volatile solids), *i.e.* naphthalene, 1,2,4,5-TMB, decanoic acid, 1-naphthol, anthracene, 1,3-DMOB, 9-methylanthracene, 12-crown-4, PhNMe$_2$ and tributylamine, have been investigated.

First, the analytes were sampled by dipping the closed end of a melting-point capillary directly into the solid. Approximately 0.1 mg of solid was sampled in this way, and similar TMEs to those when approximately 1 μL solvents were analyzed were observed. Next, the amount of solid sample was reduced in order to assess the changes in the TME. The compounds were dissolved in a solvent, *e.g.* toluene, at individual concentration of 10 mg mL^{-1}, 100 μg mL^{-1} and 1 μg mL^{-1}. They were sampled by dipping the closed end of a melting-point capillary directly into the solutions of the compounds and then air dried for approximately 3 min. Approximately 10 μg, 100 ng and 1 ng of compounds, which were dried from approximately 1 μL of solution, were analyzed. The results indicated that approximately 10 μg of solid was required to generate an efficient TME, *i.e.* both $M^{+\bullet}$ and $[M+H]^+$ ions are abundant for naphthalene. If liquid instead of solid was used, the required volume should be 10 nL, assuming a 1 mg mL^{-1} density.

Table 3.7 lists the observed ions by positive-ion DART for approximately 10 μg of individual analyte. The generation of $[M-H]^+$, $M^{+\bullet}$ and $[M+H]^+$ ions were mostly through reactions (C), (D) and (E) in Scheme 3.7 that were also used to interpret the generation of similar ions from the solvents. It should be noted that no $M^{+\bullet}$ ion was observed for decanoic acid and 12-crown-4, which is probably due to reaction (D) in Scheme 3.7, although the PAs of the corresponding (M–H) radicals were not available. Other ions were also detected as shown in Table 3.7 due to gas-phase ion–molecule reactions.

3.2.7.4   Ionization under the TMEs of Volatile Solids

The ionization of impurities in volatile solids should take place through gas-phase ion/molecule reactions with the ions of these volatile solids. Two such samples, *i.e.* 1 ng naphthol in 10 μg naphthalene and 1 ng naphthalene in 10 μg naphthol (1 : 10 000), were analyzed. Abundant $M^{+\bullet}$ and $[M+H]^+$ ions of naphthol were observed for the sample of 1 ng naphthol in 10 μg

Table 3.7 Observed ion peaks with relative intensity over 10% in the positive-ion DART mass spectra of 10 μg individual solid analytes for the TMEM investigation.[a]

Analytes	$[M-H]^+$	M^+	$[M+H]^+$	$[M_2+H]^+$	Others detected ions	
Naphthalene		68%	100%			
1,2,4,5-TMB	11%	91%	100%			
Decanoic acid			61%	100%	$[M-H_2O+H]^+$	58%
1-Naphthol		43%	100%			
Anthracene		51%	100%			
1,3-DMOB		17%	100%		$[M-H+CH_3]^+$	20%
9-Methylanthracene		78%	100%		$[M+O]^+$	12%
					$[M+O_2+H]^+$	27%
12-Crown-4			100%			
PhNMe$_2$	44%	36%	100%		$[M+CH_3]^+$	20%
					$[M-CH_3+2H]^+$	<10%
Tributylamine	24%		100%		$[M-C_3H_7]^+$	83%

[a]Contribution of ion intensity from the isotopic peak of $[m/z-1]$ was subtracted. Adapted with permission from L. G. Song, S. C. Gibson, D. Bhandari, K. D. Cook and J. E. Bartmess, *Anal. Chem.*, 2009, **81**, 10080–10088. Copyright 2009 American Chemical Society.

naphthalene. As shown in Table 3.7, the TME from 10 μg naphthalene consisted of its $M^{+\bullet}$ and $[M+H]^+$ ions, which would ionize naphthol through reactions (F) and (H) in Scheme 3.7 because naphthol possesses a lower IE and stronger PA value than naphthalene (Table 3.5). No ions of naphthalene were observed for the sample of 1 ng naphthalene in 10 μg naphthol. As shown in Table 3.7, the TME from 10 μg naphthol consisted of its $M^{+\bullet}$ and $[M+H]^+$ ions, which would not ionize naphthalene through reactions (F) and (H) in Scheme 3.7 because naphthalene possesses a higher IE and weaker PA than naphthol (Table 3.5).

A critical analyte-to-matrix ratio is explored to better predict the effect of TME. When the analyte-to-matrix ratio is lower than the critical ratio value, DART ionization will be controlled by the TME. Three more samples were analyzed: 10 ng naphthalene in 10 μg naphthol (1:1000), 100 ng naphthalene in 10 μg naphthol (1:100) and 1 μg naphthalene in 10 μg naphthol (1:10). $M^{+\bullet}$ and $[M+H]^+$ ions of naphthalene were observed when naphthalene is in excess of 100 ng, which indicated that the DART ionization was no longer controlled by the TME. Therefore, naphthalene ionization in a naphthol matrix was mainly controlled by the TME in ratios below 1:100. It should be noted that the critical ratio should be dependent on the DART temperature and the boiling points of the analyte and matrix.

When a mixture of solid analytes with comparable relative amounts are ionized by positive-ion DART, ion–molecule adducts, *i.e.* $[(M_a+H)+M_b]^+$ where the ions and molecules were both from the analytes, have been observed.[45] Figure 3.11 shows representative positive-ion DART mass spectra of a simulated slurry sample corresponding to 50% reaction progress of a batch slurry reaction at 30% (w/w) slurry concentration. In Figure 3.11(a), the solvent of the slurry, *i.e.* xylenes, was evaporated, which took approximately 3 min at ambient temperature. In Figure 3.11(b), the slurry was completely

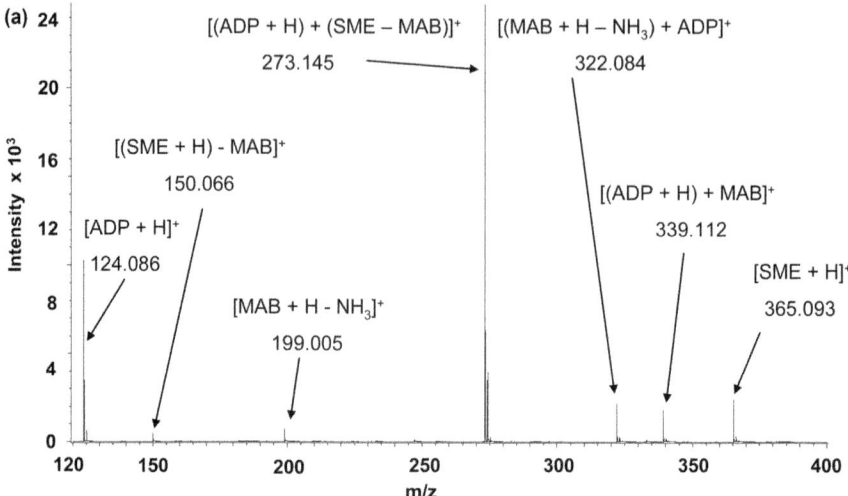

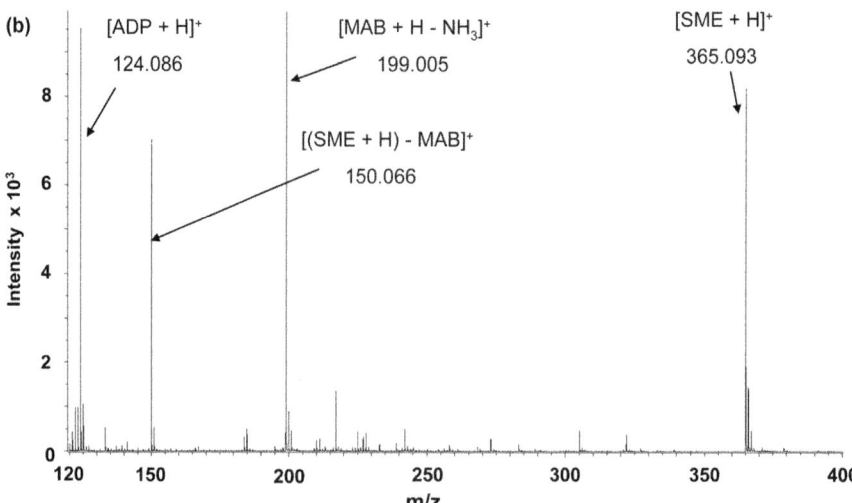

Figure 3.11   Representative positive-ion DART mass spectra of a slurry sample
corresponding to 50% reaction progress at 30% (w/w) slurry concentration: (a) after solvent, *i.e.* xylenes, was evaporated; (b) after slurry was
completely dissolved in xylenes and became a 0.08 mM solution.
Reprinted with permission from D. S. Cho, S. C. Gibson, D. Bhandari,
M. E. McNally, R. M. Hoffman, K. D. Cook and L. G. Song, *Rapid
Commun. Mass Spectrom.*, 2011, **25**, 3575–3580.

dissolved in xylenes and became a 0.08 mM solution, which took approximately 30 min with ultrasonication at ambient temperature. Protonated
molecules of 2-amino-4,6-dimethylpyrimidine (ADP) and sulfometuron
methyl ester (SME), *i.e.* $[ADP+H]^+$ and $[SME+H]^+$ and in-source

fragmentation ions of 2-(amino-sulfonyl) benzoate (MAB) and SME, *i.e.* $[MAB + H-NH_3]^+$ and $[SME + H-MAB]^+$, were observed in both Figure 3.11(a) and (b). However, additional ion–molecule adducts, *i.e.* $[(ADP + H) + MAB]^+$, $[(MAB + H-NH_3) + ADP]^+$ and $[(ADP + H) + (SME-MAB)]^+$, were obtained (Figure 3.11(a)).

The generation of uncommon ion–molecule adducts *via* DART ionization, *i.e.* $[(M_a + H) + M_b]^+$ where the ions and molecules were both from the analytes, which is shown in Figure 3.11(a), can also be interpreted by the TMEM.[12] The reagent ions generated by comparable relative amounts of analyte solids should be from the volatile analytes, *i.e.* ADP and MAB. Consequently, ion–molecule adducts of the reagent ions and analyte molecules, *i.e.* $[(ADP + H) + MAB]^+$, $[(MAB + H-NH_3) + ADP]^+$ and $[(ADP + H) + (SME-MAB)]^+$, were observed.

3.2.8 DART In-source Fragmentation

DART in-source fragmentation of labile molecules has been studied in detail in a few articles.[46–49] Systematic studies with the antifungal voriconazole by Lapthorn and Pullen[46] demonstrated the increase in the relative abundance of fragment ions when DART gas temperatures were ramped from 50 to 500 °C.[46]

Harris and Fernandez[50] used finite-element computational simulations to provide a graphical representation of the thermal gradients and fluid dynamics in the DART ionization region. Comparison of the simulations with matching experiments have demonstrated that gas velocity and temperature gradients play a central role in successfully transmitting ions generated within the DART ionization region into the atmospheric-pressure interface of the mass spectrometer. It was also observed that due to heat losses to the environment, the effective gas temperature measured at different points within the ionization region was consistently lower than the DART gas temperature set through the software.

Recently, the internal energy distributions of a series of *p*-substituted benzyl-pyridinium ions generated by both DART and ESI were measured by Fernandez and coworkers[48] using the "survival yield" method. DART mean internal energy values at gas flow rates of 2, 4, and 6 L min^{-1} and at set temperature of 175, 250, and 325 °C were in the 1.92–2.21 eV range. ESI mean internal energy at identical temperatures in aqueous and 50% methanol solutions ranged between 1.71–1.96 eV and 1.53–1.63 eV, respectively. The results indicated that ESI is a "softer" ionization technique than DART but that there was a certain degree of overlap between the two techniques for the particular TOF mass spectrometer used in the study. As a whole, when helium was used to sustain the glow discharge, there was an increase in DART internal energy with increasing gas temperatures and flow rates, indicating thermal ion activation and increased ion-source activation within the first differentially pumped region of the mass spectrometer. There was no evidence of internal energy deposition pathways from meta-stable-stimulated desorption or excess energy released from large differences in proton affinities, but fragmentation induced by high-energy helium metastable was observed at the highest gas flow rates and temperatures.

3.2.9 Nitrogen and Argon DART

3.2.9.1 Nitrogen DART

Nitrogen is inexpensive, readily available and may be substituted for helium in DART ionization. However, nitrogen DART has only been mentioned as a comparison to helium DART in a few articles so far.[8,10,51] Cody *et al.*[8] indicated that nitrogen DART primarily produces vibronically excited-state nitrogen molecules. The lowest metastable state of N_2 is at 6.2 eV, not high enough to ionize most molecules, but there are higher electronic and vibrational states that may contribute, up to *ca.* 11.88 eV.[21] Some characteristics about nitrogen DART have been described:[10] nitrogen has a higher breakdown potential than helium and a higher electric field is required to initiate an atmospheric-pressure glow discharge; it is more difficult to raise the temperature of nitrogen than helium, therefore an efficient gas heater is required to achieve comparable gas temperature; in addition, analyte oxidation is more commonly observed for some compounds.

Borges *et al.*[51] indicated that the higher internal energy of helium coupled with its higher thermal conductivity resulted in enhanced fragmentation as compared to nitrogen DART in the analysis of organometallic compounds.

At present, the ionization mechanism of positive-ion nitrogen DART has not been worked out in detail in the literature. However, there are noteworthy differences from positive-ion helium DART. First, in the present authors' hands, positive-ion nitrogen DART background spectra with no analyte present are typically at least two orders of magnitude weaker in ion intensity than for helium DART. Secondly, the ions observed in the background spectrum largely do not correspond to the expected air ions, and vary in mass from day to day. Finally, when an analyte is present, spectra roughly comparable in mass to that from helium DART are observed, at least for a wide range of polar organic molecules. This implies that the metastables from nitrogen are not ionizing the air species, but rather the analyte directly. The lowest IE for air species is that of O_2 at 12.07 eV.[16] This is above most of the known occupied metastable states of nitrogen.[21] Those close to it, such as the $E^3\Sigma_g^+$ state at 11.88 eV, are dissociative at even moderate vibronic levels[21] and known to be short lived, *ca.* 10 μs.[52] It is known that both ground state $N(^4S)$ atoms and metastable $N(^2D)$ and $N(^2P)$ atoms, 2.38 and 3.58 eV up, respectively, from the ground-state atoms, are produced in discharges in N_2.[20] These may recombine to produce a range of excited N_2 in states with at least 9.8 eV, and up to *ca.* 12.3 eV excess energy, but those are still marginal for generating O_2^+. Thus, there will be a wide range of metastable N_2 states occupied, from 6.18 eV up to *ca.* 12 eV. Analytes with ionization energies less than *ca.* 12 eV should thus be ionized. This is seen for common solvents such as methanol, ethanol, 2-propanol, acetone and ethyl acetate.[53] The corresponding ionization process should be similar as described in reactions (C) and (D) in Scheme 3.7 for positive-ion helium DART. The solvents undergo Penning ionization first to generate $S^{+\bullet}$ ions and self-protonation afterwards to transform $S^{+\bullet}$ to $[S + H]^+$ and/or $[S_2 + H]^+$ ions.

3.2.9.2 Argon DART

Argon DART was first used by Dane and Cody[54] due to its greater selectivity compared to helium DART. The long-lived argon excited state has an energy of 11.55 eV for the $^3P_2$ state and 11.72 eV for the $^3P_0$ state[55] compared to 19.8 eV for the helium $^2S_3$ state. The IE of water is 12.6 eV (Table 3.1), so water will not undergo Penning ionization in argon DART. Protonated water is a relatively nonspecific chemical ionization reagent because its proton affinity is weaker than that of most DART analytes (Table 3.3). In argon DART, a dopant can be selected for the substitution of water in helium DART so that better selectivity can be achieved.

Argon DART has been used for the selective ionization of 2,4,6-triamino-1,3,5-triazine, *i.e.* melamine, in powdered milk at the presence of 5-hydroxymethylfurfural (5-HMF), which is formed as a result of heating the milk powder.[54] Because protonated melamine (calculated *m/z* 127.0732) has the same nominal *m/z* as protonated 5-HMF (calculated *m/z* 127.0395), its analysis in powdered milk can be interfered by 5-HMF using low-resolution mass spectrometry. Argon DART is used in combination of two dopants, *i.e.* acetylacetone (AcAc) and pyridine, to selectively ionized melamine at the presence of 5-HMF. This involves a multistep reaction sequence, as shown in Scheme 3.8, in which AcAc undergoes Penning ionization followed by self-protonation. The protonated AcAc then reacts to protonate trace pyridine vapor, and protonated pyridine then selectively ionizes melamine. Although pyridine alone could be used directly as a dopant, it is desirable to minimize laboratory exposure to this compound.

The IEs of AcAc and pyridine are 8.85 and 9.26 eV, respectively. The PAs of AcAc and pyridine are 873.5 and 930 kJ mol^{-1}, respectively. The PA of melamine and 5-HMF are not available, but the amine functional groups in melamine suggest that melamine should have a high PA.

$$Ar^* + AcAc \rightarrow Ar + AcAc^{+\bullet} \tag{A}$$

$$AcAc^{+\bullet} + AcAc \rightarrow [AcAc - H]^\bullet + [AcAc + H]^+ \tag{B}$$

$$[AcAc + H]^+ + \text{pyridine} \rightarrow AcAc + [\text{pyridine} + H]^+ \tag{C}$$

$$[\text{Pyridine} + H]^+ + \text{melamine} \rightarrow \text{pyridine} + [\text{melamine} + H]^+ \tag{D}$$

Scheme 3.8 Selective ionization of melamine in powdered milk by using argon DART.
Reproduced from ref. 54.

Yang *et al.*[49] tested six organic solvents, *i.e.* methanol, ethanol, acetonitrile, acetone, fluorobenzene and dichloromethane, as dopants for argon DART. The results indicated that a solvent with low IE and low PA is good for the sensitivity of argon DART. Because acetonitrile possesses an IE of 12.2 eV (Table 3.2), which is higher than the internal energy of the long-lived argon excited state, it had no influence on the signal intensity of hypaconitine by argon DART. Dichloromethane possesses an IE of 11.33 eV (Table 3.2), which is close to the internal energy of the long-lived argon excited state, therefore not too much signal increase was observed on argon DART of hypaconitine. As for methanol, ethanol, acetone and fluorobenzene, their IEs (Table 3.2) are lower than the long-lived argon excited state, hence signal increase was observed on argon DART of hypaconitine and it was inversely proportional to the PAs of the solvents.

Argon DART is also reported to generate less fragmentation than helium DART, presumably because the long-lived excited state of argon has lower energy than that of helium.[49] A representative example is displayed in Figure 3.12 for the ionization of 2′-deoxycytidine. As shown in Figure 3.12(a), helium DART generated a high degree of fragmentation. With argon DART, however, the number and intensity of the fragment peaks are significantly reduced (Figure 3.12(b)).

3.2.10   DART and other Plasma-based Ambient Ionization Techniques

Fernandez and coworkers[7] recently listed plasma-based ambient ionization techniques to comprise DART, flowing atmospheric-pressure afterglow (FAPA), low-temperature plasma (LTP), dielectric barrier discharge ionization (DBDI), and microplasmas. All these techniques involve the generation of a direct current or radiofrequency electrical discharge between a pair of electrodes in contact with a flowing gas such as helium or nitrogen, generating a stream of ionized molecules, radicals, exited-state neutrals, and electrons. Some or all of the plasma species are directed toward the sample, with optional secondary heating of the plasma gas stream to enhance desorption. The main differences usually reside in one of the following points:[7] (a) if the plasma species are or are not removed from the flowing gas stream previous to interaction with the sample; (b) if the plasma gas stream is heated or not, and if the heating is performed by a supplementary heating element, or by Joule heating induced by the discharge current itself; (c) if the discharge is operated in dc, ac pulsed mode, *etc.*; (d) if the regime in the current–voltage curve (*I–V*) where the discharge is operated; (e) if the discharge is in the point-to-plane or annual configuration; and (f) if the plasma is established with a macro- (several millimeters or more) or micrometer-sized gap.

FAPA was first described under the name flowing afterglow atmospheric-pressure glow discharge (APGD).[56,57] DART and FAPA share many

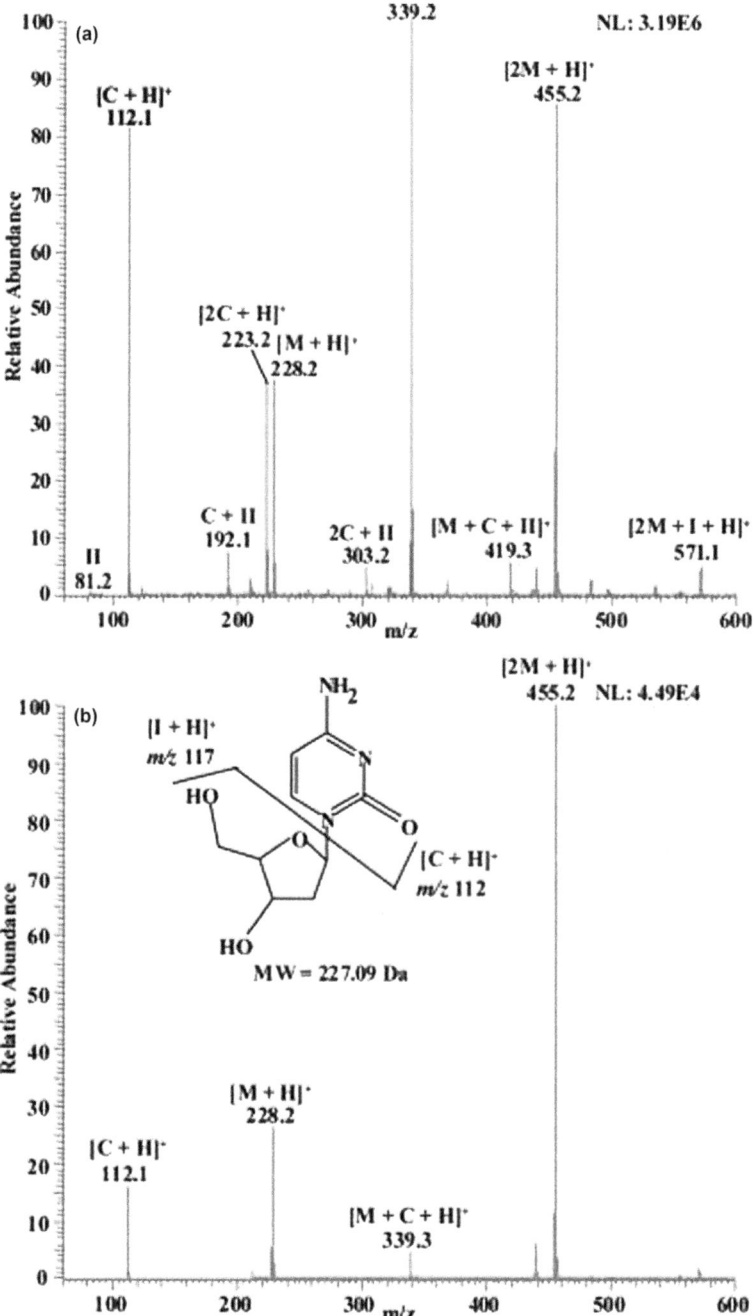

Figure 3.12 Positive-ion DART mass spectra of 50 μg mL^{-1} 2′-deoxycytidine: (a) helium DART; (b) argon DART.
Reprinted with permission from H. M. Yang, D. B. Wan, F. R. Song, Z. Q. Liu and S. Y. Liu, *Anal. Chem.*, 2013, **85**, 1305–1309. Copyright 2013 American Chemical Society.

similarities, but they have some important differences:[7] (a) in FAPA plasma species are not filtered by any electrodes before interaction with the sample, (b) FAPA is operated in the glow-to-arc (G–A) *I–V* regime (~ 25 mA) where DART is operated in the corona-to-glow (C–G) regime (2–5 mA), and (c) heating of the FAPA gas stream is not achieved with a resistive heating element as in DART but through Joule heating within the electric discharge. A study by Hieftje and coworkers[58] revealed that a FAPA source operated in a C–G discharge mainly produced protonated water clusters $[(H_2O)_m + H]^+$ with $m = 2$–7 in the background, whereas in the G–A discharge N_2^+, NO^+, O_2^+ and $[(H_2O)_m + H]^+$ with $m = 2$–3 were the main species. Additionally, a stronger reagent ion signal was observed by the G–A than the C–G discharge. It was also found that the gas exiting the discharge chamber reached a maximum of 235 and 55 °C for the G–A and C–G discharges, respectively.

Although these results are not directly transferable to DART experiments, as in DART plasma species are filtered by electrodes to remove charged species and an auxiliary heater is used, we think that it explained why an external heater is needed for DART ionization very well. However, we would like to indicate that under typical conditions DART produces different background mass spectra with predominant $[(H_2O)_2 + H]^+$ ion (Figure 3.1(a)). Moreover, we believe that the presence of charge-transfer reagent ions such as N_2^+, NO^+ and O_2^+ ions is not essential to generate $M^{+\bullet}$ ions as metastable helium atoms, which is abundant in the DART gas plume, is also able to generate $M^{+\bullet}$ ions through Penning ionization. On the contrary, high-abundance N_2^+, NO^+ and O_2^+ ions produced by FAPA indicated possible higher-energy plasma that could result in higher in-source fragmentation. Nevertheless, we do agree with the authors that a greater abundance and a wider variety of reagent ions produced by FAPA may transform to better ionization efficiency, *i.e.* better signal intensity of analyte ions.

DBDI and LTP make use of unheated plasma that interacts directly with the sample.[7] This plasma is generated by a dielectric barrier discharge produced in millimeter-sized gaps in various possible configurations. Dielectric barrier discharges (DBDs) make use of the novel feature that the electrodes are positioned outside of the discharge chamber and not in contact with the plasma.[59] These types of discharges are characterized by the generation of "cold" nonequilibrium plasmas at atmospheric pressure and the strong influence of the local field distortions caused by space charge effects. In DBDs, one or both discharge electrodes are separated by or coated with a dielectric material such as glass, ceramics, or polymers. Because this material cannot pass a direct current, the electric field in the gap has to be high enough to produce gas breakdown. At high frequencies (5–10 MHz), the dielectric material is less efficient at limiting the current; thus, this is the preferred mode of operation to limit power consumption. The initial description of DBDs for the ionization of organic compounds received the name DBDI for an ion source in a point-to-plane configuration where the sample substrate also acts as the dielectric.[60] This approach was later followed by an annular DBD configuration that received the name LTP and was allowed for directing the plasma toward specific points in space that allowed the analysis of samples of various sizes and on various substrates.[61]

Shelley and Hieftje[62] recently reported a comparison study among FAPA, DART and LTP using an identical mass spectrometer. Analytical performance of the three sources was directly compared in the analysis of acetophenone, naphthalene, methyl salicylate, and triethyl phosphate. The FAPA source yielded higher signal levels for all four analytes. In particular, FAPA produced signal levels five times higher than that from DART and LTP for 45 pmol of methyl salicylate (Figure 3.13). However, the DART source produced the least in-source fragmentation, followed by FAPA and LTP (Figure 3.13). Ionization-related matrix effects were investigated by mixing small amounts of vapor-phase matrices with a continuous stream of gaseous analyte which was introduced into each ionization source. A decrease in analyte signal upon introduction of a matrix signaled an ion suppression event. When the matrix species had a proton affinity the same as or greater than that of the analyte, all the sources suffered analytic signal suppression, even at moderate matrix-to-analyte concentration ratios. In every case, the FAPA was the least susceptible to the ion suppression process, likely owing to its higher power and ion current. In contrast, when the proton affinity of the matrix species was lower than that of the analyte, no matrix effect was observed with DART, although[19] an effect persisted for both FAPA and LTP.

3.3 Ionization Mechanism of Negative-ion DART

Negative ions – typically either $[M–H]^-$ *via* proton loss from analyte M, or $M^{-\bullet}$ ions formed by electron attachment to the analyte – allow access to additional structural information in DART mass spectra. As with chemical ionization mass spectrometry,[19] not all analytes form such ions, but for those that do, this provides useful structural identification.

3.3.1 Generation of Background Ions

The production of negative analyte ions *via* the DART mechanism starts with the same primary process as for positive ion generation: ionization of the various molecules in air by metastable neutrals, as in Scheme 3.2, reactions (A)–(C), also shown in Scheme 3.7, reaction (A). The difference for negative ion production is that it is the electrons coproduced in those processes that lead to the negative ions. In the conventional mechanism,[8] the electrons are "thermalized by collisions with the air molecules in a few nanoseconds at most." The electrons can then lead to negative analyte ions by several principal pathways:

1. The electrons can undergo direct attachment to an analyte M with a positive electron affinity, to create the radical anion $M^{-\bullet}$. (reaction (C) in Scheme 3.9) This requires that the $M^{-\bullet}$ ions be rapidly stabilized

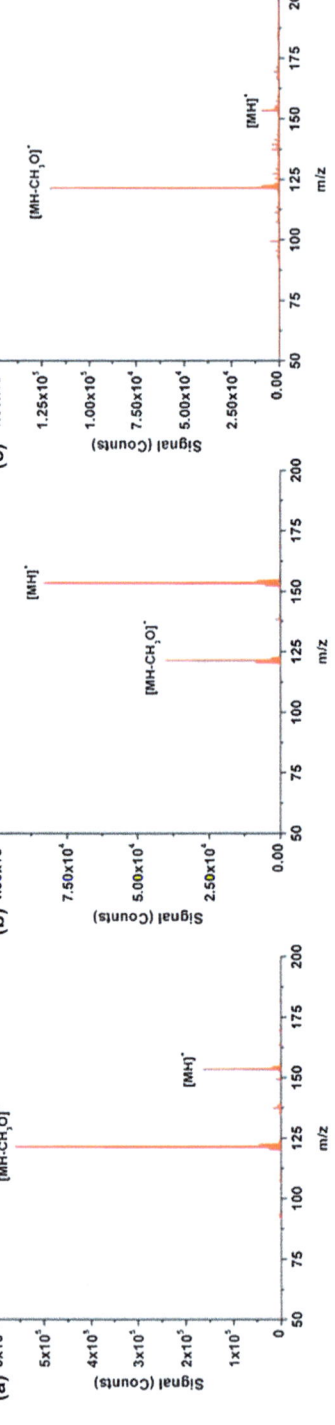

Figure 3.13 Mass spectra of methyl salicylate obtained with an ionization source of: (a) FAPA, (b) DART and (c) LTP. Reproduced from ref. 62.

by collisions, but at atmospheric pressure, this is not a problem. This also is viable only for analytes with a positive electron affinity.

2. In a similar process, electrons can also attach to O_2 to form $O_2^{-\bullet}$, as in reaction (D), Scheme 3.9. It can then act as a Brønsted base $(\Delta_{acid}H = 1477 \text{ kJ mol}^{-1})$[63] to deprotonate analytes more strongly acidic than that, creating $[M-H]^-$ ions, as per reaction (E).

3. The $O_2^{-\bullet}$ can also react with M *via* electron transfer, to create $M^{-\bullet}$, as per reaction (F), Scheme 3.9. This is limited to analytes with electron affinities (EAs) greater than that of O_2, at 0.45 eV.[63]

4. The thermalized electrons can react directly with analytes – or dopants – that undergo dissociative attachment with thermal electrons, to create fragment ions. Although this is a possibility for some analytes, the principal use here is to generate halide ions from dopants like CH_2Cl_2, which then attach to the analytes to form $[M + X]^-$ ions. This is useful in the cases where the analyte's radical anion is not bound thermochemically, and it is too weak a Brønsted acid to be deprotonated by any of the background ions.

One additional source of electrons is postulated in this Scheme 3.9, from surface collisions of metastable atoms ejecting electrons into the gas phase per reaction (B). This would create electrons with an original excess energy corresponding to the difference between the metastable atom's energy (19.8 eV for $He(2^3S)$) and the work function of the surface, roughly in the 4–6 eV range for most metals.[64] These electrons are then thermalized as above.

This scheme, however, should result in only $O_2^{-\bullet}$ as the background ion, in the absence of analyte and dopant. An examination of the background ions observed in negative-ion mode, with no analyte or dopant present, both in the original paper[8] and in further studies, as per Figure 3.14, reveals numerous additional ions, including but not limited to $O^{-\bullet}$, HO^-, HOO^-, $H_2O_3^-$, CO_3^-, HCO_3^-, CO_4^-, HCO_4^-, NO_2^-, and NO_3^-. These ions are highly

$$He^* + G \quad \rightarrow \quad G^{+\bullet} + He + e^{-*} \tag{A}$$

$$He^* + surface \rightarrow He + e^{-*} \tag{B}$$

$$e^{-*} + M \quad \rightarrow M^{+\bullet} + e^- \tag{C}$$

$$e^- + O_2 \quad \rightarrow O_2^{-\bullet} \tag{D}$$

$$O_2^{-\bullet} + M \quad \rightarrow [M-H]^- + HOO\bullet \tag{E}$$

$$\rightarrow M^{-\bullet} + O_2 \tag{F}$$

$$\rightarrow [M + O_2^-]^* \rightarrow [M+O_2]^- \tag{G}$$

Scheme 3.9   Generation of background ions in negative-ion DART.

variable in abundance from day-to-day, and with changing source conditions.[11] In Figure 3.14, the DART source potentials were set to needle $= 3500$ V, electrode $1 = -250$ V, electrode 2 (grid) $= -60$ V. This spectrum was obtained with the exit of the DART source positioned approximately 10 mm from the orifice of a JEOL AccuTOF mass spectrometer. Ions are identified within 2 mDa error. As the grid voltage changes, the relative abundance of observed ions can be changed. However, such changes in the grid voltage usually do not induce additional ions in the negative-ion

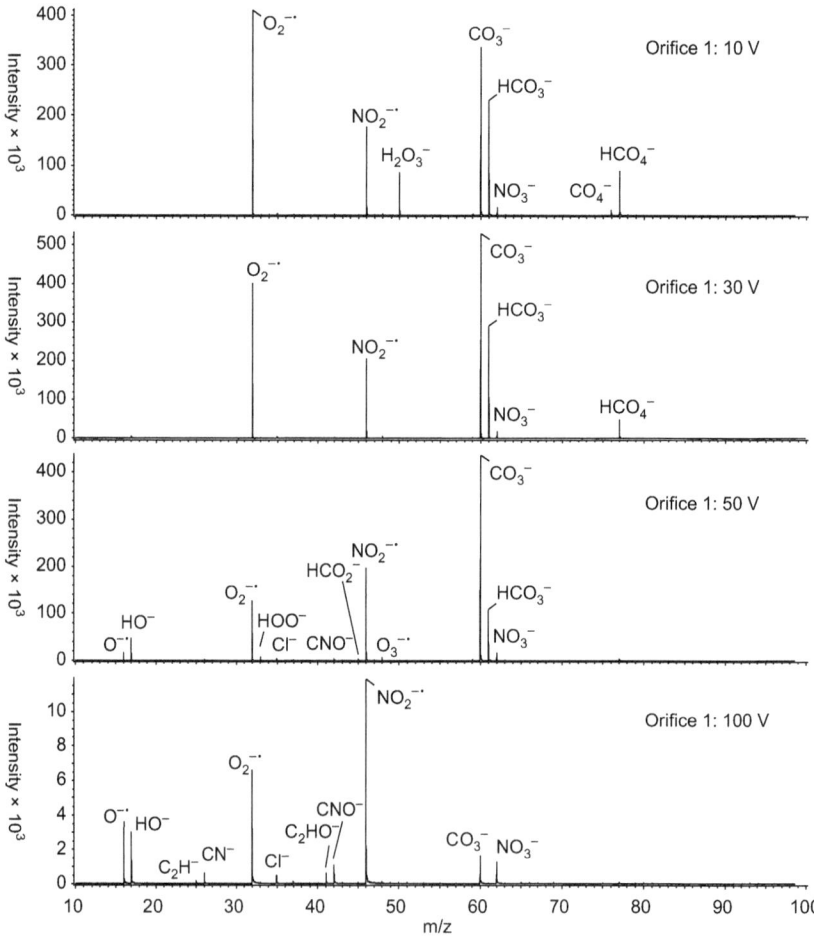

Figure 3.14 Typical negative-ion DART background mass spectra at 300 °C DART gas temperature. The DART source potentials were set to needle $= 3500$ V, electrode $1 = -250$ V, electrode 2 (grid) $= -60$ V. This spectrum was obtained with the exit of the DART source positioned approximately 10 mm from the orifice of a JEOL AccuTOF mass spectrometer. Ions are identified within 2 mDa error.

DART background mass spectra. The change of DART temperature can result in similar but less consequence as the change of orifice 1.

Due to the lack of sensitivity in the mass range below 20 *m/z* for most analyzers used with DART sources, it is difficult to quantitate ions in that range. $H_2O_3^-$ is likely the cluster ion $H_2O \cdot O_2^{-\bullet}$. The carbon-containing ions in the above list could be assigned to the reaction of background ions with ever-present atmospheric CO_2, to form these "carbonated" ions.[11] This is consistent with their decrease in intensity as the orifice 1 voltage is increased, as seen in Figure 3.14, along with an increase in the $[M-CO_2]^-$ ions corresponding to each. However, that implies that the ions O^-, HO^-, and HOO^- must be formed somehow, which is not consistent with the above scheme. Water is present in air, but is too weakly acidic by 159 kJ mol^{-1}, to be deprotonated by $O_2^{-\bullet}$.[63] Even if HOOH were present, it is also too weakly acidic by 86 kJ mol^{-1} to proton transfer to $O_2^{-\bullet}$.[63]

The origin of NO_2^- is not known, though it obviously must come from air ions. As outlined in Section 3.2.1, there are reasonable sources of neutral NO and NO_2 in the DART source. Electron attachment to NO_2 would yield the observed ion. The NO_3^- ion could be assigned to the reaction of NO_2^- with O or NO_2 with $O^{-\bullet}$. The EA of NO is so low (0.024 eV)[63] that NO^- would not survive in the source due to thermal detachment.

These background ions can be rationalized, however, by more closely examining a specific aspect of the conventional mechanism: the thermalization of the electrons produced in the primary event. These electrons when formed can have energies up to the difference between the metastable's energy and the ionization energy of the neutral molecule. This gives maximum electron energies, when using helium, of 7.7 eV from O_2, 4.2 eV from N_2, and 7.2 eV from H_2O.[16] Of course, not all electrons so formed will have this maximum energy, due to some energy being left in vibronic states of the cation that is coproduced, Nevertheless, there will be a distribution of electron energies, with some near that thermochemical limit. It is well known that O_2 and H_2O undergo dissociative attachment with large cross sections and energetic maxima right in this range.[65] Dissociative attachment of O_2 makes O^-, one of the observed background ions, with a maximum occurring at 6.4 eV.[66] H_2O undergoes dissociative attachment at a 6.6 eV maximum. Interestingly, the latter process does not yield HO^-, the most thermochemically favored species, but rather H^-, with HO^- present at about 3% of H^-, and $O^{-\bullet}$ at 0.3%.[67,68] H^-, however, reacts rapidly with H_2O to form HO^- and H_2. Conceptually, $O^{-\bullet}$ might abstract a hydrogen from water to form HO^-, but that reaction is endothermic by *ca.* 29 kJ mol^{-1}, and thus unlikely to occur.[63] $O^{-\bullet}$ does abstract a hydrogen from alkanes and alkenes to form HO^-,[69,70] a case where the presence of an analyte alters the nature of the reagent ions. It also is known to abstract a hydrogen and a proton from vinylidene groups, to create water and $[M-2H]^-$ ions.[71]

It is noted that N_2, although the most-abundant gas present, does not yield negative ions *via* dissociative attachment: neither $N_2^{-\bullet}$ or N^- are bound with respect to electron loss.[63]

Table 3.8 Thermochemical data for negative reagent ions.[a]

Ion	$\Delta_{acid}H^{b}$	$\Delta_{acid}G^{b}$	EA[c]
H^{-}	1675	1649	0.76
HO^{-}	1633	1606	1.83
O^{-}	1601	1576	1.46
HOO^{-}	1575	1546	1.08
O_2^{-}	1477	1451	0.45
NO_2^{-}	1424	1396	2.27
$HOCO_2^{-}$	1400	1366	3.68
Cl^{-}	1395	1373	3.61
$CO_3^{-\bullet}$	1386	1351	4.05

[a]From ref. 63.
[b]In units of kJ mol^{-1}.
[c]In units of eV.

At first glance, the most obvious source of HOO^{-} would seem to be hydrogen abstraction by $O_2^{-\bullet}$ from some suitable structure. Examination of the thermochemistry involved, however,[63] reveals that the bond dissociation energy of the hydrogen atom donor species would have to be less than 268 kJ mol^{-1} for such a reaction to be exothermic, and there are very few such molecules known. Simple H^{-} addition to the abundant O_2 species, stabilized by gas collisions, could provide this ion, though at present this is not proved.

The reagent ions so formed are given in Table 3.8, along with their thermochemical parameters. It is noted that the primary ions formed by electron dissociative attachment or simple attachment, are H^{-}, O^{-}, $O_2^{-\bullet}$, and some of the HO^{-}; the others must come from ion–molecule reactions of these in air.

3.3.2 Generation of $[M-H]^{-}$ Ions

It is evident from Table 3.8 that analytes considerably weaker in acidity than $O_2^{-\bullet}$ would be expected to yield $[M-H]^{-}$ ions based on the other observed reagent ions' basicity. However, because production of the primary ions H^{-} and $O^{-\bullet}$ involve resonance attachment of electrons in specific energy ranges,[67,68] anything that affects the electron-energy distribution in the source will alter the relative amounts of these ions. This includes the cleanliness of the surfaces in the source, *via* reaction (B) in Scheme 3.9, and the electric fields in the source. These will vary from day-to-day in real operation, and thus it is not surprising that negative-ion DART spectra might be less reproducible than positive-ion spectra. Due to the concentration of water in air, typically 1–4% due to water's vapor pressure and the relative humidity in a lab, it is likely that the most basic ion H^{-} will react away quickly with water and O_2. Thus, the expected upper limit on $[M-H]^{-}$ ions is likely those that can be deprotonated by HO^{-}. This is seen in that in our lab's methanol, at 42 kJ mol^{-1} stronger as an acid than water, yields a small but real 31^{-} signal when used neat in the DART source. In parallel to CIMS,[25] this upper limit to deprotonation includes all –OH compounds, anilines and carboxamides, many enolizable carbon acids, but not simple amines or alkanes.

3.3.3 Generation of M⁻• Ions

It is also obvious from Table 3.8, that anything with an electron affinity greater than that of O_2, at 0.45 eV, can undergo electron attachment by transfer from $O_2^{-\bullet}$. Although analytes with weaker EAs could attach free thermal electrons, such M⁻• ions would then be likely to transfer that electron to O_2 upon collision, and thus yield no or very little direct signal.

3.3.4 Generation of [M+X]⁻ Ions

What of attachment of anions to analytes? Use of dichloromethane as the solvent for introducing the analyte results in production of Cl^-, *via* dissociative attachment of thermal electrons.[72] This has $\Delta_{acid}G = 1356$ kJ mol⁻¹, so analytes of stronger acidity than that, such as halogenated carboxylic acids and the most strongly electron-withdrawing phenols, should still deprotonate, rather than yield attachment ions. Weaker acids that are good hydrogen-bond donors, such as carboxylic acids, alcohols, thiols, and anilines, should yield stable chloride-attached species.[25,63] Sufficiently weak hydrogen-bond donors will likely not form attachment ions, because the high temperatures in the DART source will disfavor those, for entropic reasons.

Cody and Dane[73] have observed anion attachment of $O_2^{-\bullet}$ to large alkanes, alcohols, and fatty esters. The first two classes are too weakly acidic to yield [M–H]⁻ ions from proton transfer to $O_2^{-\bullet}$, and should have negative electron affinities, so electron transfer to form M⁻• is not viable. Only alkanes beyond C18 were observed to give attachment ions, so polarizability was invoked as the reason for such species' existence. For the alcohols, even methanol gave an attachment ion, albeit in low intensity, so hydrogen bonding is likely to be the stabilizing force there. Polychlorinated alkanes gave some [M+O₂]⁻ but [M+Cl]⁻ predominated, with the Cl^- arising from self-fragmentation. The fatty acids yielded [M–H]⁻ as expected from their acidity.[25,63] Increasing the cone voltage resulted in loss of signal for the attachment ions, likely from collision dissociation due to their relatively weak bonding.

3.3.5 Nitrogen DART for Negative Ions

Very little has appeared in regard to this mode of ionization. In the present authors' lab, the background spectrum of nitrogen DART in the negative-ion mode, shows signals for $O^{-\bullet}$, HO^-, $O_2^{-\bullet}$, CO_3^-, and HCO_3^-, but the peaks for NO_2^- and NO_3^- can be the intense ones. As postulated with helium DART, the former likely arises from electron attachment to NO_2 formed by neutral atom chemistry per reactions (C) and (D) in Scheme 3.3.

3.4 Conclusions

The main reactions of the chemistry of the DART source appear to be reasonably well established, and appear to follow the general Scheme proposed originally by Cody *et al.*[8] Modes of ionization that have not been closely examined experimentally include (1) the chemistry from using nitrogen as a DART gas, (2) some of the background ion chemistry in helium negative ion DART, and (3) the source of NO_x ions.

References

1. A. Venter, M. Nefliu and R. G. Cooks, *TrAC, Trends Anal. Chem.*, 2008, **27**, 284–290.
2. G. J. Van Berkel, S. P. Pasilis and O. Ovchinnikova, *J. Mass Spectrom.*, 2008, **43**, 1161–1180.
3. G. A. Harris, L. Nyadong and F. M. Fernandez, *Analyst*, 2008, **133**, 1297–1301.
4. H. W. Chen, G. Gamez and R. Zenobi, *J. Am. Soc. Mass Spectrom.*, 2009, **20**, 1947–1963.
5. R. M. Alberici, R. C. Simas, G. B. Sanvido, W. Romao, P. M. Lalli, M. Benassi, I. B. S. Cunha and M. N. Eberlin, *Anal. Bioanal. Chem.*, 2010, **398**, 265–294.
6. D. J. Weston, *Analyst*, 2010, **135**, 661–668.
7. G. A. Harris, A. S. Galhena and F. M. Fernandez, *Anal. Chem.*, 2011, **83**, 4508–4538.
8. R. B. Cody, J. A. Laramee and H. D. Durst, *Anal. Chem.*, 2005, 77, 2297–2302.
9. Z. Takats, J. M. Wiseman, B. Gologan and R. G. Cooks, *Science*, 2004, **306**, 471–473.
10. R. B. Cody, *Anal. Chem.*, 2009, **81**, 1101–1107.
11. L. G. Song, A. B. Dykstra, H. F. Yao and J. E. Bartmess, *J. Am. Soc. Mass Spectrom.*, 2009, **20**, 42–50.
12. L. G. Song, S. C. Gibson, D. Bhandari, K. D. Cook and J. E. Bartmess, *Anal. Chem.*, 2009, **81**, 10080–10088.
13. E. S. Chernetsova and G. E. Morlock, *Mass Spectrom. Rev.*, 2011, **30**, 875–883.
14. J. Hajslova, T. Cajka and L. Vaclavik, *TrAC, Trends Anal. Chem.*, 2011, **30**, 204–218.
15. F. M. Penning, *Naturwissenschaften*, 1927, **15**, 818.
16. E. P. Hunter and S. G. Lias, in *Proton affinity evaluation, NIST Standard Reference Database Number 69*, ed. P. J. Linstrom and W. G. Mallard, National Institute of Standards and Technology, Gaithersburg MD 20899, Editon edn., 2011, http://webbook.nist.gov.
17. D. Smith, N. G. Adams and T. M. Miller, *J. Chem. Phys.*, 1978, **69**, 308–318.
18. M. A. Sutton, J. W. Erisman, F. Dentener and D. Moller, *Environ. Pollut.*, 2008, **156**, 583–604.
19. A. G. Harrison, *Chemical ionization mass spectrometry*, CRC Press, Boca Raton, FL, 1983.
20. S. N. Foner and R. L. Hudson, *J. Chem. Phys.*, 1962, **37**, 1662.

21. F. R. Gilmore, *J. Quant. Spectrosc. Radiat. Transfer*, 1965, **5**, 369–390.

22. R. A. Sultanov and N. Balakrishnan, *J. Chem. Phys.*, 2006, **124**, 124321.

23. J. E. Bartmess, *NIST Chemistry WebBook* (http://webbook.nist.gov).

24. J. E. Bartmess, J. L. Pittman, J. A. Aeschleman and C. A. Deakyne, *Int. J. Mass Spectrom.*, 2000, **195**, 215–223.

25. J. E. Bartmess, *Mass Spectrom. Rev.*, 1989, **8**, 297–343.

26. M. A. Domin, B. D. Steinberg, J. M. Quimby, N. J. Smith, A. K. Greene and L. T. Scott, *Analyst*, 2010, **135**, 700–704.

27. JEOL, *Rapid Analysis of p-Phenylenediamine Antioxidants in Rubber*, http://www.jeolusa.com/PRODUCTS/AnalyticalInstruments/MassSpectrometers/AccuTOF%E2%84%A2DART%C2%AE/tabid/230/Default.aspx#175917-industrial-materials

28. F. M. Fernandez, R. B. Cody, M. D. Green, C. Y. Hampton, R. McGready, S. Sengaloundeth, N. J. White and P. N. Newton, *ChemMedChem*, 2006, **1**, 702.

29. J. Schurek, L. Vaclavik, H. Hooijerink, O. Lacina, J. Poustka, M. Sharman, M. Caldow, M. W. F. Nielen and J. Hajslova, *Anal. Chem.*, 2008, **80**, 9567–9575.

30. S. Banerjee, K. P. Madhusudanan, S. P. S. Khanuja and S. K. Chattopadhyay, *Biomed. Chromatogr.*, 2008, **22**, 250–253.

31. B. Roschek, R. C. Fink, M. D. McMichael, D. Li and R. S. Alberte, *Phytochemistry*, 2009, **70**, 1255–1261.

32. A. H. Grange and G. W. Sovocool, *Rapid Commun. Mass Spectrom.*, 2008, **22**, 2375–2390.

33. J. A. Laramee, H. D. Durst, J. M. Nilles and T. R. Connell, *Am. Lab.*, 2009, **41**, 25.

34. E. S. Chernetsova, M. V. Ovcharov, G. V. Zatonskii, R. A. Abramovich and I. A. Revelskii, *J. Anal. Chem.*, 2011, **66**, 1348–1351.

35. JEOL, *Detection of Oleocanthal in Freshly Pressed Extra-Virgin Olive Oil*, http://www.jeolusa.com/PRODUCTS/AnalyticalInstruments/MassSpectrometers/AccuTOF%E2%84%A2DART%C2%AE/tabid/230/Default.aspx#175914-food-flavors-fragrances

36. JEOL, *"No-prep" Analysis of Lipids in Cooking Oils and Detection of Adulterated Olive Oil*, http://www.jeolusa.com/PRODUCTS/AnalyticalInstruments/MassSpectrometers/AccuTOF%E2%84%A2DART%C2%AE/tabid/230/Default.aspx#175914-food-flavors-fragrances

37. JEOL, *Detection of the Peroxide Explosives TATP and HMTD*, http://www.jeolusa.com/PRODUCTS/AnalyticalInstruments/MassSpectrometers/AccuTOF%E2%84%A2DART%C2%AE/tabid/230/Default.aspx#175916-homeland-security

38. JEOL, *Direct Analysis of Adhesives*, http://www.jeolusa.com/PRODUCTS/AnalyticalInstruments/MassSpectrometers/AccuTOF%E2%84%A2DART%C2%AE/tabid/230/Default.aspx#175917-industrial-materials

39. JEOL, *Rapid Analysis of Glues, Cements, and Resins*, http://www.jeolusa.com/PRODUCTS/AnalyticalInstruments/MassSpectrometers/

AccuTOF%E2%84%A2DART%C2%AE/tabid/230/Default.aspx#175917-industrial-materials

40. Y. P. Zhao, M. Lam, D. L. Wu and R. Mak, *Rapid Commun. Mass Spectrom.*, 2008, **22**, 3217–3224.
41. Z. Q. Zhu, L. G. Song and J. E. Bartmess, *Rapid Commun. Mass Spectrom.*, 2012, **26**, 1320–1328.
42. D. B. Robb, D. R. Smith and M. W. Blades, *J. Am. Soc. Mass Spectrom.*, 2008, **19**, 955–963.
43. J. M. S. Henis, *J. Am. Chem. Soc.*, 1968, **90**, 844.
44. S. G. Lias, J. E. Bartmess, J. F. Liebman, J. L. Holmes, R. D. Levin and W. G. Mallard, *J. Phys. Chem. Ref. Data*, 1988, **17**, 18–25.
45. D. S. Cho, S. C. Gibson, D. Bhandari, M. E. McNally, R. M. Hoffman, K. D. Cook and L. G. Song, *Rapid Commun. Mass Spectrom.*, 2011, **25**, 3575–3580.
46. C. Lapthorn and F. Pullen, *Eur. J. Mass Spectrom.*, 2009, **15**, 587–593.
47. M. Curtis, M. A. Minier, P. Chitranshi, O. D. Sparkman, P. R. Jones and L. A. Xue, *J. Am. Soc. Mass Spectrom.*, 2010, **21**, 1371–1381.
48. G. A. Harris, D. M. Hostetler, C. Y. Hampton and F. M. Fernandez, *J. Am. Soc. Mass Spectrom.*, 2010, **21**, 855–863.
49. H. M. Yang, D. B. Wan, F. R. Song, Z. Q. Liu and S. Y. Liu, *Anal. Chem.*, 2013, **85**, 1305–1309.
50. G. A. Harris and F. M. Fernandez, *Anal. Chem.*, 2009, **81**, 322–329.
51. D. L. G. Borges, R. E. Sturgeon, B. Welz, A. J. Curtius and Z. Mester, *Anal. Chem.*, 2009, **81**, 9834–9839.
52. T. G. Slanger, in *Reactions of Small Transient Species, Kinetics and Energetics*, ed. A. Fontijn and M. A. A. Clyne, Academic Press, New York, 1983, pp. 231–301.
53. P. Dwivedi, D. Gazda, W. Wallace, T. Limero, A. Macatangay and F. Fernandez, in *61st ASMS Conference on Mass Spectrometry and Allied Topics*, Minneapolis, MN, Editon edn., 2013.
54. A. J. Dane and R. B. Cody, *Analyst*, 2010, **135**, 696–699.
55. C. K. Fagerquist and M. K. Hellerstein, *J. Am. Soc. Mass Spectrom.*, 2001, **12**, 754–761.
56. F. J. Andrade, J. T. Shelley, W. C. Wetzel, M. R. Webb, G. Gamez, S. J. Ray and G. M. Hieftje, *Anal. Chem.*, 2008, **80**, 2654–2663.
57. F. J. Andrade, J. T. Shelley, W. C. Wetzel, M. R. Webb, G. Gamez, S. J. Ray and G. M. Hieftje, *Anal. Chem.*, 2008, **80**, 2646–2653.
58. J. T. Shelley, J. S. Wiley, G. C. Y. Chan, G. D. Schilling, S. J. Ray and G. M. Hieftje, *J. Am. Soc. Mass Spectrom.*, 2009, **20**, 837–844.
59. U. Kogelschatz, *Plasma Chem. Plasma Process.*, 2003, **23**, 1–46.
60. N. Na, M. X. Zhao, S. C. Zhang, C. D. Yang and X. R. Zhang, *J. Am. Soc. Mass Spectrom.*, 2007, **18**, 1859–1862.
61. J. D. Harper, N. A. Charipar, C. C. Mulligan, X. R. Zhang, R. G. Cooks and Z. Ouyang, *Anal. Chem.*, 2008, **80**, 9097–9104.
62. J. T. Shelley and G. M. Hieftje, *J. Anal. At. Spectrom.*, 2010, **25**, 345–350.

63. J. E. Bartmess, in *Negative Ion Energetics Data, NIST Standard Reference Database Number 69*, ed. P. J. Linstrom and W. G. Mallard, National Institute of Standards and Technology, Gaithersburg MD 20899, 2011, http://webbook.nist.gov

64. *CRC Handbook of Chemistry and Physics*, ed. R. C. Weast, CRC Press, Boca Raton, FL, 1983, pp. E76–77.

65. H. Massey, *Negative Ions*, Cambridge University Press, Cambridge, 1976.

66. D. Rapp and D. D. Briglia, *J. Chem. Phys.*, 1965, **43**, 1480–1489.

67. P. J. Chantry and G. J. Schulz, *Phys. Rev.*, 1967, **156**, 134–141.

68. M. Doumont, A. Henglein and K. Jager, *Z. Naturforsch., A: Astrophys., Phys. Phys. Chem.*, 1969, **24**, 683.

69. D. K. Bohme and F. c. Fehsenfe, *Can. J. Chem.*, 1969, **47**, 2717–2719.

70. D. K. Bohme and L. B. Young, *J. Am. Chem. Soc.*, 1970, **92**, 3301–3309.

71. J. H. J. Dawson and K. R. Jennings, *J. Chem. Soc., Faraday Trans. 2*, 1976, **72**, 700–706.

72. L. g. Christop and J. a. Stockdal, *J. Chem. Phys.*, 1968, **48**, 1956–1960.

73. R. B. Cody and A. J. Dane, *J. Am. Soc. Mass Spectrom.*, 2013, **24**, 329–334.

CHAPTER 4

Atmospheric Samples Analysis Probe (ASAP) Mass Spectrometry

CHARLES N. McEWEN,*[a] TAM LIEU,[a] SARAH SAYLOR,[a]
MARIAN TWOHIG[b] AND MICHAEL P. BALOGH[b]

[a] University of the Sciences, Philadelphia, PA 19104, USA; [b] Waters
Corporation, Milford, MA, USA
*Email: c.mcewen@usciences.edu

4.1 Introduction

In the 1970s, Horning's group at Baylor introduced atmospheric-pressure
ionization (API).[1–4] The group demonstrated the efficacy of API for directly
vaporized compounds as well as for effluents from gas chromatographs and
vaporized effluents from liquid chromatographs.[3–5] Instrumentation was
developed for gas chromatography (GC) interfaced with API mass spectrometry (MS), and the high sensitivity of negative ion GC/API-MS was
demonstrated for electronegative compounds.[6–8] To effect direct ionization,
Horning's group vaporized volatile compounds into a stream of hot nitrogen
and introduced the gaseous mix through a heated tube used for GC/MS into
the API source.[4] At the time these techniques were introduced, mass spectrometers capable of API were not commercially available. The introduction
of electrospray ionization (ESI) MS by Fenn and coworkers[9,10] reignited
interest in API mass spectrometers. The liquid-introduction method pion-
eered by Horning's group was improved[11] and later referred to as atmospheric-pressure chemical ionization (APCI) to distinguish it from ESI, which
is also an API method. The popularity of liquid chromatography coupled

New Developments in Mass Spectrometry No. 2
Ambient Ionization Mass Spectrometry
Edited by Marek Domin and Robert Cody
© The Royal Society of Chemistry 2015
Published by the Royal Society of Chemistry, www.rsc.org

with ESI and APCI resulted in the latter's popular characterization as a solely liquid-chromatography (LC) effluent ionization method.

The first experiments leading to the development of the ASAP method followed John Fenn's visit to the DuPont Experimental Station in 2002 shortly before he received the Nobel Prize in Chemistry.[12] Professor Fenn was excited by the potential for self-regulated electrospray from a wick.[13] Paper spray[14] and leaf spray[15] are prominent examples of applying this technology in an analytically useful manner. Following Fenn's visit, researchers at DuPont began to explore the utility of "wick spray" to deliver solutions, such as fragrances, into the gas phase in a controlled manner.[16] An undesirable discharge accompanied the wick-spray technique because of protruding stray fibers at the wick's tip producing a high local electric field. Gaseous compounds in the air were readily ionized in the discharge.[17] Because API LC/MS instruments were by then commonly available, the potential for interfacing GC to a mass spectrometer designed for LC/MS was recognized,[18,19] and Waters Corporation now lists GC/API-MS systems among its regular product offering. During the development of these systems, a crude form of the ASAP method was used to rapidly analyze samples. The atmospheric samples analysis probe (ASAP) method uses APCI to generate ions from compounds introduced directly into the atmospheric-pressure ion source and vaporized using the heated gas available with APCI and ESI ion sources. The potential analytical utility of the method was not fully recognized until observation of a demonstration of direct analysis in real time (DART)[20] by Robert ("Chip") Cody at the 2005 Pittsburgh Conference.

The ASAP method has proven to be a powerful, simple, and reliable method for analyzing a wide array of compounds. It offers high sensitivity directly from a wide range of materials at atmospheric pressure. This method is one of several ambient ionization methods spearheaded by desorption electrospray ionization (DESI)[21] and several reviews of these ionization technologies have been published.[22–24] A clear advantage of the ASAP method lies in its cost and practicality. Only minor modifications to commercial ESI or APCI ion sources are necessary to configure it, and for most sources the ASAP configuration does not interfere with the source's normal operation. The commercial versions of the ASAP method available from Waters Corporation and M&M Mass Spec, through IonSense Corporation, are now used worldwide. Here, we discuss fundamentals as well as areas in which the method has been applied and new developments that allow extending the ASAP method to compounds that cannot be vaporized.

4.2 Discussion

4.2.1 Fundamentals

Sample is usually introduced into the API source *via* a disposable melting-point tube inserted into the ionization region using an ASAP probe. A flange

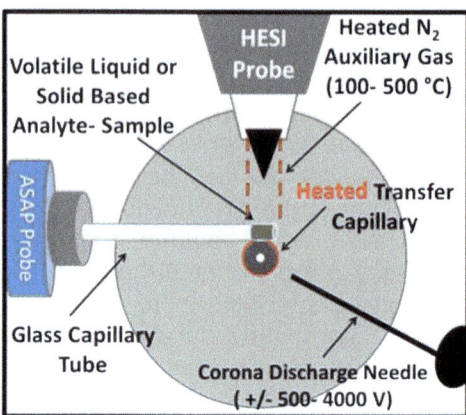

Figure 4.1   Schematic of an ion source showing the ASAP probe arrangement. The ion-inlet entrance is shown as an inlet tube that can be heated, but heat is not required and a skimmer inlet also works.

fitted onto the source guides the tube into the heated gas stream emanating from an APCI probe or heated ESI probe (Figure 4.1).[25] Except for gases, the ASAP method can introduce samples in almost any form into an API source. The heated gas used with ESI or APCI for nebulization and desolvation converts vaporizable compounds placed on the melting-point tube to gas-phase molecules, and an electric discharge or photoionization process subsequently ionizes them for mass spectrometric analysis.[22] Thus, ionization occurs, in the case of an electric discharge by APCI, or using a light source by photoionization. Therefore, the ASAP method operates within the controlled environment of the ion-source housing, which improves reproducibility and safety but makes automation difficult.

Samples analyzed using the ASAP method can be synthesized or biological compounds, tissue, organisms, or polymers. Materials, or pure compounds, can be rubbed against the closed end of the melting-point tube to transfer surface compounds to the melting-point tube for analysis so long as excess material or compound is wiped off. Alternatively, small bits of biological or polymeric sample can be inserted into the open end of tube that is then inserted into the heated gas stream.[17] Other means can be used to hold a sample for ASAP analysis, including using a pyroprobe for pyrolysis experiments.[17,26]

As the final step before they are analyzed, the vaporized molecules are ionized by reagent ions produced in a cascade of ion–molecule reactions that begin with ionization of nitrogen gas by an electric discharge (Scheme 4.1).[17,27,28] This cascade of ionization events is the same for chemical ionization and APCI. Typically, the analyte molecules are ionized by proton transfer from H_3O^+, the most-abundant reagent ion in a clean ion source, in the positive-ion mode. This process is equivalent to a chemical ionization source in which water vapor serves as the reagent gas. As with chemical ionization, negative ions are produced as well as positive ones. Moreover, depending on the analyte molecule, negative ions are produced by electron capture, dissociative electron capture, and proton abstraction.[28]

$$N_2 + e \rightarrow N_2^+ + 2e$$
$$N_2^+ + 2N_2 \rightarrow N_4^+ + N_2$$
$$N_4^+ + H_2O \rightarrow H_2O^+ + 2N_2$$
$$H_2O^+ + H_2O \rightarrow H_3O^+ + OH\bullet$$
$$H_3O^+ + n(H_2O) + N_2 \rightarrow H^+(H_2O)_n + N_2$$
$$H^+(H_2O)_n + A \rightarrow AH^+ + nH_2O \ (A=\text{analyte } n= 1\text{-}4)$$
$$N_4^+ \ (\text{or } N_2^+) + A \rightarrow A^+ + xN_2$$

Scheme 4.1 Ion–molecule reactions in APCI.

In APCI analyses, negative ionization is highly sensitive for electronegative compounds (5–9).

There are differences in ionization using the ASAP method as opposed to using APCI with LC/MS. Typically, APCI is used for less-polar compounds that do not ionize well by ESI. Nevertheless, some types of compounds are too nonpolar even for APCI. Many such compounds are more effectively ionized using photoionization.[29] APCI fails to ionize many of these less-polar compounds well, chiefly because the solvent used in LC/MS provides the reagent ions, and these solvent reagent ions require a forbidden endothermic or inefficient thermoneutral proton transfer for ionization. For example, with water as the mobile phase, protonated solvent clusters, which are less acidic than H_3O^+, prevail in number. Proton transfer to certain analyte compounds, therefore, may not occur because the reaction is endothermic. Alternatively, the transfer does occur, but it occurs slowly or reversibly and thus with poor sensitivity because the reaction is nearly thermoneutral. In ASAP, the H_3O^+ ion is sufficiently acidic to ionize most organic compounds, and it often does so with sufficient exothermicity to produce fragmentation. Ionization in ASAP is identical to ionization in solvent introduction APCI, with a major exception: ASAP can analyze less-polar compounds because the absence of solvent precludes suppression of ionization by solvent. In other words, solvent in the gas phase limits ionization of molecules less basic than the solvent. The effective removal of solvents from low-flow LC/MS has also been demonstrated to provide nearly universal ionization of vaporizable compounds.[30]

Using dry nitrogen as the vaporizing gas, nonpolar compounds that API liquid-introduction methods do not ionize are readily ionized, as demonstrated in Figure 4.2. The theoretical isotope pattern of hexachlorobenzene is shown in Figure 4.2(a) and the ASAP mass spectrum of the molecular ion region in Figure 4.2(b). In this case, charge exchange from the ionized nitrogen (N_2^+ and N_4^+) produces the molecular radical cation. The ASAP method can even ionize saturated hydrocarbons,[31] as shown by the hydrocarbon hexatriacontane in Figure 4.2(c) and (d). In Figure 4.2(c), the $[M - H]^+$ ion produced by hydride abstraction is observed along with extensive fragmentation at a vaporizing-gas temperature of 250 °C. The ions observed are the same as those observed using methane as the reagent gas in chemical ionization mass spectra of saturated hydrocarbons. Interestingly, reducing the inlet temperature to 100 °C produces the odd electron molecular ion and much less fragmentation. It appears that the additional

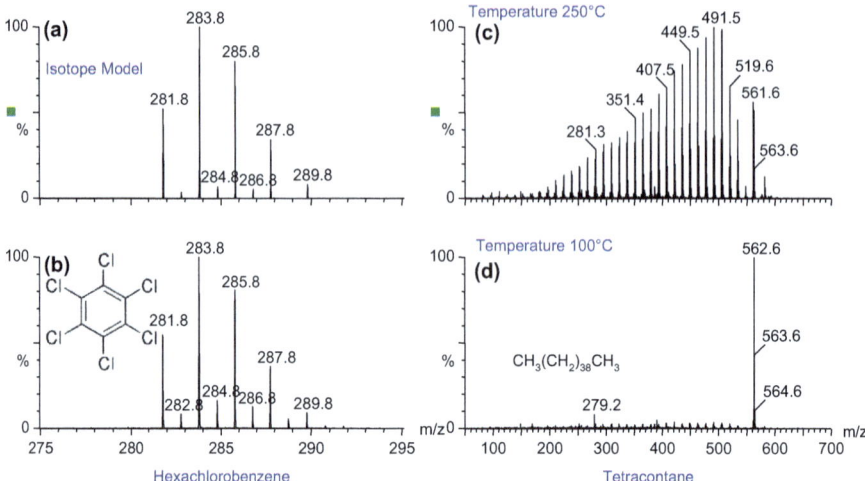

Figure 4.2   The theoretical isotope pattern for the molecular ion (a) and the ASAP, positive-ion, mass spectrum of the molecular ion region for hexachlorobenzene (b). The ASAP mass spectrum of hexatriacontane with a vaporizing-gas temperature of 250 °C (c), and 100 °C (d) show major differences in the degree of fragmentation.

thermal energy supplied by the vaporizing gas suffices to greatly enhance fragmentation of this compound.

Conversely, water vapor or ammonia gas can be added to the ionization region to reduce the exothermicity of the reaction and eliminate fragmentation for thermally stable compounds. Reducing fragmentation by using weak acids for the gas-phase protonation step also ensures that only compounds more basic than the reagent gas are observed. Reducing fragmentation, therefore, yields the benefit of a more-selective ionization.

The ion–molecule reactions occurring at atmospheric pressure are the same as those that occur, at much lower pressure, in a CI source. There are differences, however. At atmospheric pressure, many more collisions occur than in the lower pressure of a CI source. Thus, reactions between reagent ions and analyte molecules are highly efficient, and API produces a higher percentage of ionized analyte than does CI. Nevertheless, the ion transfer from an atmospheric-pressure environment to the mass-analyzer's vacuum is less efficient than it is in the lower-pressure CI source, moderating the sensitivity gain. Moreover, the higher collision rate at atmospheric pressure makes collisional cooling more efficient, so less fragmentation is expected. Thus, the ASAP method can be viewed as a direct probe for API, similar to a direct probe for CI. Unlike CI, however, ASAP requires no vacuum lock with its attendant increase in analysis time and loss of volatile samples.

4.2.2   Applications

The original paper published on the topic of ASAP-MS demonstrated the utility of the method for analyzing liquids, solids, and biological tissue.[32]

Steroids, polymers, and polymer additives were demonstrated, as well as carotenoids, capsacins, and lipids from biological tissue. Both accurate-mass measurements and mass-selected fragmentation were demonstrated. Subsequent design of a probe and flange inlet simplified sample analysis and increased reproducibility.[25] The end of the melting-point tube, its exterior surface containing the sample, could then be precisely placed into the heated gas stream flowing through the ionization region. An interesting early example of the method's use was the direct analysis of fungal cells in the presence of various concentrations of inhibitors to the ergosterol pathway.[33] In this work, growing fungus cells were pipetted into wells of a bactiplate, and known fungicides were added in known concentrations. Inhibition of the ergosterol pathway led to buildup of sterols in the pathway prior to the blockage. The sterols were readily detected, and the concentration dependence and mode of action of each fungicide was rapidly determined. Another early study involving the ASAP method demonstrated that not only was the method useful for analysis of small-molecule drugs, it was also an easy way of following the progress of chemical reactions.[27]

The ASAP method has been highlighted in several general-readership articles[34-36] as well as by manufacturers of mass spectrometers.[37-40] Reports on the utility of this approach for a variety of problems are set forth below. Yet many laboratories, including ours, use the method principally in a manner similar to direct-probe analyses. The initial applications of the ASAP method were to obtain accurate mass of synthesized samples. The speed of analysis and the ability to eliminate fragmentation so that the purity of the samples could be verified in seconds resulted in rapid acceptance of the method using API mass spectrometers capable of accurate-mass measurement.

The advantages of ASAP relative to a vacuum direct probe analysis are that sample introduction is much faster and volatile compounds can be introduced without being pumped away. The degree of fragmentation can be controlled, as with chemical ionization. Generally, only molecular-ion information is desired, so soft ionization is important. Soft ionization is readily achievable by adding, for example, water or ammonium hydroxide solution through the ESI or APCI probe or ammonium salts with the sample. Fragmentation for characterization is often achievable using collision-induced fragmentation in conjunction with mass selection, or using dry nitrogen gas as a means of vaporizing the sample and reducing the water content in the ion source, inducing higher-energy ionization.

The ASAP method has been reported for the high-throughput analysis of urine and bile,[41] in the analysis of steroids[42] and drugs,[27] and in the identification of counterfeit, adulterated products of synthetic phosphodiesterase type-5 inhibitors.[43] The analysis of airborne particles[40-44] is a particularly interesting application. The ASAP method was assessed as a means of detecting pesticides and illegal dyes in foods and cereals.[45] Qualitative determinations of strobilurin pesticides were made simply by stirring the ASAP probe directly into grains of wheat. The method was also used in the analysis of vanilla-flavored food products.[46] Applications also include the analysis of low molecular weight synthetic polymers as well as crosslinked polymers and polymer blends,[47–50] the analysis of polymer additives,[26] crude oil analyses,[17,51,52] and the rapid screening of chemical-warfare nerve agents.[53] Additionally, the ASAP approach was used to detect impurities in 2-naphthalenamines,[54] for the analysis of nucleosides,[55] for the detection of carotenoids in diverse materials,[56] and in the analysis of natural products.[57] An unusual application of ASAP-MS was the measurement of vapor pressures and heats of sublimation of dicarboxylic acids to better understand the impact of aerosols on climate, visibility, and health.[58] The ASAP method was used in the characterization of polycyclic tetracarboxylic acids found in oilfield pipelines and equipment.[59]

Theoretically, ASAP can be interfaced to, and is useful with, any API mass spectrometer. It has proven especially powerful, however, when used with mass spectrometers capable of high-mass resolution, accurate-mass measurement,[17,31] or ion-mobility spectrometry.[48,52] Combined with tandem mass spectrometry, the ASAP method was applied to quantitate pesticide residues on vegetables.[60] A calibration curve from 7 ppb to 7000 ppb, with an R^2 of 0.994, was demonstrated for pure melamine.[18] Finally, combined with solid-phase extraction (SPE), the method was used to obtain, for 13 beta-agonists in porcine urine, a linear correlation coefficient better than 0.98, over the concentration range 0.5 ng mL^{-1} to 300 ng mL^{-1}.[61] The LOD was 200 pg mL^{-1} for MS/MS acquisition times of less than 1 min.

Recent developments in the use of ASAP-MS were displayed at the 61st annual conference of the American Society for Mass Spectrometry. Both GC/API-MS and ASAP were used to analyze hydrocarbons.[62] Combined with ASAP, high-mass resolution and accurate-mass measurement were used for forensic analysis,[63] to analyze plasticizers in glass food-jar gaskets,[64] and for characterizing insoluble materials produced in the development of cosmetics.[65] Ion mobility/mass spectrometry was combined with ASAP to characterize bacterial metabolites[66] and analyze thermoplastic copolymers.[67]

4.2.3   Analysis of Inks by ASAP-MS

As an example of the use of the ASAP method, dyes composing the different-colored inks from 36 ballpoint pens produced by different manufacturers were analyzed directly from paper by simply rubbing the melting-point tube over the ink, obtaining a mass spectrum, and subtracting the mass spectrum of the paper background obtained by rubbing a second melting-point tube over the paper in areas where ink was not present. The results from both positive and negative ion analyses were compared with the ASAP mass spectra of synthetic dyes commonly used in inks, and against the mass spectra of inks directly from the pens. Fresh and old ballpoint pen writings were analyzed. To better identify dyes in some of the inks, thin layer chromatography (TLC) was used to separate mixtures of dyes present in an ink sample. Colored areas on the TLC plate were scraped into a vial. The dyes were then extracted into methanol for analysis by ASAP-MS. This approach would associate the accurate mass of the dye with its color in a rapid analysis.

Figure 4.3 shows the mass spectrum of solid rhodamine B obtained by touching the closed end of the melting-point tube against the dye powder and wiping the tube using a Kim Wipe before analysis. The temperature of the nitrogen gas used to vaporize the sample was 400 °C. The ion at *m/z* 443.2283, the cation portion of the dye, was obtained using a Thermo Scientific Orbitrap Exactive mass spectrometer at 100 000 mass resolution. The relative abundance of the ions at *m/z* 443.2283, 415.1975, and 387.1665 in the positive-ion mass spectrum approximates that seen in the negative-ion mode. In the latter mode, however, the ions appear 2 Daltons lower in mass, as expected for proton abstraction relative to protonation.

Analyzing red ink from a BIC multicolor ballpoint pen shows two major dyes (Figure 4.4). One of the dyes is rhodamine B and the other auramine O, with a cation *m/z* of 268.1802. The much-less abundant loss of C_2H_5 at *m/z* 415.2008, shown in Figure 4.5, distinguishes rhodamine B in this ink from rhodamine 6G in the red ink of a Pental ballpoint pen. After analyzing the ink of all 36 pens, it was possible to distinguish, in a blind analysis, writing on paper for all of the inks that showed differences. This was true even though the dyes used in all the inks were not readily identified using the direct-analysis approach. Figure 4.6 shows the use of TLC to separate the

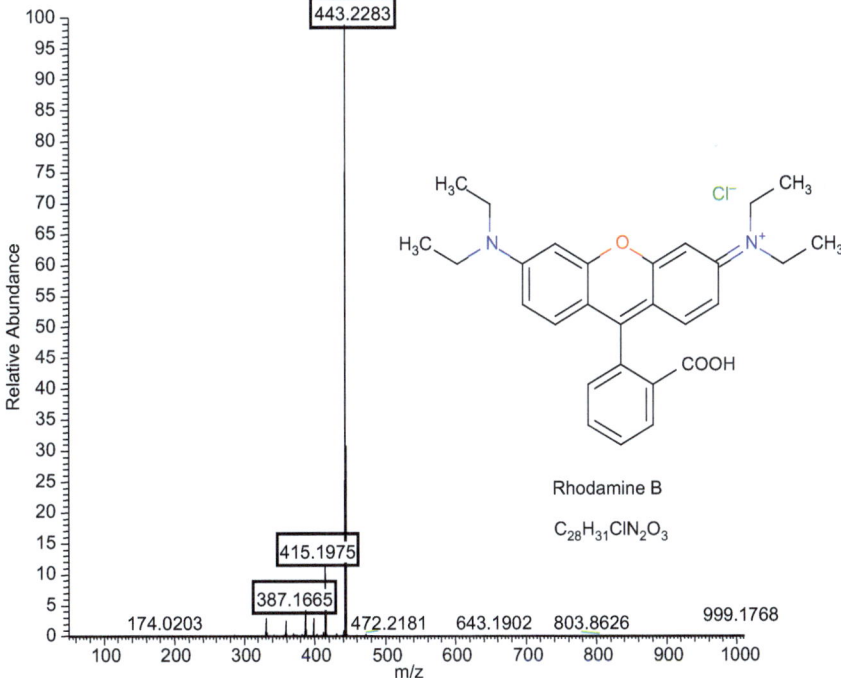

Figure 4.3 ASAP mass spectrum of solid rhodamine B and the structure of rhodamine B.

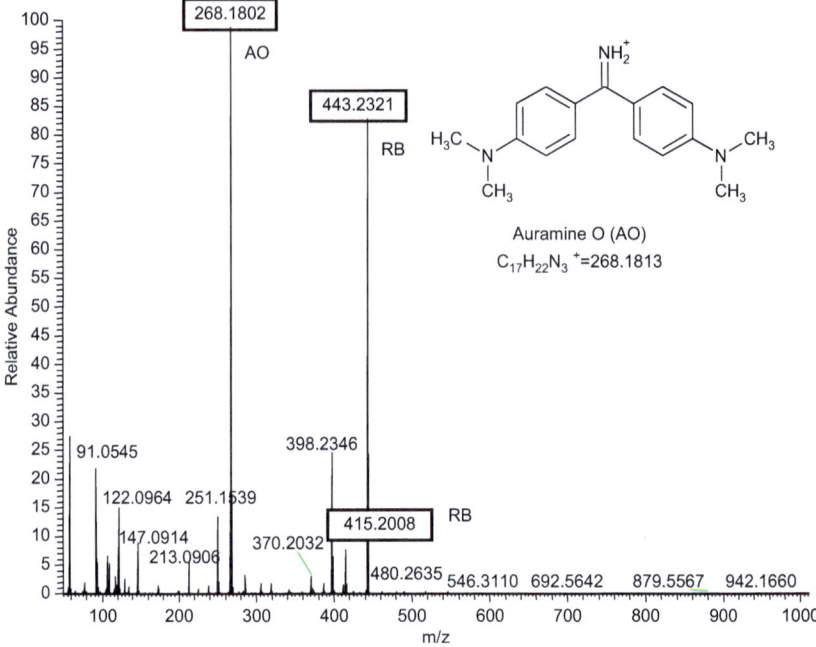

Figure 4.4 ASAP mass spectrum of red BIC multicolor ballpoint pen and the structure of auramine O.

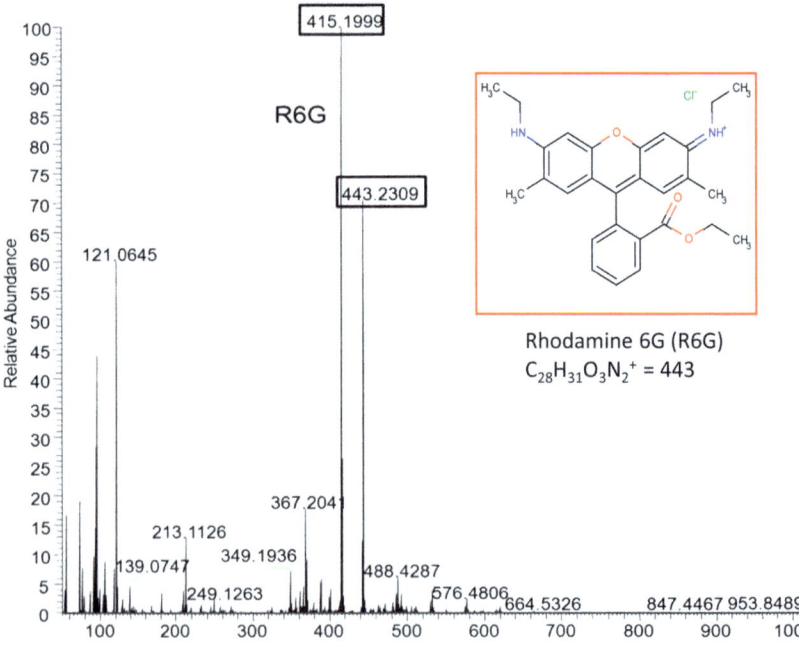

Figure 4.5 ASAP-MS analysis of red Pental ballpoint pen ink, from paper, and the structure of rhodamine 6G.

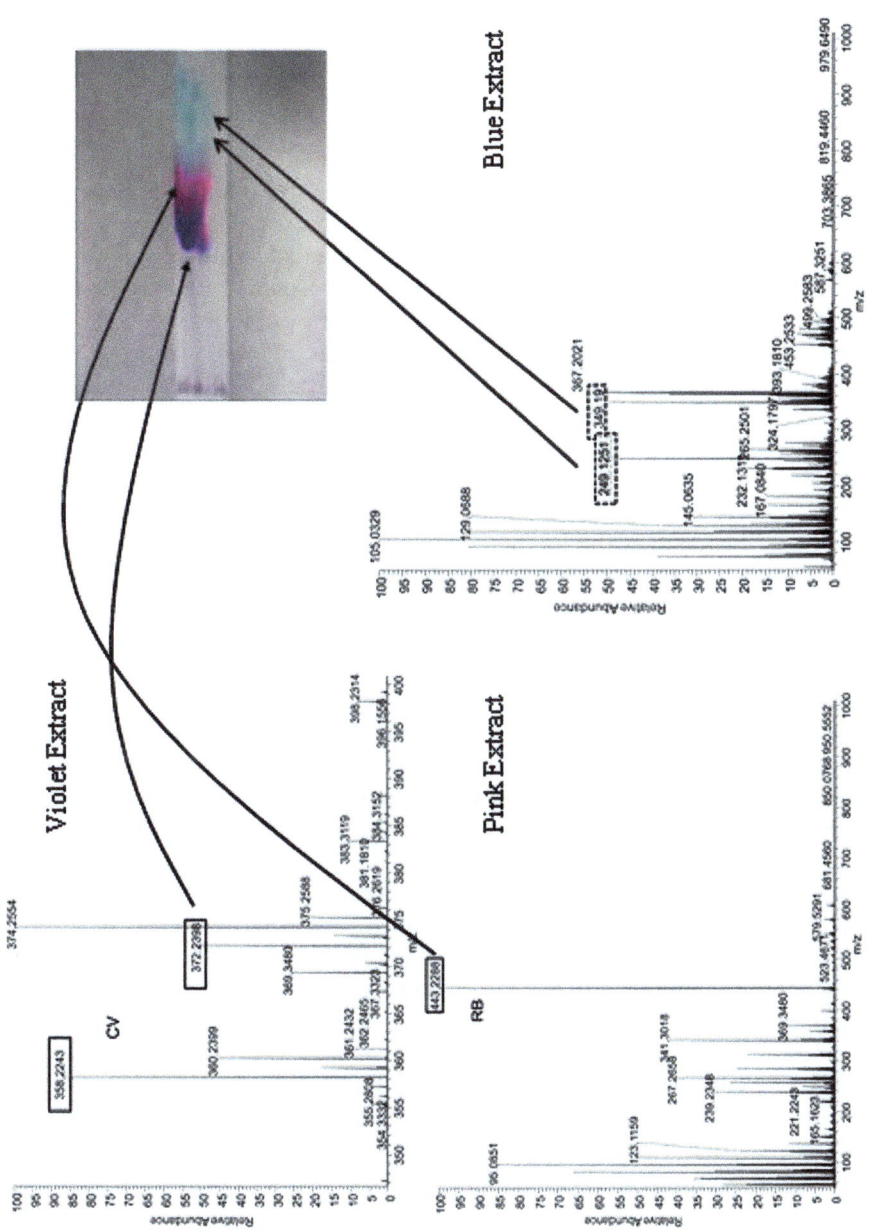

Figure 4.6 Separation of violet ink from a Pental R.S.V.P. ballpoint pen using thin-layer chromatography and mass spectra of some of the separated components.

various dyes in a Pental R.S.V.P. violet-ink ballpoint pen and the ASAP mass spectra of three of the separated dyes. It is clear from the analysis that rhodamine B is used rather than rhodamine 6G in this ink.

4.2.4 Extension of the ASAP Method to Nonvolatile Compounds

Though ASAP-MS has been very successful for analyzing compounds that vaporize or even materials after pyrolysis, it fails for nonvolatile compounds. Because the method is simple and inexpensive, in most cases requiring only minor modifications of existing ESI or APCI sources, extending its capability to analyzing nonvolatile compounds would prove useful. Initial success in doing so was demonstrated in a desorption electrospray ionization (DESI) experiment in which the ESI probe of a Thermo Scientific IonMax source was aligned with the melting-point tube from the ASAP probe.[68] Proteins as large as myoglobin were thus ionized. The sample could be placed on the melting-point tube from solution or as a solid and wiped off, leaving a trace residue. The method requires properly aligning the ESI probe and the melting-point tube so that the electrospray-charged droplets skim the edge of the tube. Another drawback is that the method is subject to sodium adduction.

Later, laserspray ionization,[69,70] a method whereby a matrix-containing analyte is ablated into a heated inlet tube to produce multiply charged ions similar to ESI (see Chapter 17), was shown applicable for nonvolatile compounds using the ASAP probe.[71] For this method, the analyte is applied to the closed end of the melting-point tube together with a matrix compound, similar to the approach used in MALDI. The laser is aligned to ablate the matrix/analyte mixture into the heated, mass-spectrometer inlet, where ions are produced. This method requires a laser and alignment of its beam with the melting-point tube. An advantage, however, is that the small area sampled by the laser beam allows multiple samples to be placed on a single melting-point tube. Mass spectra from four samples ranging in molecular weight from peptides to the protein ubiquitin could be obtained in one minute by turning the adjustment wheel of the ASAP probe to move the melting-point tube relative to the laser beam.

More recently, a method termed matrix-assisted ionization vacuum (MAIV) was introduced. The method converts volatile and nonvolatile compounds to gas-phase ions simply by exposing the analyte in a suitable matrix to the instrument's vacuum (see Chapter 17).[72,73] Using the MAIV matrix, 3-nitrobenzonitrile, it was discovered that mild heat substituted for vacuum to produce ions from volatile and nonvolatile analyte in the matrix. Thus, applying the 3-NBN matrix and analyte onto the closed end of the melting-point tube and inserting the tube into the stream of heated nitrogen (as the ASAP method would require), ions were generated from an array of compounds including the protein myoglobin.[74] Almost no sodium adduction was observed. This method, which we refer to as matrix assisted ASAP

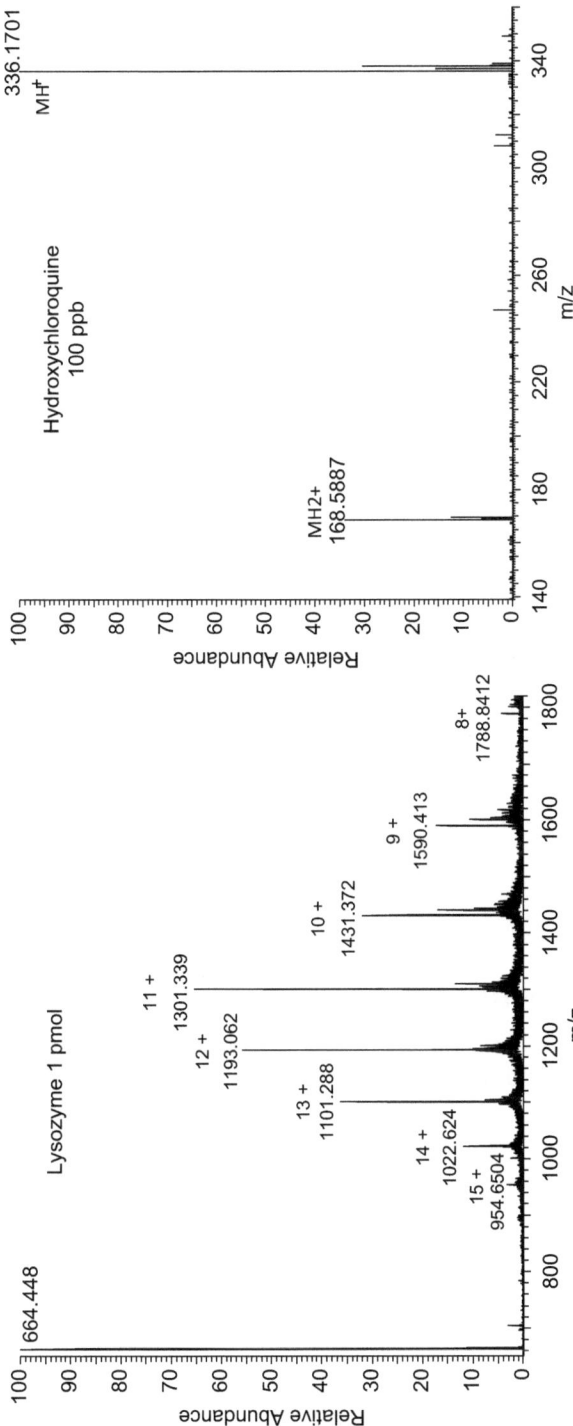

Figure 4.7 Mass spectra of 1 pmol of lysozyme (left panel) and 1 μL of a 100 ppb solution of hydroxychloroquine (right panel) using 3-NBN as matrix applied onto a melting-point tube and inserted into a heated (75 °C) nitrogen gas stream using the ASAP probe.

(MA-ASAP), requires the nitrogen from the heated ESI or APCI probe to be only about 60 °C. Even 2,5-dihydroxyacetophenone works well as a matrix, though it requires a slightly higher gas temperature ($\sim$90 °C). The position of the gas flow requires more adjustment than is necessary with the ASAP method, and MA-ASAP has been demonstrated only on a Thermo Scientific IonMax source using an M&M ASAP probe. Yet the authors demonstrated that the approach is applicable to Waters instruments by hand holding the melting-point tube in a heated gas stream. Representative mass spectra of lysozyme and the drug hexachloroquine are shown in Figure 4.7.

4.3 Conclusion

A direct probe for atmospheric-pressure ion sources has proven to be a simple yet effective means of analyzing a wide range of materials. Small molecules are readily ionized from melting-point tubes secured in an API ion source by an ASAP probe. With the assistance of matrices or solvents, the ASAP method has successfully analyzed small and large nonvolatile compounds, including proteins. The simplicity and speed of depositing sample on a disposable low-cost melting-point tube and inserting the tube into heated nitrogen gas from commercial ESI or APCI probes makes the method very attractive. The ASAP method requires no sample preparation because neat liquids, solids, and solutions can be placed directly on the closed end of a melting-point tube, wiped, and inserted into the source for analysis. Materials such as biological tissue and polymers or plastics can be inserted into the open end of the melting-point tube for analysis. Nonvolatile and insoluble materials can be pyrolyzed to obtain information, and those that are soluble and nonvolatile can be ionized with matrix assistance using the ASAP method. The method has proven so useful that, in a few years, it may become standard equipment on API mass spectrometers.

Acknowledgements

This work was made possible by funding from NSF-CHE-1112289 and the Richard Houghton endowment to the University of the Sciences to CNM.

References

1. E. C. Horning, M. G. Horning, D. I. Carroll, I. Dzidic and R. N. Stillwell, *Anal. Chem.*, 1973, **45**, 936–943.
2. E. C. Horning, D. I. Carroll, I. Dzidic, K. D. Haegele and M. G. Horning, *J. Chromatogr. Sci.*, 1974, **12**, 725–729.
3. D. I. Caroll, I. Dzidic, K. D. Haegele, R. N. Stillwell and E. C. Horning, *Anal. Chem.*, 1975, **47**, 2369–2373.
4. E. C. Horning, D. I. Carroll, I. Dzidic, K. D. Haegele and S.-N. Lin, *Clin. Chem.*, 1977, **23**, 13–21.

5. D. I. Caroll, I. Dzidic, R. N. Stillwell, M. G. Horning and E. C. Horning, *Anal. Chem.*, 1974, **46**, 706–710.
6. R. K. Mitchum, W. A. Korfmacher, G. F. Moler and D. L. Stalling, *Anal. Chem.*, 1982, **54**, 719–722.
7. R. K. Mitchum, G. F. Moler and W. A. Korfmacher, *Anal. Chem.*, 1980, **52**, 2278–2282.
8. T. Kinouchi, A. T. L. Miranda, L. G. Rushing, F. A. Beland and W. A. Korfmacher, *HRC & CC, J. High Resolut. Chromatogr. Chromatogr. Commun.*, 1990, **13**, 281–284.
9. M. Yamashita and J. B. Fenn, *J. Phys. Chem.*, 1984, **88**, 4451–4459.
10. J. B. Fenn, M. Mann, C. K. Meng, S. F. Wong and C. M. Whitehouse, *Science*, 1989, **246**, 64–67.
11. T. R. Covey, E. D. Lee, J. d. Bruins and J. D. Henion, *Anal. Chem.*, 1986, **58**, 1451A–1461A.
12. J. B. Fenn, *Angew. Chem., Int. Ed.*, 2003, **42**, 3871–3894.
13. J. B. Fenn, *US Pat.*, US6297499 B1, 2001.
14. H. Wang, J. Liu, R. G. Cooks and Z. Ouyang, *Angew. Chem.*, 2010, **122**, 889–892.
15. J. Liu, H. Wang, R. G. Cooks and Z. Ouyang, *Anal. Chem.*, 2011, **83**, 7608–7613.
16. C. N. McEwen, W. J. Herron and R. G. McKay, *US Pat.*, US6729552, 2004.
17. C. N. McEwen, Encyclopedia Analytical Chemistry, 2010, 10.1002/9780470027318.a9045.
18. C. N. McEwen and R. G. McKay, *J. Am. Soc. Mass Spectrom.*, 2005, **16**, 1730–1738.
19. C. N. McEwen, *Int. J. Mass Spectrom.*, 2007, **259**, 57–64.
20. R. B. Cody, J. A. Laramée and H. D. Durst, *Anal. Chem.*, 2005, 77, 2297–2302.
21. Z. Takats, J. M. Wiseman, B. Gologan and R. G. Cooks, *Science*, 2004, **306**, 471–473.
22. A. Venter, M. Nefliu and R. G. Cooks, *TrAC, Trends Anal. Chem.*, 2008, **27**, 284–290.
23. G. J. Van Berkel, S. P. Pasilisi and O. Ovchinnikovia, *J. Mass Spectrom.*, 2008, **43**, 1161–1180.
24. G. A. Harris, L. Nyadong and F. M. Fernandez, *Analyst*, 2008, **133**, 1297–1301.
25. C. N. McEwen and R. G. McKay, *US Pat.*, US7977629 B2, 2011.
26. S. Trimpin, K. Wijerathne and C. N. McEwen, *Anal. Chim. Acta*, 2009, **654**, 20–25.
27. C. Petucci and J. Diffendal, *J. Mass Spectrom.*, 2008, **43**, 1565–1568.
28. C. McEwen and B. S. Larsen, *J. Am. Soc. Mass Spectrom.*, 2009, **20**, 1518–1521.
29. D. B. Robb, T. R. Covey and A. P. Bruins, *Anal. Chem.*, 2000, **72**, 3653.
30. A. Cappiello, G. Famiglini, E. Pierini, P. Palma and H. Trufelli, *Anal. Chem.*, 2007, **79**, 5364–5372.

31. C. M. Williams, NMSSC Application Notes No 7, 2010, http://www.nmssc.ac.uk/documents/NMSSC-AN7-ASAP.pdf.
32. C. N. McEwen, R. G. McKay and B. S. Larsen, *Anal. Chem.*, 2005, 77, 7826–7831.
33. C. McEwen and S. Gutteridge, *J. Am. Soc. Mass Spectrom.*, 2007, **17**, 1274–1278.
34. C. H. Arnaud, *Chem. Eng. News*, 2007, **85**, 13.
35. M. P. Balogh, *LCGC North Am.*, 2007, **25**, 368.
36. M. P. Balogh, *LCGC North Am.*, 2007, **25**, 1184.
37. Agilent, http://www.chem.agilent.com/Library/flyers/Public/5990-6637en_lo%20CMS.pdf.
38. Waters, http://www.waters.com/waters/en_US/Atmospheric-Solids-Analysis-Probe-/nav.htm?cid=10099674.
39. Thermo-Fisher, https://static.thermoscientific.com/images/D21092~.pdf, 19.
40. Bruker, http://www.bruker.com/fileadmin/user_upload/8-PDF-Docs/Separations_MassSpectrometry/Literature/literature2010/appli-tech-notes/TN-39-direct-probe-ebook.pdf.
41. M. Twohig, J. P. Shockcor, I. D. Wilson, J. K. Nicholson and R. S. Plumb, *J. Proteome Res.*, 2010, **9**, 3590–3597.
42. A. D. Ray, J. Hammond and H. Majors, *Eur. J. Mass Spectrom.*, 2012, **16**, 169–174.
43. M. Twohig, S. Skilton, G. Fujimoto, N. Ellor and R. S. Plumb, *Drug Test. Anal.*, 2010, **2**, 45–50.
44. E. A. Bruns, V. Perraud, J. Greaves and B. J. Finlayson-Pitts, *Anal. Chem.*, 2010, **82**, 5922–5927.
45. R. J. Fussell, D. Chan and M. Sharman, *TrAC, Trends Anal. Chem.*, 2010, **29**, 1326–1335.
46. P. J. Lee, A. M. Ruel, M. P. Balogh and P. B. Young, *Food Anal.*, 2010, **21**, 25–30.
47. M. J. Smith, N. R. Cameron and J. A. Mosely, *Analyst*, 2012, **137**, 4524–4530.
48. C. Barrere, F. Maire, C. Afonso and P. Giusti, *Anal. Chem.*, 2012, **84**, 9349–9354.
49. S. E. Whitson, G. Erdodi, J. P. Kennedy, R. P. Lattimer and C. Wesdemiotis, *Anal. Chem.*, 2008, **80**, 7778–7785.
50. R. P. Lattimer and M. J. Polce, *J. Anal. Appl. Pyrolysis*, 2010, **92**, 355–360.
51. A. Ahmed, Y. J. Cho, M. H. No, J. Koh, N. Tomczyk, K. Giles, J. S. Yoo and S. Kim, *Anal. Chem.*, 2011, **83**, 77–83.
52. C. N. McEwen, H. Majors, M. Green, K. Giles and S. Trimpin, *IMS/MS Applied to Direct Ionization Using the Atmospheric Solids Analysis Probe Method*, CRC Press, New York, 2011.
53. F. Zydel, J. R. Smith, V. S. Pagnotti, R. J. Lawrence, C. N. McEwen and B. r. Capacio, *Drug Test. Anal.*, 2012, **4**, 308–311.
54. H. Pan and G. Lundin, *Eur. J. Mass Spectrom.*, 2011, **17**, 217–225.
55. J. Rozenski, *Int. J. Mass Spectrom.*, 2011, **304**, 204–208.
56. S. M. Rivera and R. Canela-Garayoa, *J. Chromatogr. A*, 2012, **1224**, 1–10.

57. J. Lindberg and A. DerMarderosian, *Planta Med.*, 2012, **78**, 1262–1264.

58. E. A. Bruns, J. Greaves and B. J. Finlayson-Pitts, *J. Phys. Chem. A*, 2012, **116**, 5900–5909.

59. P. A. Sutton, B. E. Smith and S. J. Rowland, *Rapid Commun. Mass Spectrom.*, 2010, **24**, 3195–3204.

60. B.-Y. Huang, X.-H. Ouyang, J. Sun, Z.-Y. Xiao and C.-P. Pan, *Chem. J. Chin. Univ.*, 2013, **34**, 1591–1597.

61. P. L. Wang, R. G. Wang, Y. Li, X. O. Su and Z. H. Ye, *Anal. Methods*, 2012, **4**, 4269–4277.

62. C. Wu, K. Qian, K. Edwards, C. Walters, A. Mennito and C. Jurtschenko, 61st Ann. ASMS Conf. Mass Spectrom. and Allied Topics, Minneapolis, MN, 2013.

63. E. Jagerdeo, J. Clark, L. Reda and J. Leibowitz, 61st Ann. Conf. Mass Spectrom. Allied Topics, Minneapolis, MN, 2013.

64. M. Driffield, D. Speck, M. Parmar, J. Leak, L. Lister, E. Bradley, D. Roberts and S. Stead, 61st Ann. Conf. Mass Spectrom. Allied Topics, Minneapolis, MN, 2013.

65. N. Budimir and G. Hussler, 61st Ann. Conf. Mass Spectrom. Allied Topics, Minneapollis, MN, 2013.

66. N. M. Lareau, C. R. Goodwin, J. C. May, R. Kurulugama, E. Darland, B. O. Bachmann and J. A. McLean, 61st Ann. Conf. Mass Spectrom. Allied Topics, Minneapollis, MN, 2013.

67. N. Alawani and C. Wesdemiotis, 61st Ann. Conf. Mass Spectrom. Allied Topics, Minneapollis, MN, 2013.

68. J. A. Lloyd, A. F. Harron and C. McEwen, *Anal. Chem.*, 2009, **81**, 9158–9162.

69. S. Trimpin, E. D. Inutan, T. N. Herath and C. N. McEwen, *Mol. Cell. Proteomics*, 2010, **9**, 362–367.

70. C. N. McEwen, B. S. Larsen and S. Trimpin, *Anal. Chem.*, 2010, **82**, 4998–5001.

71. F. Zydel, S. Trimpin and C. N. McEwen, *J. Am. Soc. Mass Spectrom.*, 2010, **21**, 1889–1892.

72. T. I. Quickenden, B. J. Selby and C. G. Freeman, *J. Phys. Chem. A*, 1998, **102**, 6713–6715.

73. S. Trimpin and E. D. Inutan, *J. Am. Soc. Mass Spectrom.*, 2013, **24**, 722–732.

74. S. Chakrabarty, V. S. Pagnotti, E. D. Inutan, S. Trimpin and C. N. McEwen, *J. Am. Soc. Mass Spectrom.*, 2013, **24**, 1102–1107.

CHAPTER 5

Ambient Analysis by Thermal Desorption Atmospheric-Pressure Photoionization

JACK SYAGE*[a] AND KAVEH JORABCHI[b]

[a] Syagen Technology Inc., a subsidiary of Morpho Detection, Inc., Santa Ana, CA, USA; [b] Department of Chemistry, Georgetown University, Washington, DC 20057, USA
*Email: jsyage@morphodetection.com

5.1   Introduction

There have been many cocktail-hour discussions about whether ambient analysis began with DART (direct analysis in real time)[1] or DESI (desorption electrospray ionization)[2] or can it trace its origins to the use of direct sampling probes with MS that have been used for decades. The direct insertion probe is one example of this; solid-phase microextraction (SPME) is another. What distinguishes these methods from the current generation of ambient analysis techniques is the region of ionization. The current generation is based on atmospheric-pressure ionization (API) so that the sample can be exposed to the ionizer without inserting into a pressure differential device such as a vacuum interlock or a GC-like septum-sealing injector. However, there is yet another API ambient analysis method that has been used extensively for years, but has attracted little notice and that is the so-called swab/desorption devices used in the security and other industries to detect explosives (*e.g.*, in aviation security) and narcotics (*e.g.*, in border security).[3] Though the ion analyzers are mostly ion-mobility spectrometer (IMS)

New Developments in Mass Spectrometry No. 2
Ambient Ionization Mass Spectrometry
Edited by Marek Domin and Robert Cody
© The Royal Society of Chemistry 2015
Published by the Royal Society of Chemistry, www.rsc.org

devices, the more recent generation of analyzers are based on mass spectrometry (MS).[4]

Notwithstanding the origins of ambient analysis, for this review on APPI (atmospheric-pressure photoionization) it is worth noting that there are examples of ambient analysis by APPI and other forms of photoionization (PI) that predate the current API generation of ambient analysis methods. APPI (along with other ionization sources, such as [63]Ni and various APCI types) has been used in IMS for many years.[5] APPI and PI with direct solid-phase sample injection has also been used for MS. Figure 5.1 shows a schematic view of a photoionization source coupled to a MS system that allows for an insertion probe. The entire assembly is heated and has been configured like a GC injector for syringe injections of liquid samples and SPME samples as well as insertion of a probe containing solid sample on the tip.[6,7] The ambient sampling probe in Figure 5.1 has been operated as an APPI source as well as a low-pressure PI source. This source offers simplicity in that no carrier gas or nebulizer/vaporizer is needed, but because of the resilience of APPI as an ionizer, room air suffices for screening operation.

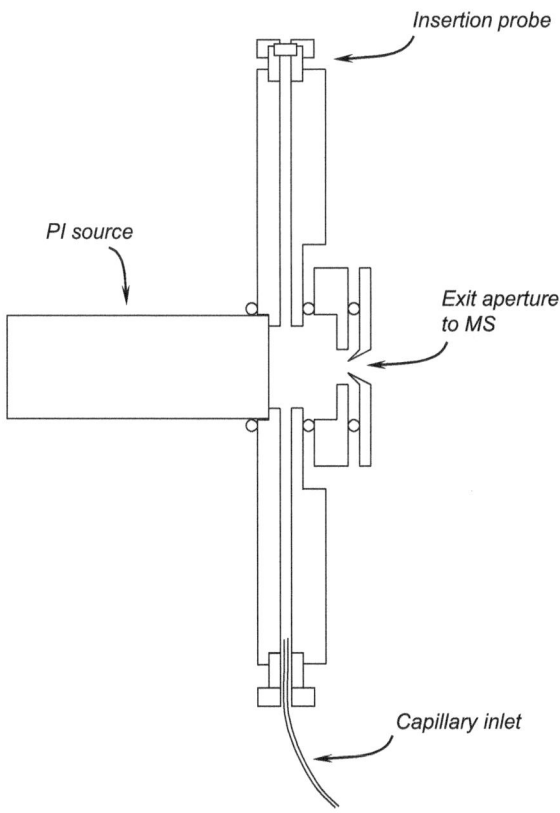

Figure 5.1 Ambient sampling photoionization source.

The source also allows for a continuous capillary flow that can serve as a calibrant, a dopant, or another sample introduction port. All flows are controlled by the pressure differential created by the vacuum behind the MS inlet.

The recent interest in ambient analysis mass spectrometry was popularized by DESI and DART. Many variations have been introduced since then with multiple reviews documenting the progress in the field.[8–12] However, DART and DESI still remain the most used techniques, partly due to their commercial availability. Ambient ion sources can be categorized based on desorption and ionization mechanisms. The APPI embodiments discussed here are based on thermal desorption followed by gas-phase ionization. The simplicity of thermal desorption is well suited for in-field analysis. Matrix effects are also less pronounced in gas-phase ionization compared to charged-droplet based methods (*e.g.* DESI), further making this approach suitable for analysis without sample preparation. Moreover, gas temperature programming may be used to effect selective desorption of analytes based on vapor pressure, creating an additional separation dimension to minimize matrix effects and to identify compounds.[13] APPI is particularly well suited to ambient analysis by thermal desorption because the heated gas can be room air rather than purified gas as is needed for other methods such as DART. Furthermore, the absence of solvent, which can absorb photons, means that the photon flux is fully available to the desorbed sample thereby enhancing sensitivity relative to its use in liquid chromatography (LC).

In plasma-based thermal desorption ionization sources, reactive species are generated by a discharge and become entrained in the gas flow prior to interaction with the sample. For example, helium metastable atoms in DART[14] and helium ions and metastable atoms in flowing atmospheric-pressure after glow[15,16] are identified as primary reactive species. These reactive species are suggested to interact with atmospheric gases (*e.g.* oxygen and water) as well as sample solvent and matrix (collectively called a microenvironment), leading to generation of ionization reagents.[17] These reactions generally produce protonation reagents, giving rise to protonated analyte ions.

For nonpolar analytes with low proton affinity, such as those encountered in the analysis of petroleum, oils, and steroid samples, the protonation mechanism is not efficient. For these compounds, charge transfer is the preferred ionization pathway. DART has been shown to generate radical molecular ions *via* reactions of analytes with helium metastable atoms and O_2^+ ions (resulting from reaction of helium metastable atoms with atmospheric oxygen).[14] However, achieving this ionization pathway requires a clean ionization region because the reactive metastable atoms and O_2^+ ions are rapidly quenched in the presence of water and trace species in air. Such conditions can be partially created by operating the DART nozzle very close to the sampling orifice of the mass spectrometer, causing the helium flow to purge the ionization region. This approach, however, reduces the residence time of neutrals in the ionization region, leading to loss of ionization efficiency. Moreover, quenching of reactive species by sample matrix (*e.g.* solvent) cannot be avoided with this approach.

5.2 Standalone Thermal Desorption Photoionization Sources

A more efficient method for generating charge-transfer reagents in ambient analysis is photoionization. This approach was first used in desorption atmospheric-pressure photoionization source (DAPPI), where a heated spray of dopant is directed onto a surface vaporizing the analytes, followed by radiation with UV photons.[18,19] Photoionization of the vaporized dopant leads to generation of reagent ions that in turn ionize the vaporized analyte. Importantly, the ionization potential for the dopants is below that of solvent and atmospheric gases. This means the reagent ions do not react with the most-abundant species (*e.g.* air) within the ambient ionization region, making the reagent ions available for ionization of the analytes and improving efficiency of ionization. However, easily ionizable matrix can still quench the reagent ions. For example, sensitivity for detection of illicit drugs using DAPPI is reduced by a factor of 2–15 in urine matrix compared to clean solvent.[20] The same matrix, however, results in a 10-fold higher signal suppression using DESI compared to DAPPI, attesting to better matrix tolerance of desorption-APPI.[20] The type of dopant impacts the ion types, particularly in positive-ion mode. Toluene and anisole readily form positive radical molecular ions and promote charge-transfer mechanisms, leading to M^+ ions from analytes, especially in the case of nonpolar compounds such as polyaromatic hydrocarbons. Mixing these dopants with a protic solvent such as methanol as well as using dopants such as acetone, hexane, and isopropanol promote proton transfer ionization, generating $[M + H]^+$ analyte ions. For most efficient ionization, one should select the dopant with a proton affinity lower than that of the analytes of interest. In negative mode, analyte ion types are not impacted significantly with the choice of dopant, however, sensitivities show dopant dependence. Acetone is frequently used as the dopant of choice for negative mode. DAPPI has been applied to screening of confiscated drugs,[21] biomass characterization,[22] and analyses of lipids[23] and cannabis samples.[24]

5.3 ASAP Adapted for APPI

We recently reported an APPI-enabled atmospheric solid analysis probe (ASAP) as another standalone APPI-based ion source for ambient mass spectrometry. In this configuration, a commercial APPI source for LC was modified to accommodate a melting-tube capillary holder with a plunger. Samples were deposited at the tip of the melting-tube capillary and were pushed into the hot gas stream from the vaporizer of the APPI source as shown in Figure 5.2.

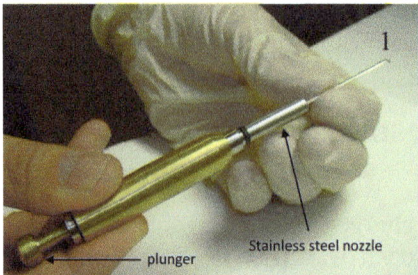
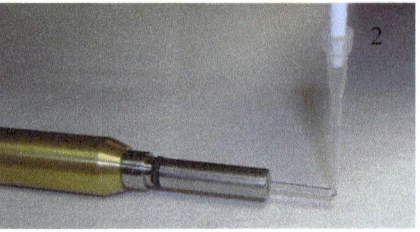
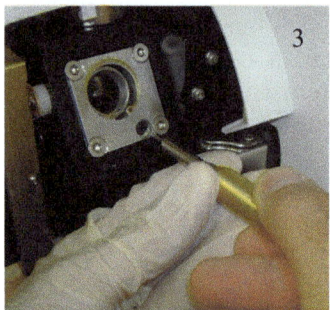

Figure 5.2 ASAP source with APPI capability. (1) The melting point capillary is installed in the probe. (2) Sample is deposited onto the capillary tip. (3) The probe is inserted into the source. (4) The capillary is pushed into the hot gas stream by the plunger.

In contrast to DAPPI, the dopant was delivered to the heated area of the nebulizer gas inlet on the source and entered the nitrogen nebulizer gas stream by vaporization rather than nebulization. The heated gas desorbed the analytes and photoionization occurred downstream using a Kr RF lamp.

Figure 5.3 compares ionization of chlorinated pesticides in positive mode for ASAP-APPI and ASAP with APCI. The charge-transfer mechanism using APPI is more efficient using APPI and leads to significant sensitivity enhancement using APPI for these hard-to-ionize compounds.

Interestingly, there is a dopant enhancement for APCI. Dopants are generally not considered for APCI in LC/MS applications; however, dopant enhancement for APCI and other corona-discharge-type ionizers are routinely used in the security industry for IMS- and MS-based instruments (as discussed in Section 5.1). Still the sensitivity for dopant-assisted APPI is significantly better than any form of APCI in Figure 5.3, at least for the class of pesticide compounds.

5.4   DART-APPI

A notable feature of photoionization is that reagent ion generation can occur after analyte desorption. A photoionization lamp can, therefore, be placed downstream of the sample, making photoionization a relatively straightforward addition to plasma-based desorption ionization techniques.

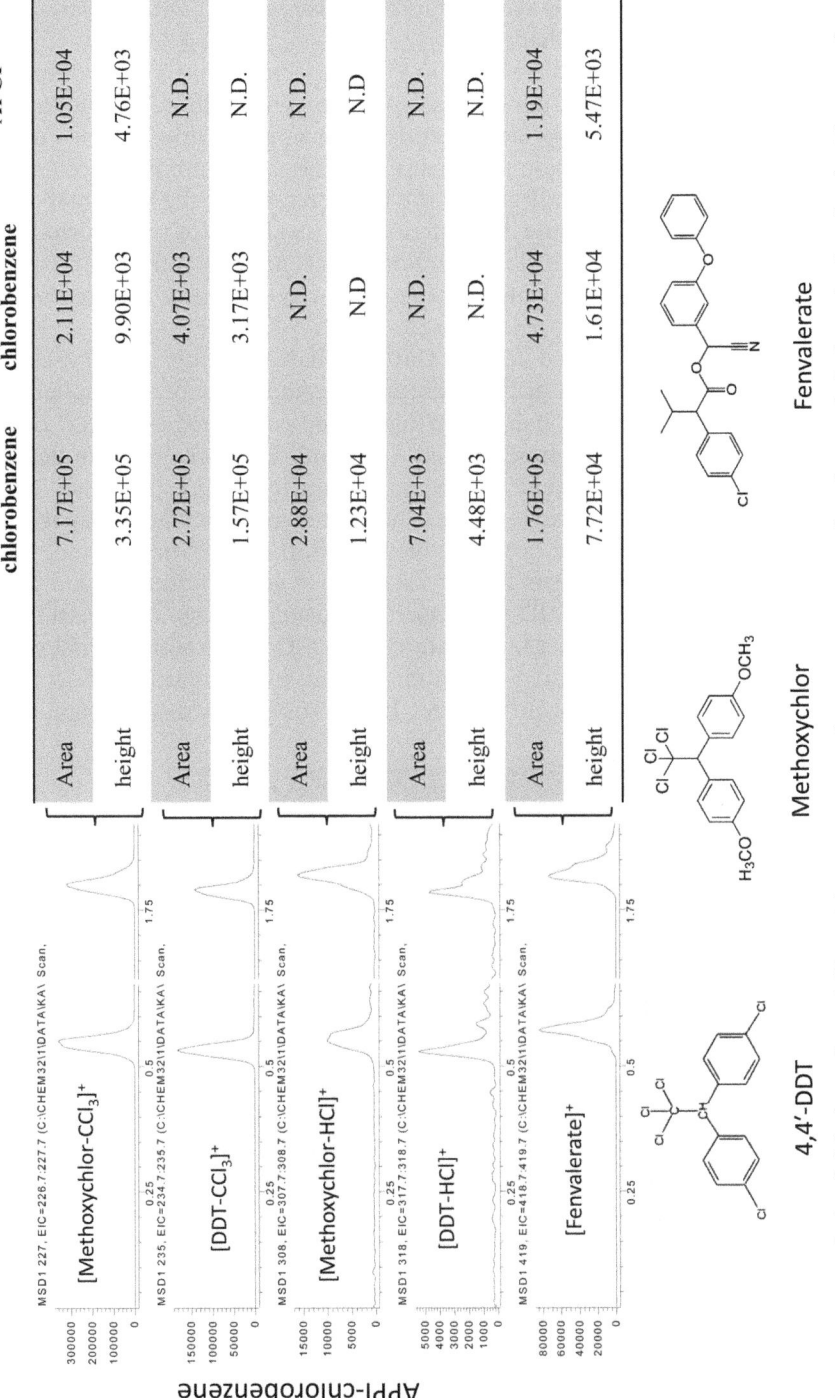

		APPI-chlorobenzene	APCI-chlorobenzene	APCI
[Methoxychlor-CCl₃]⁺	Area	7.17E+05	2.11E+04	1.05E+04
	height	3.35E+05	9.90E+03	4.76E+03
[DDT-CCl₃]⁺	Area	2.72E+05	4.07E+03	N.D.
	height	1.57E+05	3.17E+03	N.D.
[Methoxychlor-HCl]⁺	Area	2.88E+04	N.D.	N.D.
	height	1.23E+04	N.D	N.D
[DDT-HCl]⁺	Area	7.04E+03	N.D.	N.D.
	height	4.48E+03	N.D.	N.D.
[Fenvalerate]⁺	Area	1.76E+05	4.73E+04	1.19E+04
	height	7.72E+04	1.61E+04	5.47E+03

Fenvalerate

Methoxychlor

4,4'-DDT

Figure 5.3 Analysis of chlorinated pesticides with ASAP-APPI and ASAP-APCI. 1 ng of each sample is deposited at the tip. The vaporizer is set to 300 °C and chlorobenzene is used as dopant at a 500 nL min⁻¹ flow.

One example is combination of DART and APPI to augment the DART ionization capabilities. This configuration is discussed in detail below and has been previously published.[25]

The source configuration is shown in Figure 5.4. A DART-100 ion source is mounted on a single-quadrupole mass spectrometer modified with a Vapur interface.[26] The Vapur interface extends the sampling inlet of the mass spectrometer to the ionization region and provides a vacuum port to effect a flow towards the MS inlet, alleviating the high pressure in the MS due to the use of helium gas. For sample introduction, a 1 μL aliquot of the analyte mix solution is deposited at the closed end of a glass melting tube and pushed into the ionization region using a home-made rail ∼3 mm downstream of the DART outlet.

Photoionization is added to the DART unit by placing an RF-excited krypton lamp downstream of the desorption region. Dopant is supplied at flow rates of 1–5 μL min^{-1} using a syringe pump connected to a 1/32″ o.d., 100 μm i.d. stainless steel tube (placed ∼ 1 mm) downstream the outlet of the DART ion source. Toluene is used as dopant in the positive mode and acetone is used in the negative mode. The heated gas from DART vaporizes the dopant and carries the dopant molecules to the ionization region. For operation using helium, a gas flow of 2.8 L min^{-1} and a heater temperature of 280 °C are used on the DART unit. Operation using nitrogen requires a flow rate of 8 L min^{-1} and heater temperature of 450 °C. The lower thermal conductivity of nitrogen compared to helium leads to higher optimum flow rate and heater temperature for an effective heat transfer to the glass capillary.

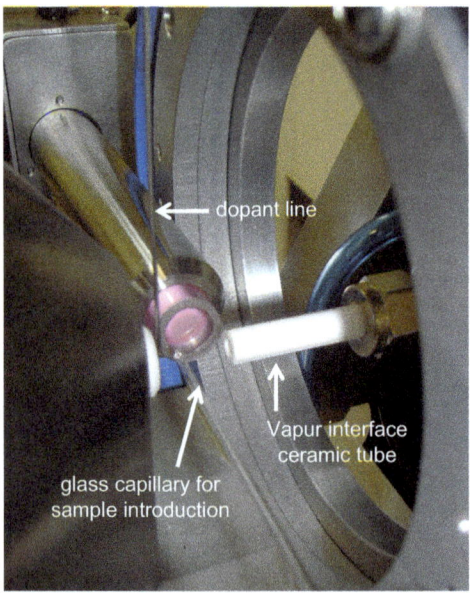

Figure 5.4   DART-APPI experimental setup.
Figure is reproduced from ref. 25 with permission.

In the results presented below, DART alone refers to operation with the discharge on, lamp off, and no dopant, whereas DART-APPI refers to discharge off, lamp on, and dopant flow on configuration.

5.4.1   Ion Types in Positive Mode

An advantage of photoionization, as noted above, is that the charge-transfer ionization pathway can be made to dominate. This characteristic is highlighted in Figure 5.5 where blank-subtracted mass spectra obtained by desorption/ionization of 6 ng anthracene (introduced as a 5 μL droplet with 1:1 water:methanol solvent) are compared for DART and DART-APPI. Note that anthracene desorption occurs simultaneously with the solvent evaporation. Therefore, ionization occurs in the presence of interfering solvent molecules. The M^+ ion for anthracene is not observed using DART alone, whereas a clear signal at m/z 178 is detected when the APPI lamp is on. In DART-APPI configuration, gas flow mainly provides the heat for desorption. Therefore, the ionization mechanism can be tuned by the choice of dopant independent of the gas. This is shown in Figure 5.5(C) where M^+ ions for anthracene were observed with DART-APPI using nitrogen for desorption. Note that DART alone operated using nitrogen did not provide any ions from anthracene at the level used above.

Figure 5.6 compares the ion types in positive mode for verapamil (1.8 ng), cortisol (6 ng), aldicarb sulfone (5 ng), and 17-α-estradiol (2 ng) using DART and DART-APPI with helium and nitrogen gases. No analyte ions were detected using DART alone with nitrogen gas; therefore the spectrum for nitrogen DART is not included. As depicted by the blue dotted lines, analytes with high proton affinity (verapamil, aldicarb sulfone, and cortisol) show similar ion types among the three ionization conditions. Protonated molecular ions are formed by photoionization either *via* hydrogen abstraction by radical molecular ions or proton transfer from protonated dopant/solvent molecules.[27]

That aldicarb sulfone forms an ammonium adduct is an interesting observation that may be attributed to higher affinity of sulfone group to ammonium ion compared to other analytes in this study. Our investigations showed that the ammonium adduct was not observed in the absence of the Vapur interface. Therefore, we attribute the formation of ammonium adducts to the release of ammonia from the ceramic tube and adduct formation inside the Vapur interface.

The largest difference in ion type among the three conditions is observed in the case of 17-α-estradiol. Helium DART alone promotes the formation of $[M - OH]^+$ for this analyte that is similar to the ion type observed using conventional APCI, suggesting protonation followed by dehydration as ion-formation mechanism. On the other hand, helium DART-APPI gives rise to formation of $M^{+\bullet}$, an indication of the charge-transfer mechanism promoted by dopant-assisted photoionization. Note that $[M - OH]^+$ for 17-α-estradiol is also observed using APPI, suggesting formation of protonated ions by photoionization *via* the mechanisms discussed above.

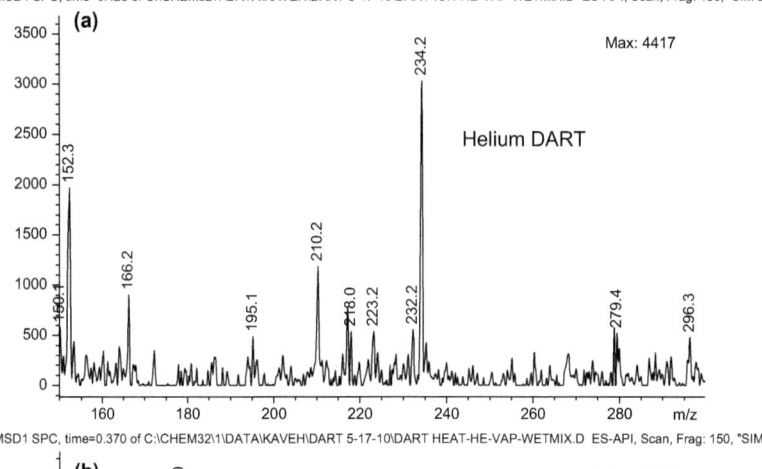

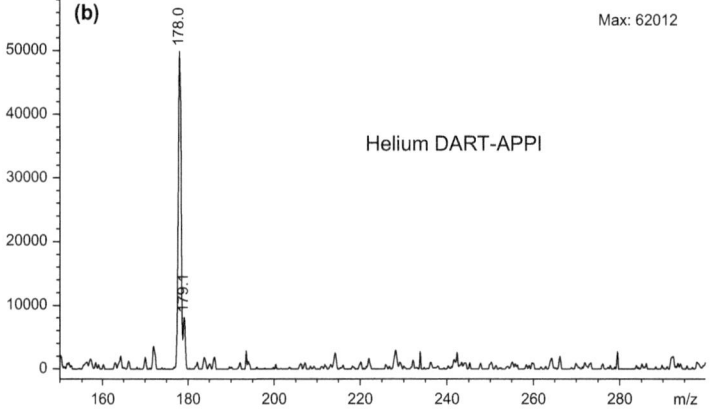

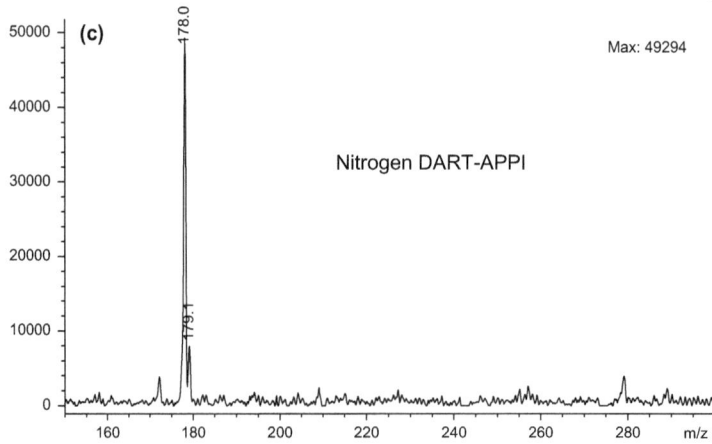

Figure 5.5   Mass spectra obtained from introduction of 6 ng anthracene in 1 : 1
water : methanol solvent using (a) DART with helium gas, (b) DART-APPI
with helium gas, and (c) DART-APPI with nitrogen gas. Anthracene
molecular ion appears at *m/z* 178.
Figure is reproduced from ref. 25 with permission.

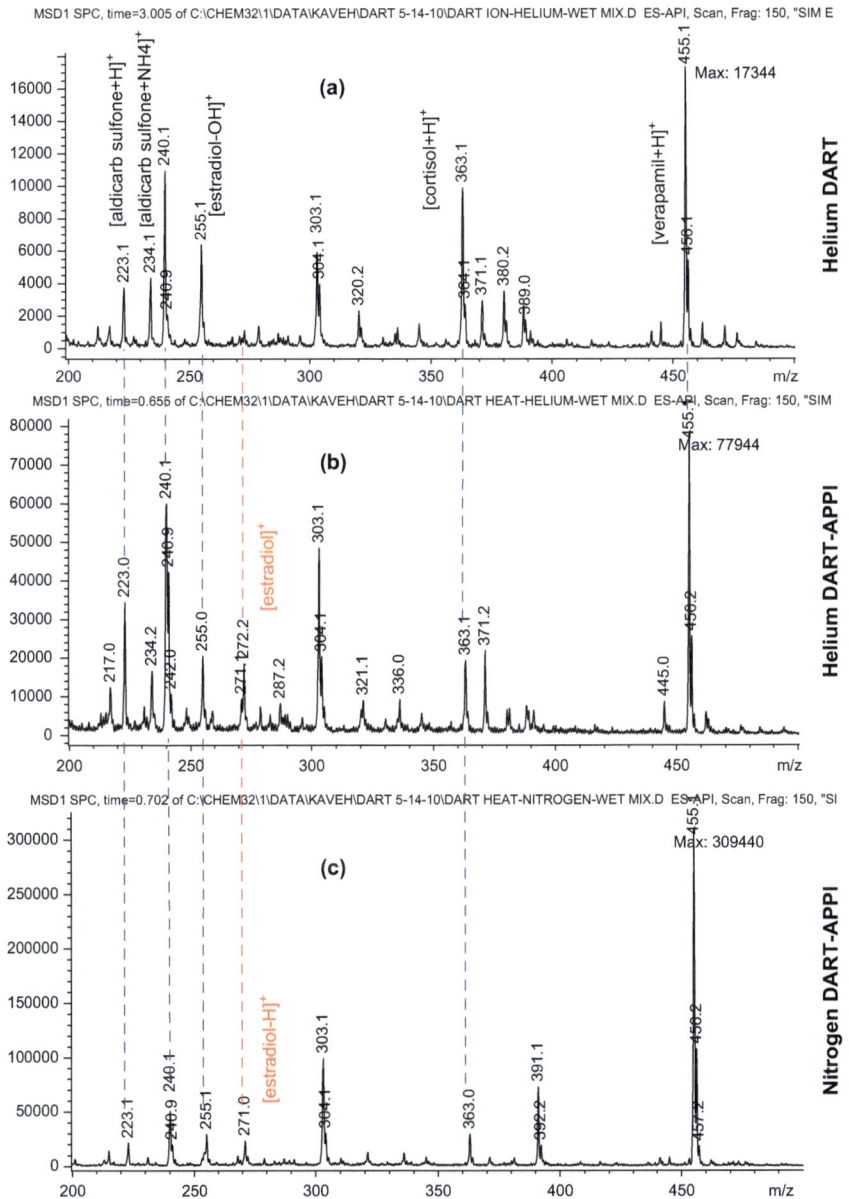

Figure 5.6 Comparison of ion types generated by (a) DART using helium gas, (b) DART-APPI using helium gas, and (c) DART-APPI using nitrogen gas. Red dashed lines show unique ions for each ionization configuration. Blue dashed lines show the common ion types.
Figure is reproduced from ref. 25 with permission.

Yet another unique feature of DART-APPI is formation of $[M - H]^+$ ion observed at *m/z* 271 for 17-α-estradiol (Figure 5.3(c)). Loss of hydrogen from radical cations can result from transfer of a hydrogen atom to background gas (*e.g.* oxygen) present in ambient ionization conditions. The availability of benzilic or allylic hydrogens is expected to increase the efficiency of hydrogen transfer. This reaction channel, however, is not favored using helium DART-APPI as evidenced by lower intensity of *m/z* 271 compared to that of *m/z* 272 (molecular ion). But the intensity at *m/z* 271 dramatically increases compared to that of *m/z* 272 using nitrogen as desorption gas, suggesting promotion of hydrogen loss channel for 17-α-estradiol at these conditions. We attribute this change to higher temperature of nitrogen gas after desorption and through the Vapur interface, improving the kinetics of hydrogen loss.

5.4.2 Sensitivities in Positive Mode

The vaporization of the analytes by thermal desorption occurs in the order of their vapor pressures. Accordingly, separation of ions over time is observed.[13] To account for this effect, we used the average area under the extracted ion chromatogram for each ion from three trials for sensitivity comparison between the desorption/ionization methods. The data are summarized in Figure 5.7 for ions commonly observed using the three ionization conditions.

DART-APPI provides a factor of 3–5 improvement over helium DART alone for the analytes studied above. Note that the *y*-axis has a logarithmic scale.

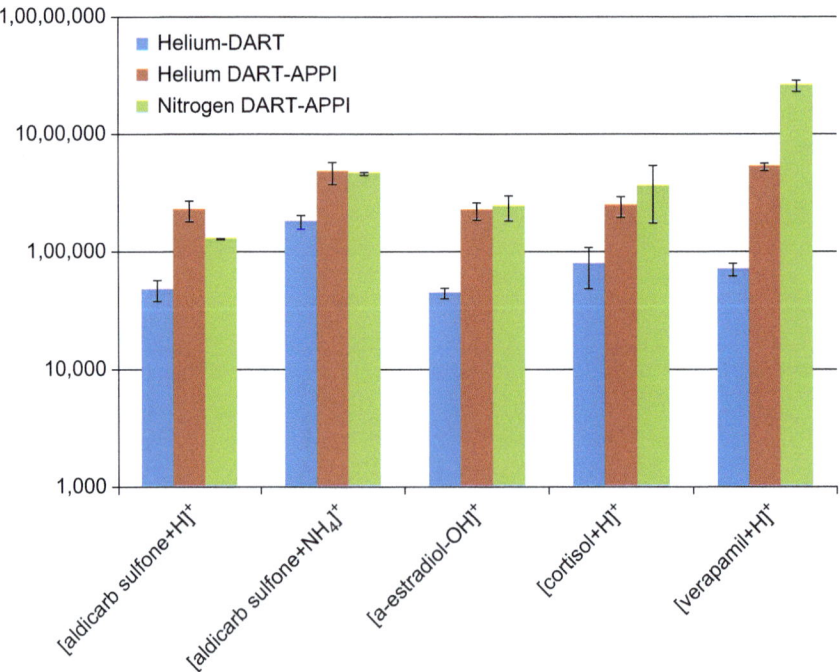

Figure 5.7 Comparison of sensitivities for common ion types using DART and DART-APPI.
Figure is reproduced from ref. 25 with permission.

More importantly, the sensitivities for helium DART-APPI and nitrogen DART-APPI are comparable. In contrast, a large loss of sensitivities is observed for nitrogen DART compared to helium DART.

5.4.3   Negative-ion Mode

The performance of helium DART alone with that of helium DART-APPI in negative mode is compared in Figure 5.8. A 1 μL sample containing 6 ng 2,5-dinitrobenzotrifluoride, 17 ng dinitrophenol, and 17 ng benzilic acid was used in these experiments. Similar ion types are observed for helium DART alone and helium DART-APPI, suggesting electron capture and dissociative

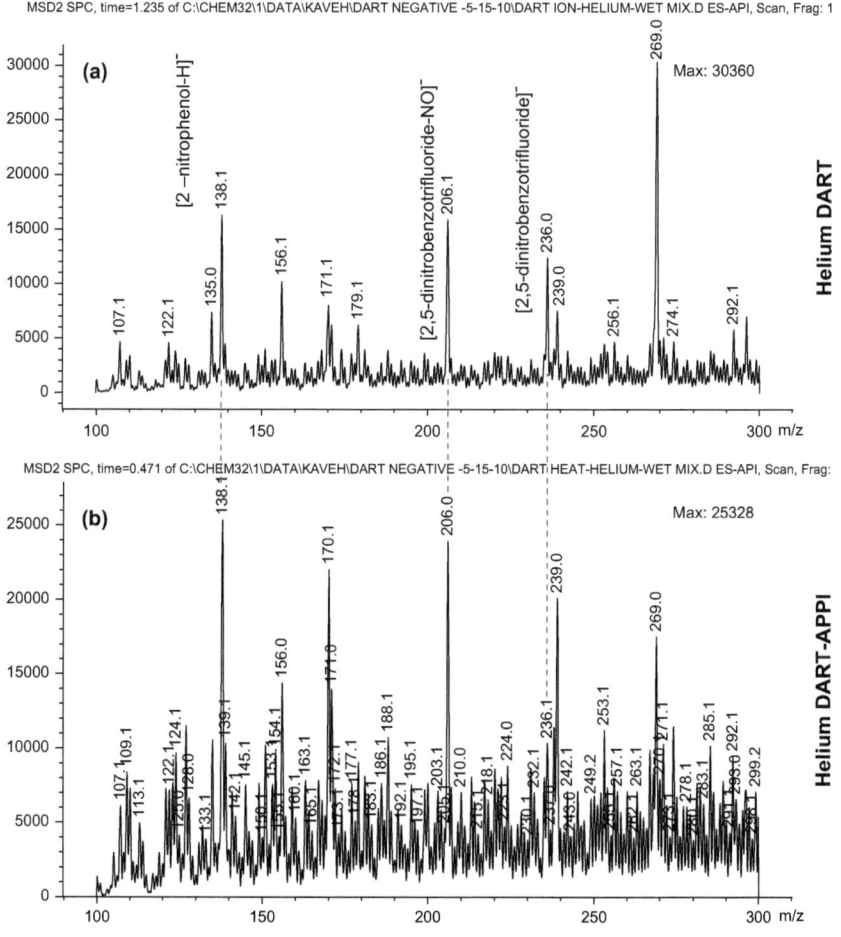

Figure 5.8   Negative-ion types generated by (a) DART using helium gas, and (b) DART-APPI using helium gas. Dashed blue lines show the common ion types from the analytes.
Figure is reproduced from ref. 25 with permission.

electron capture mechanism for 2,5-dinitrobenzotrifluoride ionization and deprotonation mechanism for 2-nitrophenol ionization. This observation is consistent with other studies reporting similar negative mode ionization mechanisms for DART and APPI.[28] However, the chemical noise in the case of DART-APPI is higher, which may be attributed to survival of toluene-related clusters in the absence of counterflow gas (see below for more discussion on the counterflow gas effect). Importantly, benzilic acid is not observed at the level introduced using either desorption/ionization conditions. We noticed a further drop in sensitivity using nitrogen DART-APPI in negative mode compared to helium DART-APPI, suggesting a faster decay of these analytes (either in the form of neutrals or ions) during the transit in the hotter nitrogen gas.

We observed an interesting trend for negative ions upon removing the Vapur interface. In this configuration, the counterflow gas is increased to 5 L min^{-1} to avoid a pressure spike in the mass spectrometer due to the use of helium gas in DART. Figure 5.9 demonstrates that the sensitivities

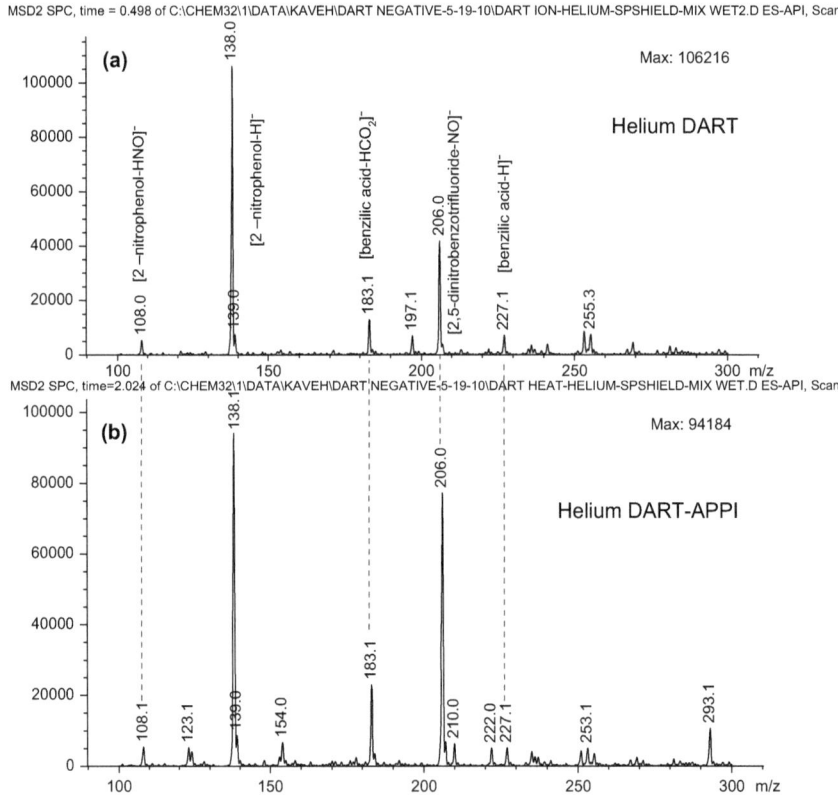

Figure 5.9   Negative-ion types in the absence of Vapur interface from (a) DART using helium gas, and (b) DART-APPI using helium gas. Dashed blue lines show the common ion types from the analytes.
Figure is reproduced from ref. 25 with permission.

for negative ions improve dramatically for both DART and DART-APPI upon removal of the Vapur interface. Such a drastic change is not observed for positive ions. We attribute this improvement in negative-ion mode to rapid cooling of the desorption gas *via* mixing with the counterflow gas. In addition, solvent clusters are removed with the use of counterflow gas, reducing the noise levels.

5.5 Conclusions

The studies above highlight the potential of photoionization as a broad-based ionization source in ambient mass spectrometry. New ion types can be generated that aid in identification of compounds. In addition, higher sensitivities are achieved with APPI. The benefits of APPI for ambient analysis are similar to those for its application for LC/MS, but with a few additional advantages:

- Wide range of ionizable compounds: Most ambient analysis methods are based on ionization sources that rely on charge (proton, cation, anion) affinity so that many compounds particularly nonpolar ones are difficult to ionize. Furthermore, the presence of high charge-affinity background compounds can steal charge and suppress the detection of analytes of interest. This problem is characteristic of ambient analysis ionization methods because the sample is generally not cleaned up or preseparated.
- Low susceptibility to ion suppression and matrix effects: This benefit of APPI is particularly important to ambient analysis because, as mentioned above, there is no clean-up or separation of the components of a sample. Photoionization relies on direct photoionization as well as dopant-assisted ionization. The latter is analogous to chemical ionization; however, the ion chemistry is much better controlled than the reliance on charge carriers in air or in solvent flows.
- High sensitivity and dynamic range: APPI is generally comparable in sensitivity to APCI for LC/MS with the edge going to APPI as the solvent flow rate decreases. Ambient analysis takes this to the limit of no flow, which greatly favors APPI over APCI for these purely thermally desorbed gas phase systems. APCI needs a charge carrier, which is usually the protic solvents used for reverse-phase LC. The solvent is generally a negative to APPI sensitivity because it can absorb photons and non-radiatively remove them from the ionization pool. APPI therefore is at its best in the absence of solvent, though very low-flow dopant introduction is still recommended. Although ambient analysis is not typically used for quantitative analysis, APPI is capable of upwards of 5 orders of linear dynamic range.
- Nitrogen (or air) can be used as desorption gas without a loss in performance when ionization is effected by APPI. This alleviates the need for additional pumping capacity required for the use of helium and

obviates the need for the Vapur interface. Moreover, this versatility makes APPI an attractive candidate for portable instruments and in-field applications where simplicity, robustness, and freedom from consumable gases are of higher importance compared to lab-based analysis.

Dopants generally improve the performance of APPI. Although used at very small quantities (1–5 µL min^{-1} in this study), one has to consider safety issues as well as the environmental impact of releasing the dopant vapor into the analysis area. Accordingly, development of safe dopants with high ionization potentials should be an area of focus to broaden the use of photoionization in ambient ionization, particularly for in-field analysis.

Acknowledgements

We thank IonSense Inc. for the loan of DART-100. We are very grateful to the California Department of Food and Agriculture for donation of pesticide standards. We also thank Sheng-Suan (Victor) Cai of Syagen for valuable discussions.

References

1. R. B. Cody, J. A. Laramee and H. D. Durst, Versatile new ion source for the analysis of materials in open air under ambient conditions, *Anal. Chem.*, 2005, 77, 2297–2302.
2. Z. Takats, J. M. Wiseman, B. Gologan and R. G. Cooks, Mass spectrometry sampling under ambient conditions with desorption electrospray ionization, *Science*, 2004, **306**, 471–473.
3. D. S. Moore, Instrumentation for trace detection of high explosives, *Rev. Sci. Instrum.*, 2004, **75**, 2499–2512, DOI: 10.1063/1.1771493.
4. J. A. Syage and K. A. Hanold, Mass spectrometry for security screening of explosives, in *Trace Chemical Sensing of Explosives*, Wiley, New York, 2007, p. 219.
5. G. E. Spangler, Theory and Technique for Measuring Mobility Using Ion Mobility Spectrometry, *Anal. Chem.*, 1993, **65**, 3010–3014, DOI: 10.1021/Ac00069a013.
6. J. A. Syage, S. S. Cai, J. W. Li and M. D. Evans, Direct sampling of chemical weapons in water by photoionization mass spectrometry, *Anal. Chem.*, 2006, **78**, 2967–2976, DOI: 10.1021/Ac0518506.
7. J. S. Wu, J. Greenlee and J. Syage, Rapid screening for chemical weapons infiltration in drinking water, *J. – Am. Water Works Assoc.*, 2004, **96**, 44.
8. D. R. Ifa, C. P. Wu, Z. Ouyang and R. G. Cooks, Desorption electrospray ionization and other ambient ionization methods: current progress and preview, *Analyst*, 2010, **135**, 669–681, DOI: 10.1039/B925257f.
9. R. M. Alberici, R. C. Simas, G. B. Sanvido, W. Romao, P. M. Lalli, M. Benassi, I. B. S. Cunha and M. N. Eberlin, Ambient mass

spectrometry: bringing MS into the "real world", *Anal. Bioanal. Chem.*, 2010, **398**, 265–294, DOI: 10.1007/s00216-010-3808-3.

10. S. C. Cheng, Y. S. Lin, M. Z. Huang and J. Shiea, Applications of electrospray laser desorption ionization mass spectrometry for document examination, *Rapid Commun. Mass Spectrom.*, 2010, **24**, 203–208, DOI: 10.1002/Rcm.4378.

11. G. A. Harris, A. S. Galhena and F. M. Fernandez, Ambient Sampling/Ionization Mass Spectrometry: Applications and Current Trends, *Anal. Chem.*, 2011, **83**, 4508–4538, DOI: 10.1021/Ac200918u.

12. M. Z. Huang, S. C. Cheng, Y. T. Cho and J. Shiea, Ambient ionization mass spectrometry: A tutorial, *Anal. Chim. Acta*, 2011, **702**, 1–15, DOI: 10.1016/j.aca.2011.06.017.

13. S. D. Maleknia, T. M. Vail, R. B. Cody, D. O. Sparkman, T. L. Bell and M. A. Adams, Temperature-dependent release of volatile organic compounds of eucalypts by direct analysis in real time (DART) mass spectrometry, *Rapid Commun. Mass Spectrom.*, 2009, **23**, 2241–2246.

14. R. B. Cody, Observation of Molecular Ions and Analysis of Nonpolar Compounds with the Direct Analysis in Real Time Ion Source, *Anal. Chem.*, 2009, **81**, 1101–1107.

15. J. T. Shelley, G. C. Y. Chan and G. M. Hieftje, Understanding the Flowing Atmospheric-Pressure Afterglow (FAPA) Ambient Ionization Source through Optical Means, *J. Am. Soc. Mass Spectrom.*, 2012, **23**, 407–417, DOI: 10.1007/s13361-011-0292-8.

16. G. C. Y. Chan, J. T. Shelley, J. S. Wiley, C. Engelhard, A. U. Jackson, R. G. Cooks and G. M. Hieftje, Elucidation of Reaction Mechanisms Responsible for Afterglow and Reagent-Ion Formation in the Low-Temperature Plasma Probe Ambient Ionization Source, *Anal. Chem.*, 2011, **83**, 3675–3686, DOI: 10.1021/Ac103224x.

17. L. G. Song, S. C. Gibson, D. Bhandari, K. D. Cook and J. E. Bartmess, Ionization Mechanism of Positive-Ion Direct Analysis in Real Time: A Transient Microenvironment Concept, *Anal. Chem.*, 2009, **81**, 10080–10088, DOI: 10.1021/Ac901122b.

18. M. Haapala, J. Pol, V. Saarela, V. Arvola, T. Kotiaho, R. A. Ketola, S. Franssila, T. J. Kauppila and R. Kostiainen, Desorption atmospheric pressure photoionization, *Anal. Chem.*, 2007, **79**, 7867–7872, DOI: 10.1021/Ac071152g.

19. L. Luosujarvi, V. Arvola, M. Haapala, J. Pol, V. Saarela, S. Franssila, T. Kotiaho, R. Kostiainen and T. J. Kauppila, Desorption and ionization mechanisms in desorption atmospheric pressure photoionization, *Anal. Chem.*, 2008, **80**, 7460–7466, DOI: 10.1021/Ac801186x.

20. N. M. Suni, P. Lindfors, O. Laine, P. Ostman, I. Ojanpera, T. Kotiaho, T. J. Kauppila and R. Kostiainen, Matrix effect in the analysis of drugs of abuse from urine with desorption atmospheric pressure photoionization-mass spectrometry (DAPPI-MS) and desorption electrospray ionization-mass spectrometry (DESI-MS), *Anal. Chim. Acta*, 2011, **699**, 73–80, DOI: 10.1016/j.aca.2011.05.004.

21. T. J. Kauppila, A. Flink, M. Haapala, U. M. Laakkonen, L. Aalberg, R. A. Ketola and R. Kostiainen, Desorption atmospheric pressure photoionization-mass spectrometry in routine analysis of confiscated drugs, *Forensic Sci. Int.*, 2011, **210**, 206–212, DOI: 10.1016/j.forsciint. 2011.03.018.

22. D. C. Podgorski, R. Hamdan, A. M. McKenna, L. Nyadong, R. P. Rodgers, A. G. Marshall and W. T. Cooper, Characterization of Pyrogenic Black Carbon by Desorption Atmospheric Pressure Photoionization Fourier Transform Ion Cyclotron Resonance Mass Spectrometry, *Anal. Chem.*, 2012, **84**, 1281–1287, DOI: 10.1021/Ac202166x.

23. N. M. Suni, H. Aalto, T. J. Kauppila, T. Kotiaho and R. Kostiainen, Analysis of lipids with desorption atmospheric pressure photoioniza-tion-mass spectrometry (DAPPI-MS) and desorption electrospray ionization-mass spectrometry (DESI-MS), *J. Mass Spectrom.*, 2012, **47**, 611–619, DOI: 10.1002/Jms.2992.

24. T. J. Kauppila, A. Flink, U. M. Laakkonen, L. Aalberg and R. A. Ketola, Direct analysis of cannabis samples by desorption atmospheric pressure photoionization-mass spectrometry, *Drug Test. Anal.*, 2013, **5**, 186–190, DOI: 10.1002/Dta.1412.

25. K. Jorabchi, K. Hanold and J. Syage, Ambient analysis by thermal desorption atmospheric pressure photoionization, *Anal. Bioanal. Chem.*, 2013, **405**, 7011–7018, DOI: 10.1007/s00216-012-6536-z.

26. S. X. Yu, E. Crawford, J. Tice, B. Musselman and J. T. Wu, Bioanalysis without Sample Cleanup or Chromatography: The Evaluation and Initial Implementation of Direct Analysis in Real Time Ionization Mass Spectrometry for the Quantification of Drugs in Biological Matrixes, *Anal. Chem.*, 2009, **81**, 193–202, DOI: 10.1021/Ac801734t.

27. J. A. Syage, Mechanism of [M + H](+) formation in photoionization mass spectrometry, *J. Am. Soc. Mass Spectrom.*, 2004, **15**, 1521–1533, DOI: 10.1016/j.jasms.2004.07.006.

28. L. G. Song, A. B. Dykstra, H. F. Yao and J. E. Bartmess, Ionization Mechanism of Negative Ion-Direct Analysis in Real Time: A Comparative Study with Negative Ion-Atmospheric Pressure Photoionization, *J. Am. Soc. Mass Spectrom.*, 2009, **20**, 42–50, DOI: 10.1016/j.jasms.2008.09.016.

CHAPTER 6

Low-Temperature Plasma Probe

JIANGJIANG LIU,[a] XIAOYU ZHOU,[a] R. GRAHAM COOKS*[b] AND
ZHENG OUYANG*[a,c]

[a] Weldon School of Biomedical Engineering, Purdue University,
West Lafayette, Indiana 47907, USA; [b] Department of Chemistry, Purdue
University, West Lafayette, Indiana 47907, USA; [c] Department of Electrical
and Computer Engineering, Purdue University, West Lafayette, Indiana
47907, USA
*Email: cooks@purdue.edu; ouyang@purdue.edu

6.1 Introduction

The low-temperature plasma (LTP) probe[1] was developed for direct sampling
ionization in chemical analysis using mass spectrometry (MS). The plasma
in an LTP probe is generated by dielectric barrier discharge (DBD) and a
discharge gas at low flow rate (<500 mL min^{-1}) and a high-voltage AC are
used to sustain the plasma in an ambient environment. Different from some
other ambient ionization methods[2] also using plasma species, the temperature of the sampling torch from an LTP probe can be adjusted over a
wide range and can be as low as 30 °C so that no damage is made to the
sample substrates, such as the skin of human fingers, fabrics on luggage, or
plastic case of a portable hard drive, *etc*. The LTP probe has been demonstrated as a powerful analytical tool for direct analysis of a wide variety of
chemicals from complex samples in their native conditions, working especially well with small organic molecules with low to moderate polarity.[3]
Qualitative and quantitative analysis using LTP probe has been widely

New Developments in Mass Spectrometry No. 2
Ambient Ionization Mass Spectrometry
Edited by Marek Domin and Robert Cody
© The Royal Society of Chemistry 2015
Published by the Royal Society of Chemistry, www.rsc.org

reported for diverse types of applications, including public safety, food safety, product quality control and forensics. Different configurations have been used for the design of LTP probes to improve their analytical performance or to satisfy various implementation requirements in different studies and applications. With the simplicity of implementation, low consumption of discharge gas and the possibility of using air as the discharge gas, LTP probe is an ideal candidate for developing portable MS analytical system for in-field chemical analysis. A handheld LTP probe was developed and tested with a miniature mass spectrometer and a coaxial LTP probe was designed for a self-sustained backpack MS analytical system. The highly reactive environment associated with the LTP has also been used to explore new chemical reactions, which produces reaction products not readily obtainable through traditional organic synthesis or lead to development of unique tools for analysis of peptides or fatty acids.

6.2 Characterization of the Low-Temperature Plasma

The original configuration of the LTP probe and the experimental setup are shown in Figure 6.1(a).[1] The sample is deposited on a solid substrate and directly sampled by the LTP probe. The LTP probe consists of a glass tube serving as a dielectric barrier and a channel for gas flow, a grounded electrode axially inserted into the glass tube, and a ring electrode outside the tube. The ring electrode is connected to an AC power supply and a discharge gas flow through the glass tube. The low temperature plasma generated through DBD extends out of the tube and is used to desorb and ionize analytes from the sample on a surface. The discharge gas can be selected from helium, argon, nitrogen or air, at a flow rate of ~ 0.4 L min^{-1} with the discharge AC set at 3 kVpp and 2.5 kHz.

The LTP probe takes advantage of a stable and low-power discharge at atmospheric pressure by the DBD, and so does another ambient ionization

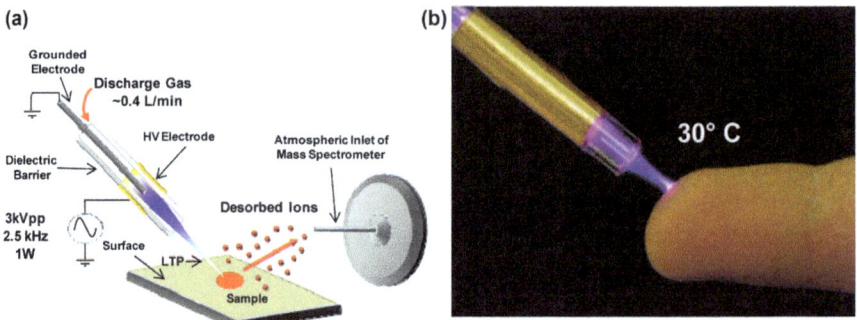

Figure 6.1 LTP probe setup for ambient ionization MS: (a) schematic of the configuration and (b) photo of the plasma torch used to sample compounds on a human finger.
Figure adapted from ref. 7 with permission.

method dielectric barrier discharge ionization (DBDI),[4] which requires the sample to be located between the two discharge electrodes. The low-temperature plasma torch from an LTP probe scans across the sample surface and the analytes are conveniently desorbed and ionized for MS analysis. It also allows an easy analysis of surfaces of large areas or bulk solutions.[1] In comparison with other plasma-based ambient ionization methods, such as direct analysis in real time (DART)[5] and flowing atmospheric-pressure afterglow (FAPA),[6] the temperature of the plasma can be significantly lower, *e.g.* 30 °C, so no damage occurs to the surfaces to be sampled (Figure 6.1(b)).[1] For the same reason, the implementation of LTP probe can be relatively simple and easy.

6.2.1 Mechanism of LTP

The LTP probe was widely applied to different analytical purposes, the mechanism of desorption and ionization involving the LTP has also been investigated,[8,9] but remains to be further understood. Preliminary studies indicated that the possible processes including thermal desorption, chemical sputtering and surface reactions, could simultaneously occur during the desorption ionization process. The LTP torch generated with helium discharge gas was characterized by optical spectroscopy using a spatially selective detection system shown in Figure 6.2(a).[9] The electron number density and rotational temperature were measured in both the LTP torch and the afterglow regions. Emission profiles of different molecular and atomic species, such as OH, N_2, N_2^+, He_I, H_I and O_I were measured with UV-visible emission spectroscopy. Some interesting findings included that the OH emission was caused by the water in the discharge gas supply, and the emissions by N_2, N_2^+ and atomic oxygen were originated from N_2 and O_2 diffusing into the LTP torch from the ambient environment. He_2^+ was found to be the dominant positive ion and performed as an energy-carrying specie to transfer the energy from the inside discharge region to the outside afterglow region. In the afterglow region, a charge transfer between He_2^+ and atmospheric nitrogen occurred and the N_2^+ was generated (Figure 6.2(b)). These findings could be useful to explain some other ionization and gas-phase reactions in mass-spectrometric study, where N_2^+ and atomic oxygen radicals play a critical role during the plasma-based ionization process.[8,9]

6.2.2 Analytical Performance

The plasma generated by LTP probe is of low temperature, approximately 30 °C measured at the spot on the surface in contact with the sampling plasma torch. Thermodegradation can be minimized and intact molecular species can be well preserved during the desorption ionization. In addition, without the discharge through the sample as for DBDI, it is feasible to perform *in vivo* analysis of biological subjects without subjecting to heat

(a)

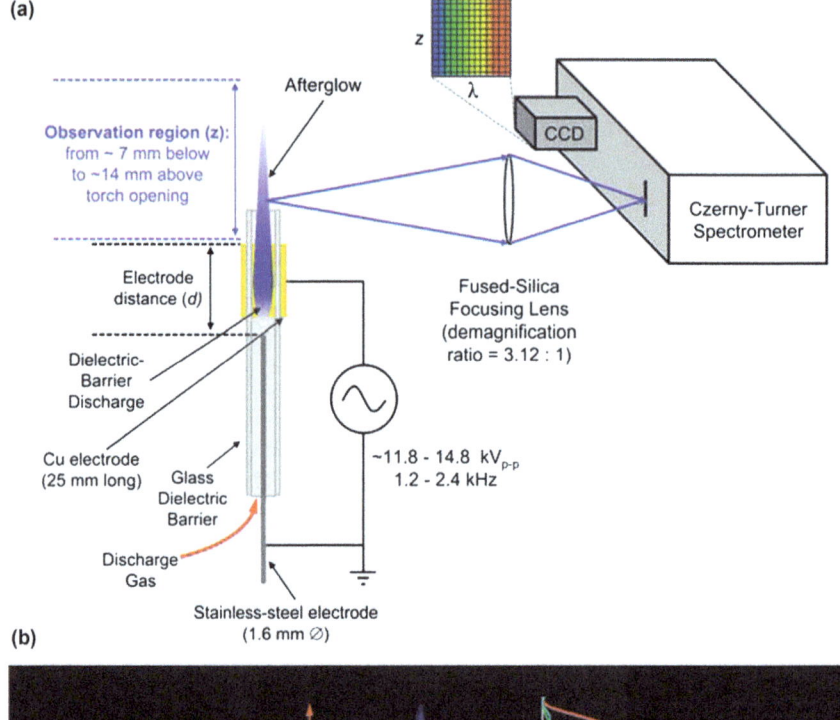

(b)

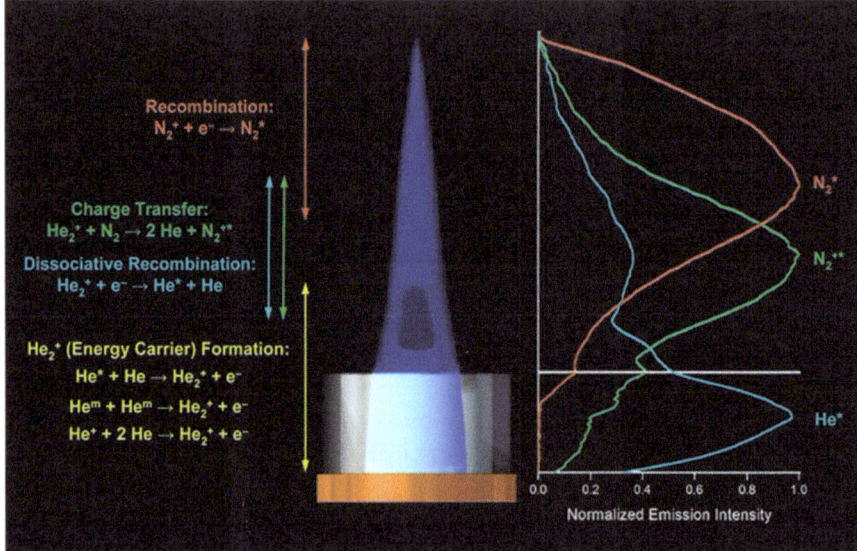

Figure 6.2   (a) Schematic diagram of the experimental setup for the measurement of electron number density and rotational temperature. (b) Schematic diagram showing the identified spatially dependent reactions for the afterglow and reagent-ion formation in the LTP probe ambient ionization source.
Figure taken from ref. 8 and 9 with permission.

damage or electric shock as shown in Figure 6.1(b). All these features make the LTP probe attractive for direct and ambient analysis of raw samples in their native conditions without sample preparation.

A comparison study using only pure sample solutions has been done with LTP, atmospheric-pressure chemical ionization (APCI) and electrospray ionization (ESI).[3] APCI and ESI are two most popular methods widely used for atmospheric-pressure ionization in analysis of purified samples. They complementarily cover different ranges in polarity and volatility of the chemicals. Although direct analysis of untreated samples might provide different study results due to an impact by matrix effect, this comparison study does provide valuable information on the applicability of the LTP, with the relative ionization efficiencies identified for compounds of different physical and chemical properties, including polycyclic aromatic hydrocarbons (PAHs) ionic species, amides, amines, imides, aldehydes, nucleoside, and pharmaceutical drug (Figure 6.3(a)).[3] The comparison results are summarized in Figure 6.3(b). As a general guidance, LTP shows more similarities to APCI than ESI in terms of the polarity and molecular weight of the analyte. LTP works well for both polar and nonpolar analytes, but not for ionic analytes. The molecular weight of analytes is also critical for chemical detection using LTP, presumably because of difficulty in desorption of intact large molecules such biopolymers from condensed phase samples.

The implementation of LTP probe for sampling analysis is relatively easy and the relative positions of LTP probe, sample and MS inlet are not critical. The susceptibility to less careful alignment was characterized by moving the glass slide around while recording the signal for 1 μg of cocaine deposited on it (Figure 6.4(a)).[1] The contour map of the recorded signal intensity (Figure 6.4(b)) shows that good signal could be obtained with the sample in a

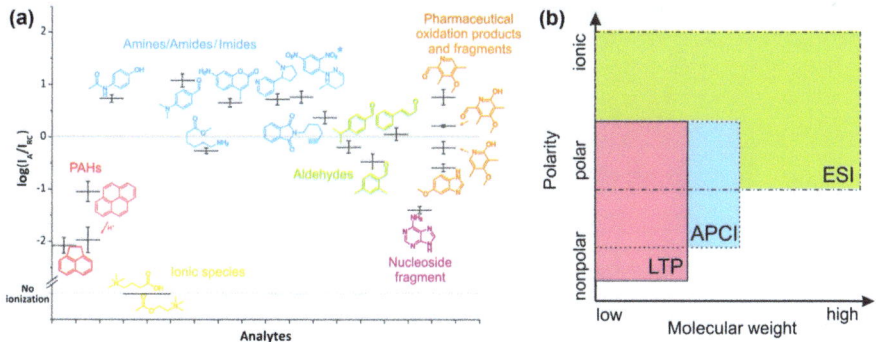

Figure 6.3 (a) Relative LTP ionization efficiencies of PAHs, ionic species, amides, amines, imides, aldehydes, a nucleoside, and a pharmaceutical. All analytes were detected in positive-ion mode except for the compound marked with an asterisk that was analyzed in negative-ion mode. (b) Classification of LTP, APCI, and ESI based on their ionization characteristics plotted against molecular weight and polarity of analytes. Figure adapted from ref. 3 with permission.

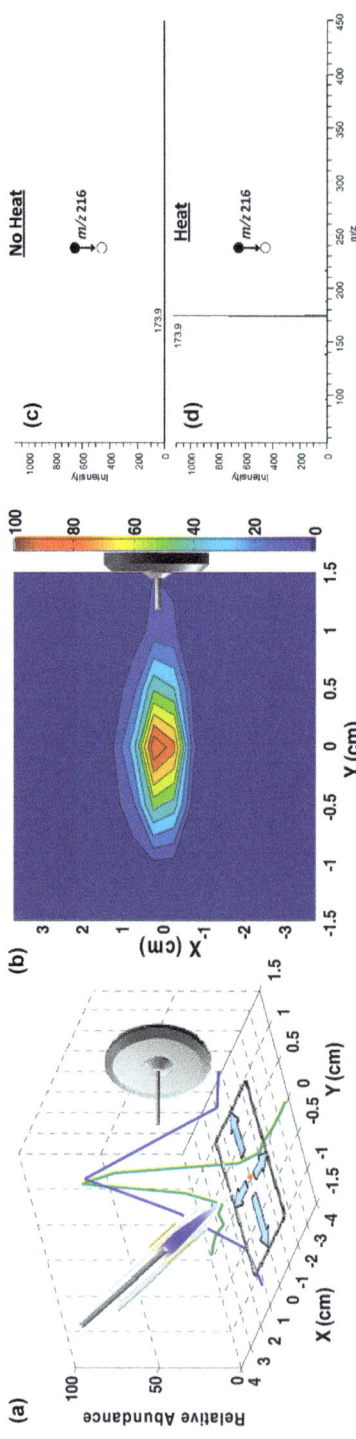

Figure 6.4   Characterization of the LTP probe sampling area using 1 μg of cocaine: (a) relative intensity of *m/z* 304 along the *x*- and *y*-axes; (b) extrapolated 2D distribution of the relative desorption efficiency. Analysis of atrazine using LTP with and without heating on sampling substrate. MS/MS product ion spectra of atrazine ([M + H] + *m/z* 216) detected from (c) an ambient substrate at room temperature and (d) a substrate heated to 150 °C. (a) and (b) Taken from ref. 1 and (c) taken from ref. 10 with permission.

1×1 cm area and analysis of samples from an area as large as 5 cm^2 could also be possible using this setup. The analytical performance of the original LTP setup is adequate for many semi- and nonvolatile compounds, such as TNT (vapor pressure $\sim 2 \times 10^{-4}$ torr at 25 °C) and cocaine ($\sim 3 \times 10^{-7}$ torr at 20 °C); studies have also been done to show that heating could facilitate the sampling ionization.[10-12] As an example for atrazine analysis with LTP, the signal intensity of protonated atrazine m/z 216 was improved by a factor of two orders of magnitude with a thermal assistance of 150 °C (Figure 6.4(c) and (d)).[10]

6.3 Designs of LTP Probe

The condition for plasma generation is critical for the analytical performance of an LTP probe. However, the actual implementation of the LTP desorption ionization can be done with different configurations. Following the first introduction of LTP probe in 2008,[1] a number of variations have been made for applying LTP-MS analysis in different applications or instrumentation packages. The LTP probe was originally demonstrated to be flexible with the type of discharge gas and has been demonstrated with helium, argon, nitrogen and air as discharge gas.[1] However, use of air is particularly attractive for in-field applications. A miniature LTP probe of 0.86 mm ID and 1.5 mm OD was designed to work with air at reduced flow rate, which can be easily supported using a miniature diaphragm pump (Figure 6.5(a) and (b)).[13] Arrays of multiple discharge probes were later used to enhance the efficiency of sampling a relative larger area (1–2 cm^2) while retaining low voltage for discharge and moderate temperature on sample surface (Figure 6.5(c)).[14]

In a typical setup using LTP probes of original configuration,[1] the LTP probe points to the surface at an angle for sampling ionization with the ions taken by the inlet capillary of the mass spectrometer (Figures 6.1(a) and 6.5(b)). During the development of integrated MS analytical systems,[15] an LTP probe with a coaxial configuration was fabricated (Figure 6.5(d)). The metal inlet capillary for the miniature mass spectrometer was inserted into the LTP glass tube and served as the ground electrode for discharge. This development enabled a "point-and-shoot" operation for sampling ionizing using LTP probe, without the need for dedicated adjustments of the angles among the sampling probe, MS inlet and the sample surface.

The softness in ionization using an LTP probe can be adjusted by varying the DBD conditions,[1] which is dependent on the discharge voltage, the distance between the electrodes, the thickness and the property of the dielectric material. For practical implementation, the DBD conditions of an LTP probe can be varied simply by pulling or pushing the ground electrode inside the tube, which results in a change in the electric field for the discharge and degree of fragmentation of the analytes during the sampling ionization.[1] An LTP source has also been designed with dual modes, switchable between a DBDI source and an LTP probe (Figure 6.6).[16]

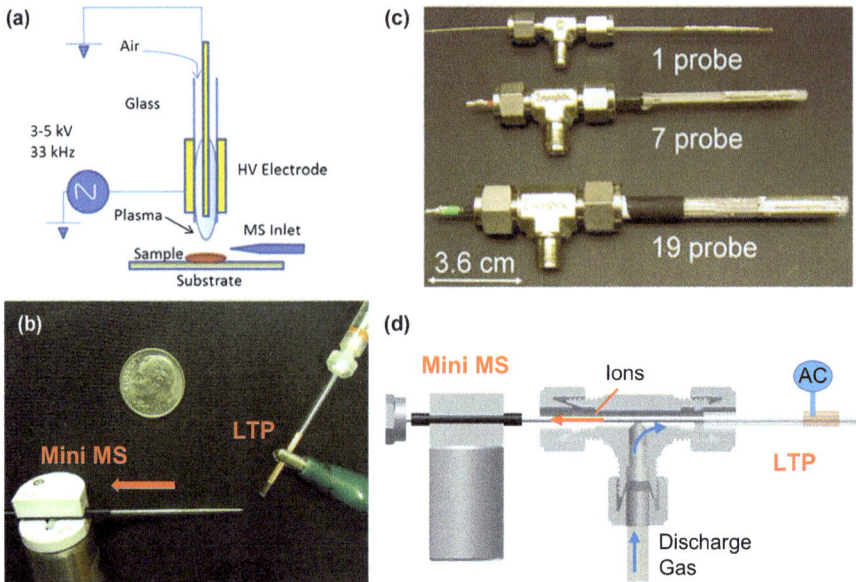

Figure 6.5 (a) Scheme of a miniature LTP probe. (b) Photography of the miniature LTP probe. (c) LTP probes with 1-, 7-, and 19-probe arrangements. (d) Structure of the LTP source designed for miniaturized mass spectrometer.
(a)–(c) Taken from ref. 13 and (d) taken from ref. 15 with permission.

A graphite rod, instead of stainless steel capillary, was used as the central electrode of the LTP probe. An additional copper plate electrode was placed underneath a glass slide. This source was demonstrated for the analysis of volatile organic compounds (VOCs). The gas sample was mixed in the argon discharge gas. In the DBDI mode the discharge occurs between the graphite rod and the copper plate. The soft ionization condition was used for the DBDI mode and molecular ions were observed as the major species (Figure 6.6(c)). The DBD condition in LTP probe mode was set harsh enough to produce extensive fragmentations of the analytes along with some interesting gas-phase reactions observed.

Other alternative designs with significantly smaller inner diameters of the tube of LTP probe were also explored for focusing the sampling torch and a better control of the temperature.[17] Fused capillaries of inner diameters as small as 100 μm were used to construct the LTP probes for imaging work of art, which will be further discussed later in this chapter.[18] An LTP probe with a tapered inner configuration of the tube was also developed to achieve a better control of the temperature and the diameter of the LTP torch (Figure 6.7(a)). The length and the internal diameter of the plasma adapter can be adjusted and a refined LTP torch of 300 μm in diameter was obtained with a 1 mm i.d. plasma adapter (Figure 6.7(b)). This potentially could improve the spatial resolution when using LTP for imaging applications.

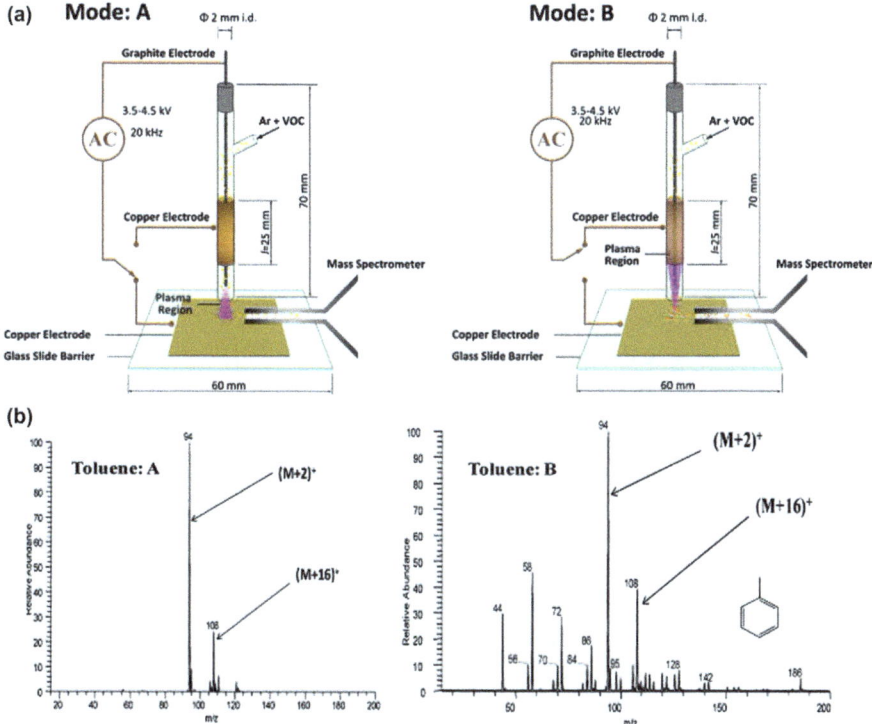

Figure 6.6 (a) Schematic of the dual-mode LTP system. (b) Spectra of toluene and ethylbenzene collected with LTP working in the two modes (mode A and mode B), respectively.
Figure taken from ref. 16 with permission.

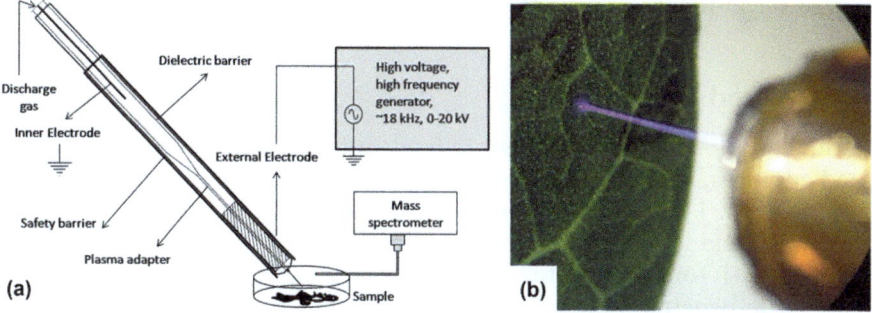

Figure 6.7 (a) Configuration of the low-temperature plasma (LTP) probe with an additional inner tube for guiding the discharge gas. (b) Photography of the LTP beam on a plant leaf. Spatial resolution studies of the LTP probe used for MS imaging application.
Figure taken from ref. 17 with permission.

6.4   Applications of LTP-MS

6.4.1   Explosive Detection

In situ detection of explosives at trace level is a major concern for public safety. Representative explosives and related compounds include hexahydro-1,3,5-trinitro-1,3,5-triazine (RDX), 2,4,6-trinitrotoluene (TNT), pentaery-thritol tetranitrate (PETN), tetryl, cyclo-1,3,5,7-tetramethylenetetranitrate (HMX), hexamethylene triperoxide diamine (HMTD) 2,4-dinitrotoluene, 1,3-dinitrobenzene, 1,3,5-trinitrobenzene, 2-amino-4,6-dinitrotoluene, 4-amino-2,6-dinitrotoluene, 2,6-dinitrotoluene, and 4-nitrotoluene. For check-point applications, the explosives are typically present as solid-phase residues on a variety of substrates, such as the surfaces of the luggage, laptop, and attire, which are made from different materials with different chemical properties. The sampling ionization method needs to be fast, effective but also minimally invasive to prevent any damage to the subjects. Using the highly reactive plasma species for desorption ionization, LTP probes have been shown to have a great potential for fast detection of the explosives using MS analysis in both positive and negative modes.

LTP-MS analysis of explosives have been done using both benchtop and miniature mass spectrometers. Figure 6.8 shows the spectra obtained with LTP probes from RDX and TNF from different surfaces (Figure 6.8(a)–(d))[1] and mixture of explosives from the glass (Figure 6.8(e)).[19] The limits of detection (LODs) for different explosives range from picograms to nanograms, and are highly dependent on the properties of the explosive compounds and the substrates.[19,20] Significant improvement in LOD was achieved by non-contact heating of the surfaces using a halogen lamp,[21] which can be easily implemented for high throughput detection required at check points. The coaxial LTP probe (Figure 6.5(d)) was integrated into a handheld sampling probe for a backpack mass spectrometer, which was designed and characterized for in-field detection of explosives (to be further discussed later).

6.4.2   Food Safety

A wide range of chemicals present in foodstuffs, such as agrochemicals and food additives, are subject to regulations. With the capability of direct surface analysis, LTP is well suited for fast analysis of foodstuffs. The agrochemical residues on fruits and vegetables can be directly analyzed by scanning LTP torch across their peels, sliced blocks or homogenized samples.[10] The penetration of the agrochemicals into the fruits could also be characterized (Figure 6.9(a) and (b)).[22] A quantitative study was carried out to analyze 13 agrochemicals from different fruits and vegetables, including ametryn, amitraz, atrazine, buprofezin, DEET, diphenylamine, ethoxyquin, imazalil, isofenphos-methyl, isoproturon, malathion, parathion-ethyl and terbuthylazine.[10] The obtained LODs ranged from 0.1 ng g^{-1} to 200 ng g^{-1}, dependent on the types of the chemicals and the samples.

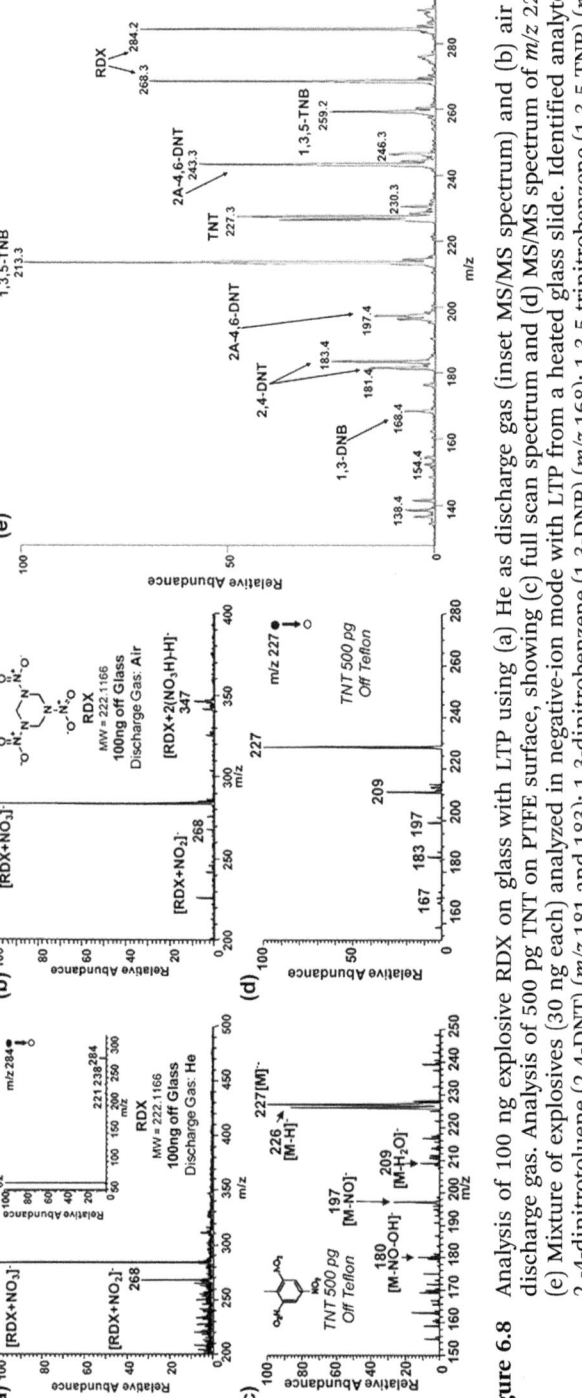

Figure 6.8 Analysis of 100 ng explosive RDX on glass with LTP using (a) He as discharge gas (inset MS/MS spectrum) and (b) air as discharge gas. Analysis of 500 pg TNT on PTFE surface, showing (c) full scan spectrum and (d) MS/MS spectrum of *m/z* 227. (e) Mixture of explosives (30 ng each) analyzed in negative-ion mode with LTP from a heated glass slide. Identified analytes: 2,-4-dinitrotoluene (2,4-DNT) (*m/z* 181 and 183); 1,3-dinitrobenzene (1,3-DNB) (*m/z* 168); 1,3,5-trinitrobenzene (1,3,5-TNB) (*m/z* 213 and 259); 2-amino-4,6.dinitrotoluene (2A-4,6.DNT) (*m/z* 197 and 243); RDX (*m/z* 268 and 284); and TNT (*m/z* 226 and 227). Figure taken from ref. 1 and 19 with permission.

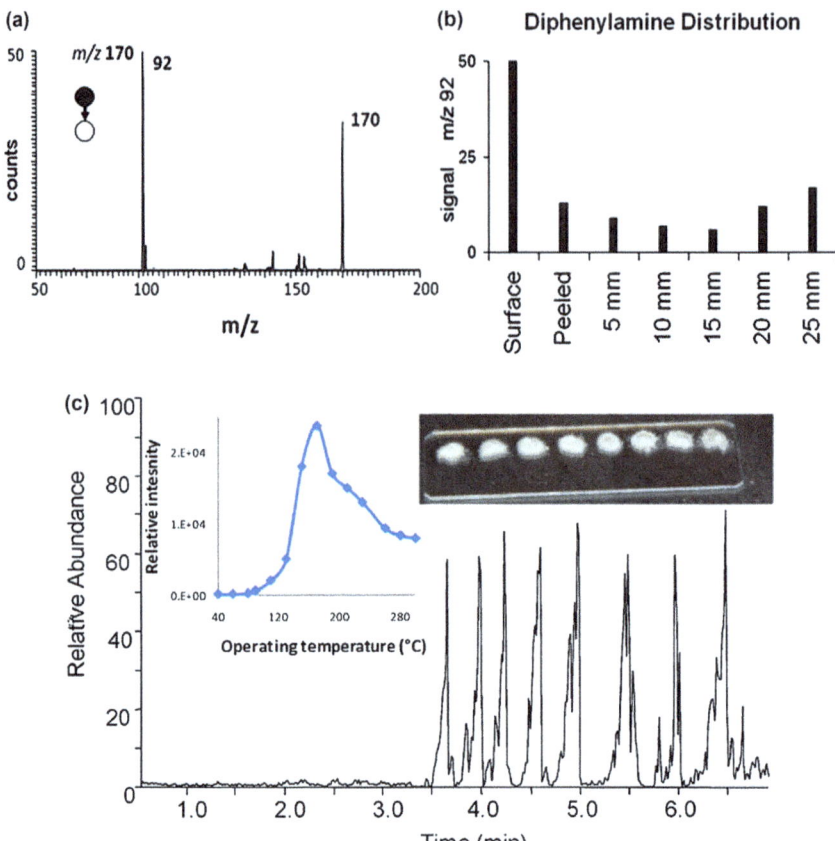

Figure 6.9   (a) MS/MS spectrum of diphenylamine (m/z 170 → m/z 92) tested with LTP. (b) Concentration profile of diphenylamine throughout the apple based on the signal intensity of m/z 92 collected with MS/MS. The apple was cut into 5 mm increment slices from a transverse section and each slice was tested with LTP individually. (c) High-throughput analysis with LTP on eight control samples (0.5 to 3.0 min) and eight 5 ng mg^{-1} milk powder samples (3.5 to 6.5 min) in single reaction monitoring mode. Samples were heated at 170 °C. Left inset: signal intensity as a function of the sample temperature. Right inset: photograph of the eight samples deposited on a glass slide.
(a) and (b) Taken from ref. 22 and (c) taken from ref. 12 with permission.

The addiction of melamine in milk and milk products resulted in a major food safety incident in 2008. Methods for direct detection of melamine in urine and milk products were developed using LTP. Melamine in milk powder at subnanogram levels can be detected using LTP-MS within a minute (Figure 6.9(c)).[12] This method has been adapted for a miniature mass spectrometer. An LOD of 250 ng mL^{-1} and a linear dynamic range of 0.5–50 µg mL^{-1} with a relative standard deviation of *ca.* 7.6–16.2% was achieved for melamine in whole milk.

Bacteria is another major concern of food safety and is the most common cause of food poisoning. Although not being able to analyze DNAs or proteins directly, an LTP probe had been previously demonstrated for fatty acid analysis in complex food samples, such as olive oil.[23] As an attempt for direct identification of bacteria, the fatty acid profiles of the bacteria were obtained by direct analysis of the bacteria in growth media using LTP-MS. Principal component analysis was subsequently carried out to differentiate bacterial species based on the data acquired with accurate mass measurement.[24] Gram-positive and gram-negative bacteria could be readily distinguished, and 11 out of 13 Salmonella strains showed distinctive patterns.

6.4.3 Pharmaceutical Tablets

The LTP probe has been demonstrated for the direct MS analysis of a wide variety of drug compounds.[1,25] As the active ingredients in the drug tablets, the concentrations of these drug compounds are typically sufficiently high, with major peaks appearing in the MS spectra obtained in full MS scans using LTP probe for sampling (Figure 6.10). Commercial drug tablets of 11 types have been analyzed in a high-throughput mode (600 samples/h), including hormone (steroidal), antipyretic analgesics (nonsteroidal anti-inflammatory), cardiovascular, digestive system, neuropsychotherapeutic, diuretic, antithyroid, sulfa anti-inflammatory, antiparastic, sedative-hypnotics, and antibacterial drugs.[25] Most of these tablets can be directly analyzed using LTP, except for some of them that needed to have the coating removed to expose the internal ingredients for the analysis. This type of analysis has a great potential for identification of counterfeit drugs. Without involving solvents in the analysis, LTP probe is also suitable for online, real-time analysis for the quality control in the processes for drug production.

6.4.4 Forensics

As a demonstration of the complex mixture analysis that is routinely needed for forensics, the LTP probe was used for examination of stomach contents of a diseased dog.[1] Direct sampling ionization was applied for MS analysis and protonated Terbufos (m/z 289) and Terbufos sulfoxide (m/z 305) were observed. These two compounds are common in Ternbufos-based insecticides, which was suspected to be the cause of the death of the dog. Analysis of chemicals in body fluids and other biological samples has also been explored using LTP. Different from spray ionization methods, the salts at high concentrations in body-fluid samples, such as urine, have relatively low impact on the ionization by LTP. Metabolites in a dried urine spot on PTFE could be easily observed.[1] The analysis of drugs of abuse in urine, saliva and hair extract samples were systematically investigated with 14 drugs or drug metabolites (listed in Table 6.1).[26] LODs of 10 ng mL^{-1} or better were obtained without any sample preparation for amphetamine, methamphetamine, caffeine, ketamine, methadone in a urine matrix (Table 6.1).[26]

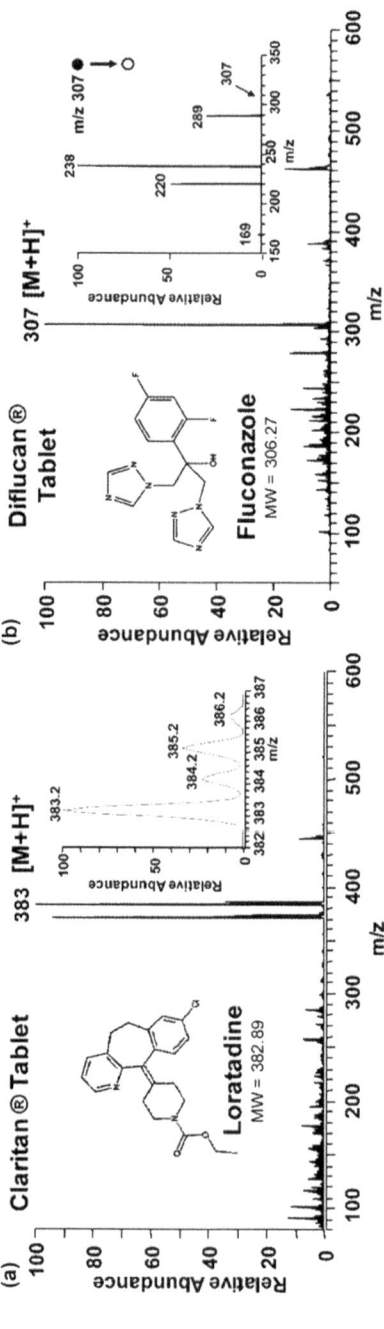

Figure 6.10 Analysis of chemicals from various matrixes using LTP probe. (a) A 10 mg Claritin tablet containing loratadine, inset: molecular ion region. (b) A 100 mg Diflucan tablet containing fluconazole, inset: MS/MS spectrum. Figure taken from ref. 1 with permission.

Table 6.1 Limits of detection of analytes in different matrices.[26] Table adapted from ref. 26 with permission.

Compound	LOD of standard/ ng ml^{-1}	LOD raw urine/ ng ml^{-1}	LOD 1:2 dilution/ ng ml^{-1}	LOD 1:5 dilution/ ng ml^{-1}
Amphetamine	1	100	10	10
Benzoylecgonine	500	1000	500	500
Caffeine	0.5	100	10	10
Cannabidiol	100	100	100	100
Cocaine	5	100	10	10
Codeine	50	1000	100	100
Diazepam	10	100	10	10
Ephedrine hydrochloride	50	1000	100	100
Heroin	1000	10×10^3	10×10^3	10×10^3
Ketamine	5	10	10	5
Methadone	5	10	10	10
Methamphetamine	0.5	100	10	10
Morphine	100	10×10^3	10×10^3	1000
Δ^9-THC	100	1000	500	500

Thermoassistance by heating the sample to 100 °C was found to be effective for improving the LODs by a factor of a hundred or more for benzoylecgonine (cocaine metabolite), codeine, diazepam, and Δ^9-THC. A dilution of the urine samples was also helpful for the analysis, which presumably is due to a reduction of matrix effects.

6.4.5 Environmental Monitoring

The convenience of applying LTP-MS analysis for environmental monitoring was demonstrated by direct analysis of atrazine in water. Atrazine is a very popular herbicide widely used in the US, leading to a significant distribution and contamination in soil and groundwater. The analysis of the water samples was simply done by sweeping the plasma torch from an LTP probe across the surface of the samples while the MS or MS/MS analysis was performed (Figure 6.11(a)).[1] Excellent S/N in an MS/MS spectrum were obtained at a concentration of 10 ppb (Figure 6.11(b)). A linear dynamic range between 1 ppb to 1 ppm was demonstrated in this experiment (Figure 6.11(c)). A similar method was applied for reaction monitoring, where the reaction products in a bulk organic solvent were examined in real time using an LTP probe.[27]

The LTP probe has also been modified and used for direct analysis of VOCs in air (Figure 6.11(d)). The indoor air was introduced downstream of the plasma region inside the modified LTP probe, where the analytes interacted with the plasmas species, were ionized and subsequently introduced into a mass spectrometer for MS analysis.[28] The VOCs tested in this study included methanol, ethanol, isopropanol, acetone, benzene, formaldehyde, butanone, toluene, ethylbenzene, ethyl acetate, butyl acetate,

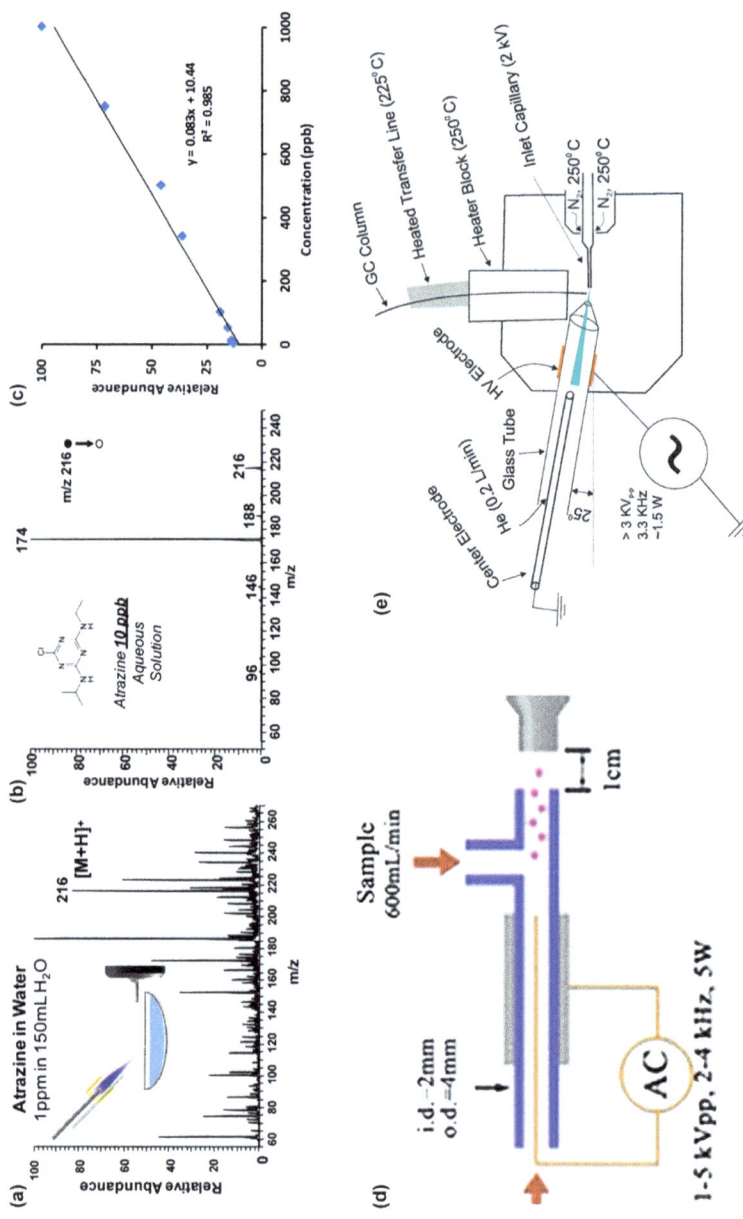

Figure 6.11 Direct analysis of atrazine with LTP from an aqueous solution. (a) Full scan spectrum of 1 ppm solution and (b) MS/MS spectrum of *m/z* 216 in 10 ppb solution. (c) Plot of relative abundance of atrazine *vs.* concentration showing the linear dynamic range between 1 ppb and 1 ppm. Figure taken from ref. 1 with permission. (d) Modified LTP for direct analysis of VOCs in indoor air. (e) LTP ionization for analysis of VOCs using GC-MS. (a)–(c) Taken from ref. 1; (d) adapted from ref. 28; (e) taken from ref. 29. All with permissions.

methyl isobutyl ketone, acetic acid and pyridine, which covered most of the category of VOCs (Figure 6.11(e)).[28] Protonated molecules and product ions due to gas-phase reactions were observed. In another experiment, an LTP probe of original configuration was used as the ionization source in a GC-MS system. Comparisons of LODs were made between LTP probe and EI (electron impact), CI (chemical ionization), APCI or PI (photo ionization) using 20 compounds within 8 functional groups.[29] A relatively harsh ionization condition was used for LTP and intense fragmentation was observed. LODs with LTP was found to be comparable to APCI and PI.

6.4.6 Material and Surface Analysis

The energy involved in the desorption ionization using LTP probes can be adjusted over a wide range. In addition to the energy variation demonstrated for generating the molecular or fragment ions,[1] the ablation of the substrate material by LTP plasma can also be avoided or enhanced based on the need in an application. In a demonstration of using LTP to authenticate paintings and calligraphy, a micro-LTP probe was used for imaging mass spectrometry.[18] The condition of the LTP desorption ionization was soft so as not to cause any damage to the rice paper, as confirmed by SEM analysis, but sufficiently energetic for desorbing and ionizing the compounds contained in the ink (Figure 6.12). The characteristic composition of the ink was used to differentiate the genuine and counterfeit art works. In another application for surface analysis, the LTP-MS was used to characterize the self-assembled monolayers (SAMs) on copper surfaces prepared with n-dodecylmercaptan (NDM) and l-phenyl-5-mercaptotetrazole (PMTA).[30] Characteristic ions of the organic compounds, mainly $[M+M-H]^+$ for NDM SAM and $[M+H-S]^+$ for PMTA SAM, were observed through a direct analysis of the SAM surfaces using an LTP probe. The same method was applied for characterization of the layer-by-layer self-assembled multilayer films on quartz plates, which were prepared using 4-aminothiophenol (4-ATP, denoted as M) capped Au particles and thioglycolic acid (TGA, denoted as F) capped Ag particles. The desorption ionization by LTP generated characteristic ions of the films, including $[M]^{+\bullet}$, $[M-NH_2]^+$, $[M-HCN-H]^+$, and $[F+H]^+$, $[F-H]^+$, $[F-OH]^+$, $[F-COOH]^+$.[31]

The energy for LTP desorption can be further increased by effective focusing of the plasma, which has been demonstrated by removing the inorganic coatings on surfaces for analysis.[32] A fused silica capillary of 150 μm i.d. was used as the discharge tube of the LTP probe (Figure 6.13(a)) and the focused plasma torch was capable of making a hole of a diameter less than 10 μm through a thin layer of Ta on silicon after a 30-s ablation (Figure 6.13(b)). The sample material was converted into aerosols that were transferred into an ICP-MS for analysis. Depth profiling was demonstrated with a 100-nm single layer and a multilayer (100 nm Al/250 nm SiO_2/100 nm Au/50 nm Cr) coatings on silicon substrates (Figure 6.13(c)). A lateral resolution of about 200 μm was obtained for surface analysis using the

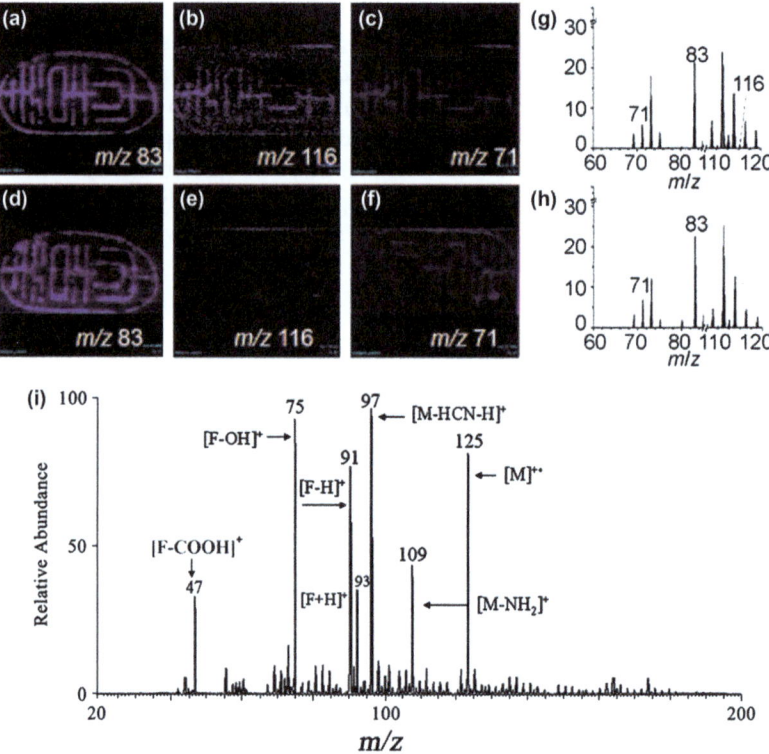

Figure 6.12 LTP imaging of inkpads of the seals on calligraphy. Ion maps generated from (a)–(c) a genuine calligraphy, and (d)–(f) a counterfeit. Mass spectra collected from (g) a genuine calligraphy and (h) a counterfeit. (i) LTP-MS spectrum of $(Au/Ag)_6$ multilayer film, M = 4-ATP and F = TGA. (a)–(h) taken from ref. 18 and (i) taken from ref. 31 with permissions.

LTP ablation. The etching of poly(lactic acid) and poly(methyl methacrylate) films by a helium LTP probe was characterized in another study[33] using secondary ion mass spectrometry (SIMS) to profile the microcraters made by the LTP ablation.

6.5 Chemical Reactions Facilitated by LTP

The LTP probe serves as an effective desorption and ionization tool, which is largely due to the generation of highly reactive species by the plasma at the ambient condition. Its potential for studying organic reactions has been recognized, while it was developed and used for chemical analysis. Using a modified LTP as shown in Figure 6.14(a), dihydrogenation of benzene and other arenes was observed (Figure 6.14(b)) by simply passing the headspace vapor of benzene through the helium LTP and analyzing the reaction products directly using a mass spectrometer.[34] The mechanistic studies

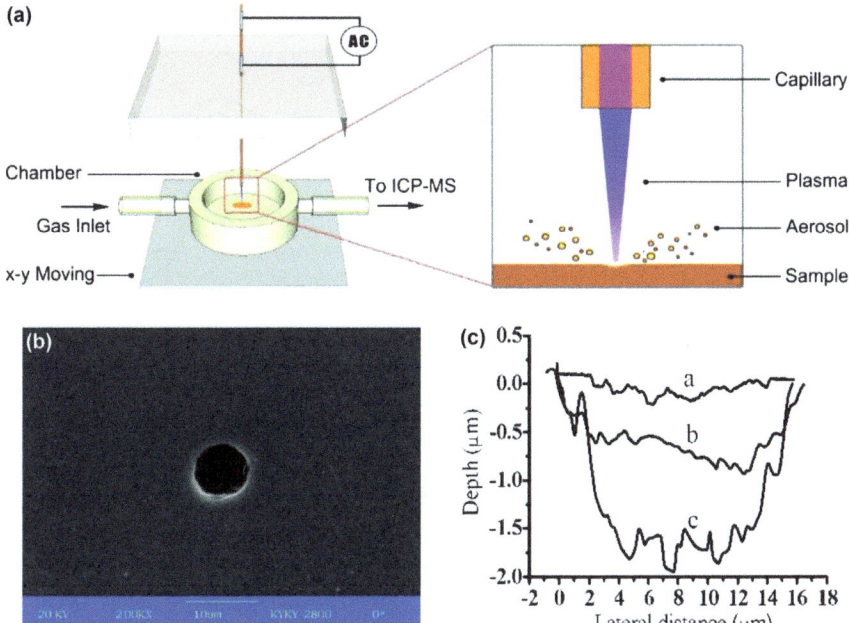

Figure 6.13 (a) Schematic diagram of the LTP probe-ICP-MS system for depth profiling. (b) SEM photograph of a Ta thin layer ablated by LTP probe for 30 s. (c) Depth profiles across the central part of the crater generated with increasing ablation time from (a) 30, (b) 90 s to (c) 240 s, respectively. Figure taken from ref. 32 with permission.

indicated that the reduction was initiated by the capture of low-energy elections from the discharge surface. Partial reduction of benzene to cyclo-hexadiene with high selectivity is very challenging using traditional organic synthesis methods and could only be achieved with the sodium/alcohol Birch reduction[35] and some well-controlled metal-catalyzed reductions.[36,37]

Using the similar experimental setup, formation of pyridine by replacing one carbon in benzene with an atomic nitrogen was observed in an LTP reaction with benzene and 1% NO in N_2 (Figure 6.14(c)).[38] A preparative apparatus was also setup with an LTP sustained by DBD in the headspace to produce the pyridine from bulk benzene solvent (Figure 6.14(d)). The reaction solvent was sampled and analyzed to monitor the yield of pyridine as a function of reaction time (Figure 6.14(e)). A mechanistic study was carried out with exact mass measurement, tandem mass spectrometry and isotope labeling methods. A reaction pathway for the pyridine formation was proposed to involve a hydrogen replacement by nitrogen-containing species produced in the LTP followed by the ring-opening and -closing reactions of the derivatized benzene. The direct synthesis of heterocyclic compounds from benzene could not be achieved using typical synthesis methods but now could be prepared through a single-step reaction using LTP. Other gas-phase reactions, including Eberlin reaction and imine formation, have also

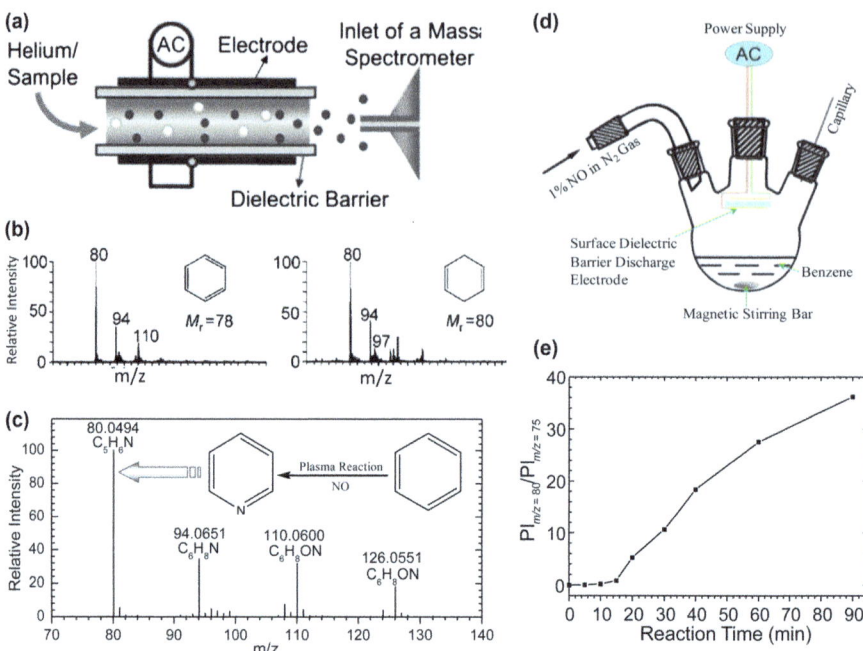

Figure 6.14 (a) Schematic view of the experimental setup with a modified LTP for organic reactions. (b) LTP mass spectra of benzene and 1,4-cyclohexadiene. (c) MS spectrum of products of an LTP reaction with benzene and 1% NO in N_2. (d) Apparatus for the preparation of pyridine from bulk benzene solvent using a LTP generated with DBD. (e) Yield of pyridine as a fuction of reaction time.
(a) and (b) Adpated from ref. 34 and (c) and (d) taken from ref. 38 with permissions.

been studied using the LTP.[39] The results from these reaction studies indicate a new potential direction for producing new materials through simple reactions involving highly reactive plasma species.

Some other unique and interesting phenomena associated with the reactive environment of LTP were also discovered and potentially could have significant impacts on the analysis of biological compounds. Exposure of nano-EIS emitter to LTP was found to turn the nano-ESI from a soft ionization to a harsh one, with extensive backbone fragmentation observed for the sprayed peptide ions (Figure 6.15(a) and (b)).[40] Mechanistic studies suggested that the electrolytes were released from the glass emitter into the spray solution; when relatively low spray voltage, *e.g.* 1.2 kV, was applied for nano-ESI, the high local electrolyte concentration possibly resulted in a direct ejection of solvated ions at high speeds, which fragmented due to the subsequent collisions with neutral molecules in gas phase. The phosphate groups on the peptides, however, were preserved during this process, which presumably was because of the protection due to the solvation. The results of this study indicate an effective, simple method for studying

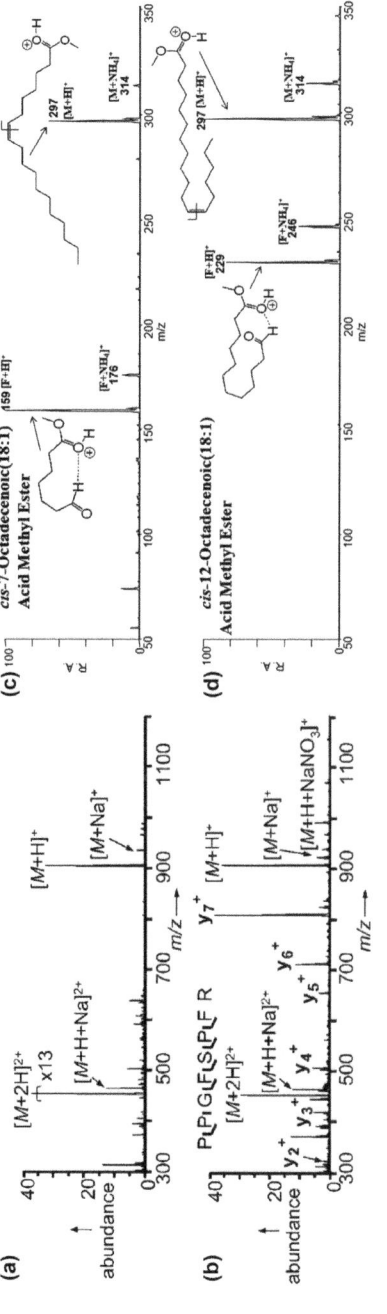

Figure 6.15 Spectra of bradykinin 2–9 (PPGFSPFR) tested with nano-ESI source (a) without and (b) with the exposure of LTP for 2 min. Spectra of methyl esters of two isomeric monounsaturated fatty acids (18:1 FA) obtained with LTP in positive-ion mode, (c) *cis*-7-Octadecenoic (18:1) acid methyl ester and (d) *cis*-12-octadecenoic (18:1) acid methyl ester, tested with LTP. F stands for the ester-containing aldehyde oxidation product. (a) and (b) Taken from ref. 40 and (c) and (d) taken from ref. 41 with permissions.

post-translational modification of proteins and solvated ions. In another study of analyzing unsaturated fatty acids and esters using LTP, the *in situ* oxidation through the ozone reactions cleaved olefins. The fragmentation information were used to assign the double-bond positions in fatty acids.[41] This method was applied for a direct analysis of the microbial fatty acid ethyl ester mixtures from complex bacterial samples.

6.6   Development for In-field Chemical Analysis

The combination of ambient ionization sources with miniature mass spectrometer is an effective way to develop self-contained analytical systems for in-field chemical analysis.[42,43] The LTP probe is of simple configuration and consumes only discharge gas at low flow rates, which suits well for instrument miniaturization and in-field analysis. Figure 6.16 shows a handheld LTP sampler operated with a battery and a miniature helium cylinder or air. A small 7.4-V Li-polymer battery was used and could sustain the LTP for 2 h continuously. One miniature helium cylinder could supply helium for 8 continuous hours. The handheld LTP probe was tested with a wide range of samples using both lab-scale and miniature mass spectrometers. Similar or slightly better analytical performance was obtained in comparison with the LTP probes of original scale.[1] For testing with lab-scale mass spectrometer, a long-distance ion transfer over 1 m was used to allow flexibility of operation (Figure 6.16(b)).

In an effort to deliver a fully integrated analytical system for in-field chemical detection, a backpack MS analytical system (Figure 6.17(a)) was developed with a sampling probe using LTP.[15] The vacuum manifold containing the mass analyzer and the LTP probe were connected through a 1 m tube to the pumping system in the backpack, where the control electronics for the mass spectrometer were also installed. This configuration allows a flexibility of scanning surfaces for in-field operation. A discontinuous

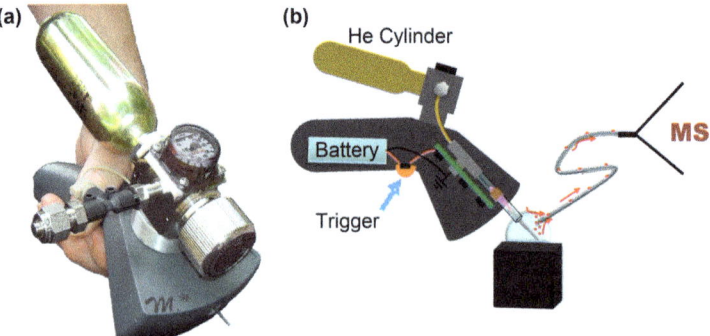

Figure 6.16   (a) Photograph and (b) schematic configuration of the handheld LTP ionization source.
Figure adapted from ref. 44 with permission.

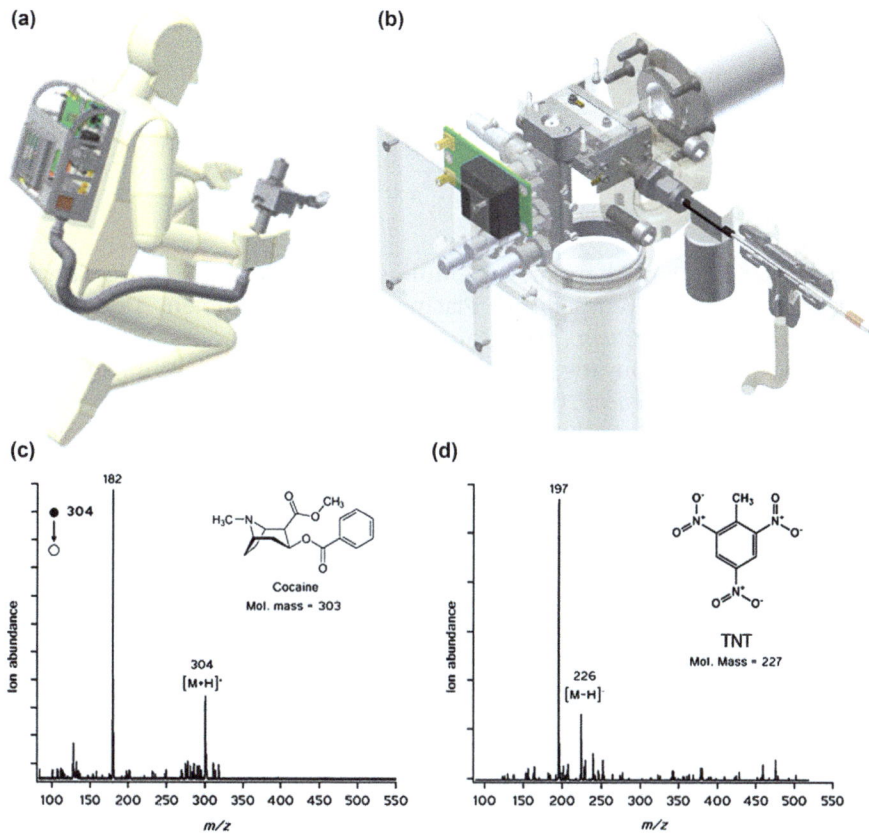

Figure 6.17 (a) The backpack mini-MS system with a LTP ionization source. (b) Structure of the handheld head unit. Spectra obtained with the backpack mini-MS system and LTP source, (c) MS/MS spectrum of cocaine (2 µg cm^{-2}) detected from human fingers under positive-ion mode; (d) MS spectrum of TNT (<1 µg cm^{-2}) detected from a glass tube under negative-ion mode.
Figure taken from ref. 15 with permission.

atmospheric-pressure interface (DAPI)[45–47] was used to transfer the ions generated by LTP into a rectilinear ion trap in vacuum for MS and MS/MS analysis. The vacuum was sustained by a miniature pumping system (Figure 6.17(b)). The LTP probe has an inline configuration as shown in Figures 6.5(d) and 6.17(b)), with a stainless steel capillary used for both MS inlet and the ground electrode for DBD. The entire system weighted about 12 kg and consumed power of an average of 65 W. A 16-cell battery pack provided a continuous working time for about 1.5 h. The performance of the backpack system was demonstrated with direct analysis of chemicals on surfaces, including chemical warfare agent simulants, illicit drugs, and explosives. LODs better than 1 µg cm^{-2} have been achieved (Figure 6.17(c) and (d)).

Acknowledgements

The research and development at Purdue University for the LTP probe and associated technologies was supported by National Science Foundation and the Department of Homeland Security through multiple projects.

References

1. J. D. Harper, N. A. Charipar, C. C. Mulligan, X. R. Zhang, R. G. Cooks and Z. Ouyang, Low-Temperature Plasma Probe for Ambient Desorption Ionization, *Anal. Chem.*, 2008, **80**, 9097–9104.
2. R. G. Cooks, Z. Ouyang, Z. Takats and J. M. Wiseman, Ambient mass spectrometry, *Science*, 2006, **311**, 1566–1570.
3. A. Albert and C. Engelhard, Characteristics of Low-Temperature Plasma Ionization for Ambient Mass Spectrometry Compared to Electrospray Ionization and Atmospheric Pressure Chemical Ionization, *Anal. Chem.*, 2012, **84**, 10657–10664.
4. N. Na, M. Zhao, S. Zhang, C. Yang and X. Zhang, Development of a dielectric barrier discharge ion source for ambient mass spectrometry, *J. Am. Soc. Mass Spectrom.*, 2007, **18**, 1859–1862.
5. R. B. Cody, J. A. Laramee and H. D. Durst, Versatile new ion source for the analysis of materials in open air under ambient conditions, *Anal. Chem.*, 2005, 77, 2297–2302.
6. F. J. Andrade, J. T. Shelley, W. C. Wetzel, M. R. Webb, G. Gamez, S. J. Ray and G. M. Hieftje, Atmospheric pressure chemical ionization source. 2. Desorption-ionization for the direct analysis of solid compounds, *Anal. Chem.*, 2008, **80**, 2654–2663.
7. X. C. Xiong, W. Xu, L. S. Eberlin, J. M. Wiseman, X. Fang, Y. Jiang, Z. J. Huang, Y. K. Zhang, R. G. Cooks and Z. Ouyang, Data Processing for 3D Mass Spectrometry Imaging, *J. Am. Soc. Mass Spectrom.*, 2012, **23**, 1147–1156.
8. G. C. Y. Chan, J. T. Shelley, J. S. Wiley, C. Engelhard, A. U. Jackson, R. G. Cooks and G. M. Hieftje, Elucidation of Reaction Mechanisms Responsible for Afterglow and Reagent-Ion Formation in the Low-Temperature Plasma Probe Ambient Ionization Source, *Anal. Chem.*, 2011, **83**, 3675–3686.
9. G. C. Y. Chan, J. T. Shelley, A. U. Jackson, J. S. Wiley, C. Engelhard, R. G. Cooks and G. M. Hieftje, Spectroscopic plasma diagnostics on a low-temperature plasma probe for ambient mass spectrometry, *J. Anal. At. Spectrom.*, 2011, **26**, 1434–1444.
10. J. S. Wiley, J. F. Garcia-Reyes, J. D. Harper, N. A. Charipar, Z. Ouyang and R. G. Cooks, Screening of agrochemicals in foodstuffs using low-temperature plasma (LTP) ambient ionization mass spectrometry, *Analyst*, 2010, **135**, 971–979.
11. G. M. Huang, W. Xu, M. A. Visbal-Onufrak, Z. Ouyang and R. G. Cooks, Direct analysis of melamine in complex matrices using a handheld mass spectrometer, *Analyst*, 2010, **135**, 705–711.

12. G. M. Huang, O. Y. Zheng and R. G. Cooks, High-throughput trace melamine analysis in complex mixtures, *Chem. Commun.*, 2009, 556–558.
13. J. K. Dalgleish, K. Y. Hou, Z. Ouyang and R. G. Cooks, *In Situ* Explosive Detection Using a Miniature Plasma Ion Source and a Portable Mass Spectrometer, *Anal. Lett.*, 2012, **45**, 1440–1446.
14. J. K. Dalgleish, M. Wleklinski, J. T. Shelley, C. C. Mulligan, Z. Ouyang and R. G. Cooks, Arrays of low-temperature plasma probes for ambient ionization mass spectrometry, *Rapid Commun. Mass Spectrom.*, 2013, **27**, 135–142.
15. P. I. Hendricks, J. K. Dalgleish, J. T. Shelley, M. A. Kirleis, M. T. McNicholas, L. Li, T.-C. Chen, C.-H. Chen, J. S. Duncan, F. Boudreau, R. J. Noll, J. P. Denton, Z. Ouyang and R. G. Cooks; Autonomous in-situ analysis and real-time chemical detection using a backpack miniature mass spectrometer: concept, instrumentation development, and performance, *Anal. Chem.*, 2014, **86**, 2900–2908.
16. M. R. Almasian, C. D. Yang, Z. Xing, S. C. Zhang and X. R. Zhang, Development of a graphite low-temperature plasma source with dual-mode in-source fragmentation for ambient mass spectrometry, *Rapid Commun. Mass Spectrom.*, 2010, **24**, 742–748.
17. S. Martinez-Jarquin and R. Winkler, Design of a low-temperature plasma (LTP) probe with adjustable output temperature and variable beam diameter for the direct detection of organic molecules, *Rapid Commun. Mass Spectrom.*, 2013, **27**, 629–634.
18. Y. Y. Liu, X. X. Ma, Z. Q. Lin, M. J. He, G. J. Han, C. D. Yang, Z. Xing, S. C. Zhang and X. R. Zhang, Imaging Mass Spectrometry with a Low-Temperature Plasma Probe for the Analysis of Works of Art, *Angew. Chem., Int. Ed.*, 2010, **49**, 4435–4437.
19. J. F. Garcia-Reyes, J. D. Harper, G. A. Salazar, N. A. Charipar, Z. Ouyang and R. G. Cooks, Detection of Explosives and Related Compounds by Low-Temperature Plasma Ambient Ionization Mass Spectrometry, *Anal. Chem.*, 2011, **83**, 1084–1092.
20. Y. Zhang, X. X. Ma, S. C. Zhang, C. D. Yang, Z. Ouyang and X. R. Zhang, Direct detection of explosives on solid surfaces by low temperature plasma desorption mass spectrometry, *Analyst*, 2009, **134**, 176–181.
21. W. Chen, K. Hou, X. Xiong, Y. Jiang, W. Zhao, L. Hua, P. Chen, Y. Xie, Z. Wanga and H. Li, Non-contact halogen lamp heating assisted LTP ionization miniature rectilinear ion trap: a platform for rapid, on-site explosives analysis, *Analyst*, 2013, **138**, 5068–5073.
22. S. Soparawalla, F. K. Tadjimukhamedov, J. S. Wiley, Z. Ouyang and R. G. Cooks, *In situ* analysis of agrochemical residues on fruit using ambient ionization on a handheld mass spectrometer, *Analyst*, 2011, **136**, 4392–4396.
23. J. F. Garcia-Reyes, F. Mazzotti, J. D. Harper, N. A. Charipar, S. Oradu, Z. Ouyang, G. Sindona and R. G. Cooks, Direct olive oil analysis by low-temperature plasma (LTP) ambient ionization mass spectrometry (vol. 23, p. 3057, 2009), *Rapid Commun. Mass Spectrom.*, 2009, **23**, 3492.

24. J. I. Zhang, A. B. Costa, W. A. Tao and R. G. Cooks, Direct detection of fatty acid ethyl esters using low temperature plasma (LTP) ambient ionization mass spectrometry for rapid bacterial differentiation, *Analyst*, 2011, **136**, 3091–3097.

25. Y. Y. Liu, Z. Q. Lin, S. C. Zhang, C. D. Yang and X. R. Zhang, Rapid screening of active ingredients in drugs by mass spectrometry with low-temperature plasma probe, *Anal. Bioanal. Chem.*, 2009, **395**, 591–599.

26. A. U. Jackson, J. F. Garcia-Reyes, J. D. Harper, J. S. Wiley, A. Molina-Diaz, Z. Ouyang and R. G. Cooks, Analysis of drugs of abuse in biofluids by low temperature plasma (LTP) ionization mass spectrometry, *Analyst*, 2010, **135**, 927–933.

27. X. X. Ma, S. C. Zhang, Z. Q. Lin, Y. Y. Liu, Z. Xing, C. D. Yang and X. R. Zhang, Real-time monitoring of chemical reactions by mass spectrometry utilizing a low-temperature plasma probe, *Analyst*, 2009, **134**, 1863–1867.

28. X. Y. Gong, X. C. Xiong, Y. E. Peng, C. D. Yang, S. C. Zhang, X. Fang and X. R. Zhang, Low-temperature plasma ionization source for the online detection of indoor volatile organic compounds, *Talanta*, 2011, **85**, 2458–2462.

29. A. W. Norgaard, V. Kofoed-Sorensen, B. Svensmark, P. Wolkoff and P. A. Clausen, Gas Chromatography Interfaced with Atmospheric Pressure Ionization-Quadrupole Time-of-Flight-Mass Spectrometry by Low-Temperature Plasma Ionization, *Anal. Chem.*, 2013, **85**, 28–32.

30. L. Ma, M. Z. Jia, J. B. Hu, J. Ouyang and N. Na, The Characterization of Self-Assembled Monolayers on Copper Surfaces by Low-Temperature Plasma Mass Spectrometry, *J. Am. Soc. Mass Spectrom.*, 2012, **23**, 1271–1278.

31. X. Y. Xu, N. Na, J. Y. Wen and J. Ouyang, Detection of layer-by-layer self-assembly multilayer films by low-temperature plasma mass spectrometry, *J. Mass Spectrom.*, 2013, **48**, 172–178.

32. Z. Xing, J. A. Wang, G. J. Han, B. Kuermaiti, S. C. Zhang and X. R. Zhang, Depth Profiling of Nanometer Coatings by Low Temperature Plasma Probe Combined with Inductively Coupled Plasma Mass Spectrometry, *Anal. Chem.*, 2010, **82**, 5872–5877.

33. S. Muramoto, M. E. Staymates, T. M. Brewer and G. Gillen, Ambient Low Temperature Plasma Etching of Polymer Films for Secondary Ion Mass Spectrometry Molecular Depth Profiling, *Anal. Chem.*, 2012, **84**, 10763–10767.

34. N. Na, Y. Xia, Z. L. Zhu, X. R. Zhang and R. G. Cooks, Birch Reduction of Benzene in a Low-Temperature Plasma, *Angew. Chem., Int. Ed.*, 2009, **48**, 2017–2019.

35. A. J. Birch, 117. Reduction by dissolving metals. Part I, *Journal of the Chemical Society (Resumed)*, 1944, 430–436.

36. G. P. Pez, I. L. Mador, J. E. Galle, R. K. Crissey and C. E. Forbes, Selective alkali metal and hydrogen reduction of benzene to cyclohexene, *J. Am. Chem. Soc*, 1985, **107**, 4098–4100.

37. C. Bianchini, K. G. Caulton, K. Folting, A. Meli, M. Peruzzini, A. Polo and F. Vizza, Stepwise metal-assisted reduction of .eta.4-coordinated benzene to cyclohexene, *J. Am. Chem. Soc.*, 1992, **114**, 7290–7291.

38. Z. Zhang, X. Gong, S. Zhang, H. Yang, Y. Shi, C. Yang, X. Zhang, X. Xiong, X. Fang and Z. Ouyang, Observation of Replacement of Carbon in Benzene with Nitrogen in a Low-Temperature Plasma, *Sci. Rep.*, 2013, **3**, 3481.

39. M. Benassi, J. F. Garcia-Reyes and B. Spengler, Ambient ion/molecule reactions in low-temperature plasmas (LTP): reactive LTP mass spectrometry, *Rapid Commun. Mass Spectrom.*, 2013, **27**, 795–804.

40. Y. Xia, Z. Ouyang and R. G. Cooks, Peptide Fragmentation Assisted by Surfaces Treated with a Low-Temperature Plasma in NanoESI, *Angew. Chem., Int. Ed.*, 2008, **47**, 8646–8649.

41. J. I. Zhang, W. A. Tao and R. G. Cooks, Facile Determination of Double Bond Position in Unsaturated Fatty Acids and Esters by Low Temperature Plasma Ionization Mass Spectrometry, *Anal. Chem.*, 2011, **83**, 4738–4744.

42. Z. Ouyang and R. G. Cooks, Miniature Mass Spectrometers, *Annu. Rev. Anal. Chem.*, 2009, **2**, 187–214.

43. X. Zhou, J. Liu, R. G. Cooks and Z. Ouyang, Development of Miniature Mass Spectrometry Systems for Bioanalysis, *Bioanalysis*, 2014, **6**, 1497–1508.

44. J. S. Wiley, J. T. Shelley and R. G. Cooks, Handheld Low-Temperature Plasma Probe for Portable "Point-and-Shoot" Ambient Ionization Mass Spectrometry, *Anal. Chem.*, 2013, **85**, 6545–6552.

45. L. Gao, R. G. Cooks and Z. Ouyang, Breaking the pumping speed barrier in mass spectrometry: Discontinuous atmospheric pressure interface, *Anal Chem*, 2008, **80**, 4026–4032.

46. L. Gao, G. T. Li, Z. X. Nie, J. Duncan, Z. Ouyang and R. G. Cooks, Characterization of a discontinuous atmospheric pressure interface. Multiple ion introduction pulses for improved performance, *Int. J. Mass Spectrom.*, 2009, **283**, 30–34.

47. W. Xu, N. Charipar, M. A. Kirleis, Y. Xia and Z. Ouyang, Study of Discontinuous Atmospheric Pressure Interfaces for Mass Spectrometry Instrumentation Development, *Anal. Chem.*, 2010, **82**, 6584–6592.

CHAPTER 7

Flowing Atmospheric-Pressure Afterglow (FAPA), the Plasma-based Source for your ADI-MS Needs

JACOB SHELLEY,[a] KEVIN PFEUFFER[b] AND GARY HIEFTJE*[b]

[a] Department of Chemistry and Biochemistry, Kent State University, Kent, OH 44242, USA; [b] Indiana University, 800 E. Kirkwood Avenue, Bloomington, IN 47405, USA
*Email: hieftje@indiana.edu

7.1 FAPA Design, Function, and Evolution

The flowing atmospheric-pressure afterglow (FAPA) source, which was initially described in the Hieftje lab at Indiana University, is based on a glow discharge at atmospheric pressure. In the initial presentations[1,2] and publications, the FAPA source was named the flowing afterglow atmospheric-pressure glow discharge (FA-APGD)[3,4] and was also referred to as helium atmospheric-pressure glow discharge ionization (HAPDGI).[5] While the first journal publications on the FAPA as a desorption/ionization source for ambient mass spectrometry did not appear until 2008,[4] it was preceded by a fundamental optical study on the helium APGD,[6] as well as multiple conference presentations.[1,2] More interesting perhaps is that the FAPA was unknowingly being developed and tested in parallel with a source similar in design, direct analysis in real time (DART).[7] While these sources may appear

New Developments in Mass Spectrometry No. 2
Ambient Ionization Mass Spectrometry
Edited by Marek Domin and Robert Cody
© The Royal Society of Chemistry 2015
Published by the Royal Society of Chemistry, www.rsc.org

to be similar at an initial glance, they behave quite differently in practice, which is due to the unique design and operating conditions of each.

The FAPA source consists of a direct-current (dc) glow discharge that is sustained at atmospheric pressure with an inert discharge gas, typically helium. Such an APGD in helium was developed and used by Lubman and coworkers for organic mass spectrometry in the early 1990s.[8,9] Lubman's configuration utilized the simplicity of direct-current power and provided good sensitivity for a broader range of compounds than atmospheric-pressure chemical ionization (ACPI) with a corona discharge, but it required that the sample be introduced directly into the discharge. This sample-introduction mode caused perturbations in the plasma as well as deposition of material onto the electrodes that led in turn to signal instabilities and memory effects. To avoid these issues, the APGD of the FAPA is physically and electrically isolated from the sample-introduction region; thus, the discharge is operated in the flowing-afterglow mode.

7.1.1 Pin-to-plate FAPA

In the first iteration of the source design,[3,4] the discharge was formed between a negatively biased pin cathode and a grounded, or near-ground potential, plate anode; the electrodes were fixed in place by and sealed to an inert, dielectric cell body (*cf.* Figure 7.1(a)). A hole in the plate allows all the ions, electrons, and excited species created in the plasma to flow into the open atmosphere and interact directly with atmospheric gases and the sample. The APGD is operated in the abnormal glow regime, with currents between 5 and 50 mA and an electrode gap between 4 and 10 mm, which requires potentials of less than 700 V. For comparison, the discharge in DART operates with voltages and currents of ~ 3000 V and ~ 2 mA, respectively, in a corona-to-glow transition region.[10] Importantly, to sustain a stable DC glow or quasiglow discharge, as is the case in FAPA or DART, a

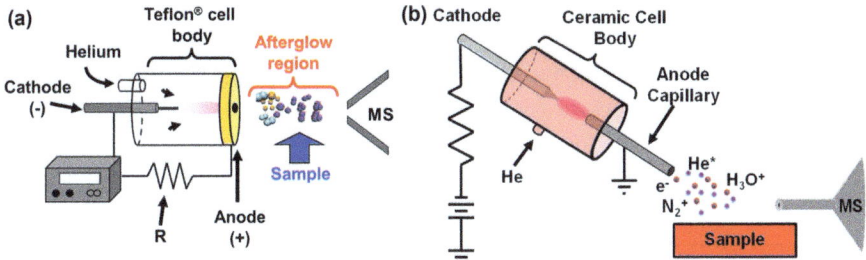

Figure 7.1 (a) Diagram of the original, pin-to-plate FAPA configuration with the discharge chamber constructed from Teflon. A hole in the plate anode allows discharge species to flow into the sample-introduction region. (b) Depiction of the newer pin-to-capillary geometry FAPA with a Mykroy ceramic discharge cell. Discharge species flow into the sampling region through the anode capillary.

ballast (or current-limiting) resistor is needed in series with the discharge to compensate for the negative dynamic resistance encountered at atmospheric pressure that would otherwise create an erratic arc.

This pin-to-plate FAPA provided simple mass spectra, consisting almost exclusively of the protonated molecular ion, and detection limits in the single femtomole range for polar and nonpolar species. The ionization mechanisms are thought to be APCI-like due to the presence of protonated water clusters, $[H_2O]_nH^+$, in the background mass spectrum.[3] This mechanism is similar to that proposed for other plasma-based ambient MS sources (*e.g.*, DART, LTP, *etc.*) except that the FAPA background spectrum exhibits a large proportion of charge-transfer species (*e.g.*, N_2^+, NO^+, O_2^+), which are largely absent in the background spectra of other plasma-based sources.[3,10] Solid analytes are desorbed from surfaces through predominantly thermal or thermally assisted mechanisms as a result of elevated temperatures in the afterglow caused by the high gas temperature of the APGD itself.[6,11] Furthermore, the gas temperature at the sample can be changed rapidly by altering the discharge current or the plasma-gas flow rate (*cf.* Figure 7.2), so no external heater is needed.

7.1.2   Pin-to-capillary FAPA

While the pin-to-plate FAPA is stable and provides good detection limits, it was observed that the source produced a significant quasicontinuum mass-spectral background as well as multiple oxidations of conjugated systems, which ultimately limited the sensitivity. To overcome these issues, a FAPA with a pin-to-capillary configuration was designed.[12] The use of a capillary prevented atmospheric oxygen from diffusing into the discharge region and

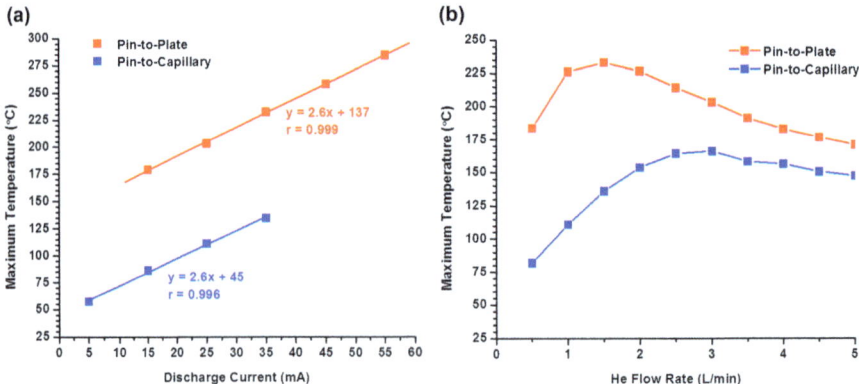

Figure 7.2   Maximum afterglow temperature for the pin-to-plate (red symbols) and pin-to-capillary (blue symbols) FAPA sources as a function of (a) APGD current and (b) helium flow rate. The flow rate in (a) was held at 1 L min^{-1}, while the discharge current in (b) was fixed at 25 mA. The decrease in temperature at higher He flow rates is caused by turbulent flow.

producing radical, atomic oxygen, which was responsible for the observed oxidation (Figure 7.1(b)). Additionally, the discharge cell body was changed to a machinable ceramic, Mykroy, which reduces the quasicontinuum background by 89% and 99% in positive- and negative-ionization modes, respectively, leading to drastically improved detection limits: 4 amol for the agrochemical ametryn. The decrease in background signal in negative-ion mode permitted detection of species that commonly form negative ions, such as nitrated explosives; the explosive PETN could readily be detected at levels of 66 fmol. This design also allows the analysis of samples with complex shapes by angling the source with respect to the sample, in a fashion similarly to DESI. An interesting type of sample that could be analyzed with this FAPA is bulk solutions. Figure 7.3(a) shows the analysis of a bulk solution of the herbicide, terbuthylazine, in neat solvent. Detection limits were obtained by scanning bulk solutions of varying concentration underneath the pin-to-capillary FAPA. Figure 7.3(b) shows a time trace for a calibration of methamphetamine solutions with a detection limit of 0.7 ppb; surprisingly, very little degradation in detection limits was observed even with complex matrices, such as raw urine.

7.1.3 Halo-FAPA

More recently, an alternative FAPA design was introduced by Pfeuffer *et al.*[13] In the halo-FAPA, the electrodes are coaxially arranged, with a cathode capillary inside an anode tube that are electrically separated by a ceramic- or polyimide-capillary spacer (*cf.* Figure 7.4). The electrodes are held in place with a Swagelok tee, which also enables introduction of the plasma gas between the electrode tubes. The close spacing of the electrodes reduces the power needed to operate the FAPA to less than 5 W without compromising

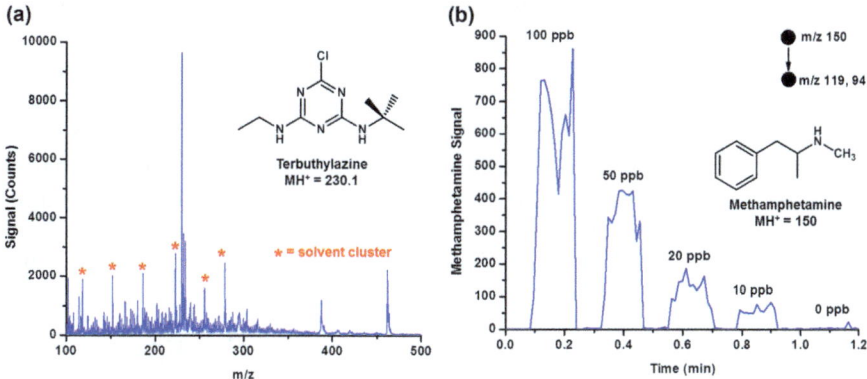

Figure 7.3 (a) Direct analysis of a bulk solution of 8 μg mL^{-1} terbuthylazine in methanol:water. The asterisk (*) denotes a solvent cluster. (b) Chronogram of methamphetamine MS/MS fragments for different concentrations in tap water. Resulting calibration curve has an r^2 of 0.999 and a LOD of 0.7 ppb.

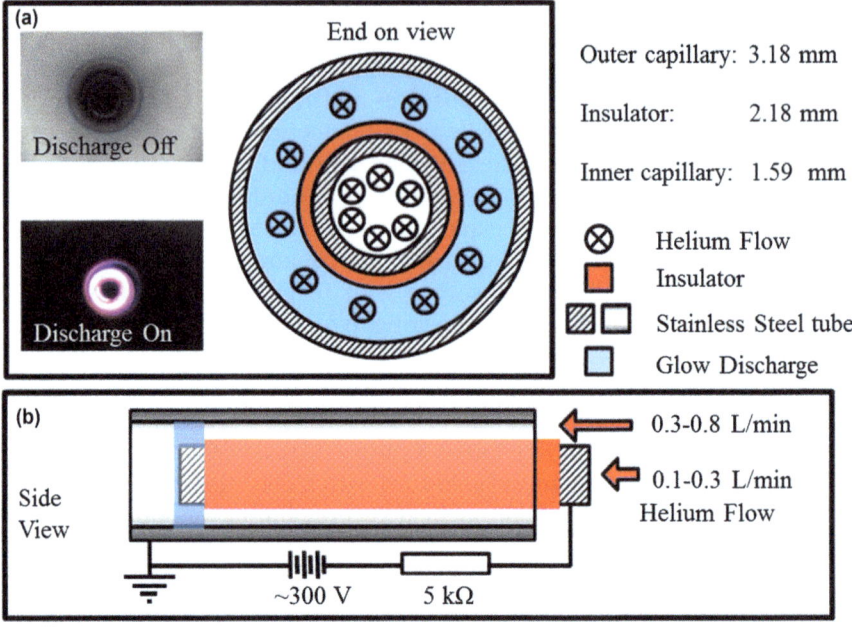

Figure 7.4 Diagram of the halo-FAPA source. (a) end-on view of cell, with schematic diagram and photographs (on and off) of the discharge. Conditions: 25 mA, 300 V, 0.4 L min^{-1} He (outer), 0.20 L min^{-1} He (inner) flows. (b) Side-on view illustrating how the concentric tubes are arranged and the placement of the insulator.

any advantages of the original FAPA. The coaxial layout of the electrodes leads to a toroidal APGD, which provides multiple additional advantages. In addition to analyzing solid samples in a conventional fashion, gases and aerosols, from nebulized liquids or laser-ablation events, can be introduced through the central channel and pass through the hole in the toroid without perturbing the plasma. Relative freedom from sample-positioning errors can also be achieved, by using the central capillary not as a means of introducing plasma gas or sample vapor into the h-FAPA discharge, but rather as a means of transporting volatilized and ionized sample species into a mass-spectrometer interface. The close proximity of the sample to the discharge provides excellent ionization efficiency for the halo-FAPA source; individual droplets containing $\sim$640 amol of lidocaine were detected with a S/N $>$ 60.

7.2   Fundamental Characterization of the FAPA

7.2.1   Optical Characterization

Efficient transfer of energy from the electrical field to electrons to, ultimately, the discharge gas, atmospheric gases, and the analytes is the primary objective of a plasma in ambient mass spectrometry source. Optical emission measurements from excited species (*e.g.*, atoms, ions, and, one of the oldest fields of "-omics", diatomics) within a discharge can produce insights into the effectiveness of this energy transfer as well as provide information about the chemistry within the plasma. A UV-visible emission spectrum can yield a short list of species formed or present in the plasma, which could react with or ionize analyte compounds. Furthermore, by characterizing the spatial structure of the emission within the discharge and the afterglow in the open atmosphere, information about the discharge structure and the source of emitting species can be determined.

Prior to the initial publication of the FAPA as an ambient mass spectrometry source, Andrade *et al.*[6] performed initial optical characterization of a sealed helium glow discharge over a range of pressures from 1 to 760 Torr. It was found that, as the pressure was increased, the negative glow, which dominates the volume and ionization processes in reduced-pressure glow discharges, is compressed to a thin disk that resides close to the cathode. Simultaneously, the positive column, which consists of slow electrons, occupies a larger region and, at atmospheric pressure, fills the majority of the discharge volume. These findings were further explained and supported through Monte Carlo modeling of the APGD.[14]

Shelley *et al.*[11] used optical-emission spectroscopy to determine excited species that were present in the FAPA discharge. Table 7.1 lists the major atomic and molecular species that were observed. In addition to high-energy helium emission lines, signals from gas impurities present in the discharge

Table 7.1 Major atomic and molecular species observed in the emission spectrum of the helium APGD from the FAPA operated at 25 mA with a helium flow rate of 1.5 L min^{-1}.

Emitting Species	Transition	Wavelength/nm
NO	$A^2\Sigma^+$–$X^2\Pi$	202–273 ($v' = 0, 1$)
CO$^+$	$B^2\Sigma$–$X^2\Sigma$	203–277 ($v' = 0, 1, 2, 3, 4, 5$)
OH	$A^2\Sigma^+$–$X^2\Pi$	281–309 ($v' = 0, 1$)
N$_2$	$C^3\Pi_u$–$B^3\Pi_g$	295–434.5 ($v' = 0, 1, 2$)
N$_2^+$	$B^2\Sigma_u^+$–$X^2\Sigma_g^+$	387–471 ($v' = 0$)
He$_2$	$e^3\Pi_g$–$a^3\Sigma_u^+$	460–475 ($v' = 0$)
	$f^3\Delta_u$–$b^3\Pi_g$	568–605 ($v' = 0$)
	$F^1\Delta_u$–$B^1\Pi_g$	605–630 ($v' = 0$)
	$d^3\Sigma_u^+$–$b^3\Pi_g$	630–650 ($v' = 0$)
H I	H_β	486.13
	H_α	656.28
He I	$3p^3P_2{}^\circ$–$2s^3S_1$	388.86
	$3p^1P_1{}^\circ$–$2s^1S_0$	501.57
	$3d^3D_{1,2,3}$–$2p^3P_{0,1,2}{}^\circ$	587.56
	$3d^1D_2$–$2p^1P_1{}^\circ$	667.82
	$3s^3S_1$–$2p^3P_{0,1,2}{}^\circ$	706.52
	$3s^1S_0$–$2p^1P_1{}^\circ$	728.14
O I	$(^4S^\circ)3p^5P_3$–$(^4S^\circ)3p^5S_2{}^\circ$	777.194
	$(^4S^\circ)3p^5P_2$–$(^4S^\circ)3p^5S_2{}^\circ$	777.417
	$(^4S^\circ)3p^5P_1$–$(^4S^\circ)3p^5S_2{}^\circ$	777.539

were also prominent. By monitoring spatially resolved emission at different gas flow rates, the presence of excited atomic oxygen was found to arise from molecular oxygen diffusing into the discharge region. Similarly, the presence of N_2^+ and OH• in the discharge was determined to come from atmospheric nitrogen and water, respectively, that diffused into the discharge.

In addition, spatially resolved emission spectra of the FAPA discharge were used to characterize the discharge structure at different discharge powers.[10,11] Specifically, two discharge currents were examined, 2 mA and 25 mA, which pertain to the operating regimes of the DART discharge and the FAPA discharge, respectively. The emission profiles for N_2^+ at 391.4 nm and O(I) at 777.19 nm, which are from gas impurities in the discharge, show that the emission and, consequently, the current density of the DART-like discharge at 2 mA is restricted to the tip of the pin electrode, which is indicative of a corona-like discharge. In contrast, the FAPA discharge produced emission along the pin electrode and throughout the positive column, which is indicative of a glow discharge. This distinction is even more apparent when the polarity of the electrodes is reversed, so the pin electrode is positively charged (*cf.* Figure 7.5). The emission maxima for species in the FAPA discharge at 25 mA exist near the cathode in the negative glow. In contrast, emission from the DART-like discharge at 2 mA resides almost exclusively at the pin electrode regardless of the polarity.

Spatially resolved emission in the FAPA afterglow region revealed that emission maxima for N_2^+ and OH• occur approximately 0.7 mm away from the plate electrode. Because this emission is occurring in a field-free region and no external energy source exists, it is likely that some plasma species is acting as a carrier to transfer energy from the discharge into the open air. It was shown with the LTP that the rotational temperature of N_2^+ can be used

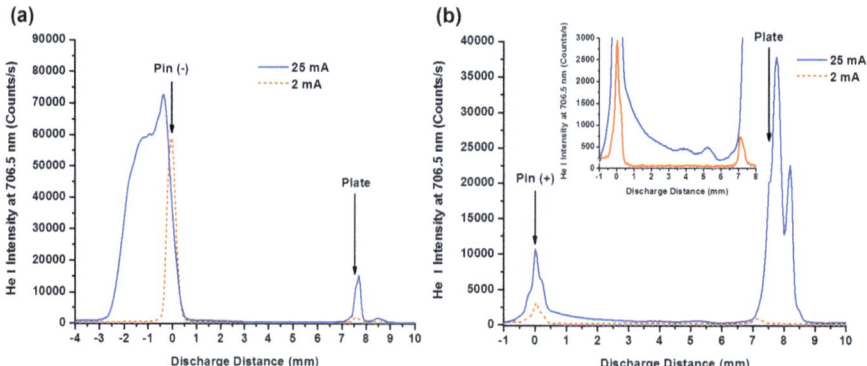

Figure 7.5   Spatially resolved emission profiles of the 706.5-nm He I line in the APGD at discharge currents of 25 mA (blue trace) and 2 mA (red trace) with a helium flow rate of 1.5 L min^{-1}. The discharge was operated in both the pin-negative (a) and pin-positive (b) polarity. The inset in (b) is on an expanded vertical axis to demonstrate helium emission in the positive column of the 25-mA APGD. In all APGD plots, a discharge distance of zero refers to the tip of the pin electrode.

to identify this energy carrier.[15] Similar to the LTP, the rotational temperature in the afterglow is artificially higher than that measured for OH or with infrared thermography. That finding, in combination with linear N_2^+ Boltzmann plots, suggests that helium dimer ion, He_2^+, and not helium metastable species, serves as the dominant carrier of energy from the discharge to the open air.

7.2.2 Mass-Spectrometric Characterization

In addition to optical methods, mass spectrometry has been a useful tool in understanding and improving the FAPA source. It has been used to identify the type and relative abundance of reagent ions produced by the source under varying operating conditions.[3] Mass spectrometry has also been used to directly compare different plasma-based sources.[10] In one case, the DART, FAPA, and LTP sources were used to ionize a relatively labile molecule, methyl salicylate (*cf.* Figure 7.6). All three mass spectra contained the protonated molecular ion as well as a commonly observed fragment ion. The ratio of the parent-ion signal to the summed signal for the analyte provides a measure termed the survival yield (SY).[16] It is an indicator of the excess energy imparted to the analyte during the ionization process and, thus, the relative softness of the source. Of course, the mass spectrometer interface contributes to the degree of fragmentation, so the survival yields can be compared only if they are obtained on the same instrument. The SY of methyl salicylate for the DART, FAPA, and LTP sources were found to be 67.3%, 22.8%, and 6.9%, respectively. These findings indicate that, of the three sources, DART is the softest, while the LTP imparts the most energy into the analyte during ionization. This trend correlates with the degree of direct interaction between the plasma and the analyte, where the DART discharge is physically isolated from the analyte, the FAPA afterglow plasma extends into the ionization region, and the LTP interacts directly with the analyte.

Mass spectrometry has been used also to measure and quantify ionization matrix effects for the FAPA, DART and LTP sources.[17] Because samples are analyzed without pretreatment or separation, matrix effects are of considerable concern to the whole ambient mass spectrometry field. Matrix effects can occur in both the desorption and ionization steps. Thus, to reduce complexity in this study, these processes were decoupled to gauge the magnitude of ionization suppression. For proton-transfer reactions, the most common ionization pathway for plasma-based sources, the suppression occurs through competitive ionization caused by variations in proton affinities (PAs). Typically, gaseous analytes, A(g), are ionized through interactions with gas-phase, protonated water clusters $[(H_2O)_nH^+]$:

$$A(g) + (H_2O)_n H^+ \rightarrow AH^+(g) + (H_2O)_n$$

$$\text{if } PA_A > PA_{(H_2O)_n H^+}$$

(7.1)

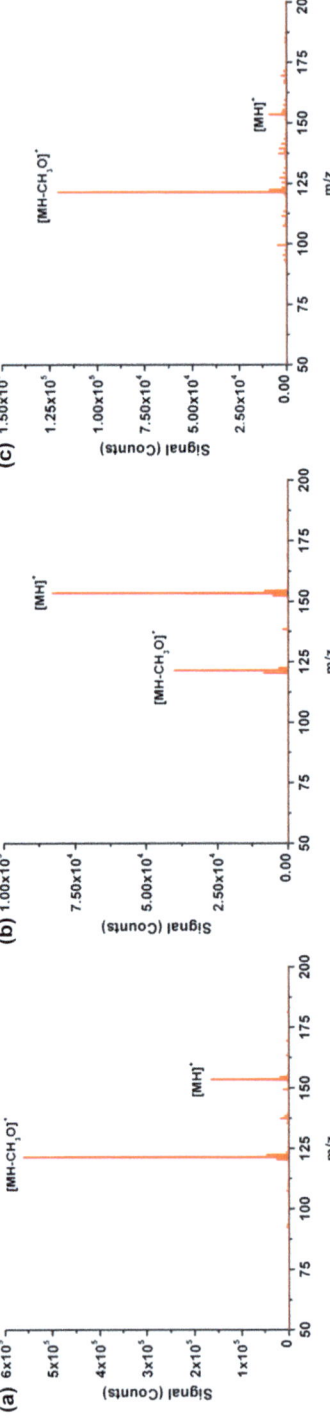

Figure 7.6 Mass spectra of methyl salicylate ($MH^+ = 153$) obtained with (a) FAPA, (b) DART, and (c) LTP ionization sources.

where PA_A and $PA_{(H_2O)_nH^+}$ are the proton affinities of the analyte and the protonated water cluster, respectively. If a matrix molecule M is also present in the gas phase, the system can undergo competitive ionization:

$$AH^+ + M^\circ \rightarrow A^\circ + MH^+$$

$$\text{if } PA_M > PA_A$$

(7.2)

where PA_M is the proton affinity of the matrix molecule.

Vapor-phase matrices were mixed with a continuous stream of gaseous analyte and introduced into each plasma-based ion source. A decrease in analyte signal upon introduction of a matrix indicated an ion-suppression event. When the matrix species had a proton affinity the same as or greater than that of the analyte, all three sources suffered analyte signal suppression, even at moderate matrix-to-analyte concentration ratios (*cf.* Table 7.2). In fact, matrix-to-analyte mole ratios of 10 were sufficient to entirely suppress analyte ion signals in the LTP and DART. The FAPA was the least susceptible to the ion-suppression process under these conditions. In contrast, when the proton affinity of the matrix species was lower than that of the analyte, no matrix effect was observed with DART, although an effect persisted for both FAPA and LTP. The ion suppression in this case is believed to be due to charge-transfer ionization between a reagent ion that is essential to water-cluster formation, such as N_2^+, and a matrix molecule.

7.2.3 Schlieren Imaging

7.2.3.1 Large-scale Schlieren Characterization[18]

Despite the numerous advantages ambient desorption/ionization sources offer, reproducibility and sample introduction still seem to be problematic. Investigations into mass transport from the source into the mass spectrometer would offer insight into how to improve the reproducibility of these analyses. Because the plasma support gas for the FAPA source (helium) is optically transparent, a method for visualizing it is necessary. Schlieren imaging[19] is just such a method; it produces images based on a gradient in refractive index, which is substantial between helium and air. A full-scale

Table 7.2 Order of susceptibility of ionization sources to matrix effects under different combinations of matrix–analyte proton affinity. A ranking of 1 indicates the source that was most prone, 3 the least, and X represents no detected matrix effect.

	$PA_A > PA_M$	$PA_A \approx PA_M$	$PA_A < PA_M$
FAPA	2	3	3
DART	X	2	2^a
LTP	1	1^a	1^a

^{a}No analyte signal detected with matrix-to-analyte mole ratio of 10.

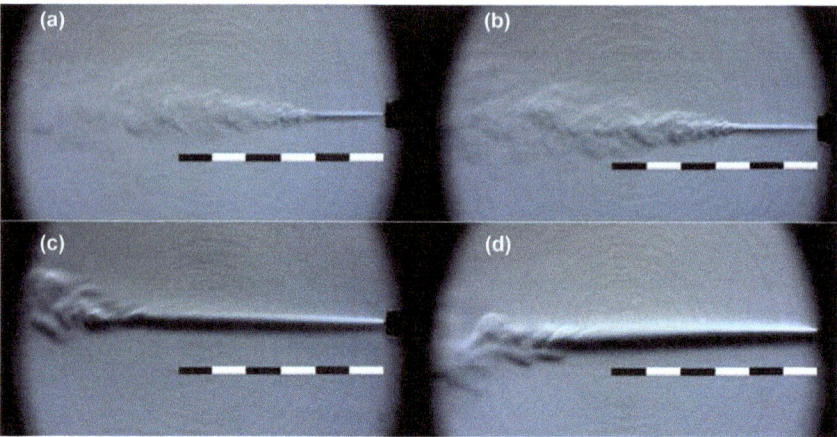

Figure 7.7   Schlieren images of a pin-to-capillary FAPA source with various capillary
inner diameters. He flow was 1.00 L min^{-1} in all cases; (a) 250 mm,
(b) 500 mm, (c) 1.0 mm and (d) 1.3 mm capillary inner diameters. Each
scale bar is 1.0 cm (total of 6 cm), oriented to capillary exit. FAPA is
positioned on right side of each frame, outside of visible zone.

Schlieren imaging setup was constructed to visualize helium flow from the
FAPA source under selected conditions. Helium flow from a pin-to-capillary
FAPA source with different capillary diameters can be seen in Figure 7.7. As
the diameter of the capillary is changed from 0.25 mm to 1.3 mm at a constant
helium flow rate, the transition from laminar to turbulent flow moves from
20 mm to 80 mm beyond the FAPA source. The displacement in the onset of
turbulence is due to a change in the Reynolds number of the helium flow.
Laminar flow is desirable as it is better characterized and more predictable
than turbulent flow; accordingly larger capillary diameters are preferred.

   Of additional interest is how the helium flow interacts with surfaces for
determining the optimal position for surface analysis. For this study an
optical flat was inserted into the Schlieren path and helium flow from the
FAPA impinged on the surface (*cf.* Figure 7.8). Angles of 30 (a–c) and 45 degrees (d–f) and flow rates of 0.50 (a,d), 1.0 (b,e) and 1.50 (c,f) 1 min^{-1} helium
were examined. At the lower angle (30°) helium always escapes the mock MS
inlet, due to its high horizontal velocity component. At a steeper angle (45°)
more helium is gathered by the mock interface and the sample interrogation
area becomes smaller. However, if the angle becomes too steep (above 60°)
backscattered helium becomes problematic and is never gathered by the
interface.

7.2.3.2   Schlieren MS Measurements

Another dimension to Schlieren analysis was performed by the addition of a
mass spectrometer to the Schlieren imaging setup. A working mass spectrometer allows realistic gas sampling and collection along with ion-identity

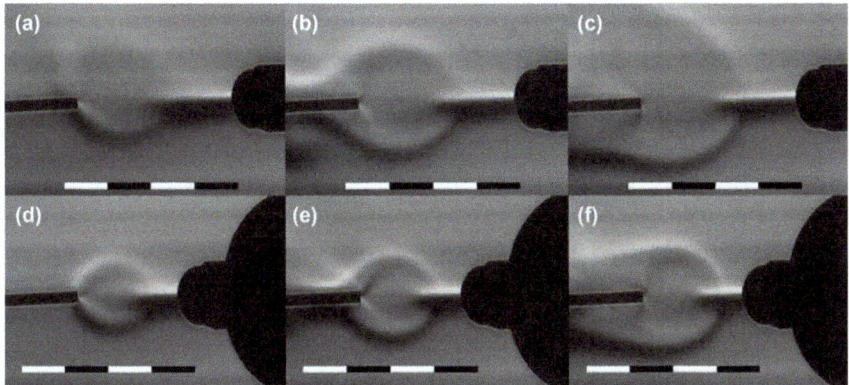

Figure 7.8 Schlieren images of FAPA beam impinging on a glass slide at 30°
(a, b and c) and 45° (d–f) degrees from perpendicular, at flow rates of
0.50 (a) and (d) 1.00 (b) and (e) and 1.50 (c) and (f) L min⁻¹. Individual
scale bars are 5 mm (20 mm total length) and aligned with capillary exit
FAPA is on right, capillary interface on left, helium flow right to left.

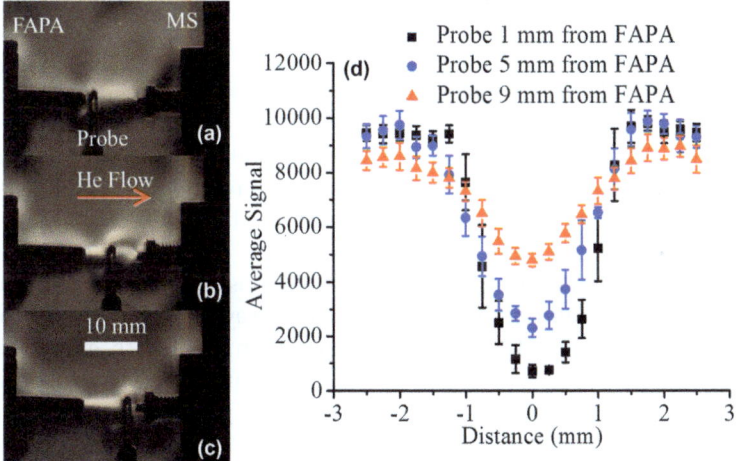

Figure 7.9 Schlieren images (a)–(c) and plot of methyl salicylate signal (d) for
melting-point capillary positions, 1, 5 and 9 mm from the FAPA source
as the MPC is moved across the MS inlet. Melting point capillary is at
distance zero for images (a)–(c). Methyl salicylate doped into the helium
flow was monitored (concentration ∼70 ppm).

information to be gathered about the FAPA source interacting with several
sample introduction arrangements. Samples can be introduced by means of
probes (melting point capillaries, MPC), employed to mechanically transfer
small amounts of sample into the afterglow of the FAPA source. Three different MPC positions were examined; it was found that probe positions close
to the MS inlet transmitted the most helium to the mass spectrometer
(*cf.* Figure 7.9). Some helium was deflected at all capillary positions, but

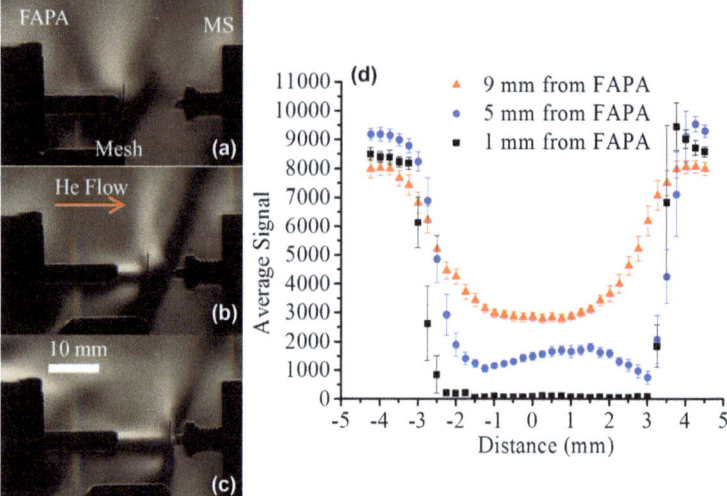

Figure 7.10 Schlieren images (a)–(c) for stainless steel mesh used for transmission-mode experiments with a plot of methyl salicylate signal (d) for mesh positions 1, 5 and 9 mm from the FAPA source as the mesh is moved across the MS interface (distance in plot d). Mesh (100×90 strands per inch) is at distance zero for images (a)–(c). Methyl salicylate doped into the helium flow was monitored (concentration ~70 ppm).

the position closest to the MS inlet suffered the least signal attenuation (50%). Additionally, a transmission-mode experiment with a stainless-steel mesh was performed to determine optimal mesh placement. Figure 7.10 shows that a position close to the FAPA source blocks all of the helium and kills the signal completely. While a position close to the MS inlet allows more helium through, some is always blocked by the mesh. A higher mesh density (100×90 strands per inch), but with greater transmission, allowed more helium through than a coarser but more obstructive mesh (70×70 strands per inch) due to the higher percent open space.

7.3 FAPA Practices

7.3.1 Laser Ablation Sample Introduction[20]

Analysis of solid samples with high spatial resolution is important for many applications and for generating molecular "maps" of a surface. By ablating a small amount of sample from a surface with laser light and subsequently transferring the particles through a gas stream to interact with the afterglow from the FAPA source, molecular information can be mapped.

Laser ablation at 266 nm was coupled to the FAPA source through 1 m of Teflon® tubing into a tee joint to ensure sufficient interaction of the afterglow with the generated aerosol (*cf.* Figure 7.11). It might at first seem surprising that intact molecules can be generated, considering the power of the laser.

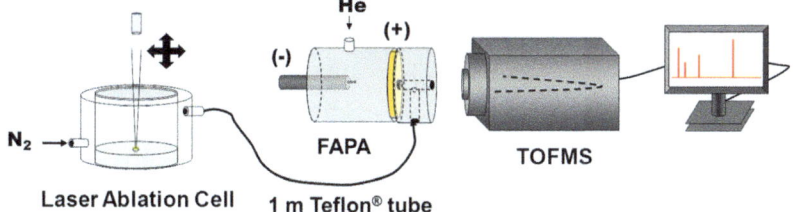

Figure 7.11 Schematic diagram of laser ablation FAPA setup. A small amount of sample is ablated with 266-nm laser light. The resulting aerosol is transferred in a stream of N_2 through a 1-m Teflon tube to the afterglow region for desorption/ionization and detection.

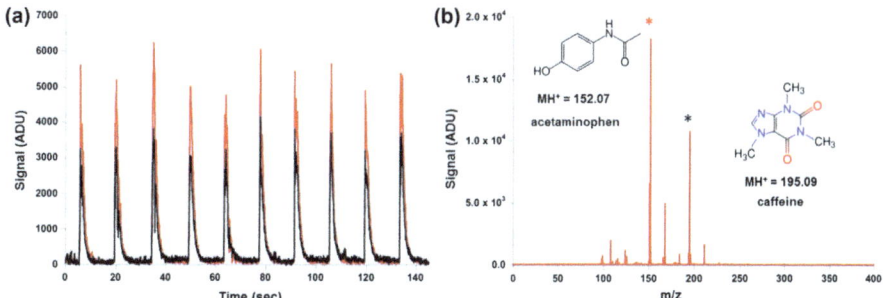

Figure 7.12 Single-shot LA-FAPA analysis of a two-component film, spin coated on glass. (a) Signal trace representing 10 single-shot laser ablation events from different positions along a sample film containing acetamino-phen (red) and caffeine (black). (b) Typical mass spectrum of a single laser ablation event.

However, it appears that nanoparticles rather than atoms or molecular fragments are formed in the ablation process.[21,22] Sample material is desorbed from the aerosol particles by the hot helium stream before they are ionized through previously described reaction mechanisms,[3] either proton or charge transfer. To assess the reproducibility of this method, a glass microscope slide was spin coated with caffeine and acetaminophen and their protonated molecular ions were monitored for 10 laser shots (*cf.* Figure 7.12). The peaks resulting from each laser pulse were ∼1.2 s long (FWHM), largely due to the size of the laser ablation chamber; a smaller chamber can reduce the washout time to less than 0.5 s.[23] The relative standard deviation of the integrated peaks for each analyte was 13%, attributable to the shot-to-shot reproducibility of the laser. Use of caffeine as an internal standard for acetaminophen and *vice versa* improved the RSD to ∼3%.

To demonstrate the imaging capability of the LA–FAPA combination, an ink-jet printer was modified to print a solution of caffeine-doped ink (*cf.* Figure 7.13). The logo for Indiana University was printed (5.6 mm across) on paper with caffeine-doped ink. Raster scanning the entire area with the

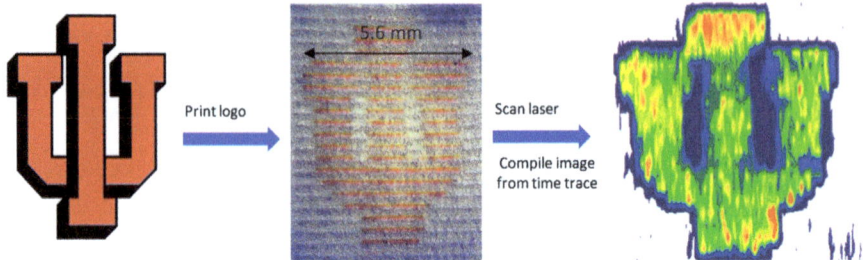

Figure 7.13   Steps in generating a chemical image of the Indiana University logo printed on paper with caffeine-doped ink. This logo is a registered trademark of Indiana University. The center image shows the logo (left) after it was laser-ablated. The path of the repetitively pulsed laser is clearly apparent. The reconstructed logo is on the right.

laser followed by reconstruction from the time domain to the space domain resulted in a map of $M + H^+$ that corresponds to the IU logo. Laser ablation can also easily be coupled to the halo-FAPA source[24] with good sensitivity. Approximately 30 ng of caffeine yielded excellent signal strength for the protonated molecular ion of caffeine.

7.3.2   Gas Chromatography/Capillary Electrophoresis (Separations)[25,26]

Despite the ability of ambient mass spectrometry to perform chemical analysis without sample pretreatment, many procedures still necessitate or benefit from the use of some separation prior to analysis. Indeed, as will be indicated later, such a separation overcomes matrix interferences that would otherwise be troublesome. Gas chromatography and capillary electrophoresis are well-established separation techniques that have successfully been coupled with the FAPA source.

A coupling tee similar to the one utilized for laser ablation was installed to couple a gas-chromatography system to the FAPA source. Two different sample mixtures were injected into the column: a mixture of 6 chemical warfare simulants (CWSs) and 13 herbicides (*cf.* Figure 7.14). The total analysis time for all CWSs was less than 5 min from extraction to detection. The CWSs were spiked in dry soil at 20 ppm by mass; the theoretical limit of detection for all but 2-CEES is 1 ppb. Coincidently, all species except 2-CEES were detected as $M + H^+$ ions, while 2-CEES was found as an odd-electron species $M^{+\bullet}$. This difference might be related to the poor detection ability as 2-CEES more readily forms a negative ion.

A 13-component neat herbicide mixture was also examined by GC-FAPA-MS. Unfortunately, only nine compounds eluted in a timely manner (<10 min); however, these exhibited excellent analytical behavior. The limits of detection for the nine components were between 2 and 6 fmol on column,

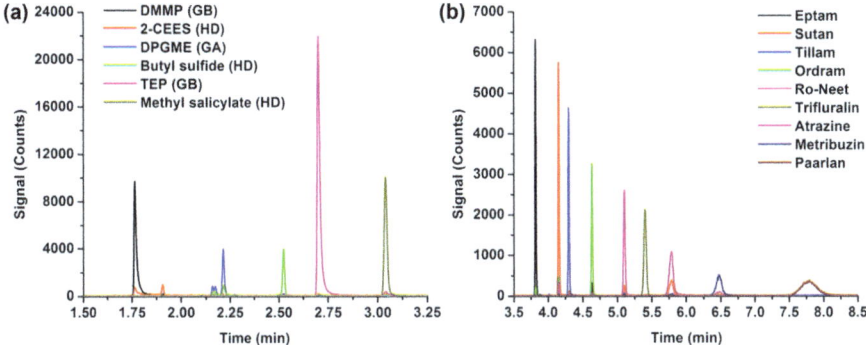

Figure 7.14 (a) Chromatogram of the GC-FAPA-MS analysis of six CWA simulants extracted from dirt. Each colored trace corresponds to the extracted ion chromatogram for the MH$^+$ ion for the analyte listed in the legend. The CWA corresponding to each simulant is given in parentheses: GB, sarin; HD, mustard gas; GA, tabun. (b) Chromatogram of the GC-FAPA-MS analysis of a mixture of 13 herbicides. Each colored trace corresponds to the extracted ion chromatogram of the MH$^+$ ion for the analytes listed in the legend. The limit of detection for every analyte was less than 6 fmol.

with at least three orders of magnitude linear working range and a relative standard deviation better than 5%.

Capillary electrophoresis was coupled to the FAPA source by nebulizing the effluent of the capillary through the afterglow of the FAPA source. A schematic can be seen in Figure 7.15. The system was optimized to give maximum signal for three model analytes (3-methylphyridine, chloraniline and tripropylamine). Parameters optimized include α, the angle between the FAPA source and the CE, the nebulizer gas flow rate, and the three distances (a, b and c) defined Figure 7.15. Selected values include a perpendicular arrangement between FAPA and CE capillary, a nebulizer gas flow rate of 4.2 L min^{-1}, and a 10-mm separation between MS inlet and CE nebulizer.

Table 7.3 summarizes results from the CE-FAPA-MS experiments. The listed limits of detection range from 74 to 610 fmol, which are comparable to those of existing CE-ESI-MS interfaces, even though the FAPA interface is much simpler to operate. To simulate complex matrices and to test the ability to detect from these complex matrices directly, 0.951 mM of *p*-chloraniline in a yeast extract, soil extract and raw urine samples was examined. The *p*-chloraniline was detected from all matrices at good S/N noise levels, as seen from the MS electropherograms in Figure 7.15. Some matrix effects are present, as the signal-to-noise ratio varies between matrices by about 4×, even though the concentration remains constant.

7.3.3 Droplet-on-demand Introduction System[27]

Introduction of microvolume samples into analytical discharges has a long history of providing both excellent analytical performance and unique

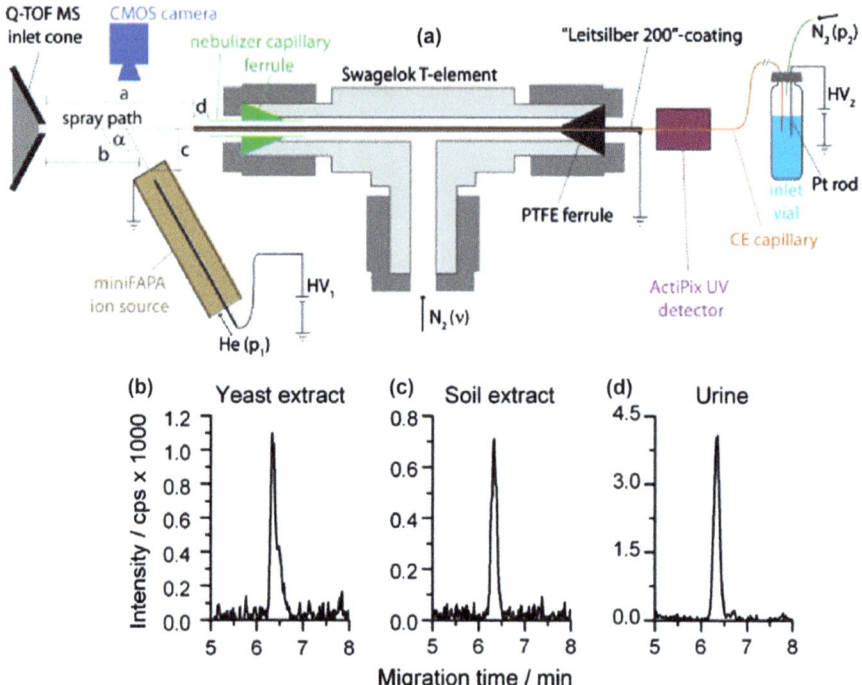

Figure 7.15 (a) Experimental setup used for coupling CE with MS using a FAPA ion source. Schematic representation showing key components of the setup including (from right to left): CE inlet vial, CE separation capillary, UV detector, nebulizer T-element assembly with the coated section of CE capillary, FAPA ion source, and MS inlet cone. The main geometrical parameters adjusted during interface optimization are marked as *a–d* and α. Figure not drawn to scale. (b)–(d) Three MS electropherograms (*m/z* 128) of *p*-chloraniline (0.951 mM) in yeast extract (b), soil extract (c), and urine sample (d). The samples were prepared by mixing 0.5 mL of filtered yeast extract, soil extract, or urine with appropriate amounts of the standards, 100 mL of 50 mM acetic acid, and adjusted to the volume of 1 mL with water.

Table 7.3 Molecular weights (M_w), dissociation constants (pKa), migration times (t_M), observed *m/z* values (protonated form), and LODs (3 *S/N* criterion) calculated according to the MS traces obtained by CE-FAPA-MS.

Compound	M_w (g mol^{-1})	pK$_a$[a]	t_M (min)	*m/z* [M + H]$^+$[b]	LOD[c] (fmol)
Pyridine	79.10	5.2	3.0	80	277 ± 22
3-Methylpyridine	93.13	5.6	3.2	94	149 ± 20
2,4,6-Trimethylpyridine	121.18	6.6	3.7	122	74 ± 37
Tripropylamine	143.27	10.6	3.9	144	610 ± 143
p-Chloraniline	127.57	4.0	5.5	128	251 ± 47
N,N-Dimethylbenzamide	149.19	n.f.	6.5	150	425 ± 7

[a]Values from various internet sources.
[b]Nominal mass-to-charge.
[c]+/− (plus or minus) 0.5×spread.
n.f. "not found".

diagnostic capabilities. In this example, a commercial ink-jet printer was modified with a custom built microcontroller to create single droplets for analysis.[28] This droplet-on-demand (DOD) device was coupled with the FAPA source through a mixing cone with a makeup flow of gas and a mixing tee in front of the pin-to-plate FAPA, similar to the one utilized for coupling the FAPA to gas chromatography. Upon introduction of analyte-containing droplets into the afterglow of the FAPA source, a signal was seen within 6 s. Simultaneous monitoring of both analyte and reagent ions shows depletion of reagent ions, concurrent with an increase in analyte signal (*cf.* Figure 7.16). The limits of detection for several drugs and drug metabolites from a spiked raw urine sample were determined with the DOD-FAPA system (*cf.* Table 7.4). LODs ranged from 0.04 µg mL^{-1} for methadone to 1.3 µg mL^{-1} for benzoylecgonine; all five analytes were detected simultaneously from raw urine, which indicates the sensitivity of this method. Improvements over probe introduction into the FAPA source of an order of magnitude were seen for benzoylecgonine, cocaethylene and cocaine, while MDMA and methadone were unable to be quantified through introduction *via* a glass probe.

This DOD system was successfully coupled to the halo-FAPA system as well.[24] The halo-FAPA provided better sensitivity for detection of cocaine and two of its metabolites (cocaethylene and benzoylecgonine) (60 cts/fg

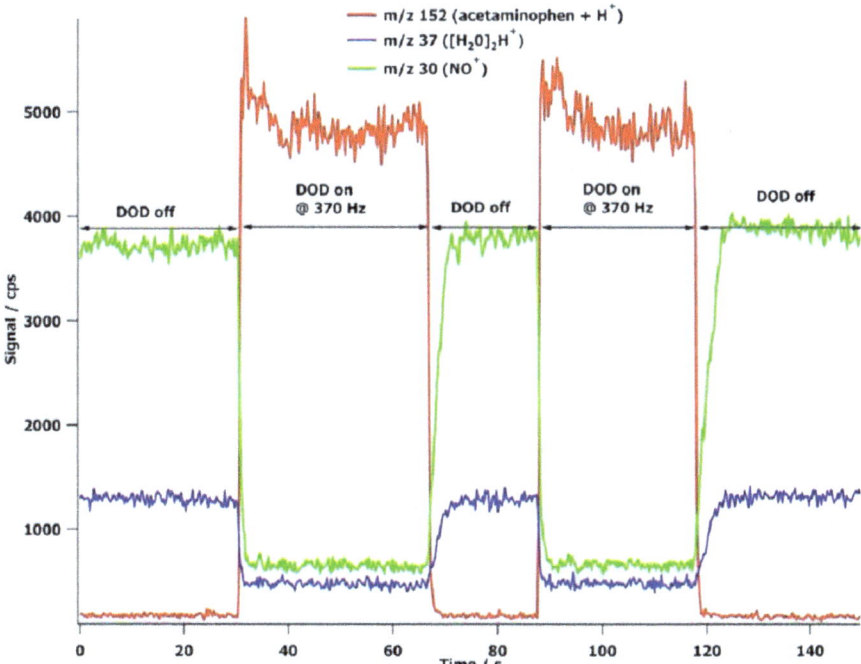

Figure 7.16 Time traces of selected ions produced within the FAPA afterglow while droplet generation was switched on and off. During droplet introduction, the analyte signal RSD is approximately 4%, neglecting the initial spike.

Table 7.4   DOD-FAPA and FAPA LODs of several drugs of abuse in doped urine without sample pretreatment and without internal standards compared to other methods reported in the literature.[a]

Analyte	MH⁺	LOD/μg mL⁻¹				
		DOD-FAPA	FAPA	LTP-MS/MS[11]	DART/ w. prec.[30]	ELISA[31]
Benzoylecgonine	290	1.3	7	1	0.024	0.3
Cocaethylene	318	0.05	1	nq[b]	0.01	nq
MDMA/Ecstasy	194	1	nq	nq	0.25[32]	1
Methadone	310	0.04	nq	0.1	nq	0.3
Cocaine	304	0.1	4	0.01	0.004	nq

[a]For FAPA and DOD-FAPA analysis, all five substances were present in the sample at the same time.
[b]nq = Not quantifiable.

compared to 38 cts/fg). Additionally, the %RSD improved due to the more stable interaction between the h-FAPA plasma and droplet, down to 0.90%, compared to 2.2% for the pin-to-plate FAPA. Finally, single droplets of 1 ppm lidocaine, corresponding to only 640 amol of analyte, could be easily detected. In comparison the pin-to-plate FAPA could detect only 0.1%.

7.3.4   Classification of Polymers[29]

Characterization of polymers is of critical importance for a variety of industrial and commercial manufacturing. The FAPA source offers a rapid solution for quality control of a wide variety of polymers from biopolymers such as sugars to synthetic homo- and copolymers. Unfortunately, due to the complexity of the resultant mass spectra, chemometric methods, such as principal component analysis (PCA) were needed to classify different polymer groups.

Mass spectra for crystal sugar (CS), *cis*-polyisoprene (IR), poly(ethylene glycol) (PEG 200), poly(ethylene terephthalate) (PET) Nylon 6/6 (PA66) and polyacrylonitrile-*co*-butadiene-*co*-styrene (ABS) can be seen in Figure 7.17(a). Due to the thermal lability of the polymers, the upper mass range was limited to ions below 400 Th. Because a variety of polymer types are present in these samples, identification based on characteristic parent ions is possible. Direct identification based on parent molecules is desirable as this has the potential for discovery analysis of unknown polymers.

A more difficult classification is of different biopolymers, specifically amylopectin, cellulose and pectin in the present case. Mass spectra of standards can be seen in Figure 7.17(b), which showcases the similarity among the three samples. Nevertheless, two-dimensional PCA is able to separate all three types of biopolymers, as is seen in Figure 7.17(c). The power of FAPA-MS coupled with three-dimensional PCA is shown in a final application[29] where different PEG chain lengths (200, 600, 900, 1000 and 1500) are successfully differentiated.

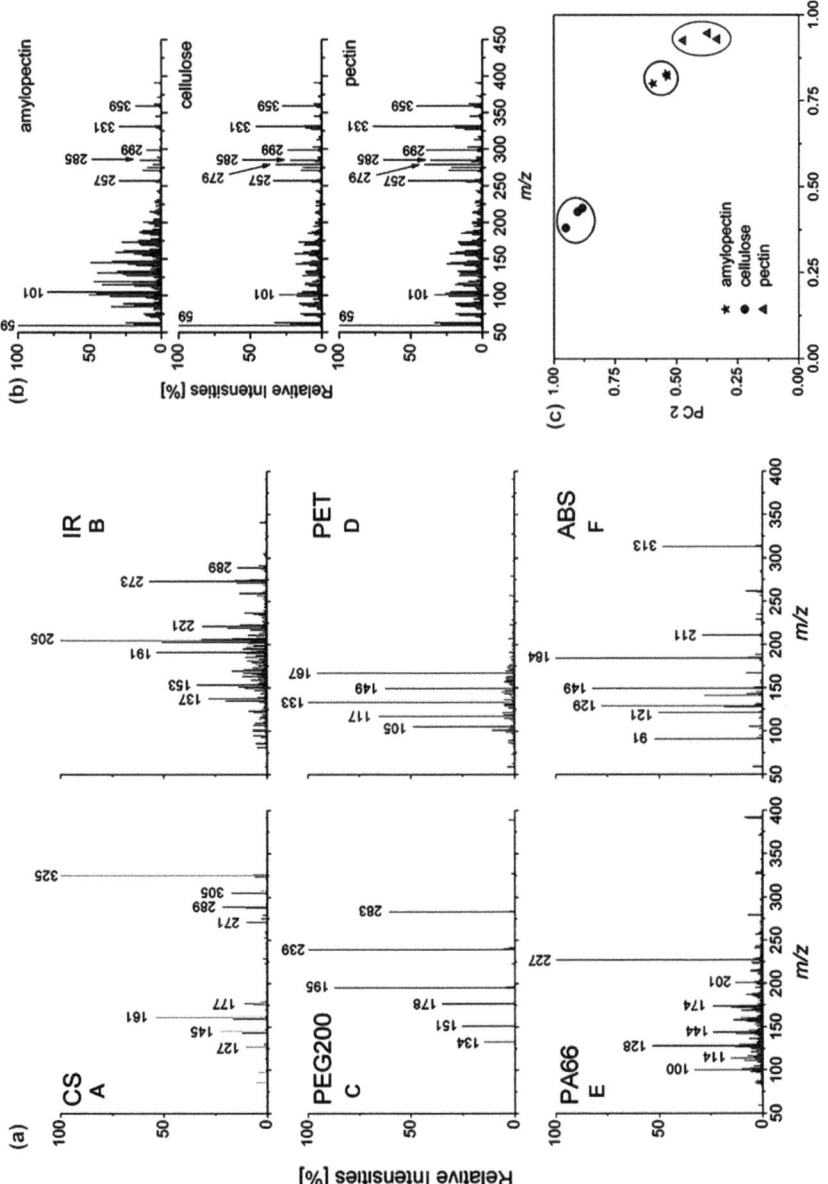

Figure 7.17 (a) Representative mass spectra (positive sampling mode) of (A) crystal sugar (CS), (B) cis-polyisoprene (IR), (C) poly(ethylene glycol) (PEG 200), (D) poly(ethylene terephthalate) (PET), (E) Nylon 6/6 (PA66) and (F) poly(acrylonitrile-co-butadiene-co-styrene) (ABS) using direct desorption/ionization FAPA-MS. (b) Representative mass spectra of amylopectin, cellulose and pectin using FAPA-MS. (c) Differentiation of three biopolymers analyzed by FAPA-MS using PCA.

7.3.5 Coupling FAPA with MHMS for Positive and Negative Ions[30]

The FAPA source has demonstrated its flexibility by being coupled with a Mattauch–Herzog geometry mass spectrograph (MHMS).[31] A schematic diagram of the coupling is shown in Figure 7.18(a). In order to get ions into the MHMS through its three-stage differentially pumped interface, a down-hill potential gradient was established from the discharge to the acceleration region (*cf.* Figure 7.18(b)).

A background spectrum was taken over the mass range 0–100 Th in both positive- and negative-ion mode, which showed the usual background features: water clusters in positive mode and oxygen and hydroxide in negative-ion mode. All ion-optic and accelerating potentials, along with the magnetic field direction were switched for negative-ion mode scans. Detection limits were determined for several organic solvent vapors (acetone,

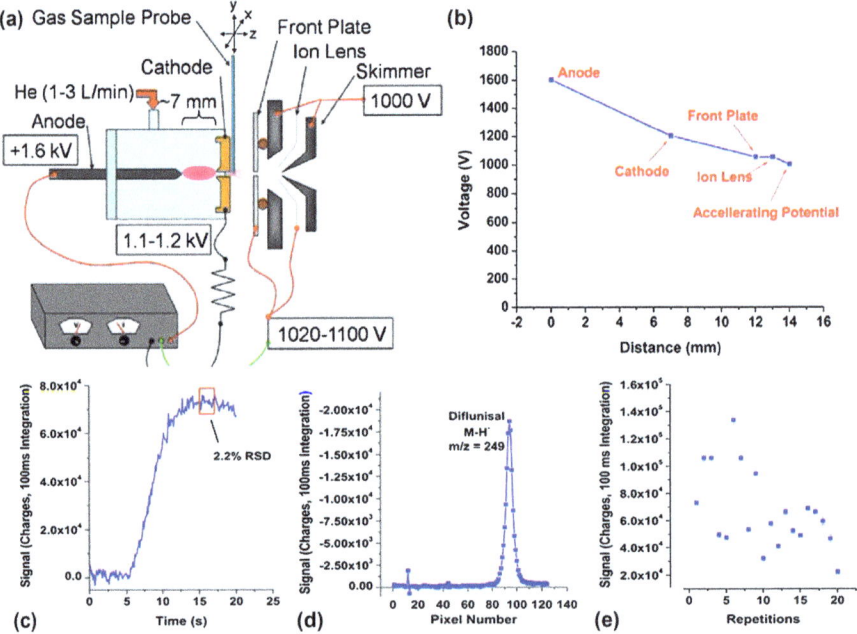

Figure 7.18   (a) Diagram of the FAPA and MHMS ion-sampling interface in positive-ionization mode. The power supply to control the FAPA source was referenced to the front plate of the MHMS for efficient ion sampling. (b) Potential profile starting from the anode and ending with the MHMS acceleration potential. (c)-(e) Analysis of diflunisal from a pharmaceutical tablet. (c) Transient produced from placing the tablet in the afterglow, (d) spectrum from the plateau region of the transient, and (e) 20 repetitions of placing material in the afterglow. Notice that the precision within a single repetition is about 2% RSD; however, precision degrades to 43% RSD (e) between runs, indicating the need for a more reproducible sample-introduction method.

tetrahydrofuran and ethanol). The limits of detection were 0.9, 24 and 13 fmol s^{-1}, respectively.

The same arrangement was employed for determination of diflunisal from a pharmaceutical tablet. A transient signal was produced after approximately 7 seconds of exposure to the afterglow (*cf.* Figure 7.18(c)), which resulted in a well-defined signal from the M–H$^-$ ion of diflunisal (*cf.* Figure 7.18(d)). However, 20 repetitions of tablet exposure to the afterglow resulted in an RSD of nearly 43% (Figure 7.18(e)), which would indicate a concentration below the detection limit, despite the S/N of the recorded spectrum (*cf.* Figure 7.18(d)). This finding indicates the poor reproducibility inherent in ambient mass spectrometry due to the need to improve sampling.

7.3.6 Detection of Pesticides in Food[32]

Rapid screening of common pesticides directly from food products is of interest for food safety. In this study, various fruit juices and fruit/vegetable skins and peels were examined for a wide variety of pesticides. Compounds were identified through analysis of neat standards with MS/MS compared with nano-ESI-MS, which showed identical results. For analysis of surfaces, solution standards were deposited on the surface and allowed to dry before analysis. For juices, filter paper was used to hold the samples.

Table 7.5 shows limits of detection for 10 pesticides on apple peel. The LODs were estimated based on a 1-cm^2 surface area interrogation and are compared against the European Union maximum residue levels (MRLs). All LODs were in the 0.01 to 5 ppb range, well below the MRLs, which showcases the ability of direct FAPA-MS to serve as a rapid screening method for pesticides.

The same batch of ten pesticides was tested in apple, cranberry, grape and orange juices. Limits of detection ranged from 1–500 ppb and do not vary widely between matrices. Lack of variance is important as it implies that this methodology could be extended to an even wider range of juices.

7.3.7 Recognition of Counterfeit Chips

Counterfeit electronics have recently been found in a number of formally trusted supply lines, which has led to a need for rapid flagging of altered electronic chips. Current methodology involves microscopy and solvent/residue testing by a trained observer who interprets the results. More advanced analytical techniques such as X-ray fluorescence have been used when necessary. Because surface adulteration is the suspected method of counterfeiting, FAPA-MS is well suited for determining if a sample has been altered from known standards. FAPA-MS coupled with chemometric techniques was utilized to differentiate counterfeit chips from known standards.

Table 7.5 Limits of detection for surface and liquid analysis. Left: The limits of detection (LODs) using direct desorption/ionization FAPA-MS of pesticides on the surface (deposited on 1 cm^2) of an apple. It was assumed that the total sample amount was desorbed and ionized. A rough estimate of the ppb concentration (mg kg^{-1}) of the pesticides in the apple is shown and compared with maximum residue levels (MRLs) designated by the EU List of the limits of determination (all values are given in ppb (ng mL^{-1})) Right: List of the limits of determination (all values are given in ppb (ng mL^{-1})) of pesticides found in different fruit juices.

Compound	LOD (ng)	Estimate in ppb	MRL EU (ppb; μg kg^{-1})	Compound	Apple juice	Cranberry juice	Grape juice	Orange juice
Carbendazim	5	4.4–5.0	50–100	Carbendazim	50	5	50	50
Carbofuran	0.02	0.018–0.02	100–300	Carbofuran	2	2	20	2
Metolcarb	0.01	0.009–0.01	not found	Metolcarb	1	100	100	50
Propoxur	0.01	0.009–0.01	50–100	Propoxur	10	100	10	10
Alachlor	1	0.88–1.0	50–100	Alachlor	50	100	500	100
Metolachlor	0.5	0.44–0.50	50–100	Metolachlor	100	50	50	10
Dinoseb	2	1.76–2.00	50–100	Dinoseb	2	20	2	20
Atrazine	0.01	0.009–0.01	50–100	Atrazine	10	10	10	100
Simazin	0.5	0.44–0.50	not settled	Simazin	5	50	50	50
Isoproturon	0.5	0.44–0.50	50–100	Isoproturon	10	10	5	100

Chemometric methods were utilized so the entire mass spectrum (*m/z* 50–500) could be used without extensive data processing. Principal component analysis (PCA) was successfully utilized to differentiate among three different counterfeit chips and a set of standard encapsulant material. The PCA plot in Figure 7.19 shows a very tight distribution for the standards. This small grouping is beneficial as it attests to the strict manufacturing requirements of the encapsulant material. Counterfeit chips (verified by a third-party tester) all group outside of the standard chips in Figure 7.19, which demonstrates the discriminatory power of chemometrics coupled with FAPA-MS.

7.3.8 Chlorpropham in Potatoes and Carbendazim in Orange Juice[33]

Detection of pesticides and fungicides from consumables is an important aspect of quality control. Two particular examples are given here. Chlorpropham is an antibudding agent typically utilized on potatoes to prevent "eyes" from forming. However, recent studies have suggested that chlorpropham is linked to antiendogenic activity in men, so rapid detection at low concentrations is very important. FAPA-MS performed

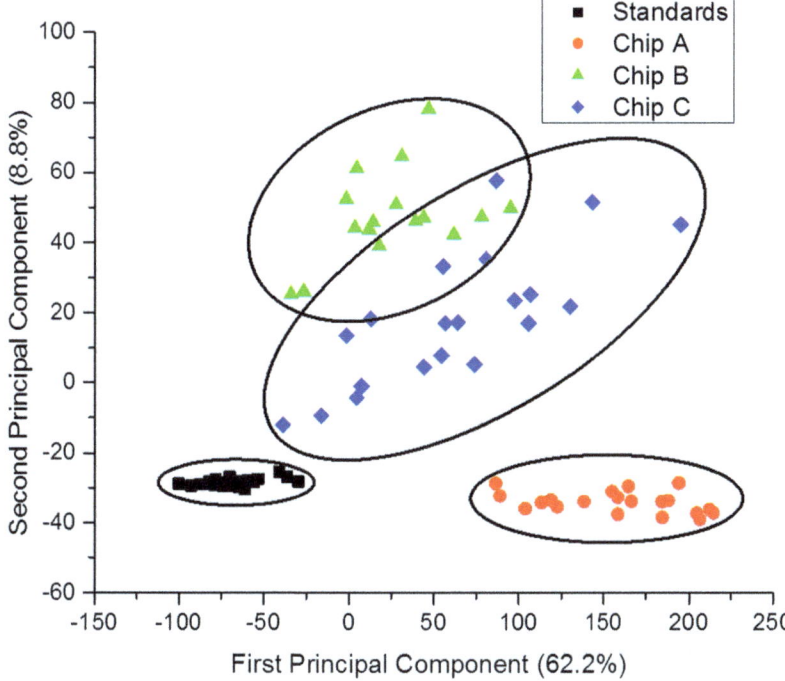

Figure 7.19   Principal component analysis of standard integrated-circuit chips compared to three counterfeit chips, which reveals the ability of PCA-FAPA-
MS to differentiate counterfeit from genuine parts.

directly on the surface of a raw potato successfully identified the
protonated molecular ion of chlorpropham, along with a characteristic
molecular fragment. More distressing to the consumer is that this
peak was seen even upon washing and cooking the potato, indicating the
persistence of the chlorpropham. Finally, FAPA-MS was performed on the
flesh inside the potato and evidence of chlorpropham was found
even there. This example shows the remarkable sensitivity of FAPA
for analysis as well as the ability to obtain some spatial selectivity for
analysis ($\sim$ 5 mm).

A recent scare of pesticide contamination in orange juice led to experiments for direct analysis of liquids with FAPA-MS. A transmission mode
geometry[34] was utilized, as it provides the fewest degrees of positional
freedom and enables rapid analysis of multiple liquid samples. An internal
standard was added to correct for any variations in matrix. Time traces
for neat standards (*cf.* Figure 7.20(a)) and spiked into orange juice
(Figure 7.20(b)) were obtained. Excellent linearity over the concentrations
tested (Figure 7.20(c)) was found with a LOD of 1 ppb and a LOQ of 10 ppb,
which is below the regulatory requirement of 17 ppb.

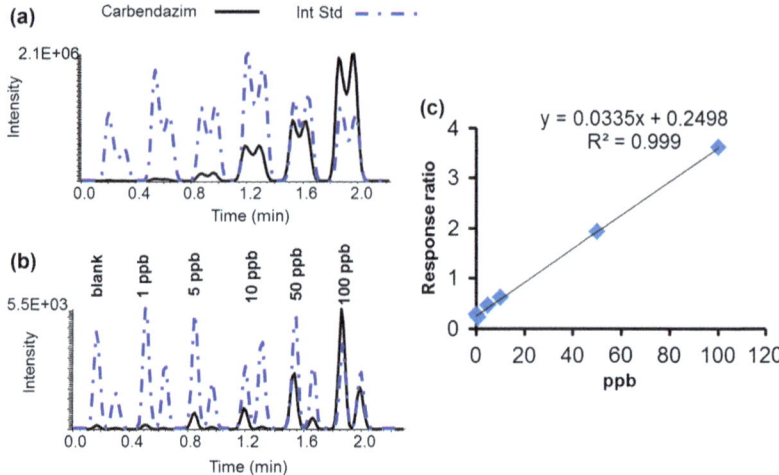

Figure 7.20 Carbendazim detection by transmission-mode FAPA experiments with an internal standard. (a) Neat solution of internal standard and carbendazim, (b) Spiked orange juice with carbendazim and internal standard, (c) Calibration plot from (a); LOD was 1 ppb.

7.3.9 Elemental Analysis through Hydride Generation[35]

Hydride generation is a common method for volatilization of certain metals and metalloids, including arsenic, germanium, antimony, selenium, tellurium, and others. The experimental setup utilized for this study was also used for detection of positive and negative ions in Section 7.3.6 (*cf.* Figure 7.18(a)). The volatile hydride is produced by reaction with an acid solution and sodium borohydride. Reaction of the acid with sodium borohydride releases hydrogen (eqn (7.3)) which reacts with the analyte metal or metalloid to form a volatile hydride (eqn (7.4)). The analyte species is A, with a valence m in solution and valence n in the hydride.

$$BH_4^- + 3H_2O + H^+ \rightarrow H_3BO_3 + 8H \tag{7.3}$$

$$A^{m+} + (m+n)H \rightarrow AH_n + mH^+ \tag{7.4}$$

Results for arsenic, antimony and selenium are displayed in Figure 7.21. Significant molecular clustering can be seen, mainly hydroxides for arsenic and water addition for germanium and antimony. These additions clutter the spectra, and spread the analyte signal over a wide range, which compromises quantitation. Additionally, spectra were obtained on the Mattauch–Herzog geometry mass spectrograph, as previously mentioned, which showed similar clustering. A calibration was performed for all analytes, but only arsenic is shown in Figure 7.22. Roll-off is seen for As^+ and AsO^+ at higher concentrations due to the greater likelihood of clustering and formation into $As_2O_3^+$, which shows a marked increase at higher concentrations. The limits of detection were 7 ppb, 2 ppb and 30 ppb for As^+, AsO^+

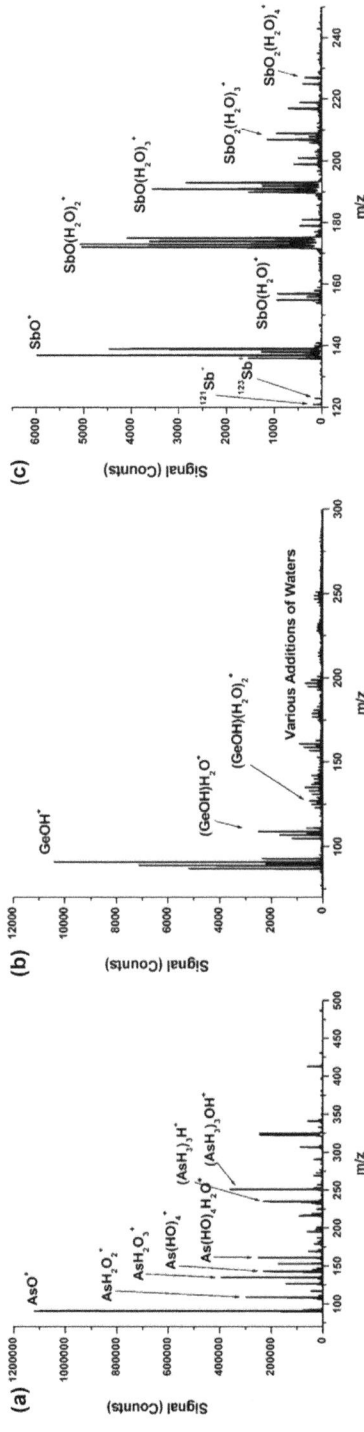

Figure 7.21 Spectra from the hydride generation of (a) As, (b) Ge, and (c) Sb taken with the FAPA and the Unique® TOFMS. Many of the same molecular clusters are observed in the two: this mass analyzer and in the Mattauch-Herzog instrument for As. The clusters observed for Ge and Sb are mainly from addition of water molecules or oxygen atoms.

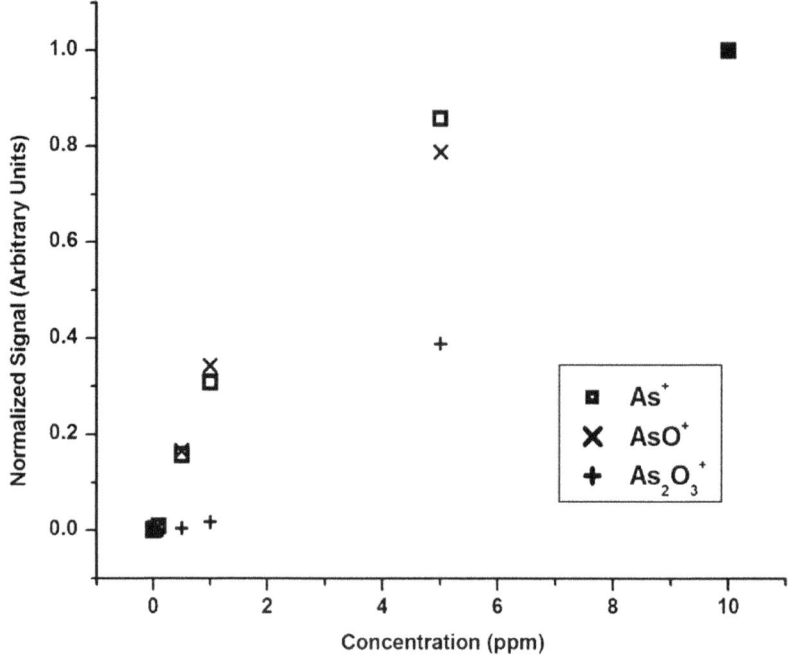

Figure 7.22 Calibration curves for three analyte peaks of arsenic. The As^+ and AsO^+ curves roll off at higher concentrations, whereas the $As_2O_3{}^+$ shows upward curvature. This behavior is due to increased formation of clusters at higher concentrations. The plots have been normalized, so slopes should not be compared quantitatively.

and $As_2O_3{}^+$, respectively. These values are 10–10 000 times worse than those achieved with an inductively coupled plasma mass spectrometer (ICPMS).[36] Despite this loss in performance, the added cost and complexity of an ICP makes the FAPA scheme attractive if sensitivity is not a primary concern.

7.3.10 Reaction Monitoring for Methcathinone[37]

Direct monitoring of chemical reactions is important to synthetic chemists for determining when a reaction has reached completion. Typically, analysis of reactions is performed by time-consuming methods such as thin-layer chromatography (TLC), gas chromatography (GC) or liquid chromatography (LC) coupled with mass spectrometry (MS) or nuclear magnetic resonance (NMR) all of which require sample preparation. The ability to analyze a reaction mixture directly from a surface with no sample preparation and to obtain detailed chemical information would be very important to synthetic chemists.

In this application, illicit synthesis of methcathinone from pseudoephedrine was monitored with a FAPA source. Two different reaction schemes were examined to compare their efficacy. The relative amount of protonated pseudoephedrine (*m/z* 166) to methcathinone (*m/z* 164) dictates how close to completion the reactions are. Resultant spectra from the two alternative schemes are shown in Figure 7.23(a) and (b). The protonated molecular ions of pseudoephedrine and methcathinone can be seen in both spectra, while the second procedure offers a much better yield of methcathinone. Dehydration of both species leads to the fragments at *m/z* 148 and 146, which is confirmed through MS/MS of methcathinone (see inset in Figure 7.23(a)).

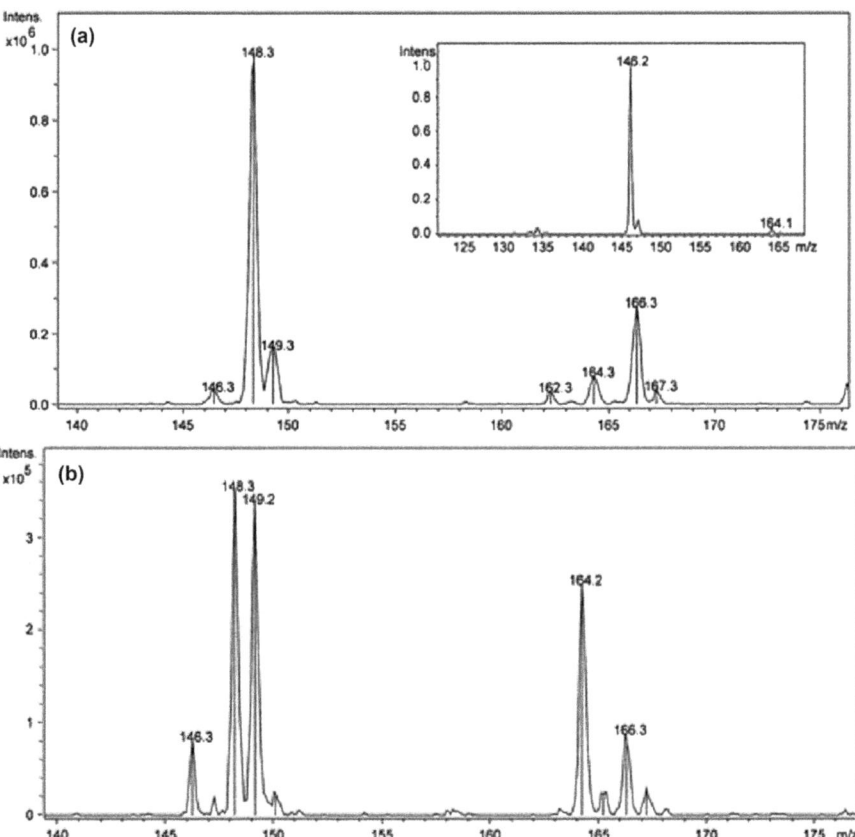

Figure 7.23 (a) Crude reaction mixture applied from the glass slide. *m/z* 164.3: protonated methcathinone; *m/z* 166.3: protonated pseudoephedrine; *m/z* 146.3: fragment ion from protonated methcathinone *m/z* 148.3: fragment ion from protonated pseudoephedrine. The insert represents the MS/MS spectrum of protonated methcathinone, confirming in-source dehydration of this ion. (b) Crude reaction mixture applied from the glass slide – modified synthesis. *m/z* 164.2: protonated methcathinone; *m/z* 166.3: protonated pseudoephedrine; *m/z* 146.3: fragment ion from protonated methcathinone; *m/z* 148.3: fragment ion from protonated pseudoephedrine. Source contamination by phthalic anhydride (plasticizer) at *m/z* 149.2 was also present in a blank run.

The results were confirmed through a liquid chromatography separation, with cleanup through ZipTip®.

7.4 Conclusions

The FAPA ambient mass spectrometry source and its alternate geometries, including the pin-to-capillary and halo-FAPA, are important sources in this developing field. Various fundamental investigations have revealed important operational characteristics for this source, with implications for other ambient mass spectrometry sources. Numerous applications for the FAPA source have been explored for far ranging fields including forensics, environmental monitoring, high-throughput analysis and reaction monitoring. The primary area of analysis is small (<500 m/z) organic molecules. Different methods of sample introduction such as laser ablation, gas chromatography, capillary electrophoresis, droplet-on-demand system and a hydride generation system can also successfully be coupled to the FAPA source. The simple geometry and construction of the FAPA along with the versatility tunable current allows makes the FAPA source very promising for ambient analysis.

References

1. F. J. Andrade, M. R. Webb, W. C. Wetzel, G. Gamez and G. M. Hieftje, Glow Discharges at Atmospheric Pressure: What? How? Why?, in: Turkey Run Analytical Chemistry Conference, Turkey Run State Park, Marshall, IN, 2005.
2. F. J. Andrade, S. J. Ray, G. Gamez, M. R. Webb, W. C. Wetzel and G. M. Hieftje, New Plasma Based Ionization Sources for Organic Mass Spectrometry, in: Pittsburgh Conference, Orlando, FL, 2006.
3. F. J. Andrade, J. T. Shelley, W. C. Wetzel, M. R. Webb, G. Gamez, S. J. Ray and G. M. Hieftje, Atmospheric pressure chemical ionization source. 1. Ionization of compounds in the gas phase, *Anal. Chem.*, 2008, **80**(8), 2646–2653.
4. F. J. Andrade, J. T. Shelley, W. C. Wetzel, M. R. Webb, G. Gamez, S. J. Ray and G. M. Hieftje, Atmospheric pressure chemical ionization source. 2. Desorption-ionization for the direct analysis of solid compounds, *Anal. Chem.*, 2008, **80**(8), 2654–2663.
5. A. Venter, M. Nefliu and R. G. Cooks, Ambient desorption ionization mass spectrometry, *TrAC, Trends Anal. Chem.*, 2008, **27**(4), 284–290.
6. F. J. Andrade, W. C. Wetzel, G. C. Y. Chan, M. R. Webb, G. Gamez, S. J. Ray and G. M. Hieftje, A new, versatile, direct-current helium atmospheric-pressure glow discharge, *J. Anal. At. Spectrom.*, 2006, **21**(11), 1175–1184.
7. R. B. Cody, J. A. Laramee and H. D. Durst, Versatile new ion source for the analysis of materials in open air under ambient conditions, *Anal. Chem.*, 2005, **77**(8), 2297–2302.

8. I. Sofer, J. Zhu, H. S. Lee, W. Antos and D. M. Lubman, An atmospheric-pressure glow discharge ionization source, *Appl. Spectrosc.*, 1990, **44**(8), 1391–1398.

9. J. Zhao, J. Zhu and D. M. Lubman, Liquid sample injection using an atmospheric pressure direct current glow discharge ionization source, *Anal. Chem.*, 1992, **64**(13), 1426–1433.

10. J. T. Shelley, J. S. Wiley, G. C. Y. Chan, G. D. Schilling, S. J. Ray and G. M. Hieftje, Characterization of Direct-Current Atmospheric-Pressure Discharges Useful for Ambient Desorption/Ionization Mass Spectrometry, *J. Am. Soc. Mass Spectrom.*, 2009, **20**(5), 837–844.

11. J. T. Shelley, G. C. Y. Chan and G. M. Hieftje, Understanding the Flowing Atmospheric-Pressure Afterglow (FAPA) Ambient Ionization Source through Optical Means, *J. Am. Soc. Mass Spectrom.*, 2012, **23**(2), 407–417.

12. J. T. Shelley, J. S. Wiley and G. M. Hieftje, Ultrasensitive Ambient Mass Spectrometric Analysis with a Pin-to-Capillary Flowing Atmospheric-Pressure Afterglow Source, *Anal. Chem.*, 2011, **83**(14), 5741–5748.

13. K. P. Pfeuffer, J. N. Schaper, J. T. Shelley, S. J. Ray, G. C. Y. Chan, N. H. Bings and G. M. Hieftje, Halo-Shaped Flowing Atmospheric Pressure Afterglow: A Heavenly Design for Simplified Sample Introduction and Improved Ionization in Ambient Mass Spectrometry, *Anal. Chem.*, 2013, **85**(15), 7512–7518.

14. T. Martens, D. Mihailova, J. van Dijk and A. Bogaerts, Theoretical Characterization of an Atmospheric Pressure Glow Discharge Used for Analytical Spectrometry, *Anal. Chem.*, 2009, **81**(21), 9096–9108.

15. G. C. Y. Chan, J. T. Shelley, J. S. Wiley, C. Engelhard, A. U. Jackson, R. G. Cooks and G. M. Hieftje, Elucidation of Reaction Mechanisms Responsible for Afterglow and Reagent-Ion Formation in the Low-Temperature Plasma Probe Ambient Ionization Source, *Anal. Chem.*, 2011, **83**(10), 3675–3686.

16. M. Nefliu, J. N. Smith, A. Venter and R. G. Cooks, Internal energy distributions in desorption electrospray ionization (DESI), *J. Am. Soc. Mass. Spectrom.*, 2008, **19**(3), 420–427.

17. J. T. Shelley and G. M. Hieftje, Ionization matrix effects in plasma-based ambient mass spectrometry sources, *J. Anal. At. Spectrom.*, 2010, **25**(3), 345–350.

18. K. P. Pfeuffer, J. T. Shelley, S. J. Ray and G. M. Hieftje, Visualization of mass transport and heat transfer in the FAPA ambient ionization source, *J. Anal. At. Spectrom.*, 2013, **28**(3), 379–387.

19. G. S. Settles, *Schlieren and Shadowgraph Techniques: Visualizing Phenomena in Transparent Media*, Springer-Verlag, New York, 2001.

20. J. T. Shelley, S. J. Ray and G. M. Hieftje, Laser Ablation Coupled to a Flowing Atmospheric Pressure Afterglow for Ambient Mass Spectral Imaging, *Anal. Chem.*, 2008, **80**(21), 8308–8313.

21. J. H. Yoo, S. H. Jeong, X. L. Mao, R. Greif and R. E. Russo, Evidence for phase-explosion and generation of large particles during high power

nanosecond laser ablation of silicon, *Appl. Phys. Lett.*, 2000, **76**(6), 783–785.

22. J. Koch, S. Schlamp, T. Rösgen, D. Fliegel and D. Günther, Visualization of aerosol particles generated by near infrared nano- and femtosecond laser ablation, *Spectrochim. Acta, Part B.*, 2007, **62**(1), 20–29.

23. A. M. Leach and G. M. Hieftje, Factors Affecting the Production of Fast Transient Signals in Single Shot Laser Ablation Inductively Coupled Plasma Mass Spectrometry, *Appl. Spectrosc.*, 2002, **56**(1), 62–69.

24. K. P. Pfeuffer, J. N. Schaper, J. T. Shelley, S. J. Ray, G. C. Y. Chan, N. H. Bings and G. M. Hieftje, Halo-Shaped Flowing Atmospheric Pressure Afterglow: A Heavenly Design for Simplified Sample Introduction and Improved Ionization in Ambient Mass Spectrometry, *Anal. Chem.*, 2013, **85**(15), 7512–7518.

25. J. T. Shelley and G. M. Hieftje, Fast transient analysis and first-stage collision-induced dissociation with the flowing atmospheric-pressure afterglow ionization source to improve analyte detection and identification, *Analyst*, 2010, **135**(4), 682–687.

26. M. C. Jecklin, S. Schmid, P. L. Urban, A. Amantonico and R. Zenobi, Miniature flowing atmospheric-pressure afterglow ion source for facile interfacing of CE with MS, *Electrophoresis*, 2010, **31**(21), 3597–3605.

27. J. N. Schaper, K. P. Pfeuffer, J. T. Shelley, N. H. Bings and G. M. Hieftje, Drop-on-Demand Sample Introduction System Coupled with the Flowing Atmospheric-Pressure Afterglow for Direct Molecular Analysis of Complex Liquid Microvolume Samples, *Anal. Chem.*, 2012, **84**(21), 9246–9252.

28. J. O. Orlandini, V. Niessen, J. N. Schaper, J. H. Petersen and N. H. Bings, Development and characterization of a thermal inkjet-based aerosol generator for micro-volume sample introduction in analytical atomic spectrometry, *J. Anal. At. Spectrom.*, 2011, **26**(9), 1781–1789.

29. M. C. Jecklin, G. Gamez and R. Zenobi, Fast polymer fingerprinting using flowing afterglow atmospheric pressure glow discharge mass spectrometry, *Analyst*, 2009, **134**(8), 1629–1636.

30. G. D. Schilling, J. T. Shelley, J. H. Barnes, R. P. Sperline, M. B. Denton, C. J. Barinaga, D. W. Koppenaal and G. M. Hieftje, Detection of Positive and Negative Ions from a Flowing Atmospheric Pressure Afterglow Using a Mattauch-Herzog Mass Spectrograph Equipped with a Faraday-Strip Array Detector, *J. Am. Soc. Mass Spectrom.*, 2010, **21**(1), 97–103.

31. D. A. Solyom, O. A. Gron, J. H. Barnes Iv and G. M. Hieftje, Analytical capabilities of an inductively coupled plasma Mattauch–Herzog mass spectrometer, *Spectrochim. Acta, Part B.*, 2001, **56**(9), 1717–1729.

32. M. C. Jecklin, G. Gamez, D. Touboul and R. Zenobi, Atmospheric pressure glow discharge desorption mass spectrometry for rapid screening of pesticides in food, *Rapid Commun. Mass Spectrom.*, 2008, **22**(18), 2791–2798.

33. J. H. Kennedy, K. P. P. G. M. Hieftje, B. C. Laughlin and J. M. Wiseman, Direct analysis of ppb Carbendazim in Orange Juice using the Flowing

Atmospheric-Pressure Afterglow (FAPA) Ambient Mass Spectrometry Source, Paper presented at the 60th American Society for Mass Spectrometry Conference on Mass Spectrometry, Springer, Vancouver, Canada, 2012, p. 151.

34. J. Chipuk and J. Brodbelt, Transmission mode desorption electrospray ionization, *J. Am. Soc. Mass Spectrom.*, 2008, **19**(11), 1612–1620.
35. G. D. Schilling, J. T. Shelley, J. A. C. Broekaert, R. P. Sperline, M. B. Denton, C. J. Barinaga, D. W. Koppenaal and G. M. Hieftje, Use of an ambient ionization flowing atmospheric-pressure afterglow source for elemental analysis through hydride generation, *J. Anal. At. Spectrom.*, 2009, **24**(1), 34–40.
36. W. C. Wetzel, J. A. C. Broekaert and G. M. Hieftje, Determination of arsenic by hydride generation coupled to time-of-flight mass spectrometry with a gas sampling glow discharge, *Spectrochim. Acta, Part B.*, 2002, **57**(6), 1009–1023.
37. M. Smoluch, E. Reszke, A. Ramsza, K. Labuz and J. Silberring, Direct analysis of methcathinone from crude reaction mixture by flowing atmospheric-pressure afterglow mass spectrometry, *Rapid Commun. Mass Spectrom.*, 2012, **26**(13), 1577–1580.

Spray Desorption Collection and DESI Mechanisms

ANDRE R. VENTER,* KEVIN A. DOUGLASS AND
GREGG HASMAN, JR.

Department of Chemistry, Western Michigan University, 1903 W Michigan
Ave, Kalamazoo, MI 49008-5413, USA
*Email: andre.venter@wmich.edu

8.1 Introduction

The so-called *ambient ionization* methods have changed the way samples are analyzed by mass spectrometry. These techniques are frequently said to require "no sample preparation",[1–3] but it would be more precise to say that these methods frequently require no *additional* sample preparation since the sample processing takes place *during* the analysis. Various sample processing steps have been implemented and are coupled in real time with well-known ionization methods, such as electrospray ionization and various forms of atmospheric-pressure chemical ionization. Sample processing steps such as *liquid–solid extraction* (DESI,[1] nano-DESI,[4] SSP[5,6]), *liquid–liquid extraction* (liquid DESI,[7] EESI[8]), *thermal desorption* (DART[9]), *energy sudden desorption* (LAESI,[10] MALDESI,[11] LA-FAPA[12]), and others are frequently used.

There are many advantages to the real-time sample processing of the ambient methods including: (1) The sample processing is incorporated into the analysis step, often leading to a simplified workflow and easier operation, useful especially for analysts without extensive training in analytical chemistry. (2) These processing steps are frequently localized, enabling the spatial distribution of analytes to be determined for chemical microscopy by

New Developments in Mass Spectrometry No. 2
Ambient Ionization Mass Spectrometry
Edited by Marek Domin and Robert Cody
Published by the Royal Society of Chemistry, www.rsc.org

imaging mass spectrometry. (3) They also reduce the potential for interface contamination and carryover problems because samples often remain outside of the analytical system right up to the point of concurrent processing and ionization, which makes these methods eminently suitable for high-throughput analysis.

Soon after the ambient ionization methods were introduced, it was realized that these new desorption sampling procedures are generally useful for sample processing for purposes other than direct mass spectrometric analysis. Liquid–solid extraction by spray desorption collection (SDC)[13] and energy sudden processing by laser-ablation transfer into a flowing stream for LC-MS[14] are examples of the application of ambient processing for purposes other than direct mass spectrometric analysis. There are many advantages to this microlocalized "ambient" sample processing both over ambient ionization mass spectrometry, as well as over more traditional sample preparation procedures. In the remainder of this chapter some of these advantages will be highlighted by focusing only on the liquid–solid extraction technique known as *spray desorption collection* (SDC),[13,15–18] although many of these points are also relevant to other desorption sample-processing methods.

SDC is different from the ambient ionization methods, in that it (1) decouples the liquids and gasses needed for processing (extraction and desorption) from those needed for efficient ionization. The separation of these processes leads to an easy coupling of large surface area sample collection with mass spectrometry, potentially even for miniature mass spectrometers with small vacuum pumps and limited pumping capacity. (2) The ability to use other methods of analyte detection in addition to MS, such as UV-Vis, scanning electron microscopy, *etc.*, or the separation of complex mixtures by GC or HPLC before detection. (3) The absence of an applied potential during collection prevents static surface charge build-up during analysis when nonconductive surfaces are used, and the neutralization of the ionizing spray from conductive surfaces is not a concern. (4) Preconcentration of low-level surface compounds, even below the limit of detection by direct ambient ionization methods, before analysis.

Compared to traditional sample preparation methods, such as solvent rinsing and swabbing, SDC offers the advantages of: (1) Reduced use of extraction solvents to recover analyte, *e.g.* from swabs, (2) the prevention of losses due to inefficient recovery from swabs, (3) sample collection from very large surface areas, (4) prevention of transfer of collected material back onto the sample surface, (5) and a simplified sample-preparation workflow, especially when the collection surface is chosen to be the analytical device needed for the next step, such as when collecting into a UV cuvette, *etc*.

In this chapter these advantages and various aspects of SDC are discussed in detail, including the use of SDC to study the mechanisms of spray desorption methods such as DESI and the generally useful applications of spray desorption for sample collection for off-line analysis by a variety of detection methods.

8.2 Using SDC and other Methods to Study the DESI Mechanism

The mechanistic aspects of DESI are described in detail in Chapter 12. In this section, we illustrate that by decoupling the extraction and desorption events from ionization and ion transport, it is possible to study these two processes separately. The techniques used for this comparative study are illustrated in Figure 8.1. The extraction and desorption relevant conditions can be studied using SDC, where the material extracted and desorbed is collected on a specifically chosen collection surface, or other collector, that can be analyzed off-line to quantify the amount of material removed from the surface. The ionization and ion-transport conditions can, in some fashion, be studied by reflective electrospray ionization (RESI-MS), where a prepared solution of analyte is electrosprayed and the spray is bounced off a clean, DESI-appropriate surface before being analyzed in real time by the mass spectrometer. The sum of these two sets of processes represent the DESI experiment.[16] By using SDC and RESI-MS together, the relative contribution to signal intensities of the different processes in DESI-MS can be studied in a comparative way.

When comparing SDC and RESI-MS to DESI-MS, it is important to ensure that all geometries, distances, and other operating conditions that are not being investigated are matched exactly. In this way, several parameters critical to the optimization of the DESI source have been investigated by SDC-ESI-MS, RESI-MS, and DESI-MS to determine their relative contributions to the spectral signal intensity.[16]

The MS intensity of rhodamine 6G collected from a glass surface by SDC was monitored by extraction and UV-Vis analysis, or analyzed directly from the glass surface by DESI-MS or RESI-MS, while ensuring source conditions were as identical as possible for the similar modes of analysis. For several of the curves, large standard errors were observed for some data points.

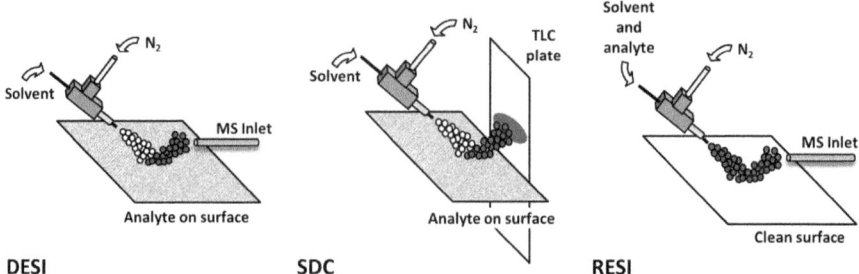

Figure 8.1 During DESI, analyte present on the surface is desorbed and analyzed immediately by MS. During SDC, analyte is present on the surface and collected *via* spray desorption onto a secondary surface (collector). During RESI, a solution containing analyte is reflected off a clean surface and analyzed by MS.

With kind permission from Springer Science and Business Media.

It is well known that signal intensities are highly variable during DESI, even when using optimized conditions,[19–25] with RSDs of <10% typically only achieved by the inclusion of an internal standard.[22] Since a large parameter space was investigated, with many data points acquired distant from the optimum source configuration, large standard errors are to be expected. In addition, many parameters are interrelated so that an adjustment of one parameter requires an adjustment of others to achieve the optimum source configuration. Regardless, by comparing each mode of analysis side-by-side, relative differences in the response to parameter changes on signal intensities can be interpreted for their contributions to either desorption or ionization processes because of their isolation by SDC and RESI, respectively.

Here, such an analysis was used to study the applied potential, gas flow rates, and distances between sprayer, surface and MS ion-sampling orifice.

8.2.1 Applied Potential

Electrosprays are typically formed by directly applying a high potential between the solvent (sample) emitter and a ground electrode, usually the mass-spectrometer inlet. During DESI-MS analysis, instrumental response increases with increasing spray potential,[21,26–28] though a signal can be obtained even without the application of a high potential.[19,27–30] Increasing the potential produces smaller droplets with higher velocities,[26] which results in higher ionization efficiency. However, high potentials can also result in an increased production of oxidative species within the electrospray, causing oxidation products to appear in the spectra,[31–33] which increases spectral complexity.

When the applied potential is optimized between +0 and +8 kV (Figure 8.2), it is apparent from the level SDC curve that desorption does not significantly depend on a high potential applied during DESI-MS. In contrast, both the RESI and DESI curves follow a trend of increasing signal intensity with increasing applied potential. The SDC result confirms that material is desorbing equally at all potential values, while the RESI curve's close match to the DESI curve indicates that the ion signal is limited by either ionization or ion-transport processes. According to the droplet-pickup model,[34] the desorption process occurs (1) by the formation of a thin liquid film into which analyte on the surface is dissolved followed by (2) a momentum-transfer event through which charged secondary droplets containing analyte are ejected from the film towards the mass-spectrometer inlet. Neither of these processes is aided significantly by applying a potential. In fact, some studies have indicated that surface charging poses negative effects to desorption.[29] These results are in good agreement with previous studies showing that an electrostatic contribution is not necessary to replicate the droplet impact characteristics observed during DESI[35] and that efficient desorption occurs during desorption sonic spray ionization,[19] which utilizes no applied potential during analysis.

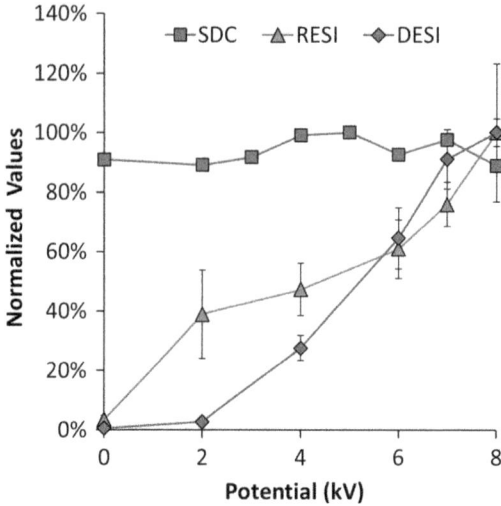

Figure 8.2 Changes in internally normalized signal intensities to changes in applied spray potential for SDC + ESI-MS, RESI-MS, and DESI-MS. With kind permission from Springer Science and Business Media.

8.2.2 Gas Flow Rate

In a typical DESI sprayer, the sheath gas exits around the inner solvent capillary, nebulizing the eluting solvent to form a heterogeneous population of charged droplets.[26] The velocity of the primary charged droplet plume increases with increasing sheath gas flow rate,[26] which is proportional to the sheath gas back pressure and the relative diameters of the capillaries. It is usually observed for DESI that an increase in the signal intensity is observed with increasing gas pressure up to an optimum value, after which instrumental response remains constant or begins to decrease.[21,23,26,27]

The sheath gas pressure was optimized[16] between 100 psi and 180 psi (Figure 8.3), which corresponded to measured nebulizing gas volumetric flow rates of 0.96 L min^{-1} and 1.72 L min^{-1} at the emitter. The DESI curve follows the expected trend of a maximal increase, with a maximum observed at 160 psi. A similar, though less marked, trend is observed for SDC. The curve for RESI is at a maximum at the lowest gas flow rate when the gas pressure is too low to direct the droplets to the surface and instead the spray is sampled directly by the inlet, without surface collision. Once the pressure is increased sufficiently, the droplets are mostly directed towards the surface, and a sudden decrease in signal is observed due to surface interaction leading to spray-diffusion, and possible analyte losses to the surface. The curve then increases linearly with increasing gas flow rate. The similar trends observed for SDC and DESI, along with the disparate trend observed for RESI, suggest that the gas flow rate has a larger impact on desorption processes during DESI than it does for the ionization or ion-transport processes. However, the sharper increase observed for the DESI curve, as

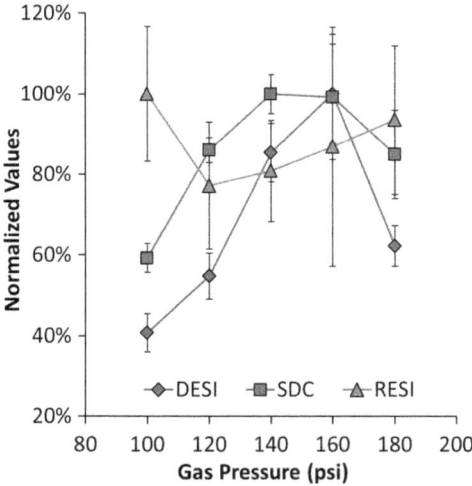

Figure 8.3 Changes in internally normalized signal intensities to changes in nebulizing gas pressures for SDC + ESI-MS, RESI-MS, and DESI-MS.
With kind permission from Springer Science and Business Media.

compared to the SDC curve, can be explained by the steady increase in intensity observed for the RESI curve, demonstrating how these independent
processes combine to influence the DESI experiment as a whole.

The effect of gas flow rate on desorption (Figure 8.3, SDC curve) can be
explained by the decreasing primary droplet velocity at low gas flow rates
resulting in poor momentum transfer. A threshold velocity is required for
the efficient production of secondary droplets after surface collision.[26,34,35]
On the other hand, at gas flow rates higher than the optimum value focusing
of the spray plume and increased droplet evaporation reduce the sample
spot size[26,36] and hence the amount of desorbed material. In addition,
sample redeposition can occur,[24] reducing the DESI-MS signal.

The ionization efficiency, characterized by the RESI curve, appears
anomalously high at the lowest gas flow rate as explained above. For all other
data points, ionization efficiency increases slightly with increasing gas flow
rate,[16] which could be due to the formation of smaller secondary droplets
with higher ionization efficiencies or a change in the geometrical aspects of
the spray, such as a change in the secondary droplet take-off trajectory.[26]
This slight increase in ionization efficiency with increasing gas pressure
does not appear enough to compensate for the decrease in desorption efficiency at high gas flow rates though.

8.2.3 Spray Emitter to Impact Zone Distance

Primary droplets leave the spray emitter at velocities in excess of 100 m s^{-1},
but drag forces slow the droplets down and decrease the kinetic energy of
droplets impacting the surface, which also effects the velocity of droplets

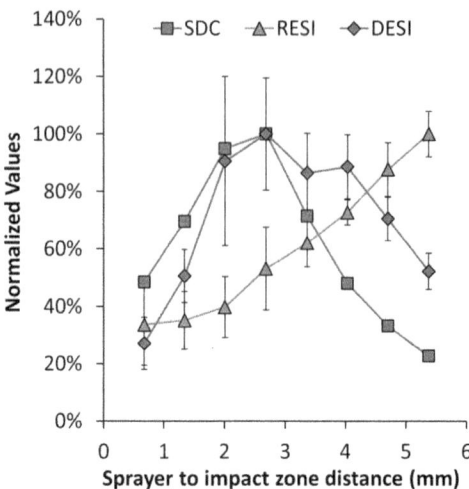

Figure 8.4   Changes in internally normalized signal intensities to changes in sprayer to impact zone distances for SDC + ESI-MS, RESI-MS, and DESI-MS. With kind permission from Springer Science and Business Media.

leaving the surface.[26] To monitor the resulting impact on the DESI de-sorption and ionization processes, the distance between the impact zone and mass inlet or collection surface was kept constant while the distance between the sprayer and the impact zone was varied from 1 mm to 8 mm for SDC, RESI and DESI as shown in Figure 8.4.[16]

For DESI-MS, instrumental response initially increases with increasing spray emitter-to-inlet distance up to 3 mm, followed by a ridge at 4 mm, after which signal intensity decreases. This interesting dependence of the DESI-MS signal on the spray emitter-to-impact zone distance can be explained by the combined contributions of both desorption and ionization efficiencies.[16]

The amount of analyte collected by SDC initially increases to a maximum at 3 mm sprayer-to-impact zone distance, after which the curve continuously decreases. At short distances, the nebulizing gas and droplet velocities are high, which can result in the sheeting of solvent along the surface instead of efficient secondary droplet formation. As the droplet velocities decrease with increasing distance, sheeting is reduced and desorption efficiency increases until the optimum value is achieved. Beyond this distance, droplet size and velocity along with the desorption efficiency continue to decrease, since at these large distances the droplets may contain too little kinetic energy once they reach the surface to effectively remove material. Some desorbed material may also redeposit between the impact zone and mass-spectrometer inlet.

Unlike for SDC, the RESI-MS signal steadily increases with increasing spray emitter-to-impact zone distance. The ionization efficiency from smaller droplets is known to be better, especially for lower molecular weight ions such as rhodamine-6G, which are assumed to be leaving droplets by the ion evaporation model.[37] As the distance between the emitter and inlet increases, the droplet size decreases due to solvent evaporation. This presumably leads to an increase in ion production and ion sampling efficiency across the entire range of spray emitter-to-impact zone distances typical for DESI-MS analysis.

The dependence of the DESI-MS response on the spray emitter-to-impact zone distance is a composite between desorption and ionization processes. Initially, the response is low due to low desorption efficiency, as shown by the SDC results. The DESI-MS response increases as desorption efficiency increases, although it increases more steeply due to the concurrent increase in ionization efficiency shown by the RESI-MS curve. DESI-MS response continues to remain high after 3 mm despite the decrease in desorption efficiency due to the increasing ionization and ion transfer efficiencies. After 4 mm, however, the benefit of increasing ionization efficiency becomes irrelevant as the decrease in desorption efficiency leads to there being little analyte available to ionize.

8.2.4 Impact Zone to Inlet Distance

Secondary droplets leave the impact zone close to the surface.[35] Aerodynamic drag decreases their velocity as they travel from the impact zone and they begin to increase in height due to dispersion.[26] This change in velocity and dispersion is not expected to substantially change the SDC response, and as can be seen in Figure 8.5[16] the SDC response falls by less than 20% from its optimal value at the shortest distance that could practically be tested. At the shortest distance, the collected band is smallest and it then gets larger with distance as nebulizing gas begins to disperse. This is not a problem for SDC when the entire collection band is analyzed subsequently. During DESI-MS, this dispersion reduces the density of droplets and ions near the MS sampling orifice and collection efficiency of the produced progeny droplets may be reduced, although sampling is assisted by the pressure differential between ambient conditions and the mass-spectrometer inlet.[36]

The RESI-MS curve in Figure 8.5 increases with increasing distance until a maximum is achieved between 3 and 4 mm, after which the curve decreases. At very short distances between the sprayer and inlet, the secondary droplets might not achieve the height necessary to avoid hitting the bottom rim of the inlet capillary. Kertesz and Van Berkel have shown that the "edge sampling" geometry gives the best response for this reason.[38] The dispersion of the secondary droplets increases with increasing distance from the impact zone,[26] so the sampling of progeny droplets and ions by the inlet also increases as they gain height from the surface with increasing distance. In addition, longer flight times allow more time for evaporation and ion production from the progeny droplets. The final decrease in the RESI curve is likely due to the decreasing sampling efficiency by mass-spectrometer inlet as the dispersion of the plume increases.

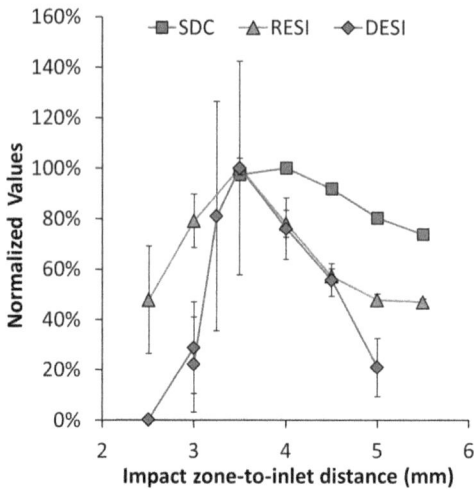

Figure 8.5   Changes in internally normalized signal intensities to changes in the impact-zone-to-inlet distances for SDC + ESI-MS, RESI-MS, and DESI-MS. With kind permission from Springer Science and Business Media.

The DESI curve follows the same trend as the RESI curve, although it is initially lower. For both techniques, a portion of the charged droplets bypass the impact zone and fly directly toward the inlet due to the high electric field at short distances.[25] This results in a decreased desorption efficiency and low response during DESI-MS while for RESI the response is higher due to the presence of analyte ions in the spray solution. The impact zone-to-inlet distance appears to affect the DESI-MS response mostly because of the ion transport and dispersion processes occurring after desorption.

In conclusion, the examples presented in this section illustrate how the separation of desorption and ionization by tools such as SDC and RESI can be used to better understand the effects that various operational parameters have on the overall DESI-MS response. It is by no means an exhaustive study of all the operating conditions, but it is interesting to observe how desorption and ionization events respond differently to changes in analysis conditions. This level of understanding should help researchers to respond appropriately to specific analysis conditions. In the following section, the problematic analysis of proteins by DESI-MS is investigated using a similar approach.

8.3   Investigating Problems Encountered with Protein Analysis by DESI-MS using SDC and RESI-MS

Protein analysis by DESI-MS displays a mass-dependent response curve, where instrumental response decreases with increasing molecular mass (Figure 8.6).[27,39] Originally, this mass dependence was attributed to the inability of larger proteins to effectively desorb from the sample surface.[39–41]

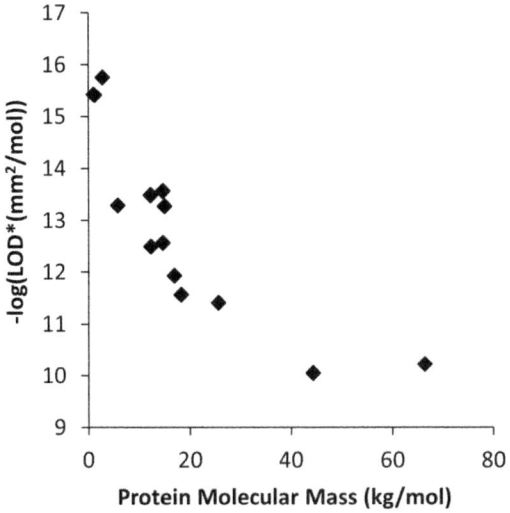

Figure 8.6   A logarithmic plot of the DESI-MS detection limits for proteins spanning a range of molecular masses shows a trend of decreasing instrumental sensitivity with increasing molecular mass.
With kind permission from Wiley.

The SDC and RESI-MS tools introduced in the previous section can also be used to investigate this mass-dependent response relative to desorption efficiency.[15] For these experiments, equimolar mixtures of proteins were characterized directly by ESI-MS. The same sample was deposited onto a porous PTFE surface and collected by SDC to study the desorption process independently of ionization. After SDC, the collected sample was again analyzed under identical ESI-MS conditions. To study the ionization process independently of desorption, RESI was used along with traditional ESI, both of which are independent of desorption. ESI alone cannot be used to study the ionization process of DESI since the distributions of size and velocity, as well as the spatial distribution, of droplets from which ionization occurs during DESI are markedly different from those of droplets generated by an ESI source.[26] During DESI, a droplet's collision with the surface also reduces the amount of charge it carries due to neutralization on the surface.[36] RESI mimics the production of secondary droplets in DESI without requiring desorption of the analyte. The relative abundances of the individual proteins are compared between the spectra for all the techniques: ESI, SDC, RESI, and DESI in Figure 8.7.

When these samples, collected by SDC, are analyzed with ESI-MS, the same relative intensities of the proteins are observed as in the direct ESI-MS experiment. This clearly indicates that the equimolar ratio of proteins is conserved during the desorption and collection processes and that the efficiency of protein *desorption* is independent of protein size for the range of proteins which were investigated, from 12 to 66 kDa.[15] Other reasons are required to explain the lack of sensitivity observed for larger proteins by DESI-MS. To investigate whether ionization plays a part, RESI-MS

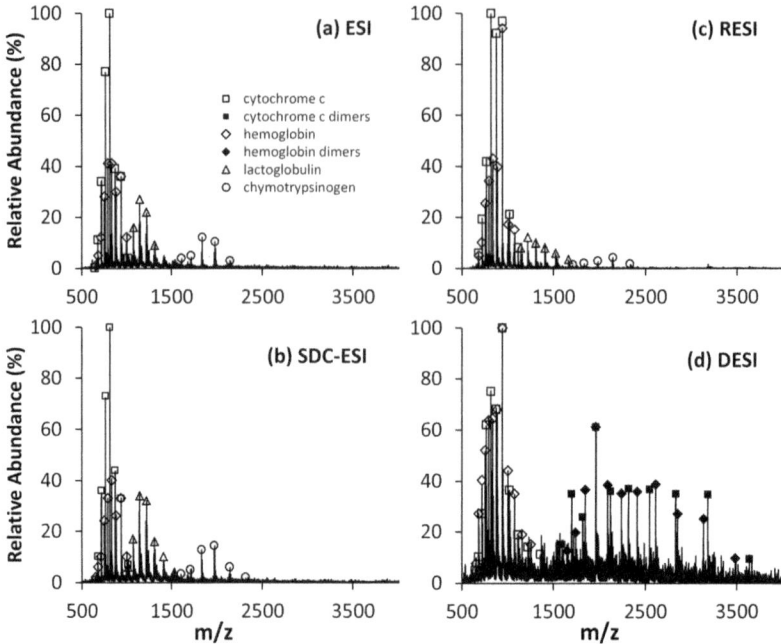

Figure 8.7 Spectra of an equimolar mixture of cytochrome c, hemoglobin, β-lactoglobulin, and α-chymotrypsinogen analyzed by (a) ESI-MS, (b) SDC + ESI-MS, (c) RESI-MS, and (d) DESI-MS. With kind permission from Wiley.

can be performed. The RESI-MS spectra are very similar to the direct ESI-MS spectra. A slight shift in charge-state distributions is observed, consistent with a small amount of charge neutralization of the droplets following their collisions with the surface. The slight loss in charging observed by RESI is not enough to account for the decrease in detection limits for DESI-MS by itself. Since the droplet interactions with the surface and the postcollisional spray distribution of DESI are mimicked by RESI, ionization and ion-transmission events occurring postdesorption are also ruled out as a leading cause of the loss in sensitivity for larger proteins. However, the DESI-MS spectra are clearly different from the direct ESI-MS spectra: no peaks corresponding to β-lactogloblulin and α-chymotrypsinogen are observed, while peaks corresponding to dimers of cytochrome c and hemoglobin are present. There must be a component of DESI that is not captured by the SDC and RESI tests, *i.e.* a process intrinsic to the overall process that is not solely due to protein removal, ionization, or ion transfer.

A major difference between DESI-MS and the SDC-ESI-MS and RESI-MS experiments is the time in which the analyte is present in the liquid phase before ionization. During RESI, the analyte is prepared in the spray solution prior to collision with the surface. Similarly, the ESI-MS analysis of samples prepared from SDC experiments requires the collected analyte to be predissolved in the spray solution. Neither method captures the transition of the analyte from a dry solid present on the surface to an aqueous solution on the short time-scale of DESI-MS. Other ambient methods that use an electrospray to generate ions, but use extraction methods allowing more time for analyte dissolution, have been found to be more amenable to the study of proteins compared to traditional DESI-MS. For example, DESI-MS becomes amenable to the study of proteins up to 150 kDa if a liquid sample is analyzed directly instead of a dry one.[7,42]

Similarly, the liquid microjunction surface sampling probe (LMJ-SSP)[6] and the nano-DESI source[4] allow longer dissolution times prior to ionization and are able to desorb proteins for successful analysis by mass spectrometry.[4,6,43] Instead of directing an electrospray at a surface as in DESI, these methods use liquid microjunctions to dissolve and desorb material. Unlike during DESI-MS, where the time between desorption and analysis is short and the sample is usually rapidly consumed,[24] there is a considerable delay between extraction and ionization of the analyte during analysis by the LMJ-SSP and nano-DESI techniques while the dissolved material traverses through the capillary *en route* to the exit where ionization occurs. When Laskin and coworkers monitored the time-dependent intensity of rhodamine 6G (443 g mol^{-1}), reserpine (608 g mol^{-1}), and cytochrome c ($\sim$12 300 g mol^{-1}) by nano-DESI-MS,[4] the times required to reach maximum instrumental response were 7, 30, and 90 s, respectively. For comparison, the maximum instrumental response for rhodamine 6G from glass was observed immediately at the onset of DESI-MS by Green *et al.*[24] It is clear that sufficient desolvation time prior to the ionization event is necessary for the efficient ionization of proteins by ambient methods using electrospray ionization.

This phenomenon can be explained by considering the typical DESI-MS sample, where material has been deposited on a surface and allowed to dry, or is naturally in a dry state. As protein samples dry, the proteins form protein–protein and protein–surface interactions, as well as interactions with salts and other impurities that are present in the protein samples and LC-MS grade solvents used to prepare them. During DESI-MS, there is little time for solvation to occur, compounded by the rapid removal of the sample from the surface. The protein–protein and protein–contaminant interactions can survive after material is removed from the surface, leading to inefficient ionization and a loss of signal. Some proteins, such as cytochrome c, appear to have a low propensity to form adduction with contaminant species during DESI-MS for reasons possibly dependent on individual protein characteristics, but it appears that there is a trend where larger proteins suffer more from this effect.[15]

8.4 Spray Desorption Collection for Noncontact Spray Swabbing

Originally purposed for the direct analysis by MS when performing DESI analysis,[1] the spray desorption method of sample processing offers many generally useful advantages over other more traditional implementations of solid–liquid extraction and surface sampling. In a traditional analytical protocol, after liquid extraction, excess solvent is removed to reconcentrate analyte by time-consuming solvent evaporation or solid-phase extraction procedures. When material needs to be sampled from surfaces, collection by swabbing also has to take place before solid–liquid extraction.

On the other hand, when spray swabbing by SDC, these actions take place concurrently with surface sampling and as the analyte is removed from sample surfaces and collected, it is concentrated into a narrow, well-defined band on a solid substrate. In the ideal case, the collection surface is chosen to be the analytical device needed for the subsequent detection or quantification method so that no further sample processing is needed.

Moreover, the original sample remains intact, without the need to grind or homogenize it before extraction. Since analyte material is collected onto a solid substrate, the choice of this collection substrate needs careful consideration to prevent sample recovery issues, similar to those experienced when extracting collected material from traditional wipes or swabs. Various collection surfaces are discussed below, both for the recovery of material from the collection surface and for when collecting onto a surface directly useable in the subsequent detection step.

8.4.1 Analysis after Spray Swab Collection

8.4.1.1 Quantitation of Analyte Collected by SDC

During collection by SDC, the desorbed material is usually deposited in a narrow band near the horizon of the sample surface. This well-defined morphology of the collected zone makes direct analysis by a number of techniques straightforward. For example, the collected band can be analyzed by aligning the desorption band with the fiber of a solid phase microextraction (SPME) needle,[17] by collecting onto the tip of a paper triangle for analysis by paper spray ionization,[18] or by many of the ambient desorption methods, such as DART- or DESI-MS.[13] In the latter case one can either scan through the narrow width, as shown in the background of Figure 8.8, or through the length because of the oblong shape.

Care should be taken when analyzing only a small portion of the collected material, however, as this can lead to a loss in quantitative information. In Figure 8.8, when scanning through the width of bands of rhodamine-6G, collected from increasingly larger areas, the calibration curve soon reaches saturation. However, when the bands are analyzed by DESI-MS while scanning through the length, or when the bands are physically removed, extracted, and analyzed by UV-Vis, a linear calibration curve of collection area to amount of detected material is observed for analyte evenly distributed over the sample surface.[13] Compared to swabs, the collection surface does not touch the sample surface and so analyte transfer can be in the collection direction only, eliminating the possibility of back-transfer of collected

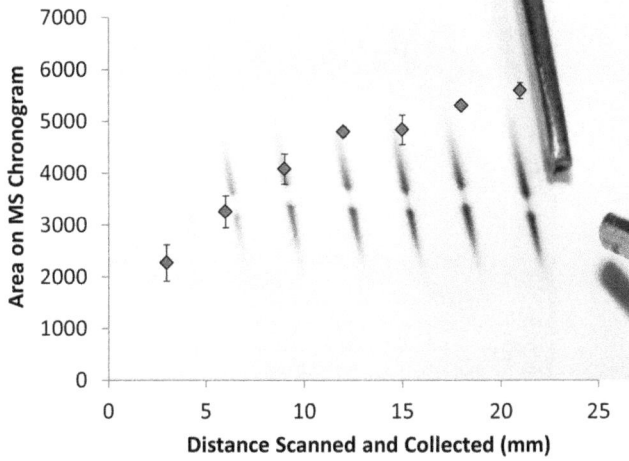

Figure 8.8 With an increase in the collection area, the amount of material collected increases linearly (see text). However, when a small portion of the collection band is analyzed, the results may not always be quantitative. Here, rhodamine was collected over increasing distances and the bands were analyzed by scanning through the width using DESI-MS. With kind permission from Wiley.

material onto the sample surface. This feature enables one to collect and concentrate material from very large surface areas onto a single narrow band, and to determine the average surface concentration, even for widely distributed compounds.[18]

8.4.1.2 Methods of Detection

Because spray swabbing by SDC is a general sample-collection technique, the potential methods for detection are unlimited. The most appropriate detection method is dictated by sample requirements such as desired limits of detection, analyte properties, availability of instrumentation, and other factors unrelated to SDC.

Once collected, the material can be extracted, leaving a liquid sample that can be handled in the same way as any other liquid sample, and a few applications are discussed below. However, by extracting the collected sample, one loses some of the benefits of spray swabbing by reincorporating additional steps prior to analysis and by diluting the collected material.

The full potential of SDC is realized when analyte is spray swabbed onto a target that is already the object needed for sample introduction for the chosen method of detection. Illustrating this, it was shown that collection can take place into sample vials and UV-cuvettes, or onto UV-cuvette surfaces, SPME fibers for separation and detection by GC-MS, PTFE targets for thermal desorption units attached to ion-mobility spectrometers, *etc.* Many other potential applications, as yet undemonstrated, are possible.

Figure 8.9   SDC with colorimetric detection. Ascorbic acid is collected onto blue
litmus paper, causing a color change in the indicating collector.

Examples could include collection onto MALDI plates, microfluidic devices,
XRF targets, screens for DART-MS, transmission DESI-MS analysis, *etc.*

   Another interesting detection approach is to make the collection surface a
sensor in itself. These sensors can be sophisticated detectors, such as a mass
spectrometer, in which case SDC reverts back to its parent technique, DESI-
MS. However, it is also possible to spray-swab directly onto other types of
sensors. A simple example of spray desorption detection (SDD) is demonstrated in Figure 8.9, where the nonvolatile analyte, ascorbic acid, is spray-
swabbed from the sample surface onto a colorimetric sensor (blue litmus
paper) causing an immediate response and an indication of the acidity of the
sample surface. Other more sophisticated SDD approaches are under
development, such as spray swabbing onto more specific and selective
chemical and electronic sensors.

8.4.2   Collectors

With SDC, after the solvent impacts the surface, chemical compounds are
extracted, desorbed and captured onto a collector. The specific requirements
for an ideal collector depend to a large extent on the choice in detection
method and the properties of the chemical compounds of interest. In some
cases, the approach may require extraction of the analyte from the collector
to produce a liquid sample, which adds the need for high recovery efficiency
to the list of requirements. Examples of surfaces both for direct analyses
from collectors and for sample extraction are presented below.

8.4.2.1 Collectors for Extraction and Dilution

Chromatography paper has proven to be a very useful collector material. It is flat, chemically inactive, pure, absorbent, and for colored analyte it is easy to see the collection band on its white surface. Extraction efficiencies are also very high, and although a variety of fiber materials are available to further tailor recoveries, Whatman #1 is a good starting point.

After analyte collection by SDC, the collected band can easily be cut out, rolled into a loose spiral and extracted with minimal solvent, sometimes with the aid of sonication, directly into an autosampler sample vial or other small vessel for analysis. Porous-PTFE sheets, which possess a high surface area and superior chemical inertness, offer a similar approach. It is also possible to collect onto thin-layer chromatography (TLC) plates.[16] In this case, one can physically remove the band and sonicate the stationary phase particles to recover collected chemicals. Prior to analysis, the particles should be removed by centrifugation or filtration.

Excellent quantitative results are possible using extraction after collection by spray swabbing, and little further method development is required if traditional analysis methods for the compounds of interest are already in place. This approach was used in the mechanistic studies of DESI, presented in Section 8.2.[16]

An even simpler method than extraction after collection is the direct collection of analyte into sample vials or cuvettes. The dry, collected material is then dissolved in a minimum volume of solvent and is ready for analysis. Depending on the configuration of the sprayers and the size of the opening into the vial or cuvette, a pressure barrier may develop, preventing efficient collection into the vial.

This problem was alleviated during the collection of proteins by SDC, discussed in Section 8.3 of this chapter, by drilling two pressure-relief holes in the back of the Eppendorf tube used for collection, as shown in Figure 8.10(a). In that case, after collection, the material was washed into a second tube after collection with minimal solvent and analyzed by ESI-MS.[15] Figure 8.10(b) shows the rinsing of collected material from a flattened Eppendorf tube with two pressure-relief holes drilled into the bottom into an intact Eppendorf tube for subsequent analysis.[15]

8.4.2.2 Collectors for Direct Analysis

In the ideal case, after collection by spray swabbing, no further processing is necessary and the detection takes place directly from the collector surface. DESI-MS detection was introduced in Section 8.4.1.1, and many other direct analysis combinations with SDC are possible. In this section, two additional examples are presented: Direct analysis of collected material by paper spray-MS (PS-MS)[44] and analysis by introducing material collected onto a SPME fiber into a gas chromatograph for separation before MS detection.

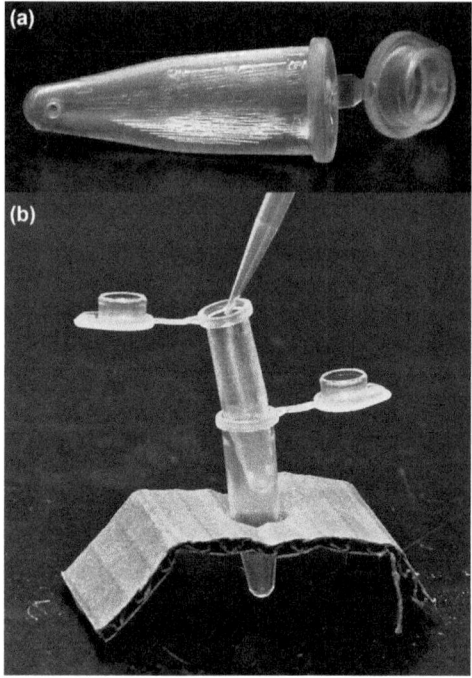

Figure 8.10   (a) An Eppendorf tube used for SDC collection showing the pressure-relief holes drilled into the bottom. (b) Sample collected into the top tube is rinsed into a new tube for subsequent analysis.

8.4.2.2.1   SDC + PS-MS.[18]   Spray desorption followed by collection onto a paper substrate was introduced above, where the collected material can be analyzed after extraction. Having collected material onto a paper substrate, yet another solid–liquid extraction-based ambient ionization method known as *paper spray ionization* can directly be used for analysis. In this case, the collected band of analyte is cut from the collector into a small triangle, to which solvent and a high voltage is applied. The solvent extracts collected analyte from the paper, and this solution flows towards the corner of the triangle pointing at the mass spectrometer where electrospray ionization occurs.

Due to the small volume of solvent and the short time for extraction, extraction efficiencies during PS-MS can suffer.[45] Nevertheless, excellent lower limits of detection are being reported.[44–46] When used in combination with SDC, PS-MS provides even greater sensitivities compared to swabbing with the paper triangle,[46] due to the way that SDC collects the desorbed material in a narrow concentrated band, which can be positioned directly at the spray tip rather than distributed over the whole paper surface.[18]

Spray swabbing before PS-MS also decouples the high volume of nebulizing gas needed for efficient analyte desorption from the conditions during the ionization process. During ionization with PS-MS, very small droplets are produced, reminiscent of those produced during nanospray ionization,[47] with subsequent improvements in ionization efficiency and without the need for a nebulizing gas. This feature makes paper spray ionization very useful for coupling to miniature mass spectrometers.[48] The combination of large-area surface sampling by SDC and convenient ionization by paper spray coupled to mini-MS will allow for *in situ* surface analysis by mass spectrometry, outside of a laboratory setting.

8.4.2.2 SDC-SPME + GC-MS.[17] Solid-phase microextraction (SPME)[49,50] and its modifications[51] are widely used to collect chemical compounds by ad- or absorption from liquid[50] or gaseous samples.[52] In combination with SDC, SPME can now also be used to collect semi- and nonvolatile compounds directly from surfaces without heating.[17] Collecting by spray swabbing onto an SPME-fiber allows easy integration of surface sampling for less volatile compounds with chromatography or other methods by using standard sample-injection procedures.

One example is the analysis of an important feedstock for biodiesel production, trap grease.[53] Here, using SDC-SPME + GC-MS significantly reduces the sample processing needed prior to analysis and prevents recurrent fouling of the GC-injector port when analyzing samples prepared by standard methods.[17] In comparison to headspace-SPME + GC-MS, SDC-SPME + GC-MS was able to provide information on both the undesirable odiferous compounds and the energy-bearing free fatty acids directly from trap grease.[17]

8.4.3 Desorption and Collection Selectivity

Every step of the process has an effect on the selectivity attainable by SDC. Selectivity during desorption is primarily influenced by solubility effects in the spray solvent, while the collection efficiency is influenced by the affinity of the analyte for the collector surface, the analyte volatility, and other effects as described below. Finally, detector selectivity depends on the choice in method of analysis, the unlimited possibilities of which fall outside of the scope of this chapter.

Analyte solubility plays a key role in the spray desorption methods, such as DESI[23,27,37] and SDC, since analyte must first be extracted from the solid substrate and then dissolve into a liquid film formed by the spray solution, before desorption can occur by momentum transfer from later arriving droplets.[34] A simple change in the concentration of the organic modifier can have a noticeable effect on the range of compounds observed in the DESI mass spectrum or collected by SDC. For example, the DESI-MS response for phenylalanine varies by several orders of magnitude when the organic fraction of the spray solvent, methanol, is varied from 0.1 to 0.9.[37] This difference can be ascribed to improved desorption and/or ionization. A similar comparison, again using the SDC and RESI tools developed in Section 8.2, would be able to differentiate between the relative effects of these two factors. An even more dramatic change in compounds observed during the DESI-MS analysis can be effected by selecting different solvents, such as chloroform and tetrahydrofuran, especially in the analysis of hydrophobic compounds.[54] Different solvent modifiers can also be used, and Cooks and coworkers demonstrated that the relative intensities of two alkaloids present in the stem of the poison hemlock plant can be inverted by switching between aqueous solutions containing 1M ammonium hydroxide or 0.2% formaldehyde.[23]

SDC has the advantage over DESI in that the solvent does not need to meet the requirements of efficient ionization after desorption. The choice of solvent for an SDC experiment is determined only by volatility and the solubility of the compounds of interest, and is not limited by solvent polarities or surface tensions. This freedom from solvent constraints should allow SDC to be even more selective in the compounds that are desorbed and collected, and allow for collection of a wider range of compounds than typically amenable to DESI-MS or ESI-MS analysis.

Analyte volatility plays an important role in collection efficiency. During SDC, the nebulizing gas is not only used for spray formation and droplet acceleration, but also to continuously remove solvent so that the compounds of interest are collected as a dry material on the collector. It stands to reason that compounds with a high volatility, approaching those of the solvent, will also be purged from the collector. This effect can be observed by comparing the relative responses for a homologous series of compounds before and after collection by SDC onto a SPME fiber.[18] Figure 8.11 shows the results for free fatty acids (FFA) between C_4 and C_{12} analyzed by GC-MS. The C_{11}-fatty acid was absent and octanoic acid-methyl ester was added.

In comparison to the direct analysis of the mixture (Figure 8.11(a)) the most volatile compound in the mixture, octanoic acid-methyl ester, is reduced by about 50% after SDC-collection (Figure 8.11(b). The other fatty acids up to decanoic acid are present in similar relative concentrations as in the original mixture. On the other end of the volatility scale, the C_{12} FFA is also absent from the postcollection sample. The absence of dodecanoic acid relates to the loss in solubility in the spray solvent as the hydrophobicity of the compounds increases, as described above.

For comparison, the result for headspace-SPME analysis by GC-MS of the same sample is also shown. The more volatile components are present at much higher relative concentrations than in the original mixture, and the signal decreases sharply with increasing carbon chain length. This highlights the complementarity of headspace- and SDC-SPME collection.

8.4.4 Large Surface Area Sampling and Sprayer Multiplexing

The drive for many ambient ionization techniques has been to reduce the area of the surface that is analyzed at any one point in time in order to improve the lateral resolution attainable during chemical microscopy experiments.[55] On the other hand, there are many applications where it would

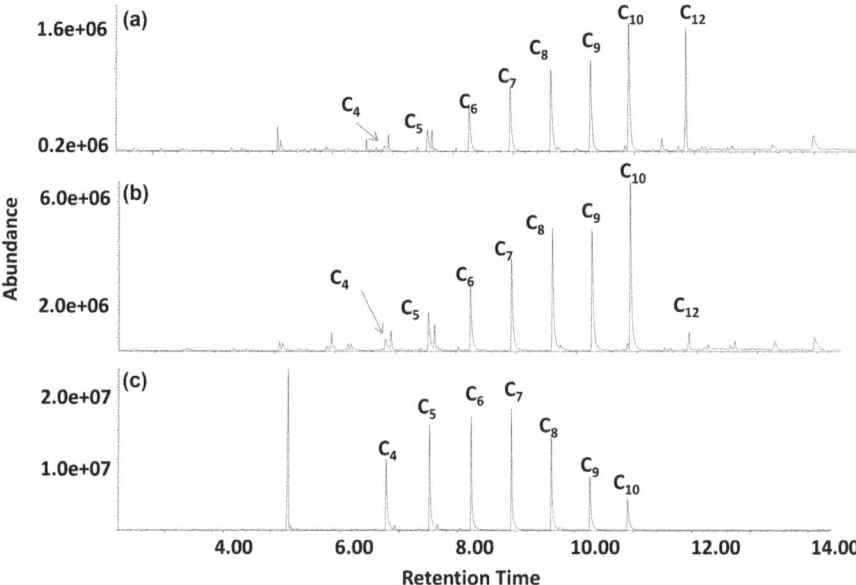

Figure 8.11 GC-MS responses for each FFA and octanoic acid methyl ester by (a) direct SPME-GCMS injection, (b) SDC-SPME + GC-MS, and (c) headspace SPME + GC-MS.

be beneficial to interrogate a large area all at once, including forensic, homeland security, environmental monitoring, and pharmaceutical cleaning validation situations.

Some ambient desorption methods, such as the laser-desorption methods, are inherently ill-suited for large-area implementation. On the other hand, when the area covered by spray-desorption methods is enlarged, a concurrent increase in spray solvent and nebulizing gas flow rates are required, placing increased demand on the atmospheric-pressure interface and MS vacuum systems. Nevertheless, a large-area DESI-MS application for pharmaceutical cleaning validation was demonstrated. Levels of detection two orders of magnitude better than the required limit were attained for certain compounds.[56]

One of the benefits of SDC is that the solvent and gas, needed for desorption, evaporates during collection, and so does not have any bearing on the subsequent analysis process. Although the two-step process means that the large area is not analyzed immediately, or in real time, the benefit of this is improved levels of detection. This stems from the concentrating effect attained when widely dispersed chemical compounds with low surface concentrations are collected in a single, narrow band on the collector and analyzed all at once.

Three iterations of SDC have been implemented to date, as demonstrated in Figure 8.12. The original design (Figure 8.12(a)) closely mimics the typical DESI sprayer, using two coaxial capillaries with the nebulizing gas

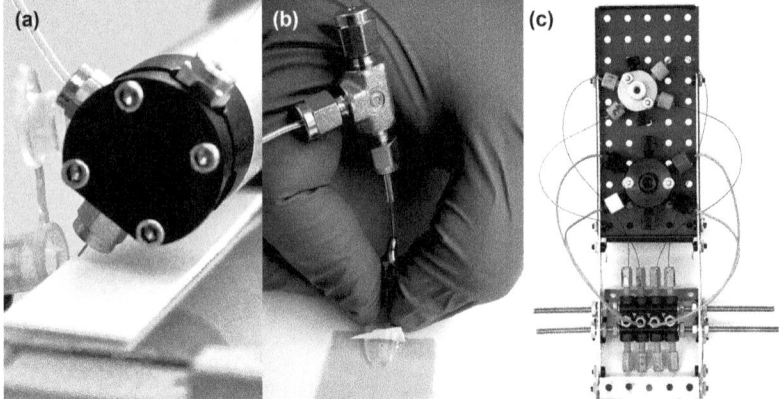

Figure 8.12 (a) A simple SDC setup, where a collector replaces the MS in front of a DESI sprayer. (b) Handheld large surface area SDC sprayer with integrated collector holder. (c) Multiplexed large surface area SDC device with four desorption channels.

surrounding a solvent line. While this design was well suited for the application of SDC to fundamental studies of DESI,[15,16] the small surface coverage of <1 mm^2 makes it impractical for timely large surface area analysis. This design was improved on by providing a turbulent flow of nebulizing gas in an enclosure to direct droplets produced from an increased solvent flow rate over a larger area of 50 mm^2 (Figure 8.12(b)).[18] Finally, this large-area sprayer was multiplexed, and when 5 sprayers are installed inline, collection can take place over an area of 5 cm^2. Figure 8.12(c) shows such a device with four desorption channels.

8.5   Conclusions

The real-time, inline and microlocalized desorption processing of compounds from surfaces was one of the key novelties of the ambient ionization methods and has proven to be extremely useful for the direct mass spectrometric analysis of compounds from complex solid surfaces at ambient conditions. These processing methods should also be considered for general-purpose sample preparation, as they offer many benefits over traditional extraction and sample-collection procedures such as swabbing or surface rinsing. Lower levels of surface concentrations can be detected, and less solvent is required. Decoupling these microlocalized desorption processing methods from real-time, inline MS analysis also allows the selection of conditions for maximum desorption efficiency without compromising the ionization and detection steps, useful especially for large surface area analysis.

In addition, decoupling the sample-processing steps such as extraction and desorption in DESI from the MS analysis allows these processes to be studied individually and much can be learned about the overall technique.

This level of understanding will aid ambient ionization practitioners to select appropriate conditions for their analysis. The tools developed to study desorption and ionization separately, SDC and RESI-MS, were demonstrated in this chapter by showing investigations in operational conditions and protein analysis. SDC and RESI-MS are tools that will also answer further questions relating to the ambient ionization methods. An example could be how each of the processes that occur during analysis, desorption or ionization, respectively is effected by changing solvents in spray desorption ambient ionization methods.

In the future it is hoped that more of these microlocalized desorption processes will be developed for improved sample-preparation procedures, and that they will be coupled to an increasing number of analysis and detection methods.

References

1. Z. Takats, J. M. Wiseman, B. Gologan and R. G. Cooks, *Science*, 2004, **306**, 471.
2. A. Venter, M. Nefliu and R. Graham Cooks, *TrAC, Trends Anal. Chem.*, 2008, **27**, 284.
3. G. A. Harris, A. S. Galhena and F. M. Fernandez, *Analyst*, 2011, **83**, 4508.
4. P. J. Roach, J. Laskin and A. Laskin, *Anal. Chem.*, 2010, **135**, 2233.
5. G. J. Van Berkel, A. D. Sanchez and J. M. E. Quirke, *Anal. Chem.*, 2002, **74**, 6216.
6. T. Wachs and J. Henion, *Anal. Chem.*, 2001, **73**, 632.
7. Z. Miao and H. Chen, *J. Am. Soc. Mass Spectrom.*, 2009, **20**, 10.
8. H. Chen, A. Venter and R. G. Cooks, *Chem. Commun.*, 2006, **42**, 2042.
9. R. B. Cody, J. A. Laramée and H. D. Durst, *Anal. Chem.*, 2005, 77, 2297.
10. P. Nemes and A. Vertes, *Anal. Chem.*, 2007, **79**, 8098.
11. J. Sampson, A. Hawkridge and D. Muddiman, *J. Am. Soc. Mass Spectrom.*, 2006, **17**, 1712.
12. J. T. Shelley, S. J. Ray and G. M. Hieftje, *Anal. Chem.*, 2008, **80**, 8308.
13. A. R. Venter, A. Kamali, S. Jain and S. Bairu, *Anal. Chem.*, 2010, **82**, 1674.
14. S.-G. Park and K. K. Murray, *J. Mass Spectrom.*, 2012, **47**, 1322.
15. K. A. Douglass and A. R. Venter, *J. Mass Spectrom.*, 2013, **48**, 553.
16. K. A. Douglass, S. Jain, W. R. Brandt and A. R. Venter, *J. Am. Soc. Mass Spectrom.*, 2012, **23**, 1896.
17. A. S. Kamali, J. G. Thompson, S. Bertman, J. B. Miller and A. R. Venter, *Anal. Methods*, 2011, **3**, 683.
18. S. Jain, A. Heiser and A. R. Venter, *Analyst*, 2011, **136**, 1298.
19. R. Haddad, R. Sparrapan and M. N. Eberlin, *Rapid Commun. Mass Spectrom.*, 2006, **20**, 2901.
20. D. R. Ifa, J. M. Wiseman, Q. Y. Song and R. G. Cooks, *Int. J. Mass Spectrom.*, 2007, **259**, 8.

21. A. Venter and R. G. Cooks, *Anal. Chem.*, 2007, **79**, 6398.
22. D. R. Ifa, N. E. Manicke, A. L. Rusine and R. G. Cooks, *Rapid Commun. Mass Spectrom.*, 2008, **22**, 503.
23. N. Talaty, Z. Takats and R. G. Cooks, *Analyst*, 2005, **130**, 1624.
24. F. M. Green, P. Stokes, C. Hopley, M. P. Seah, I. S. Gilmore and G. O'Connor, *Anal. Chem.*, 2009, **81**, 2286.
25. B. Qiu and H. Luo, *J. Mass Spectrom.*, 2009, **44**, 772.
26. A. Venter, P. E. Sojka and R. G. Cooks, *Anal. chem.*, 2006, **78**, 8549.
27. Z. Takáts, J. M. Wiseman and R. G. Cooks, *J. Mass Spectrom.*, 2005, **40**, 1261.
28. H. Chen, N. N. Talaty, Z. Takáts and R. G. Cooks, *Anal. Chem.*, 2005, **77**, 6915.
29. M. Volny, A. Venter, S. A. Smith, M. Pazzi and R. G. Cooks, *Analyst*, 2008, **133**, 525.
30. G. J. Van Berkel and V. Kertesz, *Anal. Chem.*, 2006, **78**, 4938.
31. C. Wu, K. Qian, M. Nefliu and R. G. Cooks, *J. Am. Soc. Mass Spectrom.*, 2010, **21**, 261.
32. S. P. Pasilis, V. Kertesz and G. J. Van Berkel, *Anal. Chem.*, 2008, **80**, 1208.
33. M. Benassi, C. P. Wu, M. Nefliu, D. R. Ifa, M. Volny and R. G. Cooks, *Int. J. Mass Spectrom.*, 2009, **280**, 235.
34. A. B. Costa and C. R. Graham, *Chem. Phys. Lett.*, 2008, **464**, 1.
35. A. B. Costa and R. G. Cooks, *Chem. Commun.*, 2007, **43**, 3915.
36. L. Gao, G. Li, J. Cyriac, Z. Nie and R. G. Cooks, *J. Phys. Chem. C*, 2010, **114**, 5331.
37. F. M. Green, T. L. Salter, I. S. Gilmore, P. Stokes and G. O'Connor, *Analyst*, 2010, **135**, 731.
38. V. Kertesz and G. J. Van Berkel, *Rapid Commun. Mass Spectrom.*, 2008, **22**, 3846.
39. Y.-S. Shin, B. Drolet, R. Mayer, K. Dolence and F. Basile, *Anal. Chem.*, 2007, **79**, 3514.
40. K. Heaton, C. Solazzo, M. J. Collins, J. Thomas-Oates and E. T. Bergström, *J. Archaeol. Sci.*, 2009, **36**, 2145.
41. S. Myung, J. M. Wiseman, S. J. Valentine, Z. Takáts, R. G. Cooks and D. E. Clemmer, *J. Phys. Chem. B*, 2006, **110**, 5045.
42. C. N. Ferguson, S. A. Benchaar, Z. Miao, J. A. Loo and H. Chen, *Anal. Chem.*, 2011, **83**, 6468.
43. B. G. J. Van, M. J. Ford, M. J. Doktycz and S. J. Kennel, *Rapid Commun. Mass Spectrom.*, 2006, **20**, 1144.
44. H. Wang, J. Liu, R. G. Cooks and Z. Ouyang, *Angew. Chem., Int. Ed.*, 2010, **49**, 877.
45. N. E. Manicke, Q. Yang, H. Wang, S. Oradu, Z. Ouyang and R. G. Cooks, *Int. J. Mass Spectrom.*, 2011, **300**, 123.
46. J. Liu, H. Wang, N. E. Manicke, J.-M. Lin, R. G. Cooks and Z. Ouyang, *Anal. Chem.*, 2010, **82**, 2463.
47. R. D. Espy, A. R. Muliadi, Z. Ouyang and R. G. Cooks, *Int. J. Mass Spectrom.*, 2012, **325**, 167.

48. Z. Ouyang, W. Xu, N. E. Manicke and G. R. Cooks, *Jala-J. Assoc. Lab. Aut.*, 2010, **15**, 433.
49. R. P. Belardi and J. B. Pawliszyn, *Water Pollut. Res. J. Can.*, 1989, **24**, 179.
50. C. L. Arthur and J. Pawliszyn, *Anal. Chem.*, 1990, **62**, 2145.
51. G. Ouyang and J. Pawliszyn, *Anal. Chim. Acta*, 2008, **627**, 184.
52. J. Chen and J. B. Pawliszyn, *Anal. Chem.*, 1995, **67**, 2530.
53. K. S. Tyson, J. Bozell, R. Wallace, E. Petersen and L. Moens, *Biomass Oil Analysis: Research Needs and Recommendations*, National Renewable Energy Laboratory, Golden, CO, 2004.
54. A. Badu-Tawiah, C. Bland, D. I. Campbell and R. G. Cooks, *J. Am. Soc. Mass Spectrom.*, 2010, **21**, 572.
55. C. P. Wu, A. L. Dill, L. S. Eberlin, R. G. Cooks and D. R. Ifa, *Mass Spectrom. Rev.*, 2013, **32**, 218.
56. S. Soparawalla, G. A. Salazar, R. H. Perry, M. Nicholas and R. G. Cooks, *Rapid Commun. Mass Spectrom.*, 2009, **23**, 131.

CHAPTER 9

Easy Ambient Sonic-Spray Ionization

CARLOS H. V. FIDELIS AND MARCOS N. EBERLIN

ThoMSon Mass Spectrometry Laboratory, Institute of Chemistry, University of Campinas – UNICAMP, Campinas, SP, 13083-970, Brazil
Email: eberlin@iqm.unicamp.br

9.1 MS and Easiness. A Historical Relationship

In his predictions about the merits of mass spectrometry, still in the early days of the technique, J.J. Thomson cleverly and prophetically stated: *"I feel sure that there are many problems in Chemistry which could be solved with far greater **ease** by [mass spectrometry] than by any other method"*.[1] Thomson declared therefore that a main attribute of MS ought to be easiness, or that the destiny of MS was to be easier than any other method of chemical analysis. But unfortunately throughout its history, MS quickly moved from using easy ways to ionize and measure ions, as those very simple instruments such as the cathode ray tubes and ionization techniques such as spark ionization used by the MS pioneers Thomson, Aston and Dempster, to very complicated approaches including tedious sample-preparation protocols, demanding ionization techniques or complicated, intricate and sometimes "monster" instrumentation. All of this tremendous increase in complexity and lack of easiness started to scare away beginners and potential users, so much so that soon the main drawback of MS analysis started to be summarized by many *via* single statements such as: *ultimate complexity, or lack of easiness!*

New Developments in Mass Spectrometry No. 2
Ambient Ionization Mass Spectrometry
Edited by Marek Domin and Robert Cody
© The Royal Society of Chemistry 2015
Published by the Royal Society of Chemistry, www.rsc.org

Fortunately, however, most recent developments in MS, for the last two decades or so, have moved MS back to its humble but productive origin, that is: *back to simplicity*. As for ionization, the first major revolution towards simplicity occurred in the late 1980s with the introduction of atmospheric-pressure techniques that removed the ion source from *"hell to heavens"*, that is, from the inhospitable high-vacuum environment inside mass spectrometers, placing it where we all have long hoped for but dared to dream about *thinking it possible*: at the friendly, atmospheric-pressure open ambient environment of the laboratory. Electrospray ionization (ESI) – *the Nobel laureate technique optimized and popularized by John Fenn* – is an icon of such great revolution. ESI[2] greatly simplified MS analysis, bringing MS from the restricted gas-phase environment in which only volatile, less-polar and small molecules could be handled, into the atmospheric-pressure "real world" in which now nearly all types of molecules, biomolecules, polymers and salts in solution can be analyzed by MS.

ESI has provided therefore a solid bridge[3] connecting the gas phase of MS to the solution chemistry of solvated ions and molecules and *vice versa*. ESI and other ambient ionization techniques that followed such as DESI[4] and PSI[5] also contributed to making MS analysis easier by simplifying mass spectra and its interpretation. Instead of forming multiple ions from a single molecule, arising from a sometimes disturbing combination of intact and fragment ions, or in trickier cases even by only fragment ions, ESI most commonly forms a single, intact ion (together with its corresponding and composition informative isotopologues) for each molecule, thus providing a direct "one molecule:one ion" relationship that is highly desirable and handy for mixture analysis. This direct composition:ion matching greatly simplifies the analysis of complex mixtures, and has definitively added "*separation*" as an additional MS attribute.

The easiness brought about by ESI has been summarized in an "involution" tree (Figure 9.1).[6] ESI is a spray-based technique performed directly from analyte solution at ambient conditions. As compared to the primordial electron ionization (EI) technique first applied to (small) organic molecules, ESI has eliminated therefore the need for a high vacuum (and filaments for electron beam production) but introduced the need for a cylinder of compressed and relatively inert gas, a heater, a solvent for the spraying and, in most commercial ESI sources, an electrical syringe pump.

There was therefore still room for simplification, and greater simplicity and easiness of operation have in fact been introduced *via*, for instance, a series of ambient ESI-based MS techniques such as secondary ESI (SESI,[7] Figure 9.1) and its sister technique extractive ESI (EESI).[8] These techniques, as compared to ESI, eliminated or simplified sample preparation for the analysis of gases and vapors of more volatile molecules. The pioneer ambient MS technique, DESI,[4] as well as its sisters techniques paper spray (PSI)[9] and leaf spray (LSI)[10] also contributed much to move ESI-MS further

Figure 9.1   Focusing on spray-based ionization techniques, an illustration of the simplification of ionization techniques for MS analysis *via* an "involution (from complexity to simplicity) tree". Arrows indicate new demands or elimination of parts or needs when a new technique evolves from its ancestor.
Adapted from ref. 6.

towards simplicity and easiness. These techniques have eliminated the need for preparing analyte solutions, picking up the analytes directly from auxiliary surfaces or even at the original natural surfaces in which they are naturally found and are best investigated, such as in LSI. For the involution tree illustration of Figure 9.1 therefore, in LSI ions would literally arise directly from its leafs!

9.2   The Sonic-Spray Revolution: Towards Ultimate MS Simplicity

Another less renowned but also revolutionary spray-based technique that helped MS move much closer towards simplicity was sonic-spray ionization (SSI). Introduced by Hirabayashi *et al.* in 1994,[11] SSI was found to be unique since it launched a revolutionary, much simplified and easier concept of MS ionization. For the first time in MS history, ions were produced *via* SSI without the assistance of voltage, radiation, electrical discharges or heating, requiring only a cylinder of compressed gas and a sonic-spray nebulizer.

SSI, although operating *via* an independent and distinct mechanism than ESI, may be viewed – *in terms of simplicity* – as a simpler version of ESI (Figure 9.1), for which the need for a high-voltage power supply and an electricity-demanding heater had been eliminated. In SSI, the charged droplets are produced simply by spraying, for instance, an acidified methanol solution of the analyte molecule (M) at sonic speed. Excess charge (both negative and positive) arises in the SSI droplets, not from the application of an external high voltage (as for ESI) but from a statistically imbalanced distribution of the pre-existing solution cations and anions due to the very limited excess charge capacity of the very minute droplets produced by sonic spraying. In addition, no heating was found to be necessary in SSI for droplet desolvation. Analyte ions, $[M+H]^+$ and/or $[M-H]^-$ for instance, are produced in SSI due to protonation and/or deprotonation of M depending on its basic/acidic nature.

Ion formation in SSI is therefore simpler (less-demanding instrumentation) and easier (more readily obtainable) than ESI and is based on the sole assistance of only a cylinder of compressed gas and a proper sonic-spray nebulizer. Nitrogen is normally used but even compressed or canned air can be employed due to the lack of restriction to the use of oxygen, which is a limitation for the high-voltage ESI.

Whereas ESI launched a revolutionary concept of MS ionization in which ions were produced outside the mass spectrometer, directly from salts or molecules present or preionized in solution and then simply "ejected" *via* electrospraying the solution directly into the gas phase of a mass spectrometer, sonic spray also launched the revolutionary concept of atmospheric heating and voltage-free ionization.

9.3 Easy Ambient Sonic-Spray Ionization – EASI

Using SSI as the root technique (Figure 9.2),[12] we then introduced in 2006 a new ambient desorption/ionization technique termed easy ambient sonic-spray ionization (EASI).[13,14] EASI (an acronym selected to make a direct correlation to its easiness) shares therefore most, if not all the advantageous features of SSI. SSI is used in EASI to form a stream of the very minute bipolar $(\pm)$ charged droplets from, for instance, acidified methanolic solutions. The bombardment of surfaces with such sonic-spray droplets has been shown therefore to efficiently desorb and ionize the analyte molecules resting on natural or auxiliary surfaces at ambient conditions. As for ESI and SSI, desorption and ionization in EASI is therefore also a solution process but EASI is unique in that the droplets display low charge states, either positive or negative, and the bipolar nature of EASI$(\pm)$ concomitantly forms therefore both positive and negative ions, often from the same analyte molecule. But when EASI$(+)$ is applied, the mass spectrometer is commanded to sample only the positive ions, typically $[M+H]^+$ and/or $[M+Na(K)]^+$ ions, whereas for EASI$(-)$, the negative ions, mainly $[M-H]^-$,

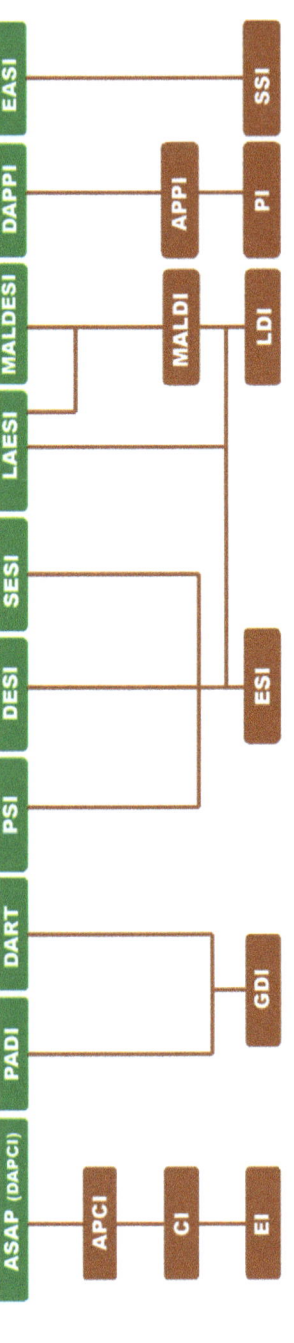

Figure 9.2 Simplified view of the "ionization tree" in ambient MS showing the root technique (in brown) and the major derived ambient MS techniques. Note that EASI is directly derived from SSI, which is in turn a unique root ionization technique that occurs *via* a distinct mechanism of unbalanced charge splitting. Adapted from ref. 6.

are sampled in a ionization process free of voltage, discharges and/or heating interferences.

The main advantages of EASI are therefore:

(a) Greatest simplicity since only a cylinder of compressed nitrogen or even air (canned air for instance) and a proper sonic-spray nebulizer are required.

(b) Ability to simultaneous produce both negatively and positively charged droplets; hence, the technique always functions in the bipolar EASI($\pm$) mode with no need for high-voltage polarity switching in the ion source when going from EASI($+$) to EASI($-$), or *vice versa*.

(c) Low charge concentration on the droplets, which seems to reduce solvent noise[12] thus favoring the analyte ions and therefore improving signal-to-noise (S/N) ratios.

(d) No thermal degradation.

(e) No electrochemical, discharge or oxidation interferences known to sometimes interfere, for instance, in ESI and DESI data.

(f) Fewer analyte ions. Due to the lack of voltage assistance, $M^{+\bullet}$ or $M^{-\bullet}$ molecular ions that are sometimes concurrently or exclusively formed in voltage-assisted techniques, have not been observed in EASI.

(g) Deeper matrix penetration, due to the high-velocity sonic-spray droplet stream, which is particularly useful for solid samples, thus providing more homogenous sampling and long-lasting ion signals. A limitation of EASI is associated with this ultrahigh-velocity sonic-spray stream that can easily blow samples away. For compact solids, samples crystallized on rough surfaces and viscous oils this blowing is not a problem. For volatile fuels, for instance, this problem can be circumvented by allowing solvent evaporation and by acquiring data for the residual polar markers.

(h) Higher sensitivity *via* enhanced signal-to-noise (S/N) ratios. Although absolute ion abundances are sometimes lower, the superior S/N ratios (which probably results from the lower charge loads of the EASI droplets) often leads to much reduced solvent and background noise due to more selective analyte ionization. This S/N advantage of EASI is directly related therefore to SSI, which for instance has been shown to provide as high as 40-fold gain in S/N in the quantitation of amino acids as compared to ESI.[15]

(i) Highly suitable for portable, low-power-demanding mass spectrometers.

(j) Reduced fragmentation. SSI is arguably the softest ionization technique, hence EASI greatly favors the detection of intact analyte ions. This softness is advantageous for the analysis of fragile molecules and complex mixtures such as, for instance, organometallic complexes, providing a more quantitative one component – one intact ion mode of detection (Figure 9.3).

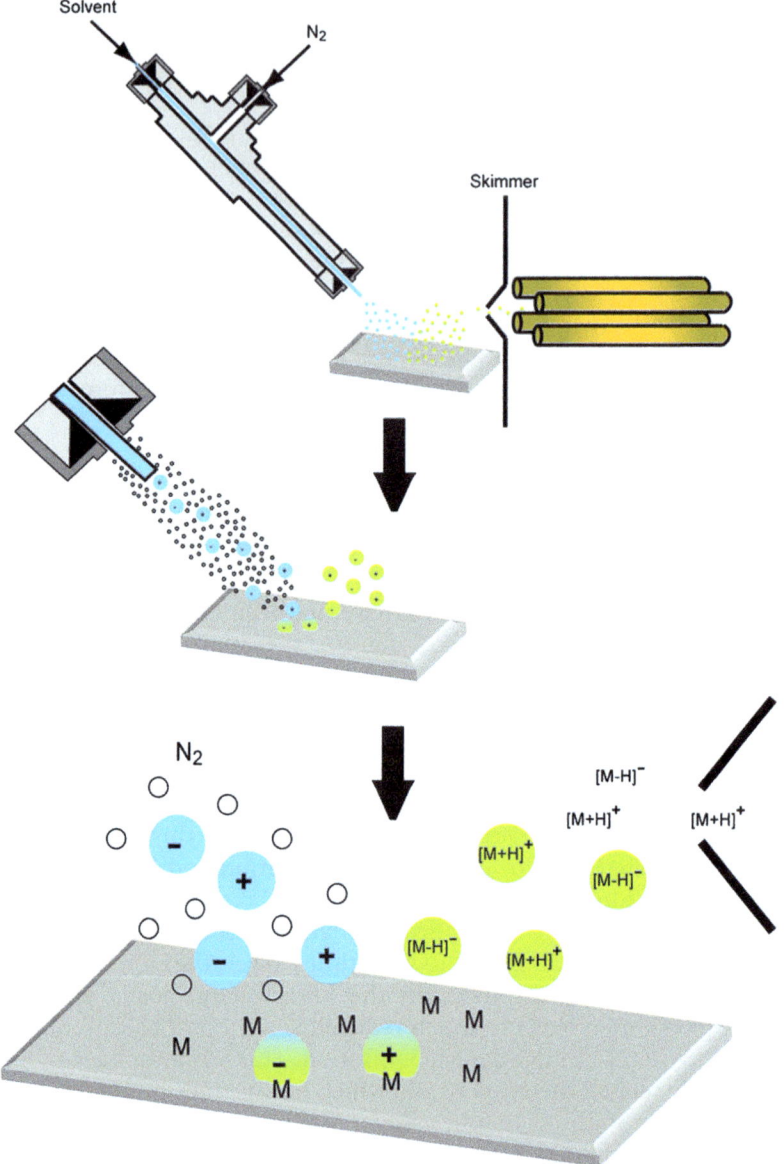

Figure 9.3 Schematic of EASI($\pm$) and its sonic-spray-based mechanism of desorption and bipolar ionization.

9.4   Hyphenated EASI

9.4.1   EASI-MIMS

EASI has been coupled to membrane introduction mass spectrometry (EASI-MIMS)[13] for the direct analysis of solution constituents (Figure 9.4). Using a

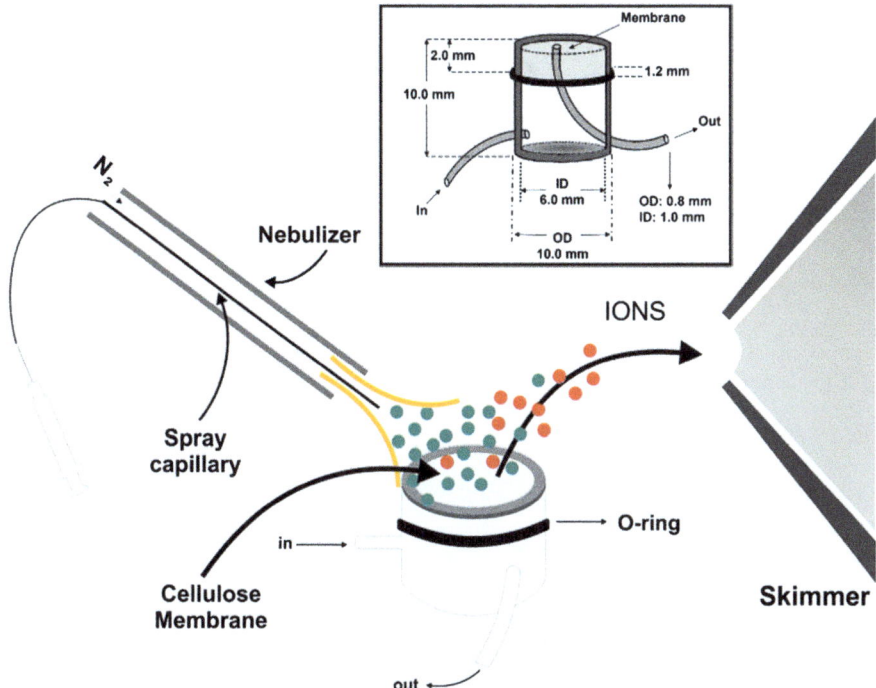

Figure 9.4 Schematics of the EASI-MIMS system. A supersonic spray of charged droplets bombards the surface of a cellulose membrane in direct contact with the aqueous analyte solution.
Adapted from ref. 13.

cellulose dialysis membrane and aqueous solutions of common drugs as proof-of-principle examples, it was demonstrated that solid but permeable and flexible membranes can be used as interfaces for the direct analysis of solution constituents *via* EASI-MS. Possible application of EASI-MIMS such as environmental analyses of effluents, online monitoring of fermentation and biotransformations and online pharmacokinetic blood analysis were discussed.

9.4.2 EASI-TLC

Direct EASI-MS chemical analysis of spots separated by thin-layer chromatography (TLC-EASI-MS)[16] and high-performance TLC (HPTLC-EASI-MS)[17] have also been performed for the more complex mixtures which (unfortunately) requires some degree of separation prior to the MS analysis. Figure 9.5 exemplifies the analysis of blends of petrodiesel/biodiesel samples by HPTLC-EASIMS.

9.4.3 EASI-MIP

Ambient MS has been performed typically on inert surfaces, but the advantageous use of active surface has been recently demonstrated in EASI-MS

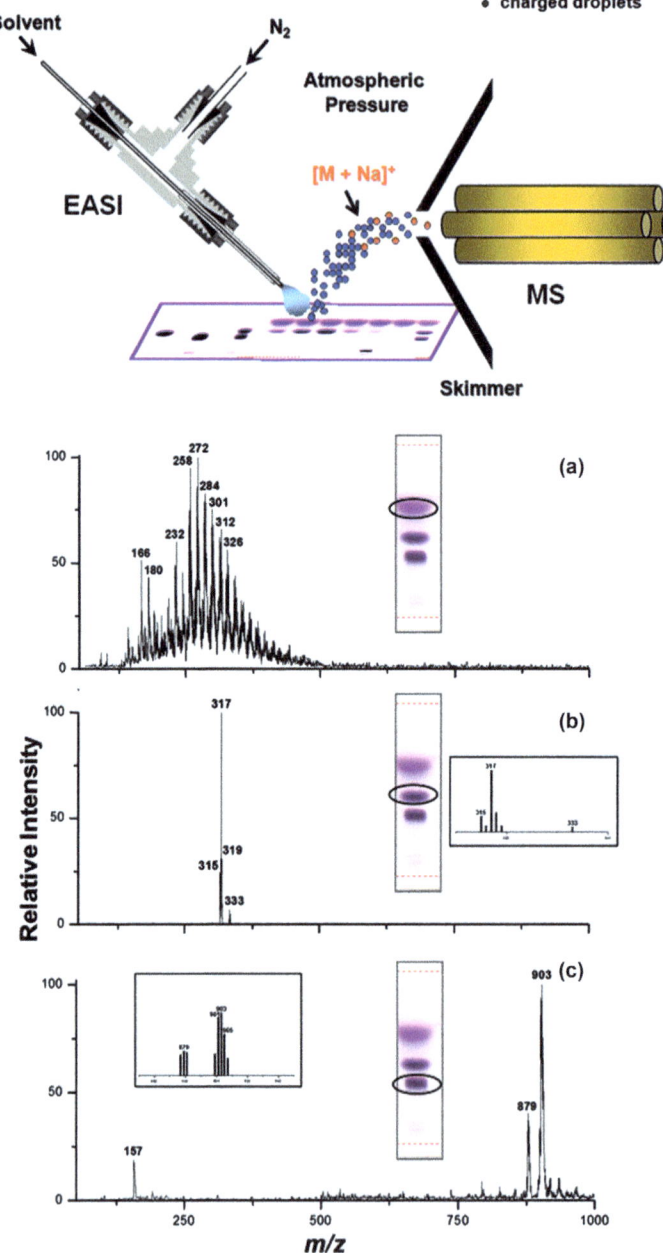

Figure 9.5 On-spot HPTLC-EASI-MS analysis of petrodiesel/biodiesel samples. Adapted from ref. 17.

experiments using molecularly imprinted polymers (MIP) were applied as a selective surface able to sequester target analytes from urine.[18] Analyte extraction was done by dipping the MIP probe to the analyte solution that was percolated through the extraction/washing cell. After the extraction period, the MIP probe was washed using a washing solution and then removed. Finally, the whole MIP surface on the probe was submitted to EASI-MS (Figure 9.6(A)). EASI desorbed the analytes from the MIP surface to the gas phase for MS analysis.

A set of 5 phenothiazines (chlorpromazine, perphenazine, triflupromazine, thioridazine and prochlorperazine) were chosen as a *proof-of-principle* class of drug samples. A chlorpromazine-imprinted methacrylic polymer was synthesized and used to prepare a MIP probe. The MIP-EASI-MS technique using acidified methanol as solvent has been shown to allow quantification of all 5 drugs in urine with LOQ of *ca.* 1 mmol L^{-1} (Figure 9.6(B) and (C)).

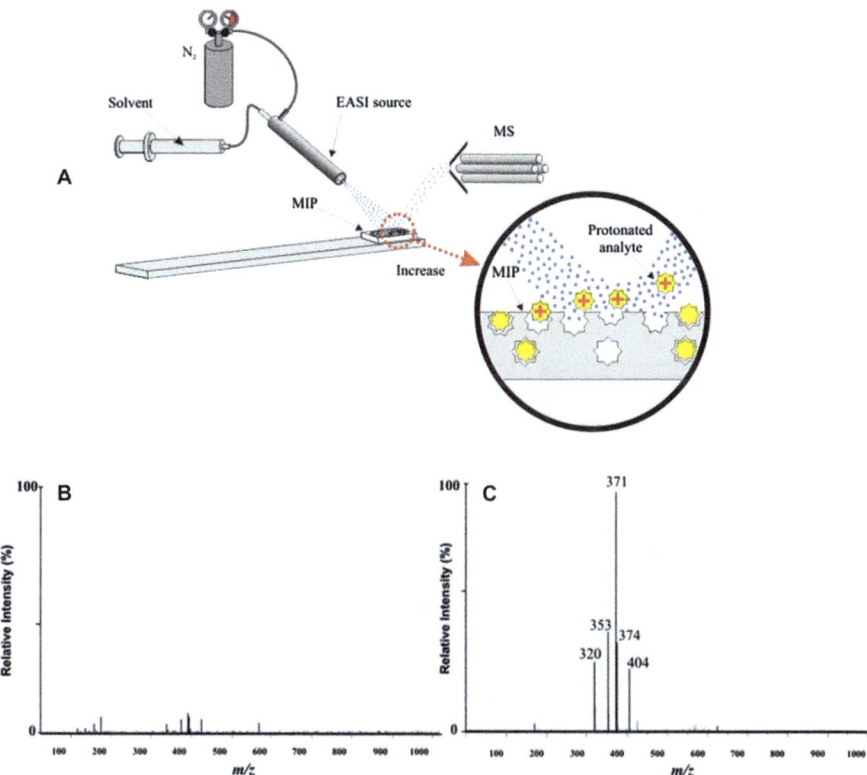

Figure 9.6 (A) Schematics of the MIP-EASI(+)-MS analysis of target molecules from solution. MIP-EASI(+)-MS of a urine sample spiked with 5 µmol L^{-1} of chlorpromazine (*m/z* 320), triflupromazine (*m/z* 353), thioridazine (*m/z* 371), prochlorperazine (*m/z* 374), perphenazine (*m/z* 404) using either a (A) nonimprinted polymer (NIP) or (B) the molecularly imprinted polymer (MIP) probe.
Adapted from ref. 18.

9.5   Sister Techniques

9.5.1   V-EASI

An even simpler and easier to assemble and use version of the EASI technique has been described.[19] The technique incorporated the classical and widely used Venturi effect and was termed Venturi easy ambient sonic-spray ionization (V-EASI). It can also operate in dual mode for both solutions and solid samples (Figure 9.7). V-EASI uses solely the forces of the sonic stream of nitrogen or air that causes sonic spray to also cause the combined result of solution pumping *via* the Venturi effect. For liquid samples, the Venturi effect of such high-velocity gas was used to pump the analyte solution to the spray region where sonic-spray ionization (SSI) that forms the intact negatively and/or positively charged molecular gaseous species takes place. For solid samples, the Venturi effect was used to pump the SSI solvent, from which the stream of very minute bipolar charged droplets formed by SSI was used to bombard the sample surface causing desorption and ionization of the analytes. V-EASI was shown to produce rather clean spectra dominated by a single-molecular species for a variety of solutions and solid samples such as drug tablets, peptides, proteins, crude oil, and cocaine. As

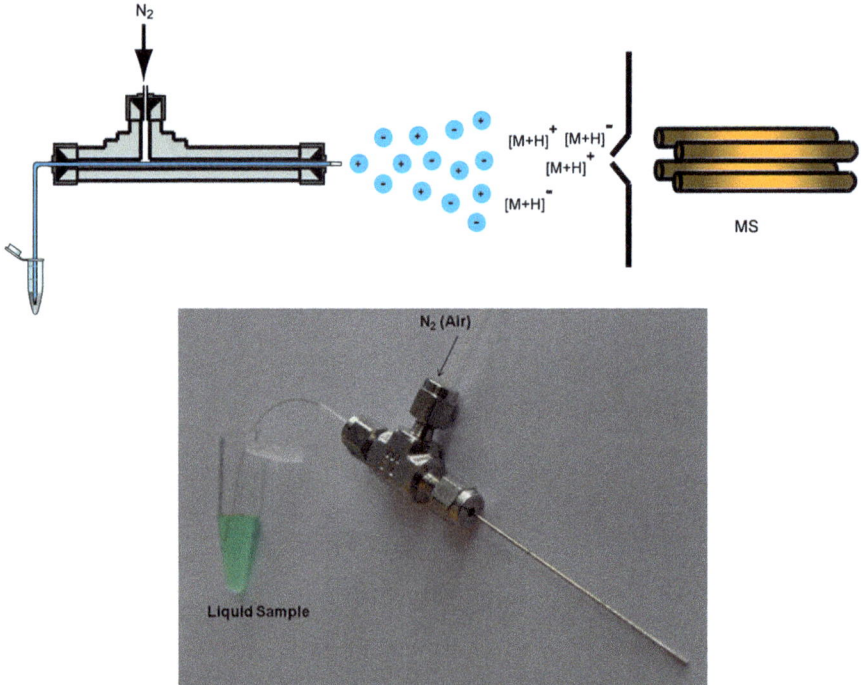

Figure 9.7   Schematic and an actual picture of the V-EASI system. Adapted from ref. 19.

for EASI, V-EASI has also the advantage of being a voltage-, heat- and radiation-free technique operating at room temperature with the sole assistance of a cylinder of compressed nitrogen (or air) and a nebulizer causing therefore no thermal, electrical or discharge interferences. The greater simplicity of V-EASI as compared to EASI with no electrical power requirement at all makes V-EASI highly suitable for its use in miniature mass spectrometers.

The V-EASI set up is also quite suitable for real-time, continuous and online monitoring of reaction solutions, as Figure 9.8 illustrates. This type of MS monitoring represents an ultimate dream for reaction monitoring since reactions could be followed in real time by the fast and highly sensitive MS technique with characterization of the changing composition of the reaction solution in terms of reactants, products and most interestingly, long-lived or even transient intermediates. Figure 9.8 shows three representative V-EASI(+) spectra acquired during the course of an Morita–Baylis–Hillman (MBH) reaction.[20] At the very beginning (Figure 9.8(A)), V-EASI(+)-MS is able to intercept the first intermediate $[M+H]^{+}$ of m/z 199. After 30 min (Figure 9.8(B)), the second key MBH intermediate was also clearly detected as $[M+H]^{+}$ of m/z 306. After 2 h (Figure 9.8(C)), the final MBH product was detected as both $[M+H]^{+}$ of m/z 194 and $[M+Na]^{+}$ of m/z 216.

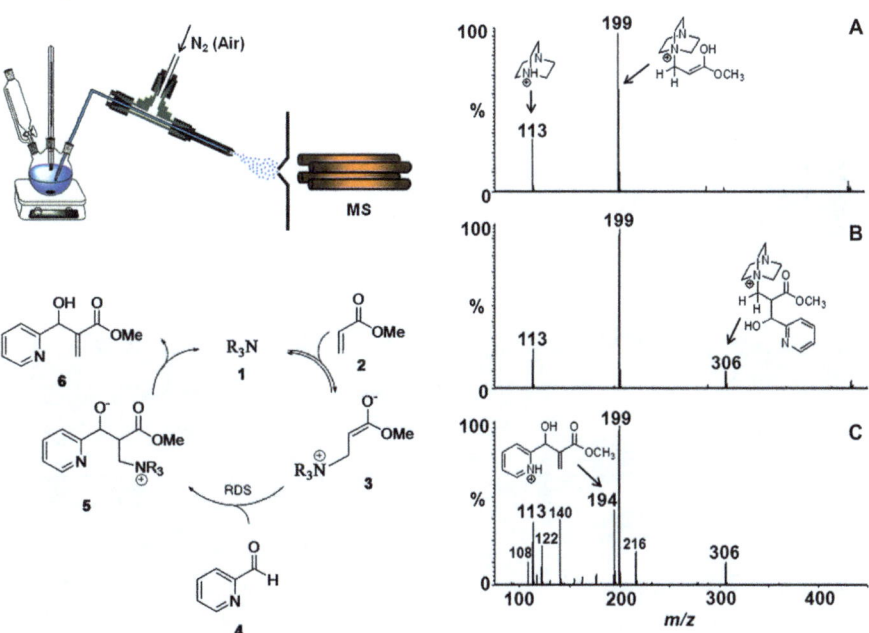

Figure 9.8 V-EASI(+)-MS online monitoring of the MBH reaction of methyl acrylate with 2-pyridinecarboxyaldehyde in the presence of DABCO. (A) $t=0$ min. (B) $t=30$ min. (C) $t=2$ h.
Adapted from ref. 6.

9.5.2 S-EASI

In the field of ambient MS, variable levels of simplicity (or lack of it) have been observed for the large and rapidly growing set of techniques. In the search for the easiest and least expensive, ideally disposable source for ambient MS, we aimed at constructing the most Spartan source design as possible, one that would ideally require no power supplies, no electrical parts at all such as syringe pumps or laser or UV light sources or glow-discharge devices and no cylinders of purified compressed and inert gases, and pressure regulators. Such a system would reduce complexity, size, cost, simplify manufacturing, operation and facilitate "in-field" analysis by portable mass spectrometers, allowing perhaps the use of disposable sources that would minimize cross contamination.

With these goals in mind, we have recently introduced a new "fully" simplified version of EASI termed "Spartan-EASI" (S-EASI).[21] It seems to constitute one of the simplest and most economical yet efficient spray-based desorption/ionization techniques for ambient MS (Figure 9.9).

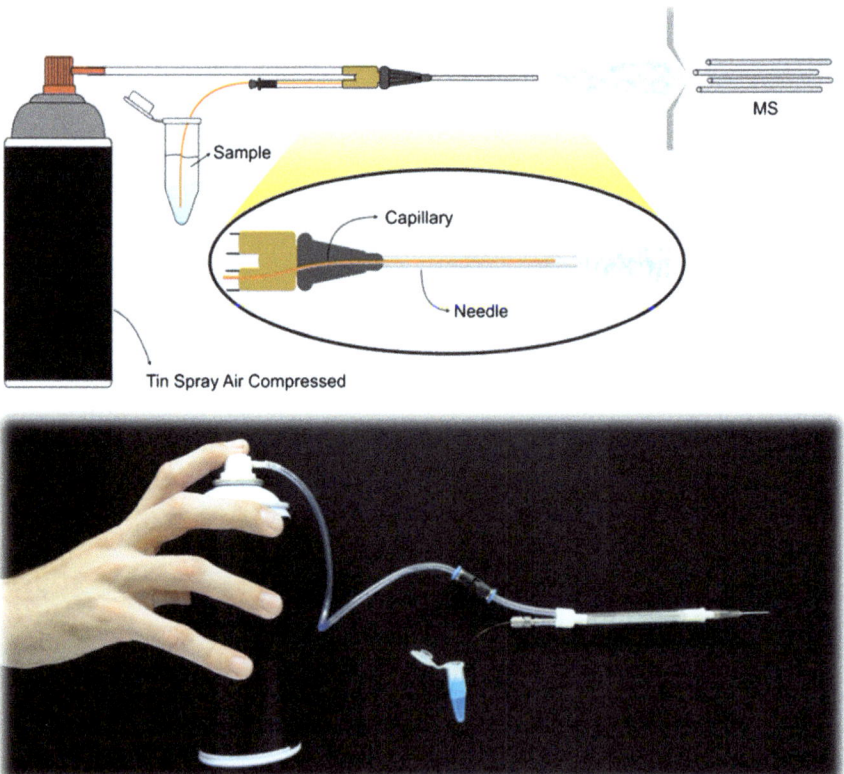

Figure 9.9 Schematics and actual picture of the Spartan EASI source. Adapted from ref. 21.

The entire fully functional S-EASI source was mounted using a surgical 2-way catheter that functions as the T-connector. To replace the cylinder of compressed nitrogen gas and gas regulators, a disposable aerosol dust cleaner can of compressed air was used and connected to one of the channels of the 2-way surgical catheter. A fused-silica capillary was introduced by the other channel of the T-connector. A simple hypodermic needle was then connected to one of the catheter ends, allowing the silica capillary to pass through its steal capillary. To simultaneously create proper self-pumping *via* the Venturi effect and efficient sonic spraying (desorption/ionization) the end of the silica capillary was positioned a few millimeters before the tip end of the needle. Pneumatic tube fittings were used for sealing. The other end of the capillary was dipped into the analyte solution for liquid samples whereas for solid samples, a proper solvent such as methanol is used as the spray solvent and the tip of the hypodermic needle is manually positioned close to the MS orifice in a nearly perpendicular angulation. For the solid samples, an angle of *ca.* 40° was used.

We have also demonstrated the efficacy of S-EASI-MS in a variety of applications as a soft, low-noise, reproducible and sensitive (high S/N ratio) technique inherently free of electrical or discharge interferences. The ease of operation combined with the simplicity of the parts and assembly should facilitate the general use and servicing of ambient MS by nonexperts such as medical doctors, nurses, police officers and soldiers.

Despite the simplicity of the S-EASI apparatus, the source showed excellent stability, reproducibility and reliability and simultaneous self-pumping and sonic-spray ionization could both be efficiently attained by the reasonably constant gas flow provided by the dust cleaner can. No voltages are applied to promote S-EASI, hence the air/HCFC mixture contained in the duster cans was found to produce no chemical interferences. Nearly constant gas flow could be sustained for 2 min with *ca.* 5 min of rest, for a total of *ca.* 40 min of MS acquisition. The disposable duster cans could be easily replaced, providing near continual use of the S-EASI source. The desorption/ionization ability of this Spartan source was also demonstrated for solid samples as for instance in the analysis of intact drug pills.

9.5.3 TI-EASI

Recently, another sister technique of EASI has been proposed – thermal imprinting EASI (TI-EASI).[22] Using this imprinting approach, the transfer of analyte molecules was performed to a paper surface, using minimal solvent amounts, followed by direct analysis of the triacylglycerol (TAG) content *via* easy ambient sonic-spray ionization mass spectrometry (EASI-MS) providing a fast method for the analysis of TAG profiles in meats, fats (Figure 9.10) and fish (Figure 9.11).

TI-EASI has been shown to be simple, fast and ecofriendly requiring no hydrolysis, derivatization or chromatographic separation. The entire TI-EASI-MS protocol is performed in a few minutes and with minimal

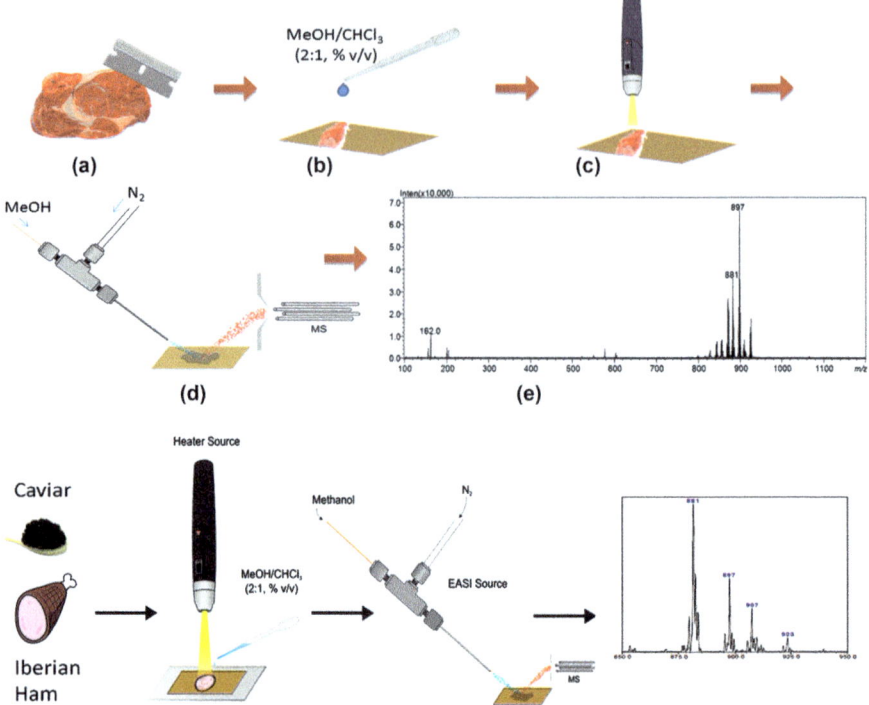

Figure 9.10 Schematics of TI-EASI for the MS analysis of TAG profiles of meats, fats, and related products such as caviar and ham.
Adapted from ref. 26.

sample handling and solvent consumption. The TAG profiles obtained *via* TI-EASI-MS have been shown to be similar to those obtained using GC and MALDI-MS analysis. Mailing of the imprinted paper was proposed for remote TI-EASI-MS analysis of meat and fat samples. Further, the TI-EASI technique was applied for the typification and quality control of Russian caviar[23] and Iberian hams[24] (Figure 9.10).

9.6 General Uses of EASI and its Sister Techniques

EASI-MS and its sister techniques have been applied with success to the analysis of different analytes and matrixes such as drug tablets,[12] perfumes,[25] surfactants,[26] vegetable oils,[27] biodiesel,[28] propolis,[29] crude oils[30] and its applications for fuel analysis has been recently reviewed.[31] Figure 9.12 displays an illustration of the "whole-world" capabilities of EASI-MS (and sister techniques) analysis, whereas Figure 9.13 displays a mosaic of such applications.

More applications of these ambient MS techniques include: (a) the direct analysis of drug pills as exemplified for sildenafil (Figure 9.14(A)), and for the characterization of TLC spots[16] as exemplified for cocaine

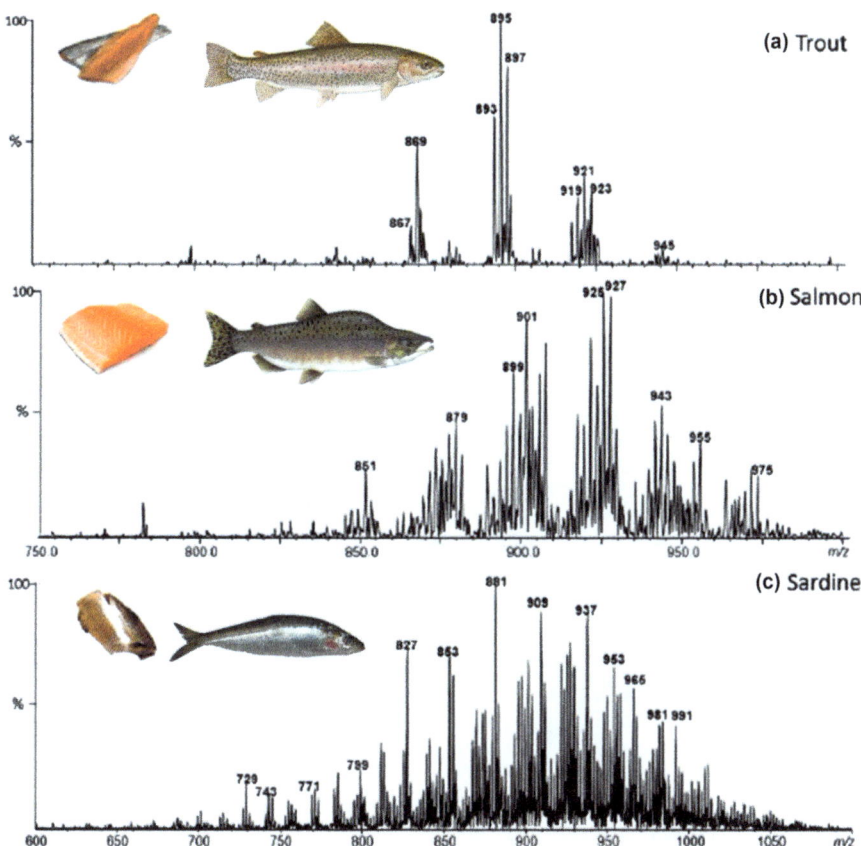

Figure 9.11 Typical TAG profiles of fish obtained by TI-EASI-MS.
Adapted from ref. 22.

(Figure 9.14(B)), for the analysis of phenothiazines in urine *via* MIP-EASI-MS (Figure 9.14(D));[18] and for the direct monitoring of known and unknown (or unexpected) drug-degradation products as a function of time and manufacturing process,[32] as exemplified for enalapril maleate (Figure 9.14(C)).

9.6.1 EASI in Forensic Chemistry

Perhaps the major area of application so far tested for EASI-MS has been in forensic chemistry. EASI-MS has been used for the analysis of ballpoint pen ink writings directly from paper surfaces[33] providing nondestructive fingerprinting identification of ink writings from different pens. Accelerated degradation showed different EASI-MS profiles for each dye. Basic Violet 3 (*m/z* 372), the most common dye in blue pens, displayed a cascade of degradation products (of *m/z* 358, 344, 330, 316 from consecutive demethylation) whose abundances increased linearly with time. This cascade of

Figure 9.12   An artistic view of the whole-world applications of chemical analysis by
EASI-MS.
Adapted from ref. 6.

degradation products functions therefore as a "chemical clock" for ink aging
(Figure 9.15). Analysis of documents with different ages confirmed the
relative ink dating capabilities of EASI-MS with applications to forgery,
superposition and crossing of lines.

EASI($\pm$)-MS has also been shown to provide comprehensive TAG and FFA
profiles for vegetable oils and these markers provide quality control, origin
certification as well as reveal adulteration – for instance the common illegal
admixture of soybean oil in the more expensive extra virgin olive oil.[34] The
main markers are either $[\text{TAG} + \text{Na}]^+$ or $[\text{FFA–H}]^-$ ions detected by EASI($\pm$)-
MS from a single droplet of the oil (Figure 9.16).

EASI(+)-MS was also shown to cause no fragmentation of TAG ions, hence
DAG and MAG profiles and contents could be concomitantly measured. The
EASI($\pm$)-MS profiles of TAG and FFA permit authentication and quality
control and was proposed to access levels of adulteration, acidity, oxidation
or hydrolysis of vegetable oils in general. EASI(+)-MS was also used to determine the level of oil oxidation, and proposed for single-shot biodiesel
analysis.[28] Methyl esters (FAME) are detected as $[\text{FAME} + \text{Na}]^+$ of m/z 317

Figure 9.13 Mosaic of applications of EASI-MS and its sister techniques: from top to down and from left to right, in the analysis of meats and fats *via* TI-EASI-MS, in the typification and quality control of amazonian oils by V-EASI-MS, in the dating of ink writings, in the screening for counterfeit bank notes in the fully direct petroleomic analysis of crude oils by EASI-MS, in the analysis of leaf extracts by V-EASI-MS and in the whole-world analysis by S-EASI-MS.

corresponding to linoleic acid ester (predominant), *m/z* 319 of oleic acid ester, and *m/z* 315 of linoleic acid ester. $[\text{FAME} + \text{K}]^+$ ions of *m/z* 331, 333 and 335 are also typically observed. DAG and TAG contaminants can also be detected as $[\text{DAG} + \text{Na}]^+$ ions around *m/z* 639 and $[\text{TAG} + \text{Na}]^+$ ions around *m/z* 903 for low-quality biodiesel samples.

A unique forensic application of V-EASI-MS has been in the typification of nobel, endangered wood and trees. It has been shown to provide nearly immediate and secure typification of the Mahogany tree,[35] a most valuable type of tropical wood. This reddish wood display unique phytochemical markers (phragmalin-type limonoids) that are rapidly detected from the wood surface by V-EASI-MS or from a simple methanol extract of a tiny wood chip *via* V-EASI-MS (Figure 9.17). Genuine samples of six other types of woods (Cedar, "Jequitibá", "Currupixá", Red Angelim, Yellow Angelim and Stone Angelim), which are commonly falsified by artificial coloring and commercialized as Mahogany, also display typical but dissimilar photochemical profiles as compared to that of the authentic wood. Variable and atypical chemical profiles were observed for artificially colored woods.

Secure chemical characterization *via* V-EASI-MS should help to control the illegal logging and trade of Mahogany and other endangered woods and their falsification. The same V-EASI protocol has also been therefore applied

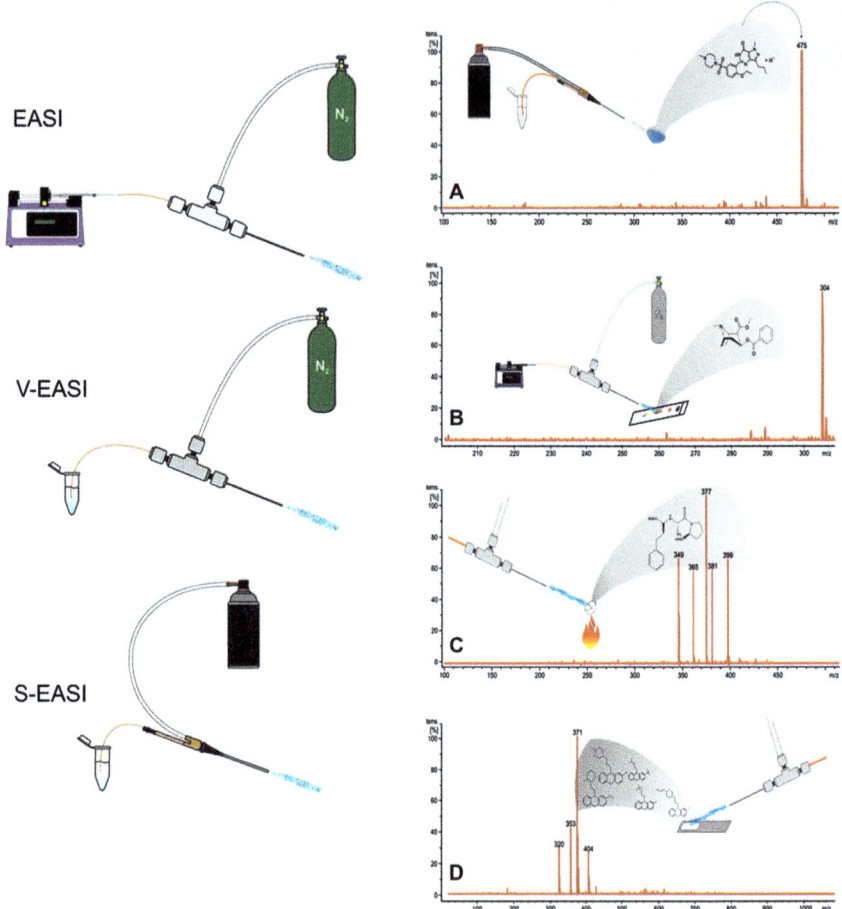

Figure 9.14   (Left) The simpler designs for EASI and its sister techniques V-EASI and S-EASI. V-EASI further simplified the EASI system by eliminating the need for electrical syringe pumping, whereas S-EASI eliminated the need for cylinders of relatively inert gases (N_2) using a can of compressed air and very simple, low-cost and disposable parts. (Right) Applications of EASI-MS (or one of its sister techniques) for the "sample-prep free" analysis of drugs (sildenafil) – (A), characterization of the cocaine spot in a TLC plate – (B), the direct degradation monitoring of enalapril maleate – (C), and (D), in combination with selective sequesters based on molecularly imprinted polymers (MIP) for the analysis of phenothiazines in urine.
Adapted from ref. 6.

recently to simple leaf extracts of *Aniba rosaeodora* Ducke ("pau rosa") and *Aniba parviflora* ("macacaporanga").[36] These are morphologically very similar Amazonian trees especially during the stage of plantation. *Aniba rosaeodora* Ducke provides valued materials to the perfume industry and is at risk of extinction. The increasing interest in plantation and the counterfeit

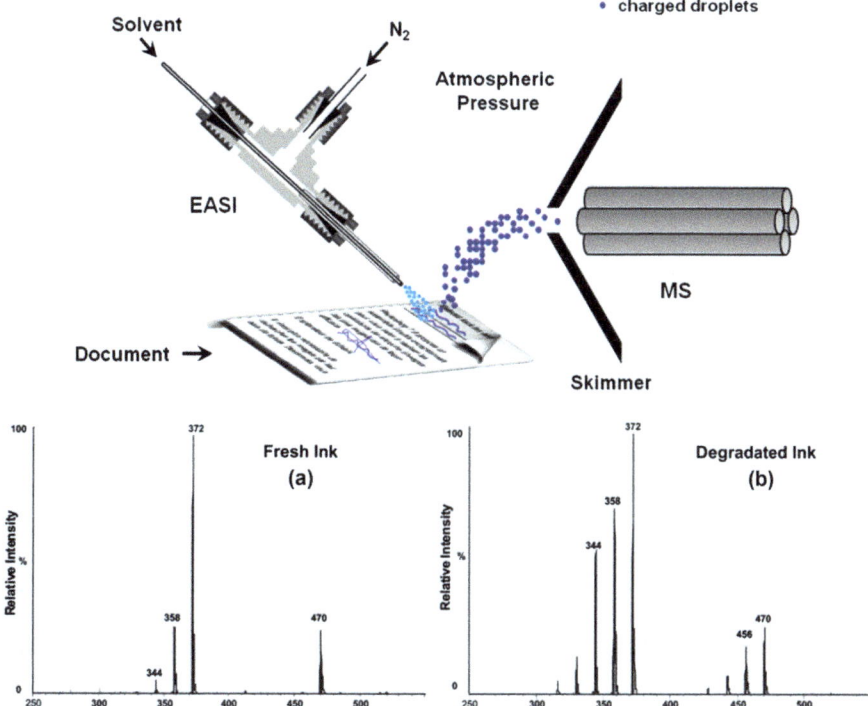

Figure 9.15 EASI-MS analysis of ink writings.
Adapted from ref. 33.

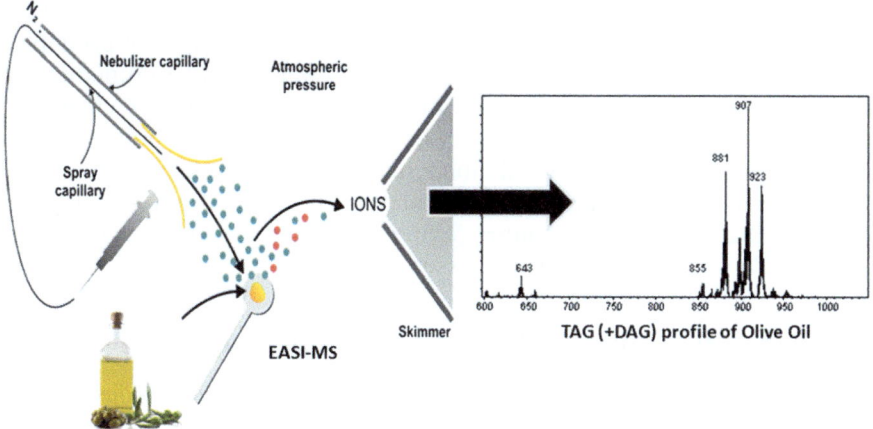

Figure 9.16 Schematics of the EASI-MS analysis of vegetable oils, and a typical TAG profile obtained.
Adapted from ref. 34.

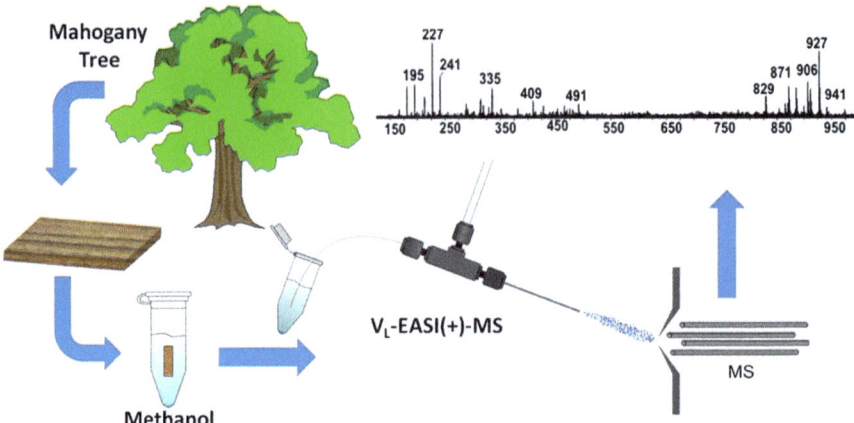

Figure 9.17   Schematics of the V-EASI-MS typification protocols applied to woods of endengerous nobel trees or a single leaf.
Adapted from ref. 35.

commercialization of *Aniba parviflora* as if it was *Aniba rosaeodora* Ducke the much more valuable and morphologically similar tree have turned the rapid and unequivocal differentiation of these species an important task. The V-EASI-MS of the extracts were shown to rapidly differentiate a single leaf of each tree. Very characteristic profiles of characteristic phytomarkers allow unequivocal differentiation of the two species. Leaf fingerprinting *via* MS analysis seems therefore to provide a rapid and secure tool for plant identification.

Detection of bank-note counterfeiting was also tested using EASI as well as DESI.[37] Brazilian real, dollar and euro bills were used as proof-of-principle samples, and both direct EASI-MS and DESI-MS were shown to function as an instantaneous, reproducible and nondestructive method for chemical analysis of banknotes (Figure 9.18). Characteristic chemical profiles were also observed for counterfeit bills made using different printing processes (inkjet, laserjet, phaser and off-set printers). The general applicability of ambient MS analysis for anticounterfeiting strategies particularly *via* the use of "invisible ink" markers was discussed.

9.6.2   Simplicity of EASI but at the Cost of Sensitivity?

As for most, if not all ambient MS techniques, due to the need to desorb the analytes directly from surfaces, direct analysis without any preconcentration or derivatization protocols, sensitivity is not always the best attribute. But the superb ion transmission and detection efficiency of modern mass spectrometers (even for miniature versions) compensate for these limitations and therefore no serious compromises of sensitivity are normally observed. For EASI and its sister techniques V-EASI and S-EASI, as well as its most recent variation thermal imprinting EASI (TI-EASI), for instance,

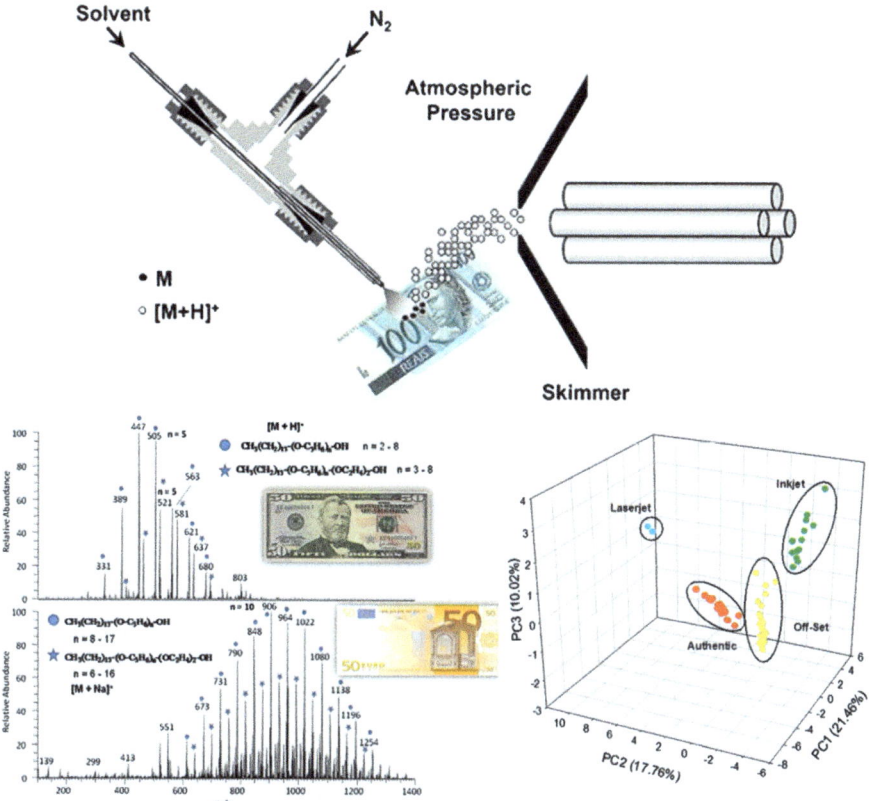

Figure 9.18 EASI-MS screening of counterfeit bank notes.
Adapted from ref. 37.

we have systematically observed less-abundant ions (as measured by absolute ion counts) but superior results in terms of S/N ratios. This gain has been associated with the reduced charge states of the SSI droplets hence with more selective ionization. The superior S/N ratio is a welcome feature of EASI, which greatly benefits detectability in ambient MS analysis.

9.6.3 Quantitation by EASI?

Despite its somewhat less-controlled fashion and the first impression that EASI would not have quantitation as one of its best figures-of-merits, quantitation has been proved to work as well. We showed, for instance, that FFA can be quantitated in crude vegetable oils with external standards by EASI-MS with reasonable linearity $(r = 0.98)$.[38] Recently, we have also showed the use of EASI-MS for the simple, fast and reliable qualitative analysis of TAG in vegetable oils.[39] In the study, TAG in edible vegetable oils, hydrogenated vegetable oils and cocoa butter were quantitated directly using EASI-MS. The results were compared with those obtained by GC-FID and mathematically by a software projection, which uses a mathematical algorithm of distribution of the FA in the triacylglycerol molecule. Good correlation coefficients were observed between the three methods for vegetable and hydrogenated oils. EASI-MS, therefore, seems to offer a promising substitute for more demanding and time-consuming gas chromatographic protocols.

9.7 Best EASI (and Sister Techniques) Performance

9.7.1 Best Sonic-Spray Nebulizers

Recently, the development of a novel and high-performance nebulizer configuration for V-EASI was reported.[40] The developed nebulizer configuration was based on a commercially available pneumatic glass nebulizer (Figure 9.19) that has been extensively used for aerosol formation in atomic spectrometry. The modified nebulizer was modified and optimal V-EASI-MS operation was achieved.

It was then demonstrated that this improved V-EASI-MS system offers distinct advantages for the analysis of coordination compounds and redox active inorganic compounds over the predominantly used ESI-based techniques. The authors also note that since V-EASI employs a "milder" ionization process (than ESI), and requires no electrical potentials for gas-phase ion formation, it was able to eliminate unwanted redox transformations, allowing also for the "simultaneous" detection of negative and positive ions (bipolar analysis) without the need to change source-ionization conditions, and also not requiring the use of syringes and delivery pumps. The authors also found that source temperature and accelerating voltage did not affect labile compounds to the extent they do in ESI-based techniques. In addition,

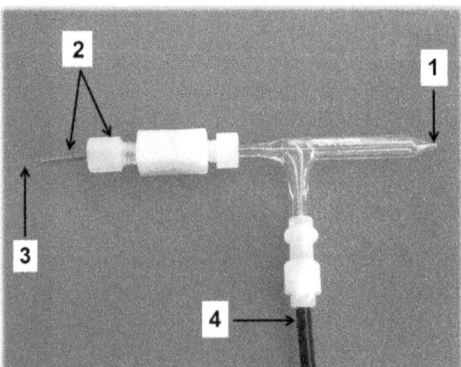

Figure 9.19 Modified Meinhard® pneumatic glass nebulizer used for optimal V-EASI-MS: (1) fused silica capillary located at glass nebulizer tip; (2) nut and PEEK sleeve used to secure fused silica capillary in place; (3) fused silica capillary going to sample; (4) 40–60 psi N_2 gas line. Adapted from ref. 40.

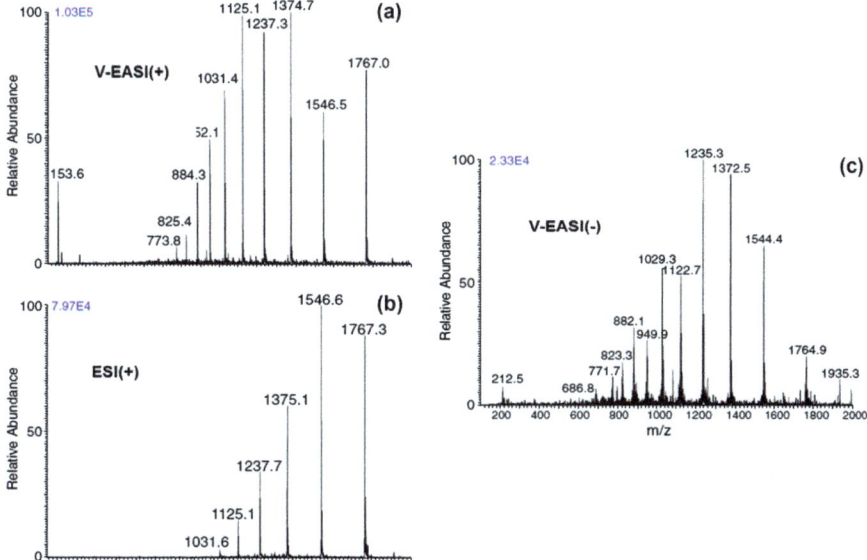

Figure 9.20 (a) V-EASI(+)-MS, (b) ESI(+)-MS, and (c) V-EASI(−)-MS of a 100 μg mL^{-1} solution of cytochrome c in 1% vol/vol DMSO. The V-EASI(±)-MS were acquired using a commercially available pneumatic glass nebulizer that provided superior performance.
Adapted from ref. 40.

bipolar analysis of proteins was demonstrated (Figure 9.20) by acquiring both positive- and negative-ion mass spectra from the same protein solutions, without the need to independently adjust solution and source conditions in each mode. Finally, the simple and efficient operation of a dual-nebulizer configuration was demonstrated for V-EASI-MS.

9.7.2 Increasing EASI Sensitivity with Dopants

One of the main advantages of EASI as compared to ESI-based ambient techniques is not absolute ion abundances but superior S/N ratios. However, although not essential for MS analysis, where the S/N ratio is the key parameter for improved sensitivity, it remains to be shown that EASI *via* SSI can also provide similar or even superior absolute ion abundances. Very recently, this ability of EASI was demonstrated in that absolute ion abundances as high or even higher than ESI were obtained by SSI (hence by EASI and its sister techniques) by the use of a SSI dopant.[41]

The dopant was 3-nitrobenzonitrile (3-NBN), which provided significant gains in ion abundances (from 1.6 to 10 fold) for dopant-SSI-MS as compared to conventional bare SSI-MS for almost all compounds tested. When compared to ESI-MS, dopant-SSI-MS was found to provide equal or better absolute intensities, although noise seemed to increase concomitantly (S/N ratios were not reported).

9.8   EASI (and Sister Techniques) in Life Science

9.8.1   An artificial EASI-MS Tongue

Very recently[42] a new artificial tongue based on an electro-version of V-EASI, namely Venturi electrosonic-spray ionization (V-ESSI) coupled to a cataluminescence (CTL) sensor array was reported for discriminating various saccharides in solution (Figure 9.21). The system integrates an electrosonic-spray ionization (ESSI) source, a liquid system of Venturi self-pumping injection for the CTL reaction, and was fabricated for enhancing CTL reactivity of aqueous samples. Comparing with simple Venturi injection by air and V-EASI without electric assistance, a remarkable enhancement of CTL signals was observed from V-ESSI. This system gave different signals for a given saccharide on different catalysts and different responses for different saccharides on the same catalyst. Then, a 4×2 CTL sensor array was used for obtaining "fingerprints" of distinct CTL response patterns. Analyzed by linear discriminant analysis (LDA), the V-ESSI CTL sensor array discriminated different saccharides (99.9% of total variation) as well as four groups of urine sugar-level for urine samples from diabetic patients (98.1% of discrimination accuracy). This new "artificial tongue" was shown therefore to provide a simple, rapid, low-cost, low energy consumption, and environmentally friendly protocol for aqueous sample discrimination, expanding applications of the CTL-based senor array and being applicable to clinical diagnoses, environment monitoring, industrial controls, food industry, and various marine monitoring.

9.9   MS Imaging by EASI

In recent work, the abilities of EASI and DESI to provide reliable MS images of a number of samples, including sections of rat brain and imprints of plant material on porous Teflon, were compared (Figure 9.22).[43] Images were recorded with the individual rows alternating between EASI and DESI, yielding a separate image for each technique recorded under what have been considered "perfectly similar conditions" on the same sample.

We note, however, that to achieve best EASI performance clearly substantial changes in ESI spray systems and the optimization of the nebulizer for best sonic-spray EASI performance would be required. This optimization can be performed on ESI nebulizers but best EASI or V-EASI conditions, as shown by the study of the optimal V-EASI nebulizer of Pergantis and coworkers,[40] would require special tubing as well as system tuning for best EASI spray conditions in order to optimize droplet size and solvent evaporation, for instance. Despite the not ideal comparison, certainly biased towards better DESI performance, the comparative MS imaging study still found that EASI worked quite remarkably for imaging of all samples. The choice of spray solvent and flow rate was more critical in tissue imaging with EASI than with DESI.

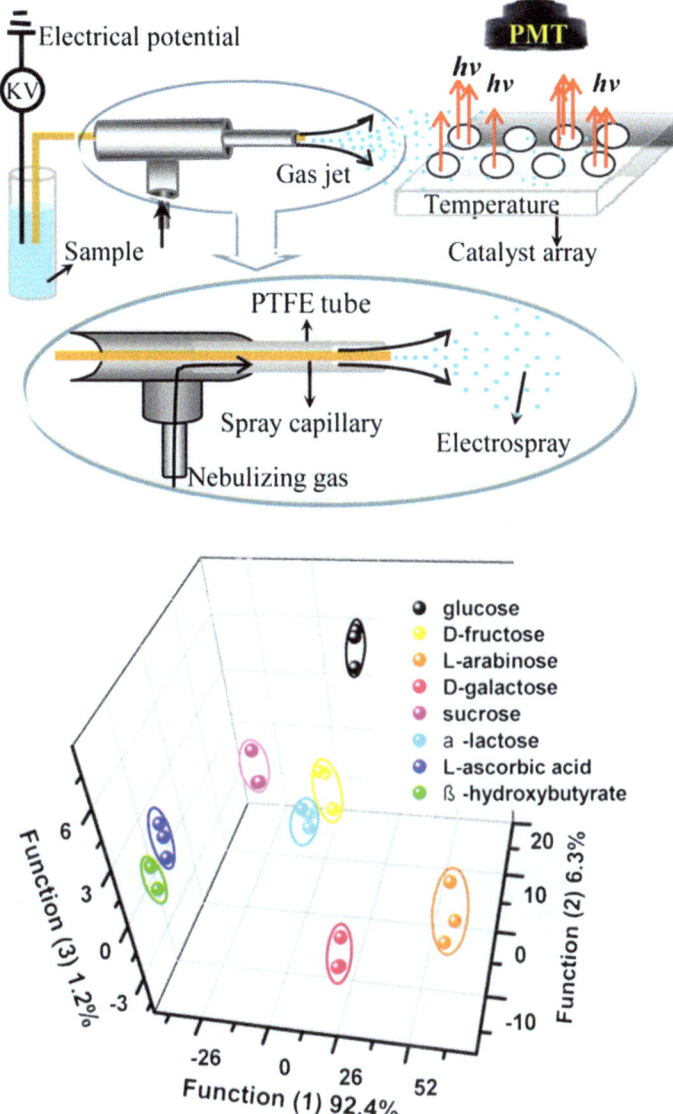

Figure 9.21 (top) Schematics of the new artificial tongue able to discriminate saccharides in solution and (bottom) the level of discrimination obtained as revealed by chemometric data treatment.
Adapted from ref. 42.

The overall sensitivity (based on both absolute ion abundances) of EASI (S/N were not compared) was, in general, just slightly lower than that of DESI. Note that the use of proper EASI nebulizers and spray conditions as well as probably dopants as discussed above could lead to much enhanced EASI images. An obvious advantage of EASI as compared to DESI for MS

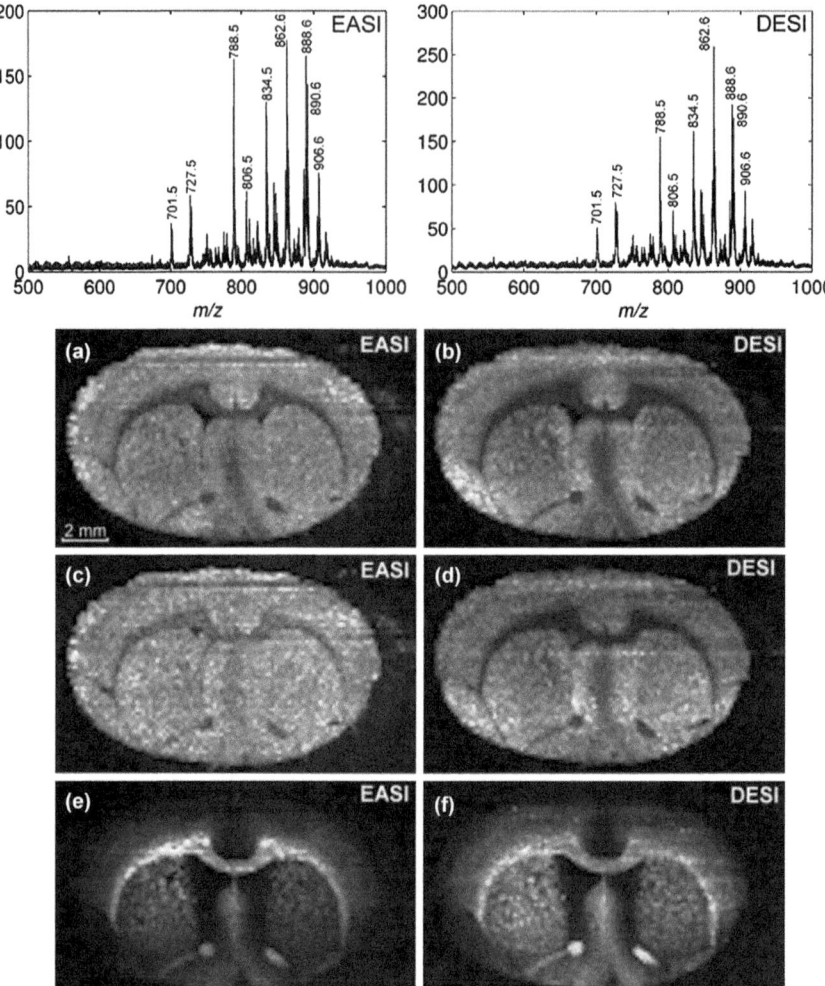

Figure 9.22 Representative EASI(−)-MS and DESI(−)-MS of sections of rat brain recorded in the negative-ion mode showing nearly identical data and S/N ratios, and comparison of the resulting EASI(−)-MS and DESI(−)-MS images.
Adapted from ref. 43.

imaging is that it can be used in cases where the application of high voltage is impractical or undesirable.

9.10 Biomarker Screening *via* an "Intelligent V-EASI Knife"

Very recently also, an "intelligent knife" that can tell surgeons immediately whether the tissue they are cutting is cancerous or not has been developed

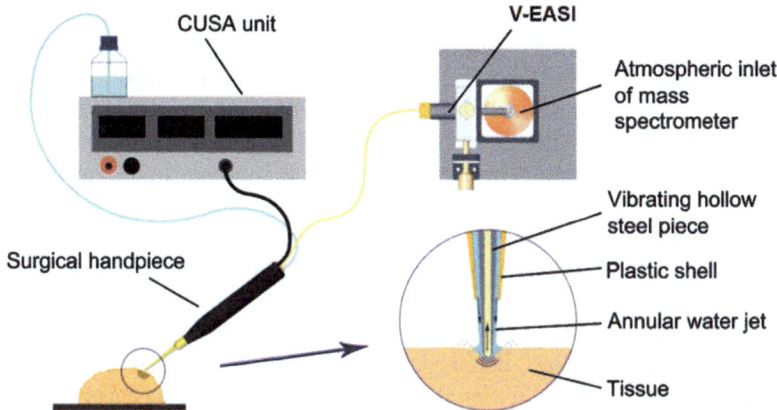

Figure 9.23 Intelligent knife use to perform *in situ* screening of cancer biomarkers
via V-EASI-MS.
Adapted from ref. 44.

based on V-EASI.[44] The "intelligent knife" was constructed to screen for
cancer biomarkers during cancer surgery by coupling a commercially available ultrasonic surgical device to a V-EASI source (Figure 9.23). Liquefied
tissue debris were sucked by the V-EASI apparatus, which could efficiently
nebulize the suspended tissue material for gas phase ion production.
V-EASI(+)-MS and V-EASI(−)-MS data were obtained from brain and liver
samples reflecting the primary application areas of the surgical device. Lipid
and peptides fingerprints of the tissues were collected in both ion-polarity
modes. The MS spectra of the intact tissue specimens were found to be
highly specific to the histological tissue type. The method has been successfully tested on postmortem and *ex vivo* human samples including
astrocytomas, meningiomas, metastatic brain tumors, and healthy brain
tissue. The technology has been already transferred to the operating theatre
to perform real-time analysis during surgery. Tests have shown that the
tissue type identified by the iKnife matched the postoperative diagnosis
based on traditional methods.

9.11 Conclusions and Perspectives

Ambient MS is still a very juvenile field but has already experienced an explosive growth in terms of many new variants, combinations, hybridization
and applications. Ambient MS has provoked a revolution in MS analysis,
bringing it definitively "back to easiness" as pointed out by Thomson. The
field has been very successful, and has tremendously simplified MS analysis, as compared to the tough high-vacuum early days of mass spectrometry:
little or literally no sample preseparation, preparation or derivatization is
required and the mass spectra are most often acquired directly for samples in
their open atmosphere real-world natural environment.

EASI and its sister techniques have therefore been developed and optimized using the ambient MS keyword in mind, that is, using "simplicity" as its main guide. But although EASI is the third most employed ambient MS techniques (third to DESI and DART) and therefore the 1st noncommercial ambient technique,[45] and certainly one of the easiest techniques in its portfolio, it still remains an underevaluated and underutilized ambient MS technique.

This limited use of EASI is so not for lack of adequate analytical figures of merit, as already commented, but most likely because of the dominance of the Nobel-laureate and widely commercially available electrospray ionization (ESI) and therefore DESI, as well as further competition for the commercially available solvent-free glow discharge-based DART technique. While recognizing the merits and great benefits of DESI and DART, as well as other suitable ambient MS techniques, we have tried hard to promote the benefits of EASI by demonstrating its potential to become a major, widely used ambient MS protocol. In particular, the milder ionization conditions affording less fragmentation of labile compounds, superior S/N ratios particularly important for EASI-MS imaging, lack of voltage, electrical discharge or heating artifacts for improved selectivity, and extreme ease of use and assembling, with the possibility demonstrated by the S-EASI source for inexpensive, disposable units, are some of the major advantages of EASI.

EASI also seems quite suitable for portable, robust, easy to operate, low cost *fit-for-purpose* mass spectrometers. Applying the "survival of the easiest" principle, we envisage a long and promising future and increasing use with several developments aimed at superior performance for EASI and its sister techniques.

References

1. J. J. Thomson Rays of Positive Electricity and Their Application to Chemical Analyses Cambridge, 1913.
2. J. B. Fenn, M. Mann, C. K. Meng, S. F. Wong and C. M. Whitehouse, *Science*, 1989, **246**, 64–71.
3. F. Coelho and M. N. Eberlin, *Angew. Chem., Int. Ed.*, 2011, **50**, 5261–5263.
4. Z. Takats, J. M. Wiseman, B. Gologan and R. G. Cooks, Mass spectrometry sampling under ambient conditions with desorption electrospray ionization, *Science*, 2004, **306**, 471–473.
5. H. Wang, J. Liu, R. G. Cooks and Z. Ouyang, Paper spray for direct analysis of complex mixture using mass spectrometry, *Angew. Chem., Int. Ed.*, 2010, **49**, 877–880.
6. N. V. Schwab and M. N Eberlin, *Drug Test. Anal.*, 2013, **5**, 3–4.
7. C. Wu, W. F. Siems and H. H. Hill, *Anal. Chem.*, 2000, **72**, 396–403.
8. H. Chen, A. Venter and R. G. Cooks, *Chem. Commun.*, 2008, 2042–2044.
9. J. Liu, H. Wang, N. E. Manicke, J. M. Lin, R. G. Cooks and Z. Ouyang, *Anal Chem.*, 2011, **83**(4), 1197–1201.

10. J. Liu, H. Wang, N. E. Manicke, J. M. Lin, R. G. Cooks and Z. Ouyang, *Anal. Chem.*, 2010, **82**, 2463–2471.
11. A. Hirabayashi, M. Sakairi and H. Koizumi, *Anal. Chem.*, 1994, **66**, 4557–4559.
12. R. M. Alberici, R. C. Simas, G. B. Sanvido, W. Romão, P. M. Lalli, M. Benassi, I. B. S. Cunha and M. N. Eberlin, *Anal. Bioanal. Chem.*, 2010, **398**, 265–294.
13. R. Haddad, R. Sparrapan and M. N. Eberlin, Desorption sonic spray ionization for (high) voltage-free ambient mass spectrometry, *Rapid Commun. Mass Spectrom.*, 2006, **20**, 2901–2905.
14. R. Haddad, R. Sparrapan, T. Kotiaho and M. N. Eberlin, Easy ambient sonic-spray ionization-membrane interface mass spectrometry for direct analysis of solution constituents, *Anal. Chem.*, 2008, **80**, 898–903.
15. M. B. Sorensen, P. Aaslo, H. Egsgaard and T. Lund, Determination of D/L-amino acids by zero needle voltage electrospray ionization, *Rapid Commun. Mass Spectrom.*, 2008, **22**, 455–461.
16. R. Haddad, H. M. S. Milagres, R. R. Catharino and M. N. Eberlin, Easy ambient sonic-spray ionization mass spectrometry combined with thin-layer chromatography, *Anal. Chem.*, 2008, **80**, 2744–2750.
17. L. S. Eberlin, P. V. Abdelnur, A. Passero, G. F. de Sá, R. J. Daroda, V. Souza and M. N. Eberlin, Analysis of biodiesel and biodiesel-petrodiesel blends by high performance thin layer chromatography combined with easy ambient sonic-spray ionization mass spectrometry, *Analyst*, 2009, **134**, 1652–1657.
18. E. C. Figueiredo, G. B. Sanvido, M. A. Zezzi and M. N. Eberlin, Molecularly imprinted polymers as analyte sequesters and selective surfaces for easy ambient sonic-spray ionization, *Analyst*, 2010, **135**, 726–730.
19. V. G. Santos, T. Regiani, F. F. G. Dias, W. Romão, J. L. P. Jara, C. F. Klitzke, F. Coelho and M. N. Eberlin, Venturi Easy Ambient Sonic-Spray Ionization, *Anal. Chem.*, 2011, **83**(4), 1375–1380.
20. G. W. Amarante, H. M. S. Milagre, B. G. Vaz, B. R. V. Ferreira and M. N. Eberlin, Dualistic Nature of the Mechanism of the Morita–Baylis–Hillman Reaction Probed by Electrospray Ionization Mass Spectrometry, *J. Org. Chem.*, 2009, **74**, 3031–3037.
21. N. V. Schwab, A. M. Porcari, M. B. Coelho, J. L. Jara, J. V. Visentainer and M. N. Eberlin, Easy dual-mode ambient mass spectrometry with Venturi self-pumping, canned air, disposable parts and voltage-free sonic-spray ionization, *Analyst*, 2012, **137**, 2537–2540.
22. A. M. Porcari, N. V. Schwab, R. M. Alberici, E. C. Cabral, D. R. Moraes, P. F. Montanher, C. R. Ferreira, M. N. Eberlin and J. V. Visentainer, *Anal. Methods*, 2012, **4**(11), 3551–3557.
23. A. M. Porcari, A. Tata, K. R. A. Belaz, N. V. Schwab, V. G. Santos, G. D. Fernandes, R. M. Alberici, V. A. Gromova, M. N. Eberlin and A. T. Lebedev, *Anal. Methods*, 2014, **6**, 2436–2443.

24. G. D. Fernandes, W. Moreda, D. B. Arellano, N. V. Schwab, G. C. N. Ruiz, P. L. Ferreira, M. N. Eberlin and R. M. Alberici, *J. Braz. Chem. Soc.*, in press.
25. R. Haddad, R. R. Catharino, L. A. Marques and M. N. Eberlin, *Rapid Commun. Mass Spectrom.*, 2008, **22**, 3662–3666.
26. A. S. Saraiva, P. V. Abdelnur, R. R. Catharino, G. Nunes and M. N. Eberlin, Fabric softeners : nearly instantaneous characterization and quality control of cationic surfactants by easy ambient sonic-spray ionization mass spectrometry, *Rapid Commun. Mass Spectrom.*, 2009, **23**, 357–362.
27. R. C. Simas, R. R. Catharino, I. B. S. Cunha, E. C. Cabral, D. Barrera-Arellano, M. N. Eberlin and R. M. Alberici, Characterization of Vegetable Oils *via* TAG and FFA Profiles by Easy Ambient Sonic-Spray Ionization Mass Spectrometry, *Analyst*, 2010, **135**, 738–744.
28. P. V. Abdelnur, L. S. Eberlin, G. F. de Sá, V. Souza and M. N. Eberlin, Single-shot biodiesel analysis: nearly instantaneous typification and quality control solely by ambient mass spectrometry, *Anal. Chem.*, 2008, **80**, 7882–7886.
29. A. C. H. F. Sawaya, P. V. Abdelnurb, M. N. Eberlin, S. Kumazawac, M. R. Ahnd, K. S. Bangd, N. Nagarajae, V. S. Bankovaf and H. Afrouzang, Fingerprinting of propolis by easy ambient sonic-spray ionization mass spectrometry, *Talanta*, 2010, **81**, 100–108.
30. Y. E. Corilo, B. G. Vaz, R. C. Simas, H. D. L. Nascimento, C. K. Klitzke, R. C. L. Pereira, W. L. Bastos, E. V. Santos Neto, R. P. Rodgers and M. N. Eberlin, Petroleomics by EASI$(+/-)$ FT-ICR MS, *Anal. Chem.*, 2010, **82**(10), 3990–3996.
31. R. M. Alberici, R. C. Simas, G. F. de Sá, R. J. Daroda, V. Souza and M. N. Eberlin, Analysis of fuels *via* easy ambient sonic-spray ionization mass spectrometry, *Anal. Chem. Acta*, 2010, **659**, 15–22.
32. P. H. Amaral, R. Fernandes, M. N. Eberlin and N. F. Hoer, *J. Mass Spectrom.*, 2011, **46**, 1269–1273.
33. P. M. Lalli, G. B. Sanvido, J. S. Garcia, R. Haddad, R. G. Cosso, D. R. J. Maia, J. J. Zacca, A. O. Maldaner and M. N. Eberlin, Fingerprinting and aging of ink by easy ambient sonic-spray ionization mass spectrometry, *Analyst*, 2010, **135**, 745–750.
34. M. F. Riccio, A. Sawaya, P. Abdelnur, S. Saraiva, R. Haddad, M. N. Eberlin and R. R. Catharino, *Anal. Lett.*, 2011, **44**, 1489–1497.
35. E. C. Cabral, R. C. Simas, V. G. Santos, C. L. Queiroga, V. S. da Cunha, G. F. de Sá, R. J. Daroda and M. N. Eberlin, *J. Mass Spectrom.*, 2012, **47**(1), 1–6.
36. R. S. Galaverna, P. T. B. Sampaio, L. E. S. Barata, M. N. Eberlin and C. H. V. Fidelis, *Phytochemistry*, 2014, submitted.
37. R. Haddad, R. C. Sarabia Neto, R. G. Cosso, D. R. J. Maia, A. O. Maldener, J. J. Zacca, G. B. Sanvido, W. Romão, B. G. Vaz, L. S. Eberlin, D. R. Ifa, N. E. Manicke, A. Dill, R. G. Cooks and M. N. Eberlin, Instantaneous Chemical Profiles of Banknotes by Ambient Mass Spectrometry, *Analyst*, 2010, **135**(10), 2533–2539.

38. R. C. Simas, R. R. Catharino, I. B. S. Cunha, E. C. Cabral, D. Barrera-Arellano, M. N. Eberlin and R. M. Alberici, *Analyst*, 2010, **135**, 738–744.

39. A. M. A. P. Fernandes, G. D. Fernandes, R. C. Simas, D. Barrera-Arellano, M. N. Eberlin and R. M. Alberici, *Anal. Methods*, 2013, 5(24), 6969–6975.

40. M. M. Antonakis, A. Tsirigotaki, K. Kanaki, C. J. Milios and S. A. Pergantis, *J. Am. Soc. Mass Spectrom.*, 2013, **24**, 1250–1259.

41. SI dopant.

42. J. Y. Han, F. F. Han, J. Ouyang, Q. M. Li and N. Na, *Anal. Chem.*, 85(16), 7738–7744.

43. C. Janfelt and A. W. Nørgaard, *J. Am. Soc. Mass Spectrom.*, 2012, **23**, 1670–1678.

44. K.-C. Schäfer, J. Balog, T. Szaniszló, D. Szalay, G. Mezey, J. Dénes, L. Bognár, M. Oertel and Z. Takáts, *Anal. Chem.*, 2011, **83**, 7729–7735.

45. G. A. Harris, A. Galhena and F. M. Fernandez, *Anal. Chem.*, 2011, **83**, 4508–4538.

CHAPTER 10

Secondary Electrospray Ionization

CHRISTIAN BERCHTOLD

ETH Zürich, Switzerland
Email: christian.berchtold@insel.ch

10.1   Introduction

A major aspect of modern analytics is the development of fast and robust methodologies to reduce the costs for each sample analysis. A variety of direct ionization methods have been developed based on classical electrospray ionization. These techniques use a pure solvent electrospray to deliver charges into the gas-phase or liquid-sample molecules (*e.g.* in aerosol droplets). These techniques, such as secondary electrospray ionization (SESI), sometimes referred to as extractive electrospray ionization (EESI), are very promising new technologies, which allow sample analysis without laborious sample preparation. An overview of the possible configurations is given in Figure 10.1. The core element of all these techniques is the electrospray introduction of clean solvent, which is either intercepted with a sample spray or sample-gas delivery. In the case of the intercepted sprays, the term extractive or fused droplet electrospray ionization may be used. In the case of the gas-delivery system, the term neutral desorption electrospray ionization (ND-EESI) or SESI are used. In order to clarify the terminology for all these techniques, it is highly recommended to use secondary electrospray ionization (SESI) since it does not suggest any mechanism such as fused droplet or extractive ESI.[1]

New Developments in Mass Spectrometry No. 2
Ambient Ionization Mass Spectrometry
Edited by Marek Domin and Robert Cody
© The Royal Society of Chemistry 2015
Published by the Royal Society of Chemistry, www.rsc.org

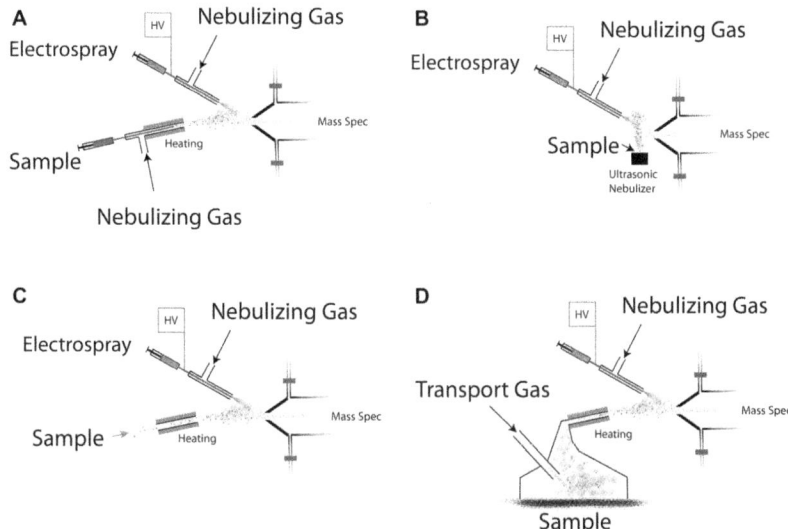

Figure 10.1 Configurations of SESI. (A) SESI in the dual spray configuration; liquid sample is transferred with a pneumatic nebulized spray into the ESI charging spray, also sometimes called EESI or FD-EESI. (B) Ultrasonic nebulizer; a liquid droplet is placed on the nebulizer and dispersed (by ultrasonic waves) in the intersection between the electrospray and mass-spectrometer inlet. (C) SESI for breath and gas analysis; breath (or another aerosol containing gas) is directed into the ESI plume. The transfer tube is heated to avoid condensation and the flow rate is potentially controlled to gain more reproducible results. (D) Neutral desorption EESI; nitrogen (or any other suitable gas) transports vaporous samples (from skin or headspace of a sample) into the electrospray plume.

In this chapter we will take a closer look at direct ionization techniques for both gas- and liquid-sample analysis, based on electrospray ionization. The recommended term SESI is used to include all possible configurations of such a setup. First, we will take a brief look at its development in a historical context in Section 10.2. The mechanistic aspects will be discussed in Section 10.3. In Section 10.4 we will discuss how to set up and optimize a SESI ion source and describe which factors have to be considered to develop a running SESI system. Finally, some practical examples of SESI applications are discussed Section 10.5. The chapter is concluded in Section 10.6. Section 10.7 acknowledges the people and institutions who influenced this article.

10.2 Development of Secondary Electrospray Ionization Technique

Yamashita and Fenn[2] recognized as early as 1984 that electrospray ionizes molecules present in the nebulizing gas and in the atmosphere. This was always regarded as a problematic issue, since such signals usually interfere with the analytes of interests. More than 10 years later, this fact was turned into an advantage by using this well-known effect on purpose. The story of electrospray-based secondary ionization techniques started in 1998 with the development of two-step electrospray ionization for the investigation of re-action products.[3] In the same year, a similar setup was used to ionize volatile organic compounds separated by GC.[4] This technique called fused droplet electrospray ionization, was introduced by Shiea and coworkers, and was further optimized in 2001[5] to ionize proteins nebulized by an ultrasonic nebulizer (such as Figure 10.1(A)). In a parallel development, secondary electrospray ionization (SESI) was introduced for forensic application and breath analysis in 2000.[6] Fernández de la Mora showed the use of SESI of breath analysis,[7] whereas Hill and coworkers showed how to use the method to analyze the gases derived from micro-organisms.[8]

Chen *et al.*[9] Introduced the term extractive ESI (EESI) for the analysis of complex matrices such as urine or milk without sample preparation in 2006. Chen's first configuration (Figure 10.1(B)) used an ultrasonic nebulizer for sample delivery. The ultrasonic nebulizer was later replaced by a second spray (comparable to the fused droplet configuration by Shiea). A liquid or dissolved sample was directly dispersed into the plume of an electrospray (containing all the matrix without precleanup). In 2007 Chen *et al.* intro-duced neutral desorption EESI (ND-EESI) as a further modification of the setup. ND-EESI is based on the transport of sample aerosol and gas by the gas flow of an inert gas[10] (Figure 10.1(D)). All of these techniques are discussed in two main review articles.[11,12]

The driving force behind the development of SESI techniques is the potential for new applications to analyze complex sample matrices without laborious pretreatment. As seen in Section 10.5, matrices such as fruits, breath or reaction mixtures have been analyzed. The development of SESI based techniques is an ongoing process, which, to the author's best knowledge, has not yet been commercialized. Ultimately, it is possible that commercialization may not even be necessary, since it is possible to modify existing ion sources and to fit this source on any instrument (see Section 10.4).

10.3 Ionization Mechanism

The major discussion about the underlying mechanism of SESI is the question of whether a gas-phase ion–molecule reaction takes place or whe-ther droplets interact in the liquid phase. The term "extractive" in EESI or "fused" in FD-ESI suggests that a physical extraction and droplet interaction step is involved in the ionization mechanism, whereas for the overall term SESI this issue is not specified. Therefore, it is a safe approximation to de-scribe all of these techniques as SESI, regardless of the actual mechanism.

Law *et al.*[13] investigated the influence of the mutual solubility of the solvents used in the sample and ionization spray. The intention of this study was to show that an extraction step is involved and therefore it is important that the droplets are able to mix as a liquid. A significant influence on the signal intensity was found to depend on the mutual solubility of sample spray solvent and ionization spray solvent. If the spray solvent and the ionization solvent are miscible, a significantly higher signal was observed. This supports the theory that liquid droplets merge prior to ionization. On the contrary, a later study by Meier *et al.*,[14] showed that the gas phase also plays an important role in the efficiency of EESI ionization. This study showed that the ionization for volatile molecules is about 100 times more efficient if delivered as vapor than as liquid droplets. Meier comes to the conclusion that, at least for volatile molecules (such as the tested amines), the gas-phase mechanism (Figure 10.2(C)) must be applied. Wang *et al.*[15] investigated the influence between two sprays in a two-spray SESI (sample spray and ionization spray liquid) configuration. In this article it was shown that the droplet fragmentation is the most dominant effect between the intercepted sprays, which causes smaller droplets. Coalescence of solvent droplets (Figure 10.2(A)) was not observed, although the process was not completely excluded. Nevertheless, a dominant liquid–liquid extraction step seems to be unlikely. Therefore, a mechanism as shown in Figure 10.2(B) is better supported by the recent articles. Fernández de la Mora has shown by calculations that the ion–molecule reaction as well as the extraction process could result in the same signal intensity.[7] Although there is a strong indication for a nonextractive mechanism, there is no proof for it. Further investigations may be needed to clarify this issue.

Although the mechanistic aspects appear to be not completely resolved, it is obvious that the charges are delivered by the primary electrospray and that these charges are transferred to the sample, permitting the use of this ionization method for the applications shown.

10.4   Experimental Aspects

SESI does not require laborious sample preparation and it is therefore easy to handle from the point of view of sample analysis. Nevertheless, its proper optimization and configuration are of major importance to analyze any sample with suitable efficiency. Since the configuration is more complex than for a standard electrospray, it is important to optimize the setup for successful SESI analyses. This section introduces all the necessary factors in order to successfully construct, optimize and use an SESI ion source. In general, such an ion source is easily made and fitted on any ambient mass spectrometer available. Chen and Zenobi described the details of a proper SESI and EESI configuration in a *Nature Protocols* article.[16] Figure 10.3 shows the most important factors, which are indicated with respect to dual-spray EESI or SESI configurations.

The core element of an SESI source is the electrospray, which delivers the charges for ionization. Its proper configuration is comparable to the configuration of a standard ESI source. Major differences to classical ESI are the generally lower flow rates of the charging spray (below 100 µL per minute) as well as the orthogonal sample introduction. Electrospray and sample delivery

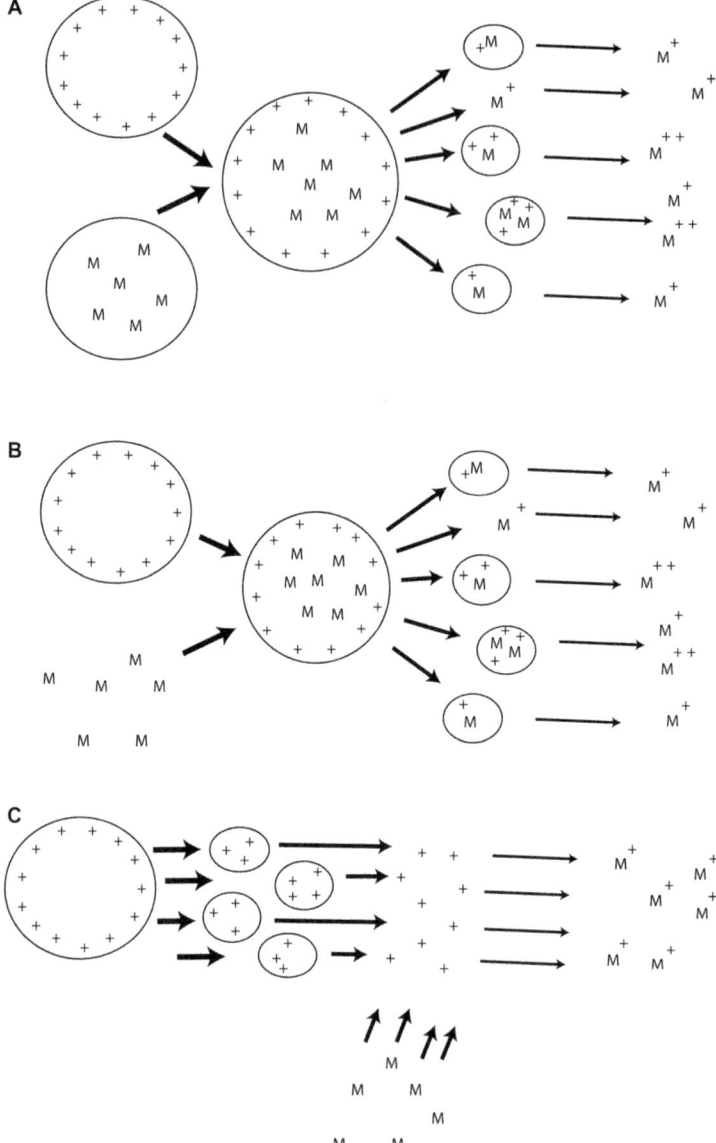

Figure 10.2 Potential mechanism of secondary extractive electrospray ionization. (A) A charged droplet formed by an electrospray source (containing only acid, water and solvent) are merged with sample droplets. After the merging, the mechanism is similar to ESI. (B) Neutral gas-phase molecules are picked up by a charged droplet. The charged droplets undergo the same mechanism as found in electrospray. This mechanism could also appear in the opposite direction if gas-phase ions interacted with neutral sample droplets. (C) Gas-phase ion–molecule reaction (comparable to atmospheric-pressure chemical ionization),[13] which involves no droplet–droplet or droplet–molecule interaction.

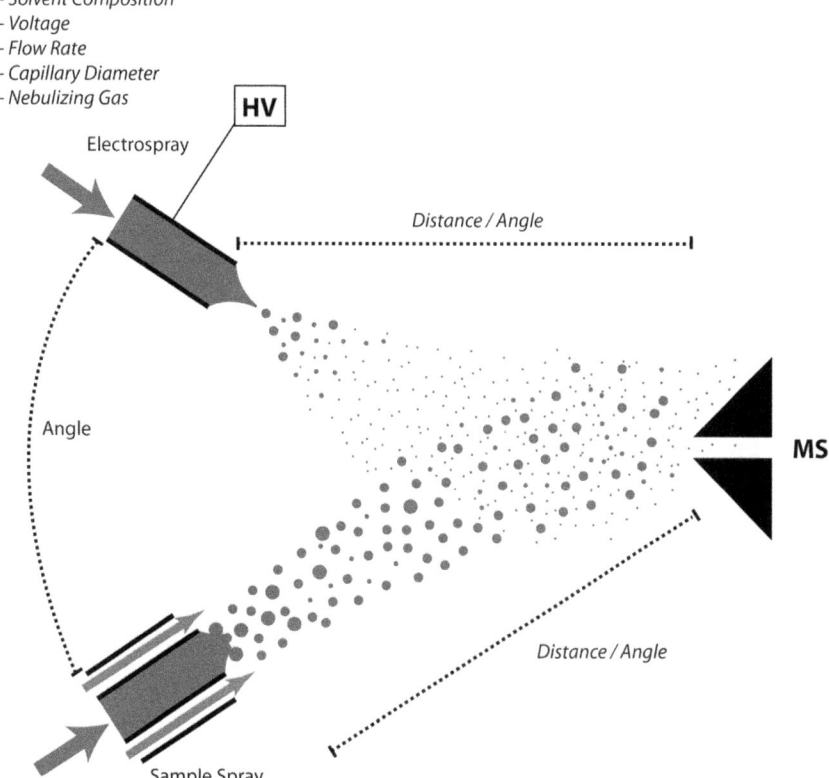

- Solvent Composition
- Voltage
- Flow Rate
- Capillary Diameter
- Nebulizing Gas

Electrospray

HV

Distance / Angle

Angle

MS

Distance / Angle

Sample Spray

- Solvent composition
- Flow Rate
- Nebulizing Gas
- Sample Delivery Rate
- Temperature
- Capillary Diameter

Figure 10.3 Factors to set up a SESI source. An electrospray, containing water, methanol and acid (or base) delivers the charges to the neutral sample spray. The ionization spray is a classical electrospray, and its performance depends on the solvent composition, capillary parameters, the voltage (electric field), and liquid flow rate, nebulizing gas flow as well as angle and distance. The sample spray, which is in a dynamic equilibrium with the electrospray, depends on the solvent composition, flow rate (nebulizing gas and sample liquid, gas, or aerosol), temperature, distance and angle. All these factors need to be optimized iteratively, whereas first the hardware is set (capillary, MS and so on), then the solvent and matrix and finally flow rates, voltages, angle and distance. For other configurations (Figure 10.1) the factors may have to be adapted, especially in case of sample introduction by ultrasonic nebulization (Figure 10.1(B)) or direct breath analysis, which is a non-stable process.

are in a dynamic equilibrium. Therefore, they cannot be optimized independently. Basically, it is a good strategy to configure the electrospray first, add the sample delivery and to optimize the whole setup in an iterative process.

10.4.1   Configure the Charging (Electro-) Spray

The charging spray or electrospray, which delivers the initial charges is no different from any other "classical" electrospray. The spray capillary may be constructed out of fused silica or a stainless steel capillary of about 5 to 150 micrometer inner diameter (average about 75 micrometer). As for every electrospray, the solvent composition, conductivity, surface tension, and pH, as well as the applied electric field, geometry and nebulizing (sheath gas) will influence performance. The solvent may be composed of water or methanol or any conductive mixture (*e.g.* methanol/water in a 1 : 1 ratio). It is beneficial to use a protic solvent mixture to provide additional proton donors and acceptors. The conductivity should be high, which means that conductive buffers may be added. These buffers need to be suitable for mass spectrometry use such as formate or acetate buffers. The pH should be adjusted low for positive mode (*e.g.* with formic acid) or high for negative mode (*e.g.* with a base such as ammonia). In general, an acid concentration in the range of 0.05% up to 30%[17] (usually 0.1 to 1% acid or base are sufficient) has been applied. The electric field, depending on the geometry (distance, capillary diameter) and voltage should be adjusted to provide a stable spray without causing discharges (kV range). The solvent flow rate of the spray may be adjusted in the range of 0.05 up to 100 μL min^{-1} and the nebulizing gas should support the spray without destabilizing it (depending on geometry about 1 L h^{-1} up to 1 L min^{-1}. All these factors are well known for ESI experiments and do not differ between different configurations and setups.[2] Since the electrospray is influenced by the sample delivery, it is not necessary to optimize the electrospray perfectly prior to the installation of the orthogonal sample delivery. It is sufficient to have a stable signal in order to optimize the entire setup.

10.4.2   Setting up the Sample Delivery

The hardware for sample delivery should be set up first. An inert material, such as fused silica or stainless steel is suitable but PTFE or a similar material may be used as long as it does not interact with the sample and the mass spectrometer (*e.g.* no plasticizers). There are two methods of sample delivery: continuous and discontinuous. In the case of the continuous liquid spray delivery, the distance, the overlap between the charging spray, and flow rate (liquid and nebulization gas) are the main factors to optimize. This is done in an interactive process, in the best case with a standard compound in the sample. It may also be helpful to use an additional standard in the electrospray solvent to compare the performance of both sprays based on a well-known compound. In this stage, the voltage, the distance (geometry) and the flow rates (sample, electrospray, nebulizing gas, and sample delivery gas) are optimized. The solvent and buffer composition as well as the hardware should be kept stable at this stage. To avoid heavy contamination or carryover, it is highly recommended to avoid directing the sample spray towards the mass spec inlet even though the overall signal may appear higher.

With regards to discontinuous sample delivery (ultrasonic nebulizer, ND-EESI, breath analysis), the optimization becomes even more challenging. The dynamic setup is more complex and less stable. Since background and sample signals vary during the optimization process, it is difficult to reach perfect conditions. This is only possible with a large number of repeated experiments, by adjusting each potential factor manually. To avoid handling a discontinuous sample delivery, it is sometimes possible to use a continuous and well-controlled gas flow to introduce sample repeatedly. For example, this may be achieved by using a constant gas flow for neutral desorption or using a mass flow controller in the sample delivery line combined with a constant nitrogen flow for breath analysis.[18] In this way, the sample gas flow rate will not vary, while its composition changes during sample introduction.

Finally, it is an important aspect of a properly optimized SESI source to set up a stable configuration in front of the ambient mass spectrometer interface. There are no important recommendations for the mass spectrometer and its interface as long as it is at ambient pressure. It may take time to set up such a system correctly, but it can be used on every instrument available and the parts needed (capillaries, gas tubes and voltage supplies) are relatively cheaply available.

10.5   Applications

One of the most important and exciting aspects of SESI is its capability to measure untreated complex matrices. The measurement of chemical reactions, food, biological matrices and cosmetic samples were addressed in various articles. The most important are mentioned in this section.

10.5.1   Reaction Monitoring

One of the first applications of SESI-like techniques was the monitoring of reaction products from the thermally unstable Wittig reaction intermediates.[3] The frozen reaction mixture was transferred after immediate melting, as a pure liquid into the electrospray plume. This enabled the detection of unstable reaction intermediates. In this early study it was shown that SESI is capable of operating with highly complex matrix. Reaction mixtures usually contain highly concentrated analytes, even if the intermediates have a fairy short lifetime. Therefore, the method does not provide a very high sensitivity. Nevertheless, it proved that SESI is a suitable new strategy for reaction monitoring.

The use of SESI for reaction monitoring has been recently reviewed.[19] Applications of SESI for reaction monitoring appear rarely in literature, and seem to be of minor usage. Reaction monitoring has to be considered as a special application of SESI with good utility for specific cases.

10.5.2   Food and Cosmetics

Food and chemicals in common use, such as cosmetics, are the applications that are most commonly investigated using SESI. These kind of samples are usually composed of highly complex matrices and with minimal sample preparation is very helpful to reduce analysis time and overall cost. Moreover, it is very helpful for researchers that these samples are easily available.

10.5.2.1   Milk

One of the early examples of food analysis by SESI was the detection of melamine in milk without sample preparation.[20] A single droplet of milk was nebulized with an ultrasonic nebulizer in front of the mass-spectrometer inlet. The electrospray was aimed through the cloud of nebulized milk to produce ions directly. This method was very fast (<1 s per sample) but not very sensitive (detection limit around 1 ppm or 1 mg L^{-1}). However, this is sufficient to detect milk that was deliberately contaminated with melamine. Melamine is added in high concentration to trick the Kjehldal method (which is the standard method to determine protein content by measuring the concentration of nitrogen). High protein content suggests a high milk quality and allows the seller to obtain a higher price for their product. Although the SESI method is very fast, it causes strong contamination of the mass spectrometer because the raw milk is basically spread over the mass-spectrometer inlet. Therefore, there is no commercial adaption of this research method. Nevertheless, it shows how SESI can deal with a complex matrix.

10.5.2.2   Ripening and Spoiling of Food

The spoiling of meat and the ripening of fruits has also been analyzed by ND-EESI. The vapor above the sample was analyzed to detect free amines and other signature compounds in the gas phase above the sample, with amines being detected from fish. The fingerprint pattern of all signals was used to classify the samples according to quality, by using principal component analysis (PCA or pattern recognition) to differentiate between rotten and fresh products. These patterns seemed to be very specific even if the sample was frozen. The advantage of the ND-EESI compared to the SESI approach for milk, that the instrument is not contaminated. The analysis of the headspace introduces much less contamination into the system. Moreover, the ability to analyze food even in the frozen state is very promising for future investigations. Unfortunately, there is no commercial application out of this SESI modification, although it could be implemented on every available instrument.[21]

10.5.2.3   Olive Oil

Olive oil was identified by analyzing vapor released from the surface (by blowing nitrogen over the surface of a small amount of oil). The vapor was directed into the electrospray by a setup comparable to the configuration used to determine the maturity of food. By the use of PCA it was possible to follow the aging of the oil as well as to identify its origin.[22] The most important aspect of olive-oil quality is its place of production, and especially the technique of production. SESI in this configuration allows a fast and distinct identification.

10.5.2.4 Toothpaste and Sunscreen

The fatty basis of sunscreen ages by the same mechanism as olive oil ages. Therefore, the same approach can be used to measure sunscreen. Toothpaste was analyzed with the same experimental setup in order to identify brand and quality.[23,24] All of these techniques work with headspace transported by a constant gas stream into the electrospray plume. Olive oil, sunscreen and tooth paste are basically analyzed with the same setup (Figure 10.4). This demonstrates the utility of SESI for creamy and oily compounds without sample preparation.

10.5.2.5 Perfumes

Perfumes were identified using SESI by aiming the electrospray into the vapors above perfume test strips. The tests strips were wetted with a typical test sample as is done for trials in a shop. SESI provided specific fingerprint patterns in the same way as was observed for olive oil. For example, it was possible to distinguish between fake and genuine perfumes.[25] This technique is potentially applicable on a standard electrospray setup without any further modification (Figure 10.5).

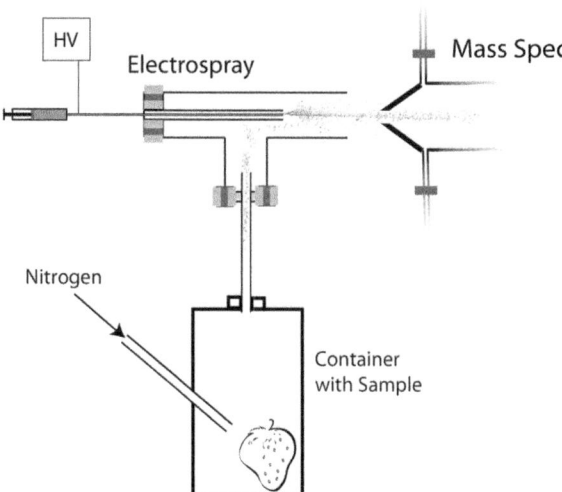

Figure 10.4 Fruits or meat are stored in containers and the released vapour is directed into the electrospray plume. As soon as the fruits start to ripen or the meat spoils the signals observed changes. This works even before a change is visible and even if the sample is frozen.

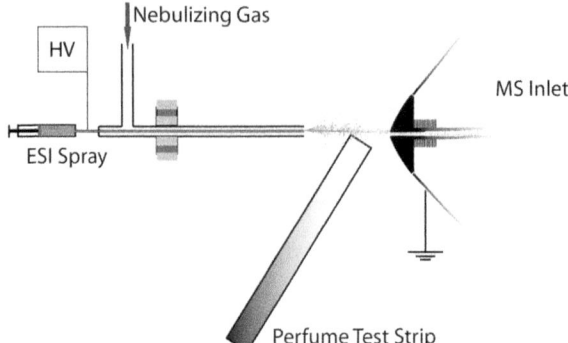

Figure 10.5 Test strips are put into the ESI plume, causing a specific pattern, which allows the identification of the perfume used.

10.5.3 Biological and Medical Matrices

Biological matrices are the most promising field of application for SESI and surely the area with the highest potential for future applications. The complexity of biological matrices is even higher than for food (generally) and its accessibility is more complicated since ethical approvals is often needed for this work. A recent review shows several important applications of SESI in this field in detail.[26]

10.5.3.1 Breath Analysis

Breath is one of the major matrices that can be analyzed with SESI, and breath analysis is a field that could lead to various new medical applications. Breath is a fairly noncomplex matrix that is directly related to the actual health status of an individual. Moreover, it is not very complicated and noninvasive to sample breath. The capability of SESI to access even molecules of low volatility, such as caffeine, is very promising. This supports the potential of the method to detect unknown and important metabolites and drugs in breath as was first shown in 2007 by Chen *et al.*[27] by detecting caffeine in exhaled breath after drinking coffee. In the same year and in parallel, Martinez-Lozano Sinues[28] showed how to use SESI to ionize biomarkers in breath and to detect specific exhaled fatty acids in a later study.[29] Other applications such as the detection and quantification of nicotine after smoking[30,31] or the detection of valproic acid (a drug used in epilepsy therapy) have been shown.[32] The individualized and time-dependent metabolic profiles of different test subjects was also analyzed. This study shows nicely how breath analysis is used to profile individuals and follow their metabolic changes over the day in real time.[18] A promising application is the use of breath analysis by SESI to detect lung infections.[33–35] These applications are still in the early phases of investigation and the setups shown are still not completely optimized. The use of

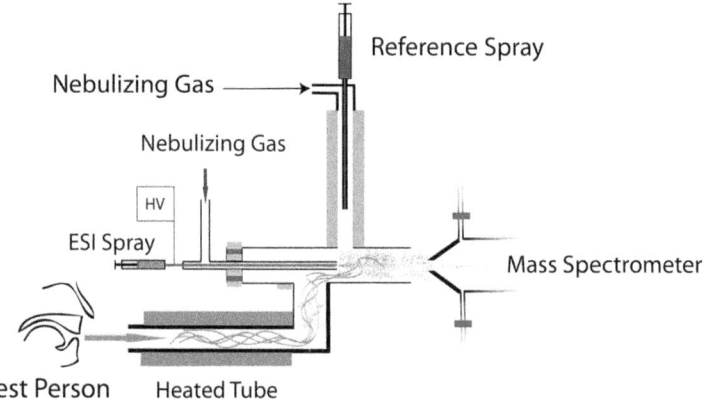

Figure 10.6 Breath is directed into the plume of the electrospray. The transfer tube is heated to avoid condensation. The pressure control allows standardization of the amount (volume) of breath delivered and the additional spray allows to calibrate and standardize with a reference standard. Such a setup allows quantification of exhaled drugs. In the case of untargeted metabolomics investigations the reference system may not be needed.

an ion funnel to enhance the sensitivity of SESI and its adoption for untargeted metabolomics in breath are only the first steps towards a new field of medical investigations (Figure 10.6).

10.5.3.2 Urine

After blood, urine is one of the most challenging biological matrices. Urine is not only very complex but its composition varies widely between different patients and even during the day, depending on the patient's constitution and consumption of food, drugs, and beverages. Therefore a complex sample preparation is usually needed to analyze such samples. Moreover, it is necessary to analyze several parameters in parallel to correct the varying water content (*e.g.* urea). SESI can reduce this workup significantly by analyzing the headspace above a urine sample. Compounds like atrazine,[36] creatinine[37] or gamma hydroxy butyrate,[38] have all been detected in urine by EESI without any sample preparation.

10.5.3.3 Skin

The human skin, as one of the largest organs, is a very important interface between the human body and the environment. Using SESI in the ND-EESI configuration, enabled the analysis of compounds directly released from the skin by blowing nitrogen over the skin surface. This enabled measurements of metabolic changes by analyzing the pattern of released fatty acids[39,40] and detection of exogenous compounds such as explosives (after exposure) or even caffeine after drinking coffee.[41]

10.6 Conclusions

SESI is a powerful technique, being able to ionize many different types of samples without any sample preparation. Although the underlying mechanism is still not fully understood, its numerous applications show that it is a very potent technique, suitable for the analysis of many different sample types with mass spectrometry. Although setup is not trivial, SESI is readily implemented for many different applications. The analysis of biological matrices such as skin or breath volatiles are especially promising since SESI can detect compounds of low volatility. Such metabolites or exhaled drugs are of major interest for therapeutic drug monitoring and diagnostics. Statistical pattern recognition, as has been demonstrated for the identification of olive oil or perfumes, is another field with a great potential. Pattern recognition can be a statistical challenge, especially if an ion source is used that is exposed to the atmosphere (such as SESI). It is therefore always important to control the environment around the source and to critically identify as many compounds as possible.

The vision of SESI is to provide a methodology that is highly sensitive and that allows analysis of many different kinds of samples without sample preparation. A very promising development in this context is the introduction of an atmospheric-pressure ion funnel in order to achieve even higher levels of sensitivity and to access lowly concentrated exhaled metabolites in real time (Figure 10.7).[42–44]

Although there is no commercial application of SESI yet, it is of interest in research and ion-source development and it has great potential for certain applications such as breath analysis.

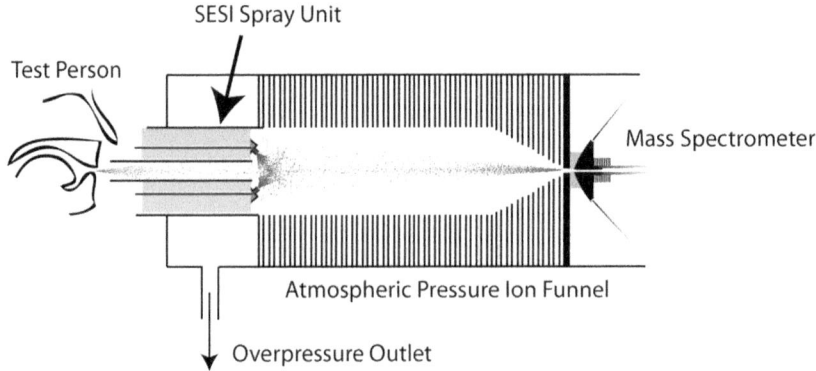

Figure 10.7   The ion funnel to optimize the efficiency of SESI. An electric field (DC and RF) supports the transport of ions into the mass spectrometer. The long distance supports the interaction of droplets or gas-phase ions and therefore leads to a 100-fold optimization of ionization efficiency. This is a recent optimization that may enhance the application of SESI for breath analysis significantly.

Acknowledgments

The knowledge gained in order to write this article is based on my Ph.D. thesis between 2009 and 2013. Therefore, I thank Prof. Dr. Renato Zenobi for his support as well as the members of the Zenobi group. Special thanks go to Dr. Pablo Martinez-Lozano Sinues as well as Dr. Lukas Meier and Dr. Stefan Schmid who supported me in many respects. Additional thanks also go to the ETH Zürich institution for providing the infrastructure for my research as well as the Swiss national foundation for founding the main part of my Ph.D. project (Grant No. K-23K1-122264).

References

1. C.-C. Lee, D.-Y. Chang, J. Jeng and J. Shiea, *J. Mass Spectrom.*, 2002, **37**, 115–117.
2. M. Yamashita and J. B. Fenn, *J. Phys. Chem.*, 1984, **88**, 4451–4459.
3. C.-H. Wang, M.-W. Huang, C.-Y. Lee, H.-L. Chei, J.-P. Huang and J. Shiea, *J. Am. Soc. Mass Spectrom.*, 1998, **9**, 1168–1174.
4. C. Y. Lee and J. Shiea, *Anal. Chem.*, 1998, **70**, 2757–2761.
5. J. Shiea, D.-Y. Chang, C.-H. Lin and S.-J. Jiang, *Anal. Chem.*, 2001, **73**, 4983–4987.
6. C. Wu, W. F. Siems and H. H. Hill, *Anal. Chem.*, 2000, **72**, 396–403.
7. J. Frenandez de la Mora, *Int. J. Mass Spectrom.*, 2011, **300**, 182–193.
8. J. Zhu, H. D. Bean, Y.-M. Kuo and J. E. Hill, *J. Clin. Microbiol.*, 2011, **49**, 769.
9. H. Chen, A. Venter and R. G. Cooks, *Chem. Commun.*, 2006, 2042–2044.
10. H. Chen, S. Yang, A. Wortmann and R. Zenobi, *Angew. Chem.*, 2007, **46**, 7591–7954.
11. H. W. Chen, G. Gamez and R. Zenobi, *J. Am. Soc. Mass Spectrom.*, 2009, **20**, 1947–1963.
12. M.-Z. Huang, S.-C. Cheng, Y.-T. Cho and J. Shiea, *Anal. Chim. Acta*, 2011, **702**, 1–15.
13. W. S. Law, R. Wang, B. Hu, C. Berchtold, L. Meier, H. W. Chen and R. Zenobi, *Anal. Chem.*, 2010, **82**, 4494–4500.
14. L. Meier, S. Schmid, C. Berchtold and R. Zenobi, *Eur. J. Mass Spectrom.*, 2011, **17**, 345–351.
15. R. Wang, A. J. Groehn, L. Zhu, R. Dietiker, K. Wegner, D. Guenther and R. Zenobi, *Anal. Bioanal. Chem.*, 2012, **402**, 2633–2643.
16. H. Chen and R. Zenobi, *Nat. Protoc.*, 2008, **3**, 1467–1475.
17. M. Li, H. Jiang, H.-M. Li and R.-F. Xu, *Prog. Biochem. Biophys.*, 2012, **39**, 194–198.
18. P. Martinez-Lozano, L. Zingaro, A. Finiguerra and S. Cristoni, *J. Breath Res.*, 2011, **5**(1), 016002.
19. X. Ma, S. Zhang and X. Zhang, *TrAC, Trends Anal. Chem.*, 2012, **35**, 50–66.
20. L. Zhu, G. Gamez, H. W. Chen, K. Chingin and R. Zenobi, *Chem. Commun.*, 2009, 559–561.

21. H. Chen, Y. Sun, A. Wortmann, H. Gu and R. Zenobi, *Anal. Chem.*, 2007, **79**, 1447–1455.
22. W. S. Law, H. W. Chen, R. Balabin, C. Berchtold, L. Meier and R. Zenobi, *Analyst*, 2010, **135**, 773–778.
23. X. Zhang, Y. Liu, J. Zhang, Z. Hu, B. Hu, L. Ding, L. Jia and H. Chen, *Talanta*, 2011, **85**, 1665–1671.
24. X. Li, B. Hu, J. Ding and H. Chen, *Nat. Protoc.*, 2011, **6**, 1010–1025.
25. K. Chingin, G. Gamez, H. Chen, L. Zhu and R. Zenobi, *Rapid Commun. Mass Spectrom.*, 2008, **22**, 2009–2014.
26. H. Gu, N. Xu and H. Chen, *Anal. Bioanal. Chem.*, 2012, **403**, 2145–2153.
27. H. Chen, A. Wortmann, W. Zhang and R. Zenobi, *Angew. Chem.*, 2007, **46**, 580–583.
28. P. Martınez-Lozano and J. F. d. l. Mora, *Int. J. Mass Spectrom.*, 2007, **265**, 68–72.
29. P. Martinez-Lozano and J. Fernandezde la Mora, *Anal. Chem.*, 2008, **80**, 8210–8215.
30. J. H. Ding, S. P. Yang, D. P. Liang, H. W. Chen, Z. Z. Wu, L. L. Zhang and Y. L. Ren, *Analyst*, 2009, **134**, 2040–2050.
31. C. Berchtold, L. Meier and R. Zenobi, *Int. J. Mass Spectrom.*, 2011, **299**, 145–150.
32. G. Gamez, L. Zhu, A. Disko, H. Chen, V. Azov, K. Chingin, G. Kraemer and R. Zenobi, *Chem. Commun.*, 2011, **47**, 4884–4886.
33. Z. Jiangjiang, J.-D. Jaime, D. B. Heather, A. D. Nirav, I. A. Minara, K. A. L. Lennart and E. H. Jane, *J. Breath Res.*, 2013, 7, 037106.
34. J. Zhu, H. D. Bean, J. Jiménez-Díaz and J. E. Hill, *J. Appl. Physiol.*, 2013, **114**, 1544–1549.
35. Z. Jiangjiang, D. B. Heather, J. W. Matthew, W. L. Laurie and E. H. Jane, *J. Breath Res.*, 2013, 7, 016003.
36. Z. Zhou, M. Jin, J. Ding, Y. Zhou, J. Zheng and H. Chen, *Chem. Commun.*, 2008.
37. X. Li, X. Fang, Z. Yu, G. Sheng, M. Wu, J. Fu and H. Chen, *Anal. Chim. Acta*, 2012, **748**, 53–57.
38. C. Berchtold, S. Schmid, L. Meier and R. Zenobi, *Anal. Methods*, 2013, **5**, 844–850.
39. M. Katona, J. Denes, R. Skoumal, M. Toth and Z. Takats, *Analyst*, 2011, **136**, 835–840.
40. P. Martinez-Lozano and J. Fernandez de la Mora, *J. Am. Soc. Mass Spectrom.*, 2009, **20**, 1060–1063.
41. H. Chen, B. Hu, Y. Hu, Y. Huan, Z. Zhou and X. Qiaoc, *J. Am. Soc. Mass Spectrom.*, 2009, **20**, 719–722.
42. L. Meier, C. Berchtold, S. Schmid and R. Zenobi, *Anal. Chem.*, 2012, **84**, 2076–2080.
43. L. Meier, C. Berchtold, S. Schmid and R. Zenobi, *J. Mass Spectrom.*, 2012, **47**, 1571–1575.
44. L. Meier, C. Berchtold, S. Schmid and R. Zenobi, *J. Mass Spectrom.*, 2012, **47**, 555–559.

CHAPTER 11

Probe Electrospray Ionization

MRIDUL KANTI MANDAL AND KENZO HIRAOKA*

Clean Energy Research Center, University of Yamanashi, 4-3-11 Takeda, Kofu, Yamanashi, 400-8511 Japan
*Email: hiraoka@yamanashi.ac.jp

11.1   Introduction

In 1917, Zeleny observed the electrospray of liquid glycerol from a capillary,[1] and the electrified liquid meniscus, which took the shape of a cone, was theoretically analyzed by Taylor.[2] Dole *et al.* described that the molecular ions might be liberated from the electrosprayed droplet,[3] and the first electrospray ionization mass spectrometry (ESI-MS) was demonstrated by Fenn and coworkers.[4,5] ESI-MS has now become one of the indispensable tools for the analysis of biomolecules. In order to cope with small volumes of liquid solution, miniaturized ESI ion sources with low flow rates have also been developed using fine glass capillaries, *i.e.* nano-ESI.[6–9] Wilm and Mann described theoretically and demonstrated experimentally that a narrower spraying capillary with a much reduced flow rate could generate smaller initial droplets, and hence better ion desorption from the liquid.[8] Due to the smaller aperture size of the capillary, needle clogging may occur during the analysis. In addition to clogging, it may also be difficult to initiate the electrospray from capillaries for aqueous solution with diameters smaller than 1 μm due to the surface tension of the liquid.

Several designs of electrospray ion sources that use noncapillary probe had been put forward to overcome the clogging problem. Shiea and coworkers succeeded in electrospraying a sample solution deposited on a copper wire ring after applying a high voltage to the copper wire, and mass spectra of

New Developments in Mass Spectrometry No. 2
Ambient Ionization Mass Spectrometry
Edited by Marek Domin and Robert Cody
© The Royal Society of Chemistry 2015
Published by the Royal Society of Chemistry, www.rsc.org

proteins similar to those of conventional ESI could be obtained.[10] The technique was further explored using optical fibers wired with copper or platinum wires,[11] a glass rod,[12] and nanostructured tungsten oxide.[13] Electrospray from the solution deposited on micropillar chips has also been reported.[14]

In 2007, probe electrospray ionization (PESI) using a sharp solid needle as an electrospray emitter was developed in our laboratory.[15] The idea of PESI was originated from field desorption (FD) because FD basically utilizes the electrospray phenomenon.[16] PESI is free of clogging problems, robust, and could be used to analyze real-world samples directly with little or no sample preparations.[17] PESI-MS is usually performed by a pick-and-spray process that differentiates it from other ESI-based ionization techniques.

This chapter deals with the fundamentals and application of PESI-MS from various aspects, namely, (a) experimental procedures, (b) sequential and exhaustive ionization, (c) proteins and peptides analysis from complex matrix, (d) imaging mass spectrometry (IMS), (e) ambient cancer diagnosis, (f) real-time reaction monitoring, and (g) related topics to PESI-MS.

11.2   Experimental Setup for PESI-MS

In a conventional ESI experiment a capillary is set in front of the ion sampling orifice of the mass spectrometer and liquid is supplied continuously by a syringe pump. Electrospray is generated by applying high voltage (HV) to the capillary. In PESI a solid needle is used instead of using a capillary. A schematic diagram of the experimental system is shown in Figure 11.1.[15] A disposable stainless steel acupuncture needle (o.d. 0.12 mm) with a tip

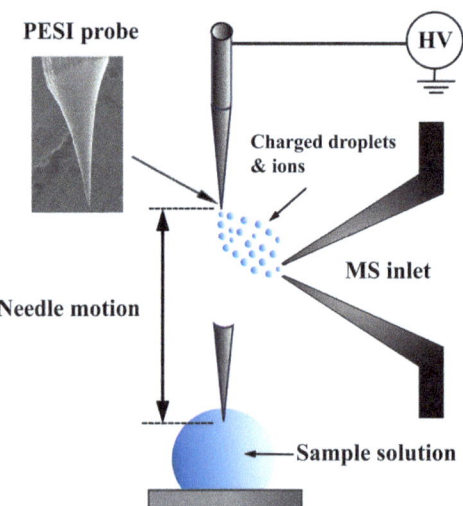

Figure 11.1   Conceptual idea of the ion source for PESI coupled with a mass spectrometer.
(Adapted with permission from ref. 30. Copyright 2010 Royal Society of Chemistry.)

diameter of ∼700 nm was used as a typical electrospray emitter. Any other kinds of electroconductive needles such as microneedles for scanning tunneling microscopy, sewing needles, W/Ni/Au/Pt wires, *etc.* can also be used. The needle was driven up and down along the vertical axis by a linear motor-actuated system. The lowest position of the needle tip was adjusted to just touch the surface of the biological or liquid sample that was mounted on the *xyz* manipulation stage. The sample volume trapped on the tip depends on the size of the needle, invasion depth, viscosity and surface tension of the sample, and hydrophobicity of the needle surface. When an acupuncture needle with a 0.12 mm diameter and ∼700 nm tip diameter is used, the sample volume is ∼0.35 pL with the invasion depth of ∼10 μm for low-viscosity samples such as urine.[18] When the needle comes into contact with the sample, both the needle and the sample are kept at ground or floating potential. To generate ES, a high voltage of ∼2 kV is applied to the needle when it is moved up to the highest position. The optimized horizontal and vertical distances from the tip of the needle to the apex of the ion sampling orifice of the mass spectrometer were 3 and 2 mm, respectively. The distance of the needle stroke was normally set at 10 mm.

11.2.1 Invasion Depth Control

A shunt resistor (2 kΩ) was connected in series with the solid needle at the high-voltage side for the electrospray current measurement, and the current signal was amplified and coupled to the low-voltage side using cascade isolation amplifiers. Postamplification and signal conditioning were performed using an RMS-DC converter, after which the 50 Hz background signal was used to determine the needle invasion depth. When the needle was not in contact with the sample, a 50 Hz background current of ∼10 nA originating from the power-line supply was observed. When the needle tip was just in contact with the sample surface, the background current increased by a factor of 2–3. This needle position was used as a reference to determine the needle invasion depth.[19]

11.2.2 Physical Characteristics of PESI

The experimental conditions were optimized based on the spray current, optical microscopy, and PESI mass spectra. After application of high voltage to the needle, a tiny Taylor cone was formed at the tip of the solid needle probe, with its volume determined by the size of the needle tip. The liquid flow rate depends on the voltage applied to the needle as well as the loaded liquid amount.

In order to examine the relationship between the aperture size and the electrospray plume, optical images of the electrospray plumes generated from the different sizes of needles were observed.[20] The tip of the solid needle was illuminated using a frequency-doubled Nd:YAG pulsed laser (532 nm) with the pulse width of 4 ns, and the electrospray generated from the tip was observed using a long working-distance lens. The forward scattering light from the electrosprayed droplets was captured using a digital camera. Figure 11.2(d)–(f) show snapshots of the electrospray generated from three solid needles, with tip diameters of 0.7, ∼10, and ∼100 μm,

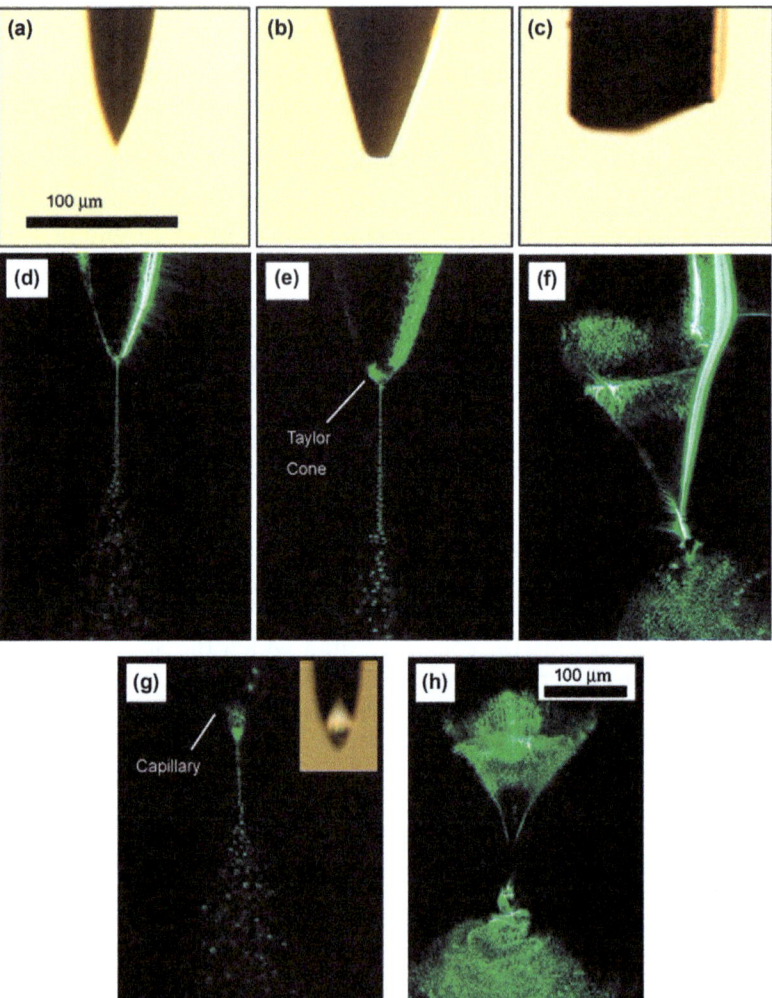

Figure 11.2 Backlight illumination photographs and electrospray images captured using pulsed laser microscopy for (a) and (d): solid needle of 700 nm tip diameter, (b) and (e): solid needle of ∼10 μm tip diameter, and (c) and (f): blunted solid needle of ∼100 μm in diameter. (g) Image for nanoelectrospray from capillary of ∼5 μm inner diameter; the shape of the capillary is shown in the inset with the same magnification. (h) Conventional electrospray generated using capillary of ∼100 μm inner diameter. The scale bar in (a) is also valid for (b)–(g).
(Adapted with permission from ref. 20. Copyright 2008 American Chemical Society.)

respectively. Backlight illumination photographs for these needles are shown respectively, in panels a, b, and c of Figure 11.2. The liquid sample was 10^{-5} M gramicidin S in 10^{-2} M ammonium acetate aqueous solution, and the needle was flashed with the pulsed laser for several tens of milliseconds after the application of high voltage.[20]

For comparison, the electrospray of 0.1% acetic acid in water/methanol (v/v = 1/1) using a nanoelectrospray capillary (i.d.: ~ 5 µm), and a conventional electrospray capillary (i.d. ~ 100 µm, liquid flow rate ~ 1 µL min^{-1} without a nebulizing gas) are shown, respectively, in Figure 11.2(g) and (h). As depicted in Figure 11.2(d), a small sample solution of ~ 1 µm in diameter was accumulated on the tip of the needle, forming a tiny Taylor cone at the tip ($\sim$ fL), and fine charged droplets were electrosprayed from the cusp, similar to that of capillary nanoelectrospray shown in Figure 11.2(g). It should be noted that, due to the diffraction limit of the optical microscopy, the exact size of the sprayed droplet cannot be determined from these images. For the solid needle with the tip diameter of ~ 10 µm shown in Figure 11.2(b) and (e), the liquid Taylor cone formed on the tip was apparently bigger, with the base diameter of the cone similar to the size of the needle tip.

The amount of liquid picked up by the needle depends on the dipping depth, and the temporal spray currents for different dipping depths into the aqueous solution of 10^{-5} M gramicidin S and 10^{-2} M ammonium acetate are shown in Figure 11.3. In this measurement, the solid needle was held at a constant voltage of 2.5 kV.[11]

As the needle was moved upward, electrospray was initiated at the highest position. For a small amount of liquid (d_0 to $d_0 + 0.5$ mm), the spray current rose to a peak value (40–50 nA) and decreased gradually due to the consumption of liquid sample. For a larger dipping depth of $d_0 + 0.8$ mm, the liquid amount loaded to the tip was enough to sustain a stable electrospray (constant electrospray current) for a short period of time before the sample was totally consumed. In the $d_0 + 1.2$ mm case, the loaded liquid could not be completely sprayed within the designated period, and was quenched when the needle moved downward for the next sampling process. When the needle was overloaded with liquid sample, *e.g.*, at the dipping depth of $d_0 + 2$ mm, the current oscillated in the range of kilohertz to tens of kHz. This current fluctuation is likely to be the spontaneous spray pulsation that is known to exist in most conventional electrosprays.[21–23] If one assumes that the liquid volume of ~ 10 pL with $d_0 + 0.5$ mm,[18] the flow rate of the liquid is roughly estimated to be ~ 10 nL min^{-1} in the spray time interval of ~ 50 ms (Figure 11.3). This estimated value is of the same order of nano-ESI with ~ 1 µm i.d. capillary.

11.3 Sequential and Exhaustive Ionization in PESI

In capillary-based ES, the main part of the charged droplet that is ejected from the Taylor cone is charge depleted by spawning the highly charged microdroplets. During this off-spring droplet fission, more surface-active ions are preferentially lost from the main part of the droplet and less

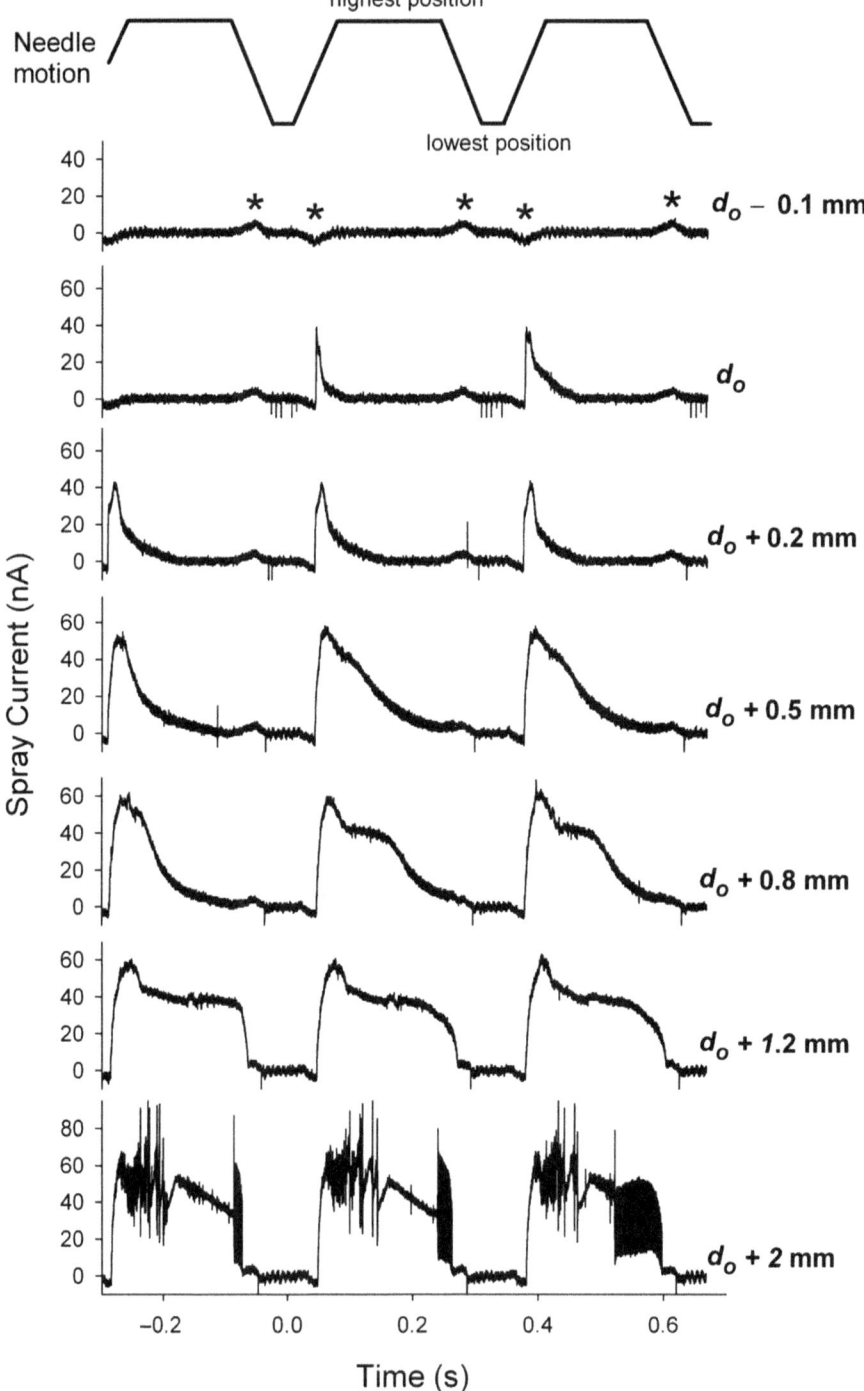

surface-active ions are apt to be left in the main droplets. This leads to the suppression effect for less surface-active components. In contrast, in PESI the main part of the liquid remains attached to the needle (*i.e.* electrode) and the excess charges are continuously supplied to the liquid until the total depletion of the sample, namely, PESI is not charge limited. PESI may be the way to analyze all ionizable components in the sample, *i.e.* sequential and exhaustive ionization.[24]

Figure 11.4(a) shows the total ion chromatogram (TIC) for the PESI experiment for 10^{-3} M Triton X100 and 10^{-5} M cytochrome c in H_2O/MeOH/ AcOH (74/25/1). In this experiment, the tangentially cut Ti wire with 0.5 mm in diameter was used as a needle probe. The liquid trapped at the tip was ~50 nL and electrospray lasted about 50 s. The high voltage (3.5 kV) was applied to the needle at 0 s. TIC shows an initial small increase followed by a low ion current, up to 30 s. Then, TIC shows a steep increase. It is evident that electrospray conditions are changing during the electrospray. Figure 11.4(b)–(d) show the extracted ion chromatograms of Triton X100 at m/z 691.6 [Triton X100 with $n = 11 + H$]$^+$, cytochrome c at m/z 941.9 [cytochrome c + 13H]$^{13+}$ and at m/z 1224.2 [cytochrome c + 10H]$^{10+}$, respectively. Figure 11.4(e)–(g) display the mass spectra measured at the times T1, T2, and T3 denoted in Figure 11.4(a). The insets in Figure 11.4(e)–(g) show the optical microscopic images of the liquid sample attached on the Ti needle at T1, T2, and T3, respectively. At T1, only protonated and sodiated Triton X-100 were detected in the mass spectrum, but no cytochrome c (Figure 11.4(e)). This trend continued just before the steep increase in TIC shown in Figure 11.4(a). The only appearance of Triton X100 in the time interval of 0–30 s suggests that Triton X-100 is being electrosprayed mainly during this period. At 30 s, the sudden appearance of cytochrome c was observed (Figure 11.4(c)).[24] Apparently, the electrospray condition changed at ~30 s, and this may be attributed to the decrease in the concentration of the surface-active Triton X-100, resulting in the availability of the liquid surface for the enrichment of the less-surface-active cytochrome c. That is, the surface excess components are being replaced from the most surface-active Triton X-100 to the less surface-active cytochrome c. This dynamic process taking place in the Taylor cone is clearly reflected in TIC (Figure 11.4(a)).

For a comparative study of PESI with capillary-based ESI, the same solution was analyzed by nano-ESI using a 1 µm i.d. capillary. The obtained nano-ESI mass spectrum was almost similar to the PESI mass spectrum

Figure 11.3 Spray current for different dipping depths of the needle into 10^{-5} M gramicidin S and 10^{-2} M ammonium acetate aqueous solution. The solid needle is held at 2.5 kV, and d_0 denotes the dipping depth where the needle tip just touched the liquid surface with detectable spray current. The exact value of d_0 is not known but is estimated to be ~100 µm or less. Asterisks (*) denote the electrical noise when the needle moved across the ion-sampling orifice of the mass spectrometer. (Adapted with permission from ref. 20. Copyright 2008 American Chemical Society.)

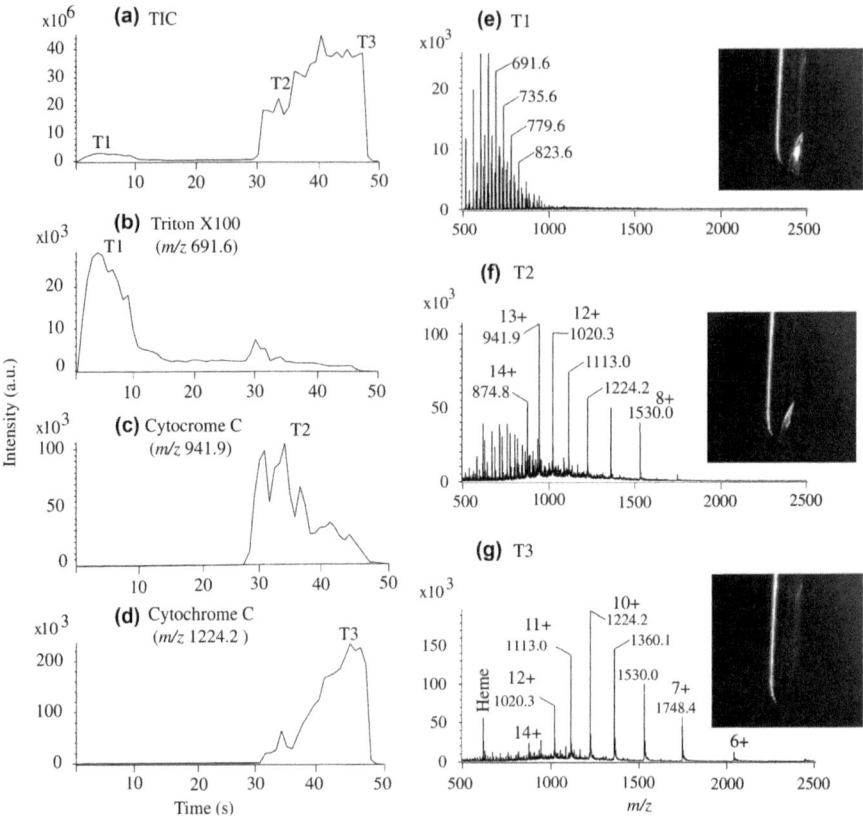

Figure 11.4   Experimental results for 10^{-3} M Triton X100 and 10^{-5} M cytochrome c in H_2O/MeOH/AcOH (74/25/1). (a) Total ion chromatogram (TIC); (b) extracted ion chronogram of Triton X100 at m/z 691.6 [Triton X100 with $n = 11 + H]^+$; (c) extracted ion chronogram of cytochrome c at m/z 941.9 [cytochrome $c + 13H]^{13+}$; (d) extracted ion chronogram of cytochrome c at m/z 1224.2 [cytochrome $c + 10H]^{10+}$; (e) Mass spectrum measured at T1; (f) mass spectrum measured at T2; (g) mass spectrum measured at T3. Insets in (e), (f), and (g) display the optical microscopic images of the tip of the Ti needle measured at T1, T2, and T3, respectively. (Adapted with permission from ref. 24. Copyright 2011 Springer.)

measured at T1 in Figure 11.4(e) and cytochrome c was totally suppressed by TritonX-100.[24] Namely, the suppression effect inherent to the capillary-based ESI is largely moderated by PESI.

Juraschek *et al.* measured nano-ESI mass spectra of insulin from NaCl solutions by using glass capillaries with orifices ≤2 μm in diameter.[21] In the mass spectrum for a 10^{-2} M NaCl and 10^{-5} M insulin in H_2O/MeOH/AcOH (48/48/4), they observed a series of $[Na_nCl_{n-1}]^+$ and $[M + H_{n-m} + Na_m]^{n+}$ where M denotes insulin. PESI mass spectra were also measured for 15×10^{-3} M NaCl and 10^{-5} M insulin in H_2O/MeOH/AcOH(74/25/1). Figure 11.5 shows the TIC (a) and mass spectra measured at T1, T2, and T3

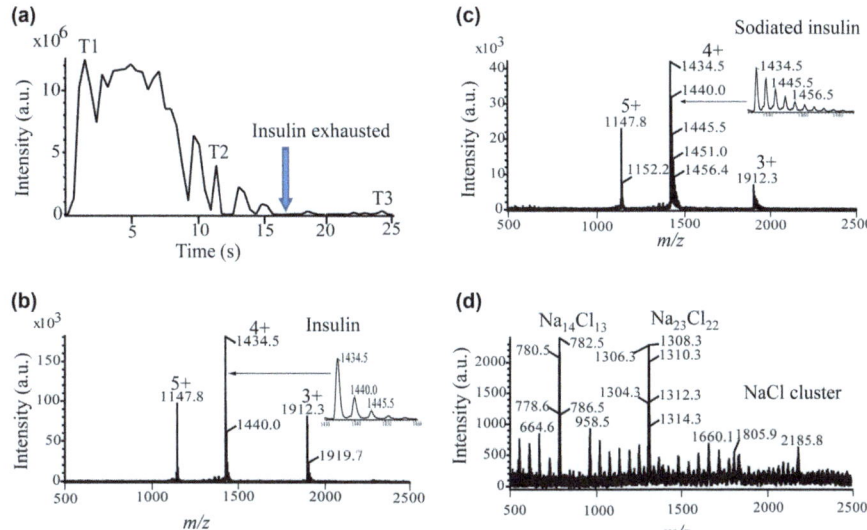

Figure 11.5 Experimental results for 15×10^{-3} M NaCl and 10^{-5} M insulin in H₂O/ MeOH/AcOH(74/25/1). (a) Total ion chromatogram (TIC); (b) mass spectrum at T1; (c) mass spectrum at T2; (d) mass spectrum at T3. The surface-modified Ti needle with an o.d. of 0.5 mm was used. (Adapted with permission from ref. 24 and modified. Copyright 2011 Springer.)

shown in Figure 11.5(a). The NaCl cluster ions $[Na_nCl_{n-1}]^+$ are absent in Figure 11.5(b) and (c), suggesting that NaCl is not electrosprayed but left in the main droplet and its concentration is being enriched with time. In fact, the value of m, the number in $[M + H_{4-m} + Na_m]^{4+}$, increases with time T1 → T2. At T3, only NaCl cluster ions but no insulin ions are observed (Figure 11.5d), *i.e.* insulin is totally exhausted at around T2. That is, almost complete separation of peptide and salt takes place in this binary sample solution. This result confirms the ionization of solution components in the order of decreasing surface activity and also the efficient ionization of the less-surface-active components, which would not occur with capillary-based electrospray.

The excess charges enriched on the surface of the Taylor cone attract the countercharge ions. The inner shell of charge and the outer atmosphere is called the electric double layer. The structure of the surface atmosphere can be described by the Debye–Hückel theory of ionic solutions, and its thickness is represented by a Debye length, λ_D,

$$\lambda_D(nm) = 0.305/C^{1/2}$$

where C stands for the electrolyte concentration in M.[25] For 15×10^{-3} M NaCl solution, λ_D is calculated to be ~ 3 nm. The appearance of only insulin without the contamination of NaCl at T1 in Figure 11.5(b) suggests that the thickness of the outer shell (*i.e.* the layer composed of insulin ions) which is stripped off at the Taylor cone is thinner than λ_D, ~ 3 nm under the experimental conditions of Figure 11.5. This result predicts that PESI is highly tolerant to the existence of salts or detergents in sample solutions.

For a real-world sample, a dissected breast cancer tissue was analyzed by PESI. The PESI mass spectra were measured immediately after 10 µL of mixed solvent of H_2O/MeOH/AcOH (50/50/1) was dropped on the surface of tumor tissue. The inset in Figure 11.6(a) shows the TIC that has three peaks. Figure 11.6(a)–(c) display the mass spectra at T1, T2, and T3 denoted in the inset in Figure 11.6(a). Ion signals due to heme, α and β chains of hemoglobin, and lipids in the range of *m/z* 700–900 were observed. The peaks denoted with asterisks and those with circles were assigned to be phosphatidylcholines (PC) and triacylglycerides (TAG), respectively. At the initial stage of electrospray (T1), heme and α and β chains of hemoglobin were observed as the major ions with weaker PC and TAG signals. With time T1 → T2, ion signals due to hemoglobin weaken and they are taken over by those of PC and TAG. Ion signals of PC are observed at T1 and T2, but they decrease to the noise level at T3. Instead, sodiated TAGs become predominant at the latest stage of electrospray. That is, separation of components takes place in the order of hemoglobin → PC → TAG.[24]

11.4 PESI as Ambient Field Desorption

When a high electric field of $\sim 10^{10}$ V m^{-1} is applied to the tip of the sharp metal needle under high-vacuum conditions, atoms or molecules approaching the tip are ionized without touching the metal surface.[26] This is the principle of field ionization (FI) and field ion microscopy is based on this. Subsequently, the ionization method for molecules deposited on fine metal needles was also developed. This method is called field desorption (FD).[27] Both of these ionization methods utilize sharp needles under high vacuum and thus are frequently discussed in the same context.

FD operated under high vacuum may be regarded as the ultimate ESI method, because single ions desorb from the sharp electrode (*e.g.*, liquid metal ion source for secondary ion mass spectrometry), making FD an excellent model to better understand ES.

Wong *et al.* observed the direct desorption of ions from a mixture of sucrose and NaCl under high vacuum.[16] Figure 11.7 shows a conceptual idea of microscopic images of the behavior of droplets from a concentrated aqueous solution of sucrose/NaCl (2:1).

On increasing the wire temperature to ~ 100 °C, the sample was elongated toward the counter electrode and almost completely carried away. After the major loss of the sample, residual samples formed. The effect of temperature on the sample droplet is to decrease the surface tension and viscosity of the sample. The latter also leads to an increase in the mobility of solvated ions, causing charging of the droplets in the high electric field. The field causes migration of the Na$^+$ and Cl$^-$ ions in opposite directions. The Cl$^-$ ions reaching the metal electrode surface are oxidized ($2Cl^- \rightarrow Cl_2$), and thus the

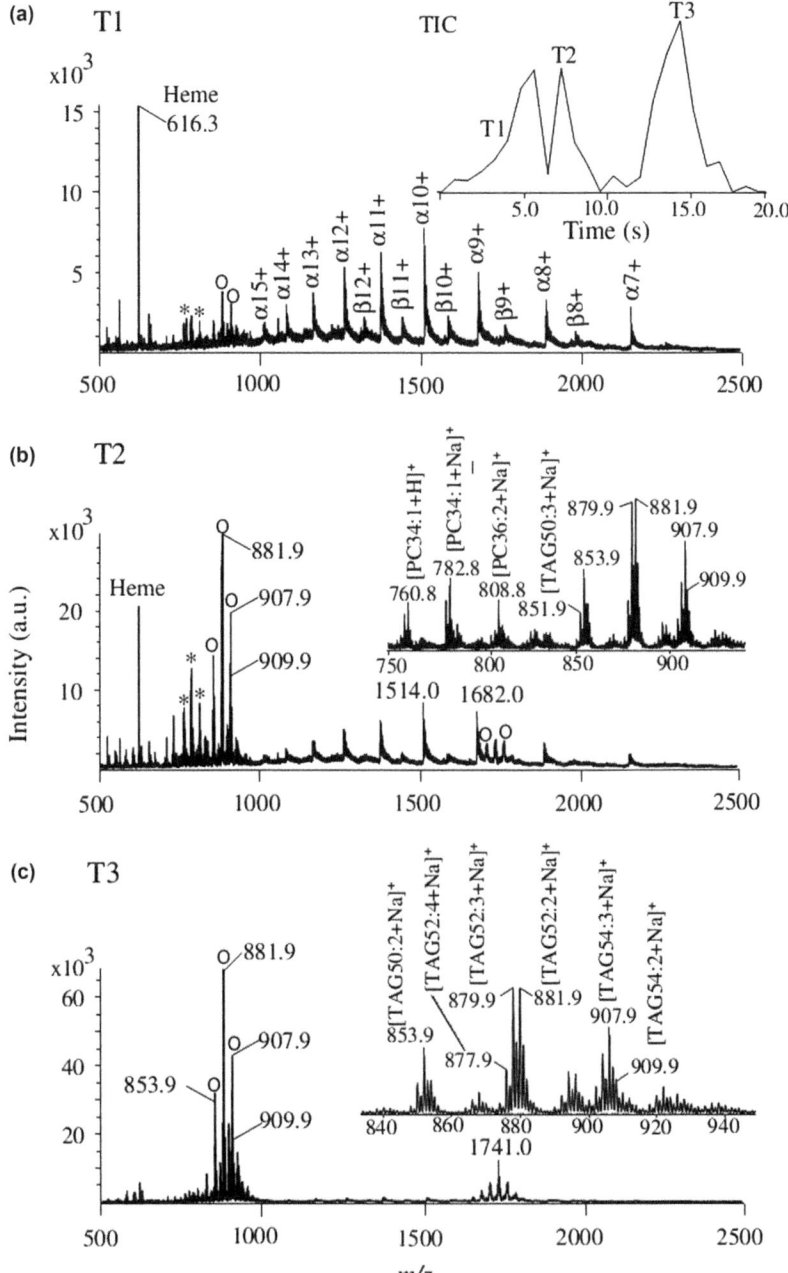

Figure 11.6 Mass spectra for a human breast cancer tissue. (a) Mass spectrum measured at T1; inset: total ion chromatogram; (b) mass spectrum measured at T2; inset: expanded mass spectrum at *m/z* 750–950; (c) mass spectrum measured at T3; inset: expanded mass spectrum at *m/z* 840–940. Asterisks (*) and open circles (○) stand for peaks from phosphatidylcholine (PC) and sodiated triacylglycerides (TAG), respectively.
(Adapted with permission from ref. 24. Copyright 2011 Springer.)

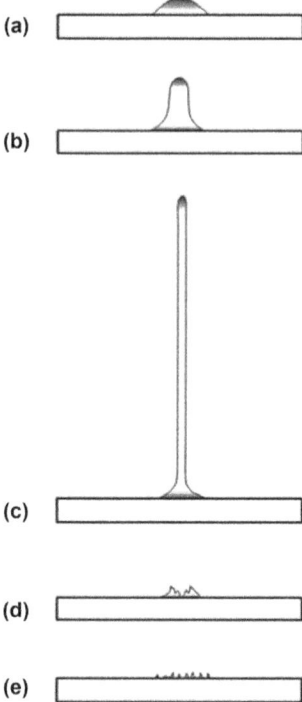

Figure 11.7 Conceptual idea of behavior for a concentrated aqueous solution of sucrose/NaCl (2 : 1) deposited on a 10 μm tungsten wire anode under the condition of FD. The anode–counter-electrode distance was 3 mm, the applied potential difference was 8 kV. The emitter heating current was slowly increased; (a) 0 mA, no change in size of the droplet, (b)-(e) current was increased from 6 mA to 18 mA: rupture of lamellae and formation of jets are observed. In (e) [M + Na]$^+$ was formed from the microcones (Taylor cones).
(Adapted with permission from ref. 16. Copyright ScienceDirect 1984).

droplets are enriched by the excess charges of Na$^+$ leading to the Rayleigh limit and droplet disintegration. At the stage of Figure 11.7(e), [sucrose + Na]$^+$ is generated as the major ion from the tip of the tiny Taylor cones. It should be noted that sodiated single molecules [sucrose + Na]$^+$, but not microdroplets are formed at the later stage of FD. Thus, FD may be regarded as the ultimate ESI that realizes the direct evaporation of single ions from the Taylor cone. This is similar to the liquid metal ion source. The electric field necessary for single-ion evaporation is estimated to be 10^8–10^9 V m^{-1}. The species that desorb from the cone are, of course, not neutral but ionic. In this sense, FD is more appropriately defined as "field-induced ion desorption". In contrast to FD, FI accompanies the electron tunneling in the ionization process, and thus FD and FI are based on the different ionization mechanism.

The electric fields generated for the conventional ESI using a 0.1 mm diameter capillary, the nano-ESI using a few-μm diameter capillary, and PESI using a needle with a 700 nm tip diameter are calculated to be $\sim 10^7$, $\sim 10^8$, and $\sim 10^9$ V m^{-1}, respectively. The value for PESI, $\sim 10^9$ V m^{-1}, is close to that for the ion-evaporation regime. In this respect, PESI may be able to realize direct single-ion desorption from the Taylor cone. However, this is not usually the case for the liquid samples, namely, fine microdroplets are formed but not the single ions (see Figure 11.2(d)), because ES for solutions is a dynamic process and is far from equilibrium, *i.e.* there is a voltage drop over the liquid cone and the cone angle becomes smaller than 99°.[8,6,28] However, in PESI, the liquid amount held on the tip becomes increasingly smaller. Actually, the liquid film becomes too thin to be recognized by the optical microscope at the last stage of spray. At this stage, strong ion signals are frequently observed (Figure 11.4(a) and inset of Figure 11.6(a)). Because PESI does not adopt a continuous sample supply but the liquid trapped on the needle tip is electrosprayed until total depletion of the sample, a phenomenon similar to FD could take place at the last stage of ES. With depletion of the sample, the sample gets thinner. This situation must result in an increase of the electric field exerted on the liquid film. At the final stage, the sample containing nonvolatile components may become viscous by solvent evaporation. The thin sample film could be enriched by the excess charges with the electric field applied to the metal needle ($\sim 10^9$ V m^{-1} with the tip diameter of ~ 700 nm), and this value is close to the ion-evaporation regime. The PESI results for human breast cancer could be regarded as ambient FD. In Figure 11.6(c), the abundant sodiated TAG molecules were observed as dominant ions, suggesting that the ion evaporation regime, *i.e.* the ambient FD, is realized at the last stage of PESI (see Figure 11.8). The appearance of [TAG + Na]$^+$ in Figure 11.6(c) is similar to the result that [sucrose + Na]$^+$ is the predominant ion at the last stage of FD (Figure 11.7(e)).[16]

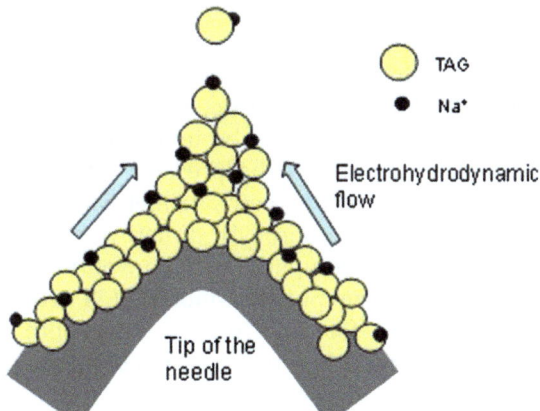

Figure 11.8 Illustration of the desorption of ions at the last stage of PESI. (Adapted with permission from ref. 29 Fundamentals of Mass Spectrometry. Copyright 2012 Springer)

11.5 Peptides and Protein Analysis from Complex Matrix

Biological samples usually contain high concentration of salts and the removal of salts and the related sample purification processes play a critical role in successful proteomics analysis with mass spectrometry.[30] Desalting treatment is mandatory before mass spectrometric analysis. Though online/inline desalting and urea purification approaches are proved to be successful, they may produce memory effect and plug the membrane frequently. Several ionization methods have been developed that allow the direct analysis of native samples without purification step. For example in 2002, Shiea's group reported a method called fused droplet electrospray ionization (FD-ESI). By fusing the charged acidic methanol droplets with sample aerosol, they successfully detected the analytes from the concentrated NaCl and sodium phosphate solution.[31] DESI[32] and extractive electrospray ionization (EESI)[33] also allow the mass spectrometric analysis to be performed without the tedious clean-up steps for salts or other contaminants. PESI was also found to be applicable to protein and peptide analysis from high concentration of salts.[30] For example, Figure 11.9 depicts the positive ion mass spectra of myoglobin with varying amounts of NaCl in H_2O/MeOH/AcOH (74/25/1) solvent obtained by PESI-MS (Figure 11.9(a)–(d)) and nano-ESI-MS (Figure 11.9(e)–(h)). As the concentration of NaCl increases, intensities of sodium adduct signals increase for both PESI-MS and nano-ESI-MS. As shown in Figure 11.9(c), PESI could detect the ion signal of myoglobin from the NaCl solution up to 250 mM or higher (estimated Debye length λ_D: ~1 nm). When the NaCl concentration exceeded 50 mM, the nano-ESI produces only salt cluster signals without any peaks from myoglobin (Figure 11.9(g)).[30]

Detergents are used as indispensible reagents in membrane biochemistry for the purpose of solubilization and extraction of proteins.[34] After the extraction, elimination of detergent is necessary for the better performance of ESI-MS. Elimination of detergents is laborious and time consuming, and also sample loss may be unavoidable.[35] However, PESI was found to be highly tolerant to the presence of detergents such as sodium dodecyl sulphate (SDS), cetyl trimethylammoniumbromide, Triton X100 and 3-[(3-cholamidopropyl) dimethylammonio]-1-propanesulfonate and also TRIS buffer compared with conventional ESI and nano-ESI.[35] Figure 11.10 shows the positive mode PESI, conventional ESI, and nano-ESI mass spectra for 2.5×10^{-5} M myoglobin and different concentrations of SDS and TRIS buffer in H_2O/MeOH/AcOH (74/25/1) solution. From the figure, it is evident that PESI is much superior to the capillary-based ESI for the samples with the contamination of detergents. Therefore, PESI can be a potential analytical tool for direct analysis of protein extracts and digests containing high-concentration of detergents.

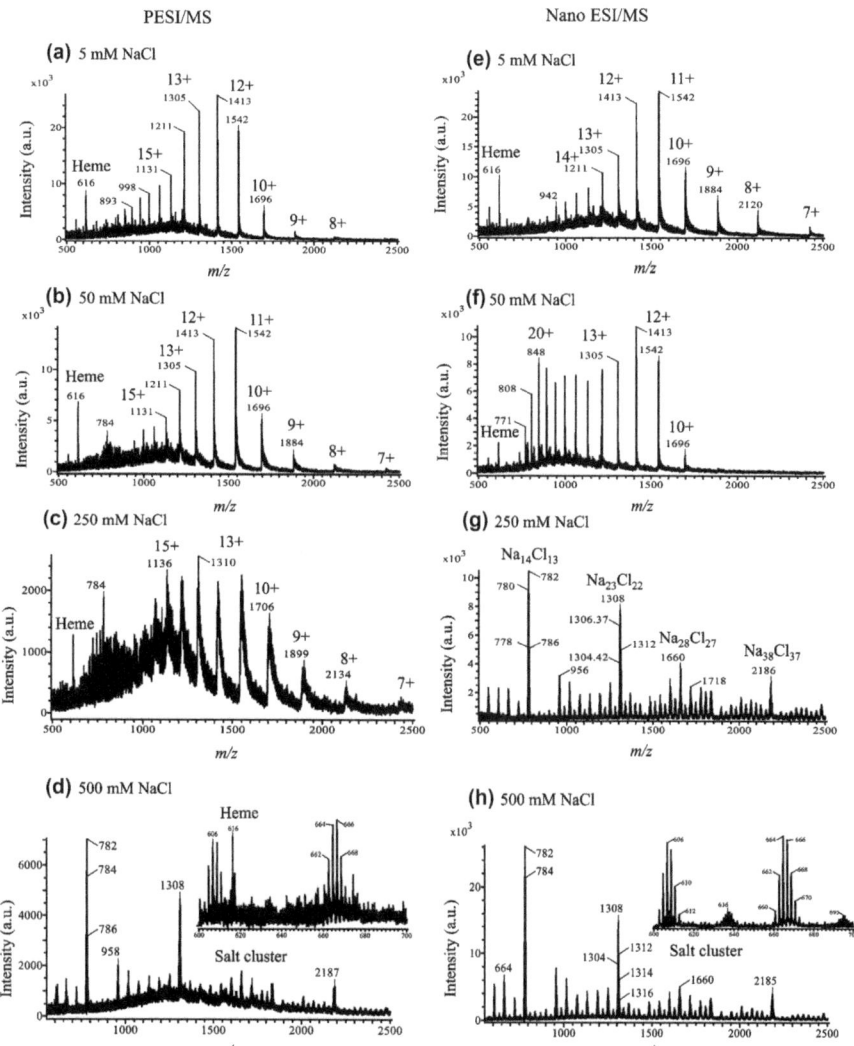

Figure 11.9 (a)–(d): Positive ion PESI mass spectra (using 100 μm titanium wire) of 10^{-5} M myoglobin in different concentrations of NaCl. (e)–(h): Positive ion nano-ESI mass spectra of 10^{-5} M myoglobin in different concentrations of NaCl. Insets show the expanded mass spectra of (d) and (h). Solvent: 74/25/1 (H_2O/MeOH/AcOH).
(Adapted with permission from ref. 30. Copyright 2011 The Royal Society of Chemistry)

11.6 Real-world Sample Analysis

PESI was found to be widely applicable to the direct analysis of biological samples such as fruits (orange, banana, *etc.*), tulips, urine, mouse brain, mouse liver, salmon egg, and living animals.[17,36–39] Since the amount of

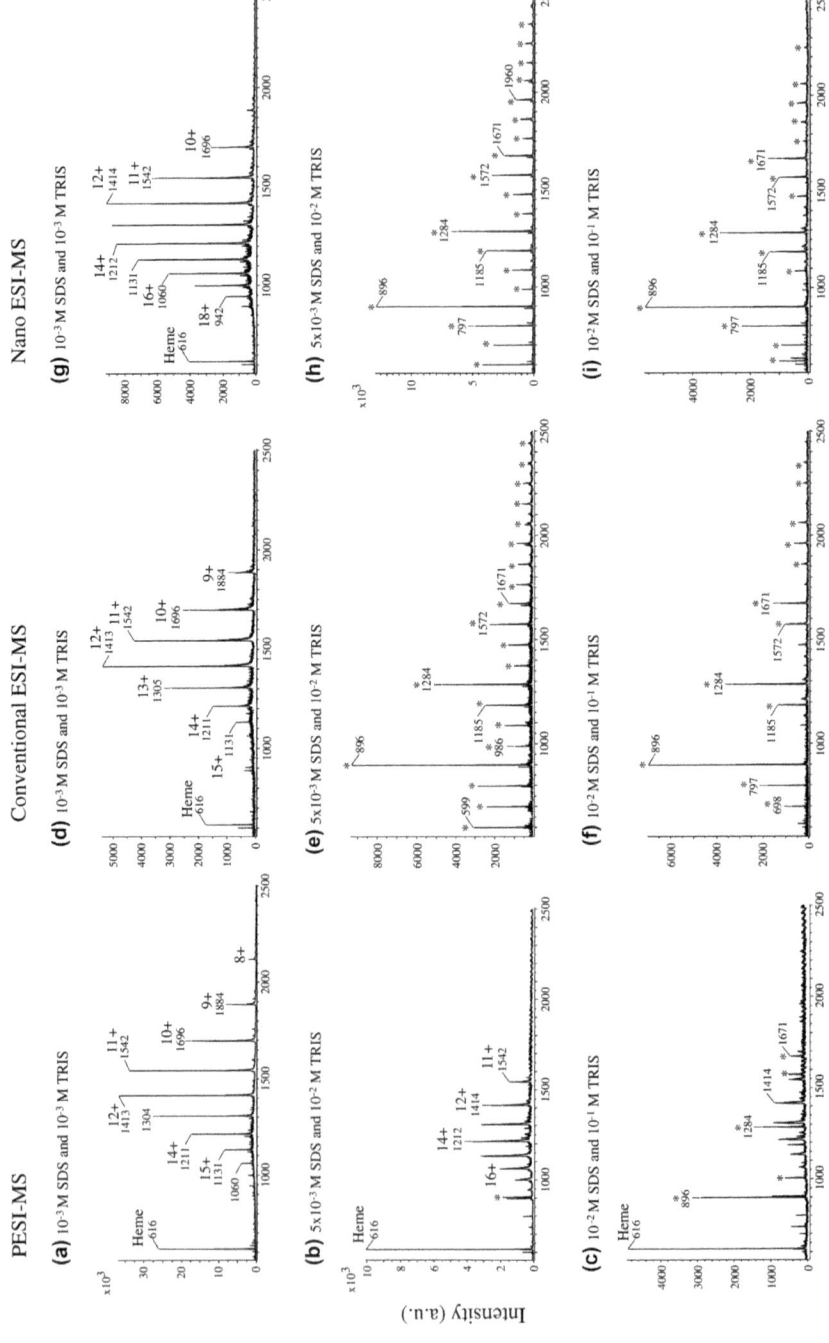

PESI-MS

Conventional ESI-MS

Nano ESI-MS

(a) 10^{-3} M SDS and 10^{-3} M TRIS

(d) 10^{-3} M SDS and 10^{-3} M TRIS

(g) 10^{-3} M SDS and 10^{-3} M TRIS

(b) 5×10^{-3} M SDS and 10^{-2} M TRIS

(e) 5×10^{-3} M SDS and 10^{-2} M TRIS

(h) 5×10^{-3} M SDS and 10^{-2} M TRIS

(c) 10^{-2} M SDS and 10^{-1} M TRIS

(f) 10^{-2} M SDS and 10^{-1} M TRIS

(i) 10^{-2} M SDS and 10^{-1} M TRIS

Intensity (a.u.)

liquid sample picked up by the needle is as small as pL order, PESI could be a promising analytical tool as a less-invasive technique for detecting bio-molecules in living systems. To show the versatility of the technique, direct profiling of tulips and human milk are demonstrated as examples.

A tulip bulb in different regions, including basal plate, outer and inner rims of scale, flower bud, petals and foliage leaves were analyzed.[38] Figure 11.11(a) and (b) show PESI mass spectra of the red part and the yel-low part of a tulip petal. Potassiated tuliposide A, tuliposide B, and di-saccharide (Hex$_2$) at *m/z* 317, 333, and 381 were detected in both spectra, respectively. Only from the red part, were some pelargonidin type pigments detected, including pelargonidin, pelargonidin 3-rutinoside, and pelargo-nidin 3-(2'''-acetylrutinoside) at *m/z* 271, 579, and 621, respectively. Figure 11.11(c) shows the PESI mass spectrum of the leave. Only four peaks at *m/z* 317, 333, 381, and 543, corresponding to potassiated tuliposide A, tuliposide B, disaccharide (Hex$_2$), and trisaccharide (Hex$_3$) were detected, respectively.[38]

As another example of real-world samples, PESI mass spectra for human and bovine milk are shown in Figure 11.12. The human milk was analyzed directly without dilution. In Figure 11.12(a), lipid ions at around *m/z* values of 800 and 1600 are observed in addition to sodiated and potassiated di-saccharide ions, $[\text{diSac} + \text{Na}]^+$ and $[\text{diSac} + \text{K}]^+$. The peaks from the human milk appeared to be more densely spaced because of the nearly equal con-tribution from sodiated and potassiated molecules, whereas the potassiated molecules in the bovine milk (Figure 11.12(b)) appeared to be stronger due to the higher concentration of potassium.[17]

To apply PESI to *in vivo* analysis, live mice were examined. Mice were anesthetized to expose the liver as shown in Figure 11.13. Then they were fixed physically on a movable sampling stage. HV was applied to the needle at its highest position and synchronized with the needle motion to generate the electrospray. A reproducible ion signal could be obtained by auto-matically repeating the sampling cycle to perform MS/MS. Mice tolerated the surgery well without showing obvious side effects. Bleeding at the analyzed

---

Figure 11.10 (a)–(c): Positive ion PESI mass spectra of 2.5×10^{-5} M myoglobin and different concentrations of SDS and TRIS buffer in H$_2$O/MeOH/AcOH (74/25/1) solution. (d)–(f): Positive ion conventional ESI mass spectra of 2.5×10^{-5} M myoglobin and different concentrations of SDS and TRIS buffer in H$_2$O/MeOH/AcOH (74/25/1) solution. (g)–(i): Positive ion nano-ESI mass spectra of 2.5×10^{-5} M myoglobin and different con-centrations of SDS and TRIS buffer in H$_2$O/MeOH/AcOH (74/25/1) solution. Peaks denoted with asterisk (*) are originated from SDS and TRIS: $[(\text{SDS})_n + \text{Na}]^+$, $[(\text{TRIS})_n(\text{HDS})_{n-1} + \text{H}]^+$, $[(\text{TRIS})_n(\text{HDS})_n + \text{Na}]^+$, $[(\text{TRIS})_n(\text{SDS})_n(\text{HDS})_n + \text{Na}]^+$, $[(\text{TRIS})_n(\text{SDS})_{n-1}(\text{HDS})_n + \text{Na}]^+$ and $[(\text{TRIS})_n(\text{SDS})_{n-2}(\text{HDS})_n + \text{Na}]^+$, where HDS stands for dodecyl hydro-gen sulfate, C$_{12}$H$_{25}$SO$_4$H.
(Adapted with permission from ref. 35 Copyright 2011 John Wiley & Sons, Ltd)

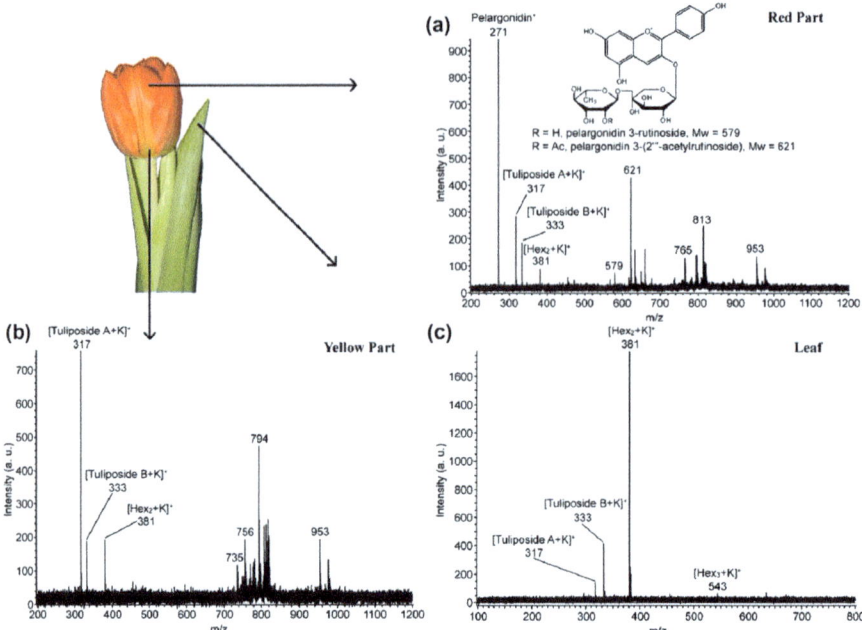

Figure 11.11   PESI mass spectra of the aerial part of a tulip plant, (a) the red part of the petal, (b) the yellow part of the petal, and (c) leaf, respectively. A penetration depth for sampling: 0.5 mm.
(Adapted with permission from ref. 38. Copyright 2009 Springer.)

region of liver was negligible and the scars on the liver surface healed after 24 h.[39]

The livers from control and steatotic mice were analyzed. As shown in Figure 11.14(A) and (B), the ion abundance for TAG in Figure 11.14(B) was much higher than those in Figure 11.13(A). Apparently, TAGs were accumulated in the liver of steatotic mice. Representative MS/MS spectra for $[PC(34:2) + H]^+$ (m/z 758.57) and $[TAG (52:3) + NH_4]^+$ (m/z 874.79) are shown in the insets of Figure 11.14(A) and (B), respectively.

11.7   Imaging Mass Spectrometry (IMS)

Elemental or molecular images of samples, including biological tissues examined under vacuum, have been reported using desorption ionization (DI) techniques, specifically secondary-ion mass spectrometry (SIMS) and matrix-assisted laser desorption ionization (MALDI). The first 2D molecular images of biological samples using ambient mass spectrometry were reported by DESI-MS in 2006.[40] Biological tissues are composed of cells that contain 70–90% water. When the tissue surface is probed by PESI, the biological fluid adhering to the needle tip can be electrosprayed directly or assisted by additional solvent added onto the needle surface. For example, an auxiliary heated solvent vapor can be used to supply

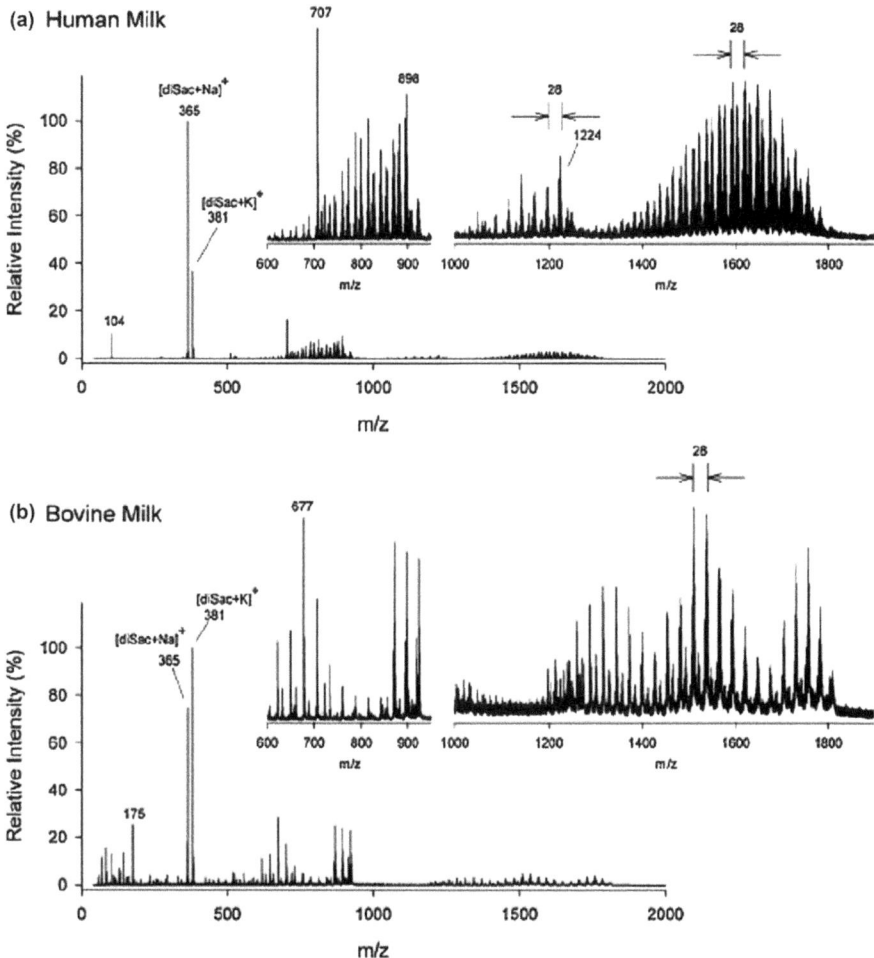

Figure 11.12 PESI mass spectra for (a) human milk and (b) local brand bovine milk formulated for new-born infants. Insets show the magnified mass spectra for each sample. Both samples were analyzed without dilution. (Adapted with permission from ref. 17. Copyright 2008 John Wiley & Sons, Ltd.)

solvent vapor to the needle tip (see Figure 11.13). The condensed vapor on the needle tip dissolves the adhered biofluid and also plays a role for the needle cleaning. Ambient imaging mass spectrometry (IMS) of mouse brain section was performed using PESI, incorporated with the solvent vapor sprayer. The histological sections were prepared by fixation using paraformaldehyde, and the spatial analysis was automated by maintaining an equal sampling depth into the sample in addition to raster scan. Figure 11.15 shows the 2D images of a mouse brain analyzed by PESI.[19] The spatial resolution was set to be 60 μm in this experiment. The microscopic

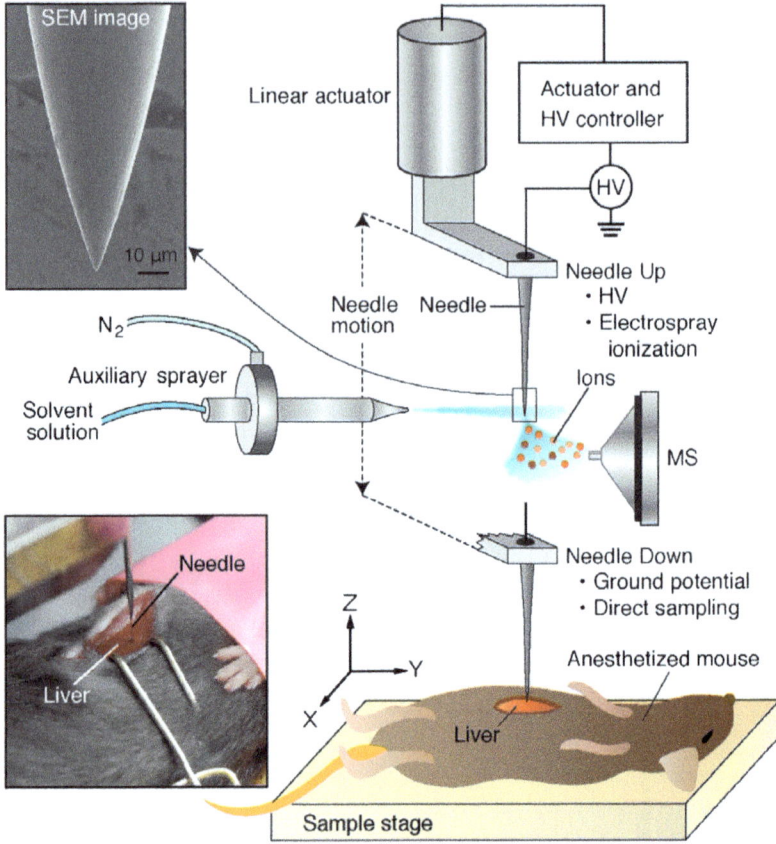

Figure 11.13 Schematic diagram of the sampling and electrospray processes of PESI–MS for the *in vivo* analysis of an anesthetized mouse. When the needle is at the sampling position (lowest position), it is electrically connected to the ground potential and the needle tip touches the surface of tissue. At the ionizing position (highest position), an appropriate solvent from a pneumatically assisted sprayer moistens the needle tip and is followed by application of HV to generate the electrospray. The upper inset shows a scanning electron micrograph of the needle probe (scale bar = 10 μm), and the lower inset shows a photograph taken during measurement in surgical intervention of a mouse.
(Adapted with permission from ref. 39. Copyright 2011 ScienceDirect.)

images of the brain section before, during and after the measurement are shown in Figure 11.16.[19]

In contrast to MALDI-MS imaging, where the mapping of phosphatidylcholines and galactosylceramides needs to be conducted separately using different matrices (*e.g.*, DHB for PC, and nanoparticles for GalCer),[41,42] both lipids could be readily detected by PESI. The ability to detect analytes nonselectively means that the suppression effect is largely moderated in PESI.

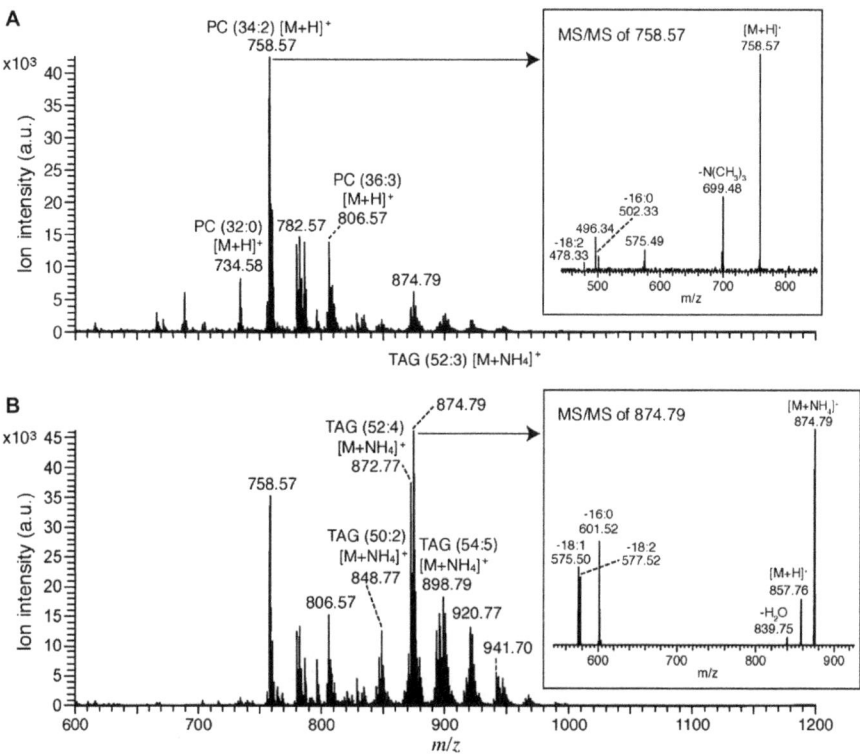

Figure 11.14 *In vivo* PESI mass spectra of livers from control (A) and steatotic (B) mice. The insets indicate MS/MS spectra of the $[M+H]^+$ ion of PC(34:2) (m/z 758.57) and the $[M+NH_4]^+$ ion of TAG(52:3) (m/z 874.79) in panels A and B, respectively. In this case, a mixed solution of chloroform/methanol (70:30, v/v) containing ammonium acetate (10 mM) was used for the auxiliary spray to efficiently ionize the lipid components.
(Adapted with permission from ref. 39. Copyright 2011 ScienceDirect.)

11.8 Cancer Diagnostics

Immediate diagnosis of human specimen is an essential prerequisite in medical routines. Direct and ambient analysis of histological tissue sections was demonstrated first by DESI-MS in 2005 where profiling of liver adenocarcinoma tissue sections revealed differences in the phospholipid profiles along a scan line extending from the tumor to nontumor regions.[43]

For cancer diagnosis by PESI, tissue samples of clear cell renal cell carcinoma (ccRCC) were freshly isolated from patients and were divided into noncancerous and cancerous regions. To confirm the histopathologic types of tumors, small pieces of tissue were set aside for paraffin sections stained with H&E (Figure 11.17(a) and (b), which were used as correlative samples to the PESI-MS. PESI mass spectra were obtained immediately after dropping 1 µL of binary solvent H_2O/CH_3OH (1/1) on the surface of the kidney tissue,

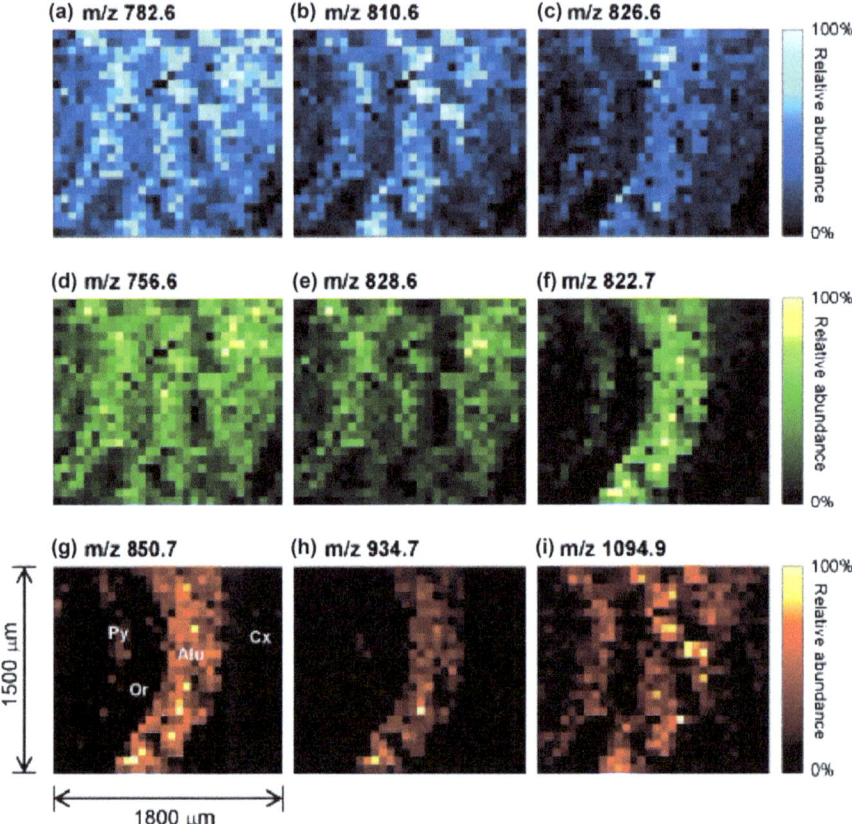

Figure 11.15   PESI-MS images of the mouse brain section for ion peaks at (a) *m/z* 782.6 ([PC(34:1) + Na]$^+$), (b) *m/z* 810.6 ([PC(36:1) + Na]$^+$), c) *m/z* 826.6 ([PC(36:1) + K]$^+$), (d) *m/z* 756.6 ([PC(32:0) + Na]$^+$), (e) *m/z* 828.6 ([PC(38:6) + Na]$^+$), (f) *m/z* 822.7 ([GalCer(22h:0) + Na]$^+$), (g) *m/z* 850.7 ([GalCer(24h:0) + Na]$^+$), (h) *m/z* 934.7 ([GalCer(30h:0) + Na]$^+$), and (i) *m/z* 1094.9. The pixel size is 60 µm and the analysis area is 1500×1800 µm^2. Cx = cerebral cortex, Alv = alveus of hippocampus, Or = oriens layer, Py = pyramidal cell layer.
(Adapted with permission from ref. 19. Copyright 2009 John Wiley & Sons, Ltd.)

which created on the tissue a spot size with a diameter of less than 2 mm.[44,45] As shown in Figure 11.17(b), the cancerous region displayed a typical honeycomb appearance, in which the cytoplasm is devoid of eosinophilic materials, showing so-called "clear cell" pattern; these microscopic findings are a hallmark of clear cell RCC (ccRCC). As shown in Figure 11.17(c) and (d), the PESI-MS could successfully generate distinct profiles of the lipid composition of cancerous region, in which TAGs were abundantly detected. Ions originating from phosphatidylcholines (PCs) (*e.g.,* PC[34:1], PC[34:2], and PC [36:2]) were shared by both noncancerous and

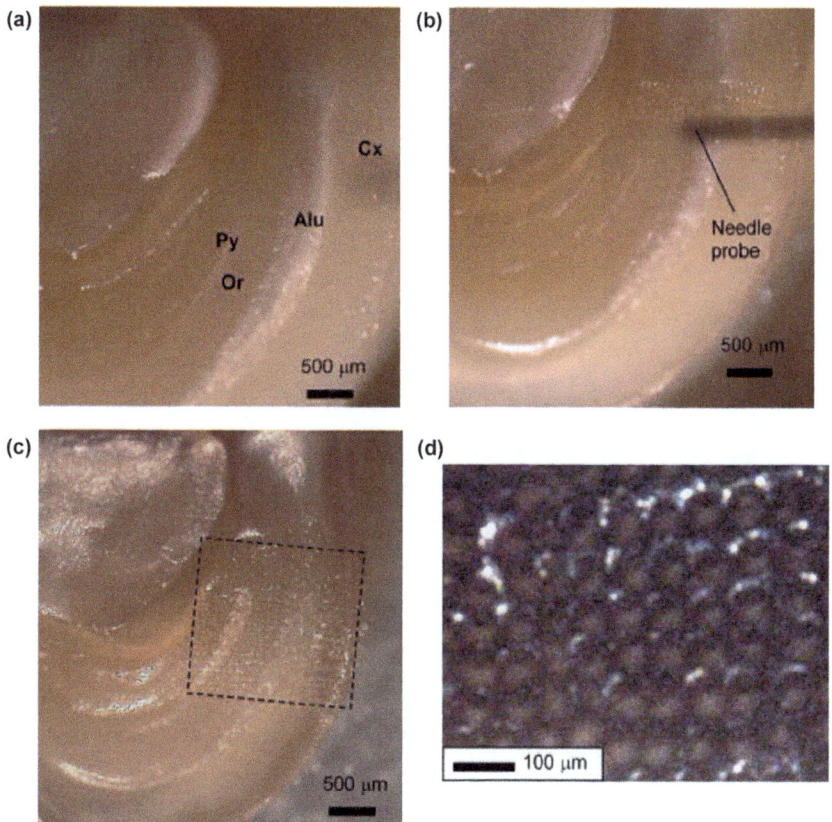

Figure 11.16 Microscopic images of the mouse brain section taken (a) before the measurement, (b) during the raster scan of the imaging PESI-MS, and (c) after the completion of the measurement. (d) Close-up inspection of the holes made by the solid needle. The scanned area is enclosed by dashed lines in (c).
(Adapted with permission from ref. 19. Copyright 2009 John Wiley & Sons, Ltd.)

cancerous regions, possibly because they are the principal components of cell membranes. It is of note that some TAGs peaks (*e.g.*, TAG [52:2], TAG [54:2], and TAG [54:3]) were detected in high abundance only in cancerous region (Figure 11.17(d)).

In addition to positive-mode PESI-MS, negative-mode PESI-MS was examined for colon cancer diagnosis.[46] In order to examine the ability of PESI for high-throughput analysis, 30 points have been selected for normal and cancerous samples as shown in Figure 11.18(a) and (b). After the needle was stopped at the ionization position, solvent vapor was provided to the needle tip from the vapor sprayer. The solvent vapor was condensed on the needle and washed away almost all the sample trapped on the tip of the needle, *i.e.* a cleaning effect. It was confirmed that the ion signals decreased

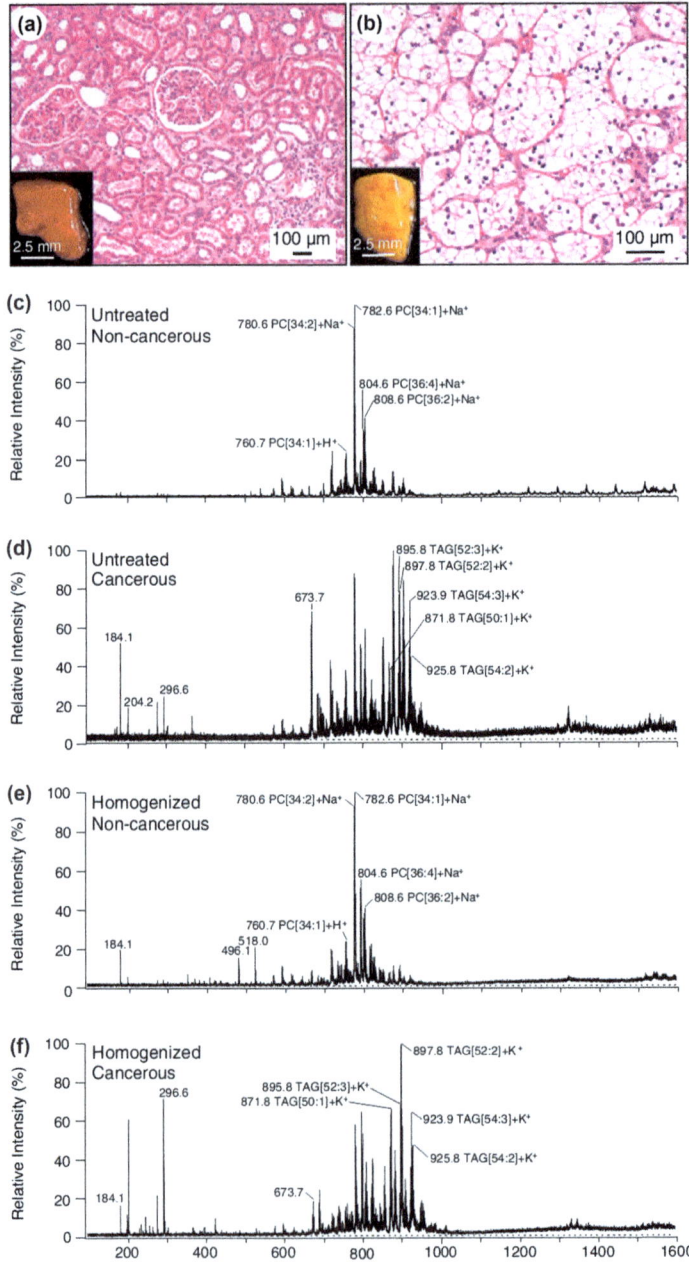

Figure 11.17 Kidney specimens were subjected to PESI-MS. Paraffin sections of kidney specimens from (a) noncancerous and (b) cancerous regions were stained with H&E. Photomicrograph (b) indicates the typical morphological appearance of ccRCC. Bars indicate 100 μm. Insets indicate the low magnification images of specimens. Bars indicate 2.5 mm. Representative patterns of mass spectra from noncancerous (c) and (e), and cancerous (d) and (f) regions.
(Adapted with permission from ref. 45. Copyright 2012 Springer.)

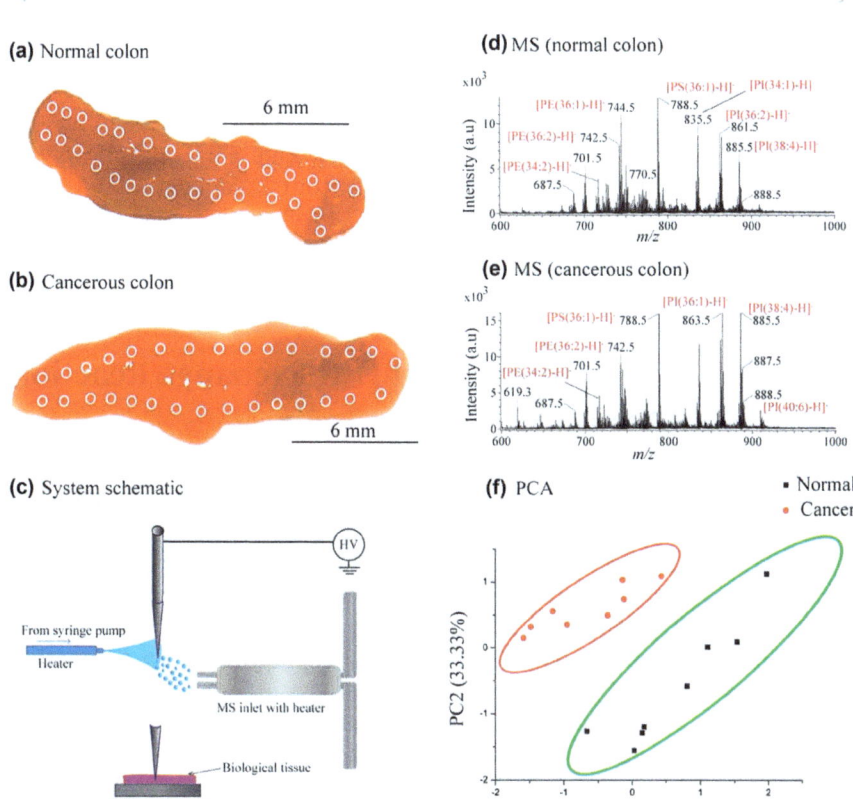

Figure 11.18 (a) and (b) Photographs of normal and cancerous human colon tissues. (c) PESI system with an auxiliary side vapor sprayer. Representative negative mode PESI mass spectra of normal (d), and cancerous (e) human colon tissues. 10 µl of isopropanol-H$_2$O (1/1) was dropped on the surface of tissue specimens prior to analysis. (f) Principal component analysis (PCA) from 8 normal and their 8 cancerous counterpart colon tissues obtained from 4 colon cancer patients. (Adapted with permission from ref. 46. Copyright 2013 The Royal Society of Chemistry.)

to zero at each shot of PESI operation. Thus the installation of the vapor sprayer could minimize the carryover of sample from one sampling location to another. The invasion depth of the needle into the sample was about 0.5 mm and with this invasion depth, the spatial resolution was 0.1 mm. The analysis of 30 points shown in Figure 11.18(a) and (b) could be made in less than 3 min with minimal carryover problem. This high-throughput analysis was attainable because the sample amount picked-up by the acupuncture needle is about a few pL and the duration of electrospray was only a few hundreds of ms. Figure 11.18(d) and (e) display the representative negative mode PESI mass spectra of cancerous tissue and their normal counterpart spots. Discriminating mass spectra were obtained for cancerous and non-cancerous spots with regard to the intensity of several lipid molecules (*e.g.*, [PI(38:4) – H]⁻). This is in good agreement with the work for colon cancer obtained by Gerbig *et al.* using DESI-MS.[47] For the proper visualization and creation of a diagnostic rule for characterizing the tissue sections, the principal component analysis (PCA) was applied. Figure 11.18(f) shows the 2D plots of PC1×PC2 for the data obtained from 8 cancerous and their 8 normal counterpart samples for 4 colon cancer patients. The spectral patterns from the normal and cancerous colon tissues were reasonably separated.[46]

11.9 Real-time Monitoring of H/D Exchange Reaction and Proteolysis

PESI-MS was used to monitor some chemical reactions in real time, such as acid-induced protein denaturation, hydrogen/deuterium exchange (HDX) of peptides, and Schiff base formation.[48] By using PESI, time-resolved mass spectra and ion chromatograms can be obtained reproducibly. Real-time PESI monitoring can give direct and detailed information on each chemical species taking part in reactions, and this is valuable for a better understanding of the whole reaction process and for the optimization of reaction parameters. PESI can be considered as a potential tool for real-time reaction monitoring due to its simplicity in instrumental setup, direct sampling and low sample consumption.

11.9.1 Hydrogen/Deuterium Exchange (HDX)

HDX mass spectrometry has been widely applied to investigate the structural dynamics and folding mechanism of proteins. In proteins, hydrogens that are not protected effectively by steric shielding and/or hydrogen bonding undergo HDX rapidly, and *vice versa*.[49] In general, sites containing more acidic hydrogen atoms undergo HDX faster. By monitoring the level of incorporated deuterium atoms, HDX can provide information on the solvent accessibility of various parts of a protein. HDX can also provide information for the identification of regions of flexibility in the protein sequence.[49] Because HDX is usually an extremely fast process, real-time monitoring of HDX is a challenge to all instrumental methods. Due to such advantages as rapid response, high sensitivity and high mass accuracy, MS is an ideal technique for the real-time monitoring of HDX kinetics because each individual exchange results in a 1-Da difference in the mass of the molecule and this can be clearly detected. Usually, two strategies are adopted for the real-time monitoring of HDX. One is to slow down the exchange rate. Under slow-exchange conditions (namely 0 °C, pH 2–3), Smith and coworkers have monitored the HDX of small peptides using continuous-flow fast-atom bombardment (CF-FAB).[50] The other strategy is fast mixing and fast

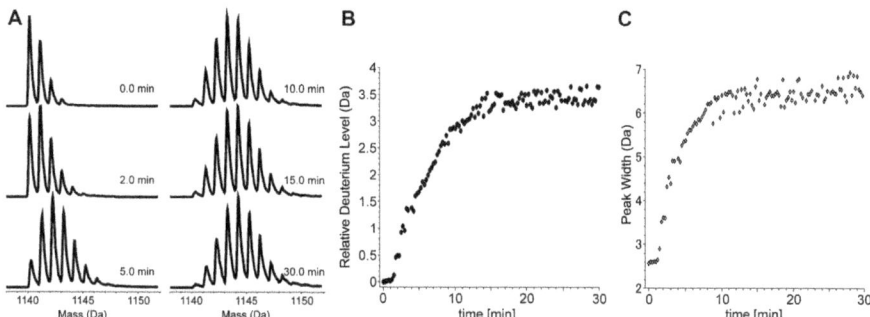

Figure 11.19 (A) Deconvoluted time-resolved mass spectra of deuterium-labeled GS monitored by PESI-MS combined with diffusive mixing strategy. Representative labeling times are shown to the right of each spectrum. Changes in relative deuterium level (B) and peak width (C) during the time course of real-time PESI-MS monitoring.
(Adapted with permission from ref. 48. Copyright 2010 John Wiley & Sons, Ltd.)

detection. By using continuous-flow capillary mixing systems, where two or more motor-driven syringes are used to mix reactant solutions rapidly in a mixing chamber, both continuous labeling and pulse-labeling can be achieved.[51,52] Herein, the combination of PESI-MS with the diffusive mixing of reactants to monitor the HDX kinetics of some peptides in real time will be described.[53] As an example, 150 mL of a 10^{-5} M gramicidin S (GS) water/methanol solution (50 : 50, v/v, containing AcOH 0.25%), and 75 mL of D-donor solution, mixture of D_2O and CH_3OD (50 : 50, v/v), were mixed. The mass-spectrometric data as a function of reaction time were deconvoluted by using Magtran34 and then analyzed using HX-Express.[53] The time-resolved PESI mass spectra and the HDX results are shown in Figure 11.19. It can be clearly seen that by adopting the method of diffusive mixing of reactants, the concentration of deuterium increased slowly that made the apparent HDX proceed slowly. Even minor changes during the HDX process could be observed clearly. For the first 1 min, no obvious changes were observed. After 1 min, the forward HDX started to dominate the whole HDX process, as shown by the increase in both the relative deuterium level and the peak width in Figure 11.19(B) and (C), respectively. At about 15 min, both the number of deuteriums exchanged and the peak width reached a plateau, which means that equilibrium of forward and reverse HDX has been reached.

HDX results for two other small peptides, bradykinin and angiotensin I, were also performed. These compounds presented different HDX behavior from that of GS with different relative deuterium levels due to their different amino acid sequences.[48]

11.9.2 Enzyme-catalyzed Reaction Monitoring

Although there are a lot of well-established methods for monitoring enzyme–catalyzed reactions, most of them are based on changes in spectroscopic properties during the conversion of substrates to products.[54] However, re-actions without optical changes are common, which are not amenable to these spectroscopic methods. As an alternative technique for enzymologic research, mass spectrometry (MS) is favored due to its specificity, sensitivity, and the ability to obtain stoichiometric information. PESI was employed to real-time monitoring of some typical protease catalyzed reactions, including pepsinolysis and trypsinolysis of cytochrome c. Due to the high electrical conductivity of each reaction system, corona discharges are likely to occur, which would decrease the intensities of mass-spectrometric signals. An ultra-fine sampling probe and an auxiliary vapor spray were adopted to prevent corona discharges. Experimental results from peptic and tryptic digestions of cytochrome c showed different and characteristic catalytic pathways.[54]

11.9.2.1   Pepsinolysis

As one of the major proteases in the human digestive system, pepsin is se-creted by chief cells in the gastric mucosa as pepsinogen, the precursor of pepsin with 44 additional amino acids.[55] Pepsin is the enzyme whose cata-lytic activity is highly dependent on the environmental pH value. Porcine pepsin, the most studied pepsin, has its maximal activity at pH 2.0 and re-tains its activity when the pH drops to 1.0, but begins to be inactive at pH 5.[56] Normally, pepsin is believed to have a broad but low specificity by preferring to cleave peptide bonds in which the carboxyl group is provided by amino acids with bulky hydrophobic or aromatic side chains.[57] The pepsinolysis is susceptible to many factors such as the ratio of substrate/enzyme, tem-perature, and pH values of solvents.[58] Hamuro *et al.* have statistically ana-lyzed large amounts of peptic digestion data, and their results showed that at low temperature (<1 °C) and low pH, the specificity of pepsin can be nar-rowed.[59] Furthermore, a recent UPLC-MS research suggested that the spe-cificity of pepsin is pH independent.[60] For the purpose of investigating the relationship between pepsin specificity and the solution pH value, peptic digestions of bovine CytC at pH 1 and 3 were monitored by PESI-MS in real time. Time-resolved mass spectra of pepsinolysis of CytC at pH 1 are shown in Figure 11.20(a). At the starting point (the time right after the mixing of pepsin and CytC solutions, noted as 0.0 min), ions of CytC with multiple charges (from $9+$ to $21+$) were detected. At 0.5 min after the starting point, the ion intensities of multicharged CytC ions decreased sharply and several multicharged fragment ions started to appear, especially the fragment 65–104 at *m/z* 583.7. At 3.0 min, the fragment 65–104 disappeared and ions of the fragment 35–101 at *m/z* 690.9 and the fragment 65–82 at *m/z* 551.7 became dominant; 60.0 min later, the most abundant peaks were ions of fragments 65–82 and 83–96. From the extracted ion chromatograms (EICs) of four typical ions shown in Figure 11.20(b), the kinetic process of pepsinolysis of the substrate could be clearly seen. CytC was consumed very rapidly to generate the fragment 65–104, a short-lived intermediate.

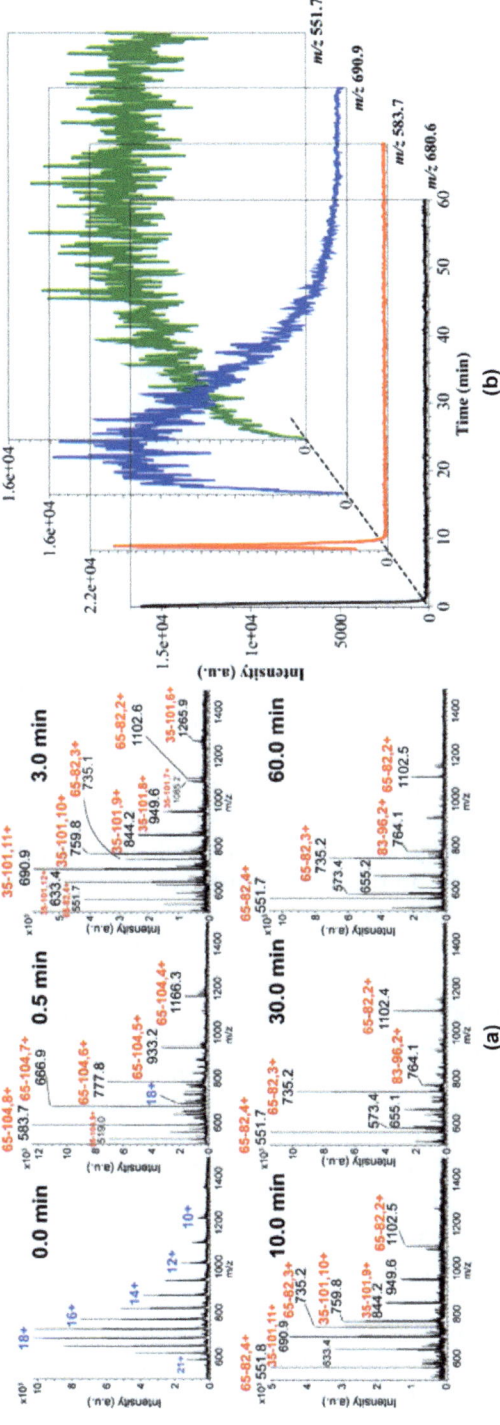

Figure 11.20 Online real-time monitoring of pepsinolysis of CytC at pH 1. (a) Time-resolved mass spectra at 0.0, 0.5, 3.0, 10.0, 30.0, and 60.0 min after the mixing of 2 μL pepsin (1 μg μL⁻¹), and 200 μL CytC (2×10⁻⁵ M); (b) EICs of ions at *m/z* 680.6 (CytC 18 +), 583.7, 690.9, and 551.7, respectively. Charge states of CytC ions are notated in blue. Adapted with permission from ref. 54. Copyright 2012 Springer.

The fragment 35–101 seemed to be generated simultaneously with the fragment 65–104 but with a slower rate. The fragment of 65–82 could be viewed as one of the final digestion products during the 60 min of pepsinolysis judged from its relatively stable ion intensity.

The pepsinolysis of CytC was also examined at pH 3 to investigate how the solution pH affected the digestion process. By comparing the peptic digestion pathways of CytC at pH 1 and 3, one significant feature of pepsinolysis is that the peptic digestion proceeds as a multistage process.[54] The final digestion products may be generated from sequentially cleaved intermediates from substrates.[61,62] As hypothesized by Belikow and Antonowa,[63] pepsin may preferentially cleave peptide bonds on the surface of a substrate. When the most-accessible peptide bonds on the surface have been hydrolyzed, peptide bonds on the inner parts will become accessible. To a pepsinolysis, the solution pH value is a vital factor because it can influence both the pepsin activity and the conformation of the substrate, which makes the whole reactions multistaged.[54]

11.9.2.2   Trypsisnolysis

Trypsin is another important proteolytic enzyme in the human digestive system. Trypsin is activated from the inactive trypsinogen secreted by the pancreas and functions in the lumen of the small intestine. Usually trypsin digests optimally at around pH 8 and 37 °C.[64] Trypsin breaks proteins down into peptides by following a strict rule of cleaving on the carboxyl side of lysine and arginine except if they are followed by proline.[65] The high specificity makes trypsin the first choice protease for most MS-based proteomics.[66] Trypsin is often notated as an opposite example to pepsin due to its high but narrow specificity, requirement of an alkaline environment, and the slowness of its catalysis. In order to compare the catalytic properties of trypsin and pepsin, trypsinolysis of CytC was performed and monitored in real time by the same method as described above. The time-resolved mass spectra at 0.0, 10.0, 30.0, and 60.0 min after the mixing of CytC and trypsin solutions are shown in Figure 11.21(a). Due to the high specificity of trypsin, most ions of digestion products could be identified according to the theoretical digestion results and the data published previously.[67–69] Compared with the pepsinolysis results shown in Figure 11.20, the detection of intact ions of CytC at 60 min could be ascribed to the slowness of trypsin catalysis. In Figure 11.21(b), the EICs of typical ions at m/z 1530.1, 779.7, 1125.8, and 617.3 corresponding to CytC 8 +, the fragment 80–86, the fragment 54–72 and heme are shown, respectively, which also indicate that trypsin has a lower catalytic activity than pepsin.

In 1994, Noda *et al.*[70] reported the existence of intermediates in the trypsinolysis of a modified lysozyme, which has a similar molecular mass to CytC. In that work, the tryptic digestion was performed at insufficient time and under suboptimal conditions. In PESI study, however, all fragment ions seemed to be generated in parallel. There was no significant observation of

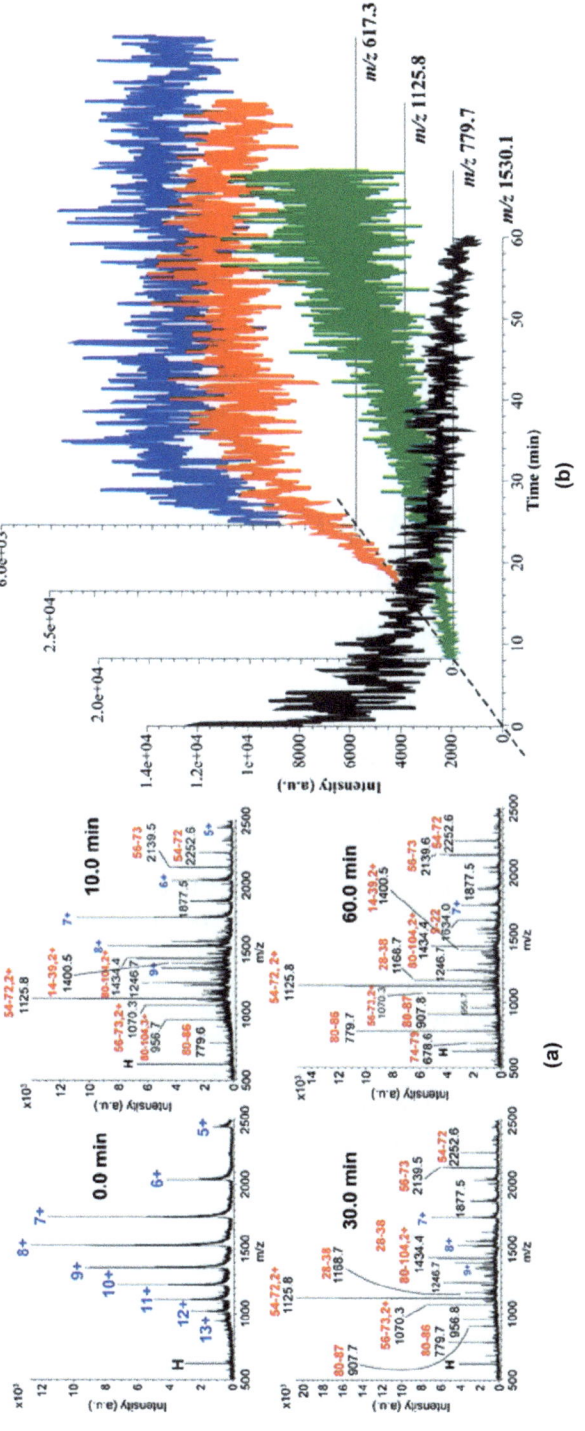

Figure 11.21 Online real-time monitoring of trypsinolysis of CytC at pH 8. (a) Time-resolved mass spectra at 0.0, 10.0, 30.0, and 60.0 min after the mixing of 2 μL trypsin gold (1 μg μL^{-1}) and 200 μL CytC (5×10^{-5} M); (b) EICs of ions at m/z 1530.1 (CytC 8+), 779.7, 1125.8, and 617.3 (heme), respectively. Charge states of CytC ions are notated in blue. H indicates heme. Adapted with permission from ref. 54. Copyright 2012 Springer.

intermediate species, whose ion intensities would comprise an increase followed by a decrease like the ion at m/z 583.7 shown in Figure 11.20(b). As shown in Figure 11.21(b), the intensities of three typical ions exhibited either a continuous increase (the ion at m/z 779.7) or a fast increase with a plateau (the ions at m/z 1125.8 and 617.3). There is much discussion concerning intermediates of trypsinolysis, where substrates with high masses were employed.[71,72] Additionally, a full tryptic digestion needs hours of incubation time. Thus, the authors assume the simultaneous observation of tryptic products of CytC, which could be regarded as a parallel but not consecutive process as far as the reaction time of 60 min was concerned in the current experiment. This may be ascribed to the relatively low molecular mass of CytC.[54] The kinetic difference between peptic and tryptic digestions of CytC reflects characteristically different stages in the human digestive system, namely, a fast and extensive digestion *versus* a slow and intensive digestion.

11.10   Related Topics with PESI-MS

11.10.1   Sheath-Flow PESI-MS

Although PESI is highly versatile, it was restricted to liquid and wet biological samples. In order to circumvent this limitation, sheath-flow PESI mass spectrometry (SF-PESI-MS) was developed.[73] SF-PESI-MS system schematic is shown in Figure 11.22. An acupuncture needle (0.12 mm o.d. with a tip diameter of 700 nm) was inserted into the gel loading pipette tips with i.d. of about 0.135 mm and o.d. of 0.2 mm. The acupuncture needle was protruded from the tip of the capillary by about 0.1–0.2 mm. Acetonitrile:H_2O (1/1) with 0.1% formic acid as a solvent was flowed through the capillary with a flow rate of 1 μL min^{-1} through the LC T-joint.

The needle was driven along the vertical axis perpendicularly to the apex of the ion-sampling orifice with a frequency of 0.5 Hz using a linear actuator with an electronic controller of regulated frequency. The stroke distance of the needle was set to be 10 mm. The needle was driven with the speed of 350 mm s^{-1}. The duration time of the needle in touch with the sample surface for sample extraction was 0.2 s. A high voltage of about 2.5 kV was applied to the needle while it was at the highest position with the duration of time of about 1.7 s. The samples were positioned on the *xyz* moving stage. The needle position was adjusted to just touch the sample surface (no invasion into the sample) under the optical microscope observation. The sample stage was moved with the speed 0.1 mm s^{-1} or faster.[73]

Analysis of cocaine on currency was first made by DART[74] using a time-of-flight (TOF) mass spectrometer and later by DESI with a handheld mass spectrometer.[75] As an example of the application of SF-PESI to real-world samples, the authors also applied SF-PESI to one US dollar bill analysis.[76] In SF-PESI experiment, several arbitrary spots of the finger-held bill were examined by touching it to the probe. The bill was examined starting from

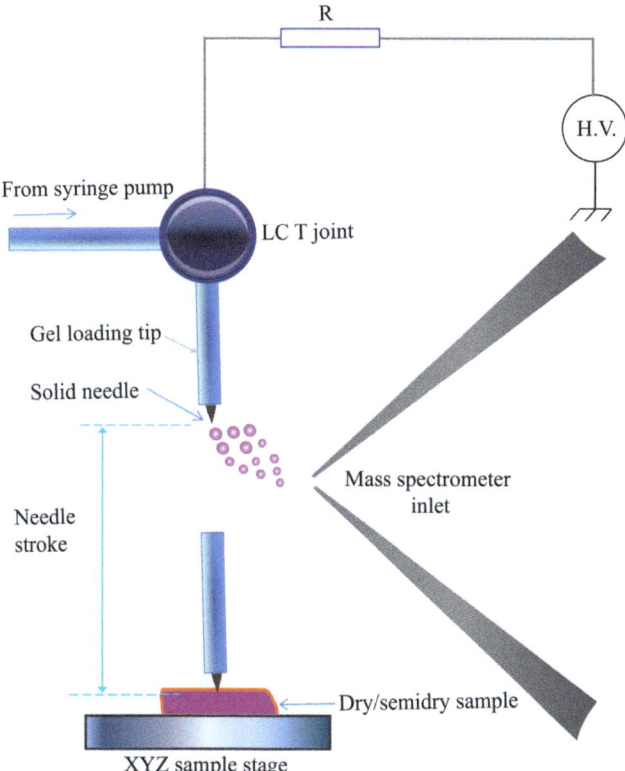

Figure 11.22 Schematic showing sheath-flow probe electrospray ionization mass spectrometry (SF-PESI-MS).
(Adapted with permission from ref. 73. Copyright 2013 American Chemical Society.)

the center to the terminus. Figure 11.23(a) represents the total ion chromatogram (TIC) scanned across the bill. Although little signal was detected in the central region, quite strong ion signals appeared when the bill was moved toward the terminus, as shown in the figure. Figure 11.23(b) shows the mass spectrum obtained at 4.5 min in Figure 11.23(a). The peak appearing at *m/z* 304 was assigned as protonated cocaine by the precise mass calibration of the TOF mass spectrometer.

By SF-PESI, 100 pg of methamphetamine and morphine deposited on a finger could also be detected easily.[76]

11.10.2 Solid-probe-assisted Nanoelectrospray Ionization

SF-PESI gave a solution to PESI for its application to dry samples. The basic concept is to supply solvent and extract analytes from the sample surface. For the alternative way to extract analytes from the sample surface, the authors developed the solid probe assisted nano-ESI (SPA-nano-ESI).[77] This

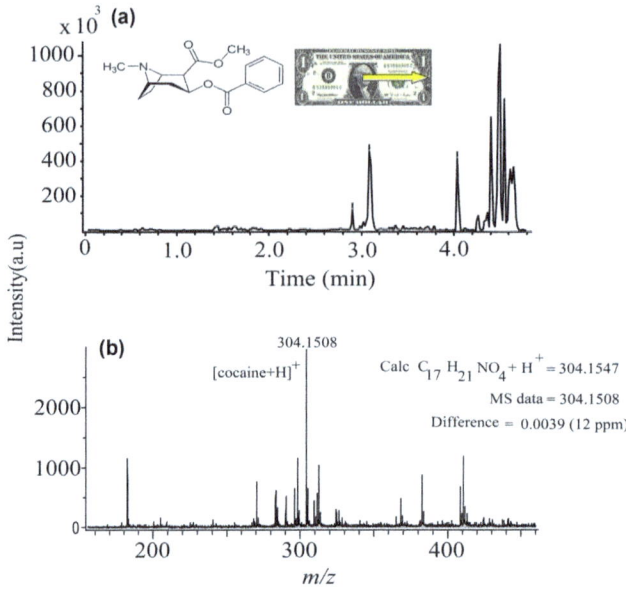

Figure 11.23 (a) Total ion chromatogram (TIC) scanned across the US one dollar banknote. (b) Mass spectrum obtained at 4.5 min in (a).
(Adapted with permission from ref. 76. Copyright 2010 John Wiley & Sons, Ltd.)

technique made it possible to perform remote sampling for mass spectrometry.[77] The procedure for SPA-PESI is shown schematically in Figure 11.24. A 0.3 mm acupuncture needle with ~700 nm tip diameter was stuck into the biological tissue with an invasion depth of about 1 mm and biofluid was sampled at the tip of the needle. After sampling, the needle was inserted into a nano-ESI capillary, in which 2 μl of organic solvent was preloaded. The nano-ESI capillary was then positioned at 5 mm from the inlet of a mass spectrometer. A high voltage (0.5–1.5 kV) was applied to a metal-coated nano-ESI capillary.

Since, in most cases, the needle normally captured only biofluids but not viscous tissues by this sampling procedure, nano-ESI capillaries were seldom clogged. The amounts of liquid trapped on the tip of the needle were about tens of pL.[18] Figure 11.25 demonstrates nano-ESI mass spectra of normal and cancerous kidney tissues, respectively, using a mixture of ACN, MeOH and isopropanol (1/1/1) as a solvent as we found that this mixture produced more abundant signal intensities than other solvents in positive mode of operation. As shown in the figure, normal and cancerous kidney tissues are clearly distinguished. For the sampling procedure that may need long-term sample preservation, we have frozen the kidney sample-loaded metal needle in glass vials at −30 °C for a few days. The obtained mass spectra before and after freezing were found to be almost identical. This suggests that dry ice, liquid nitrogen or other freezing systems could be used for preserving the

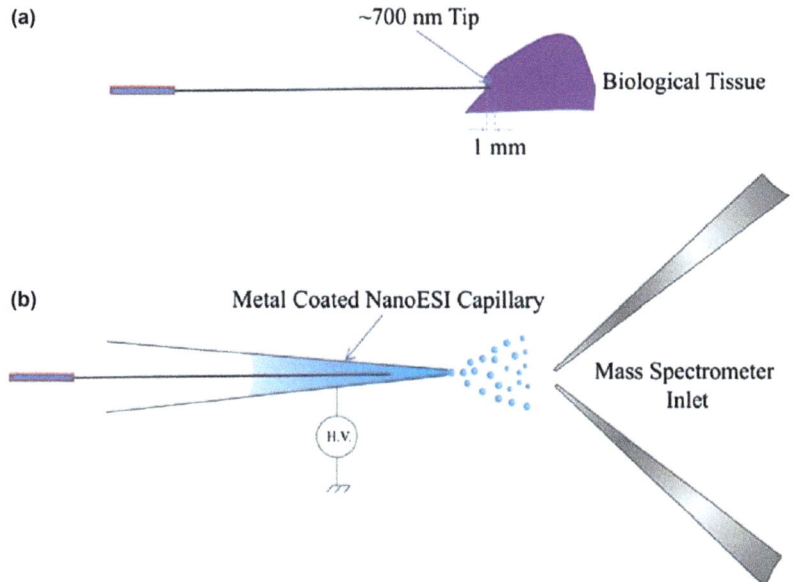

Figure 11.24 Schematic showing solid probe assisted nano-ESI (SPA-nano-ESI) using an acupuncture needle. (a) Sampling of biofluid from the tissue by sticking the acupuncture needle into the tissue. (b) Acupuncture needle inserted into the solvent-preloaded nano-ESI capillary for electrospray.
(Adapted with permission from ref. 77. Copyright 2012 The Royal Society of Chemistry.)

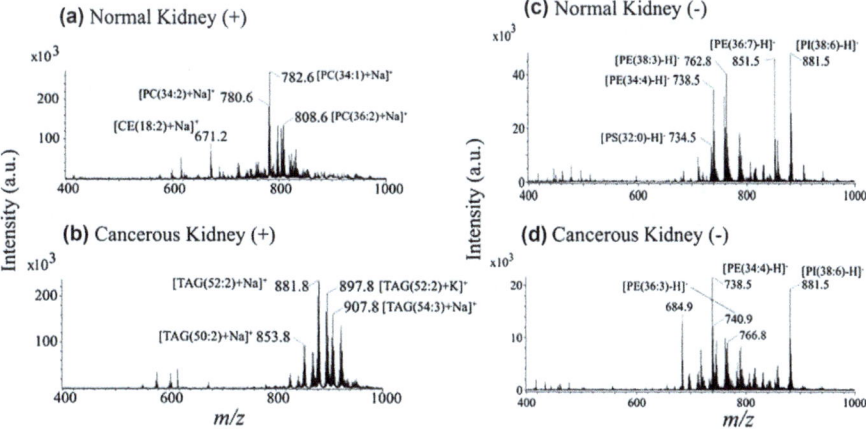

Figure 11.25 Positive mode SPA-nano-ESI mass spectra of normal (a) and cancerous (b) kidney tissues, respectively. Solvent: ACN/MeOH/isopropanol (1/1/1). Negative mode SPA-nano-ESI mass spectra of normal (c) and cancerous (d) kidney tissues, respectively. Solvent: DMF/EtOH (1/1).
(Adapted with permission from ref. 77. Copyright 2012 The Royal Society of Chemistry.)

sample when a sample-loaded needle is required to be transported from a distant place to a mass-spectrometry lab. SPA-nano-ESI succeeded in coupling the less-invasive needle biopsy with the high-sensitive nano-ESI for the off-line mass spectrometry. Dissection of organs is not a prerequisite prior to the analysis by this method. Therefore, it may fulfill the requirements of the clinical diagnosis in a doctor's room or during surgery, *e.g.*, point-of-care analysis. SPA-nano-ESI may also be applicable to relatively dry samples if they are made wet by dropping some solvents on the sample surface.

11.11   Conclusion and Perspectives

Because of adopting discontinuous sampling by a sharp needle, PESI can bring out inherent potential capabilities of electrospray. The characteristic features of PESI may be summarized as follows.

1. PESI is robust and free from clogging.
2. The gas breakdown at the needle tip is largely suppressed because the tip is wet with the liquid sample (the breakdown voltage of liquid is orders of magnitude higher than that of gas).
3. The electrochemical reactions taking place at the interface between the metal needle and the liquid supply the excess charges continuously to the liquid attached to the needle tip. This results in the sequential and exhaustive electrospray for ionizable components in the sample.[24] This is not compatible with the capillary-based ESI because the main part of the droplet is charge depleted by the spawning the highly charged microdroplets.
4. Due to the sequential and exhaustive electrospray, the suppression effect is minimized.[24]
5. Sample amounts with pL or less generate strong ion signals. In this respect, PESI is highly suitable for treating very small precious samples.[78] PESI could be applicable to biological samples with least invasiveness and high spatial resolution, *e.g.*, stethoscope, needle biopsy, cell imaging.
6. The direct pick-and-spray is suitable for real-time monitoring of chemical reactions.
7. Dry samples deposited at the needle tip can be electrosprayed by supplying additional solvent to the needle tip, *e.g.* side-vapor spray, sheath-flow PESI, SPA-nano-ESI.[73,76]
8. PESI is highly tolerable to detergents, salts and buffers. It will offer an alternative technique for the direct and high throughput analysis of biological samples without laborious sample pretreatment.[35]
9. The flexibility in instrumental configuration guarantees PESI for its application to any type of mass spectrometers that are equipped with atmospheric-pressure ion sources.
10. PESI may be extrapolated to the ambient field desorption (FD) by using a sharp needle with the field strength at the tip higher than 10^8 V m^{-1}.

One of the biggest issues in mass spectrometry is how to minimize the suppression effect. If the concentration of ionizing reagent could be higher than those of analytes, the suppression effect will be minimized or avoided. This will be partly satisfied by generating as much as possible (AMAP) ionizing reagents and introducing as small as possible (ASAP) detectable reactants in the ionizing region. PESI is in line with this demand because the excess charges are continuously supplied to the small amount of liquid sample trapped at the needle tip. In addition, a greater fraction of ions generated near the needle tip can be introduced to the mass spectrometer. PESI will pave the way for the next-generation mass spectrometry.

References

1. J. Zeleny, *Phys. Rev.*, 1917, **10**, 1–6.
2. G. Taylor, *Proc. R. Soc. A*, 1964, **280**, 383–397.
3. M. Dole, L. L. Mack, R. L. Hines, R. C. Mobley, L. D. Ferguson and M. B. Alice, *J. Chem. Phys.*, 1968, **49**, 2240–2249.
4. M. Yamashita and J. B. Fenn, *J. Phys. Chem.*, 1984, **88**, 4451–4459.
5. J. Fenn, M. Mann, C. Meng, S. Wong and C. Whitehouse, *Science*, 1989, **246**, 64–71.
6. J. H. Wahl, D. R. Goodlett, H. R. Udseth and R. D. Smith, *Anal. Chem.*, 1992, **64**, 3194–3196.
7. M. R. Emmett and R. M. Caprioli, *J. Am. Soc. Mass Spectrom.*, 1994, **5**, 605–613.
8. M. S. Wilm and M. Mann, *Int. J. Mass Spectrom. Ion Processes*, 1994, **136**, 167–180.
9. M. Wilm and M. Mann, *Anal. Chem.*, 1996, **68**, 1–8.
10. C.-M. Hong, C.-T. Lee, Y.-M. Lee, C.-P. Kuo, C.-H. Yuan and J. Shiea, *Rapid Commun. Mass Spectrom.*, 1999, **13**, 21–25.
11. C.-P. Kuo, C.-H. Yuan and J. Shiea, *J. Am. Soc. Mass Spectrom.*, 2000, **11**, 464–467.
12. J. Jeng and J. Shiea, *Rapid Commun. Mass Spectrom.*, 2003, **17**, 1709–1713.
13. J. Jeng, C.-H. Lin and J. Shiea, *Anal. Chem.*, 2005, 77, 8170–8173.
14. T. Nissilä, L. Sainiemi, T. Sikanen, T. Kotiaho, S. Franssila, R. Kostiainen and R. A. Ketola, *Rapid Commun. Mass Spectrom.*, 2007, **21**, 3677–3682.
15. K. Hiraoka, K. Nishidate, K. Mori, D. Asakawa and S. Suzuki, *Rapid Commun. Mass Spectrom.*, 2007, **21**, 3139–3144.
16. S. S. Wong, U. Giessmann, M. Karas and F. W. Röllgen, *Int. J. Mass Spectrom. Ion Processes*, 1984, **56**, 139–150.
17. L. C. Chen, K. Nishidate, Y. Saito, K. Mori, D. Asakawa, S. Takeda, T. Kubota, N. Terada, Y. Hashimoto, H. Hori and K. Hiraoka, *Rapid Commun. Mass Spectrom.*, 2008, **22**, 2366–2374.
18. K. Yoshimura, L. C. Chen, D. Asakawa, K. Hiraoka and S. Takeda, *J. Mass Spectrom.*, 2009, **44**, 978–985.

19. L. C. Chen, K. Yoshimura, Z. Yu, R. Iwata, H. Ito, H. Suzuki, K. Mori, O. Ariyada, S. Takeda, T. Kubota and K. Hiraoka, *J. Mass Spectrom.*, 2009, **44**, 1469–1477.

20. L. C. Chen, K. Nishidate, Y. Saito, K. Mori, D. Asakawa, S. Takeda, T. Kubota, H. Hori and K. Hiraoka, *J. Phys. Chem. B*, 2008, **112**, 11164–11170.

21. R. Juraschek, T. Dülcks and M. Karas, *J. Am. Soc. Mass Spectrom.*, 1999, **10**, 300–308.

22. J. Wei, W. Shui, F. Zhou, Y. Lu, K. Chen, G. Xu and P. Yang, *Mass Spectrom. Rev.*, 2002, **21**, 148–162.

23. I. Marginean, L. Parvin, L. Heffernan and A. Vertes, *Anal. Chem.*, 2004, **76**, 4202–4207.

24. M. K. Mandal, L. C. Chen and K. Hiraoka, *J. Am. Soc. Mass Spectrom.*, 2011, **22**, 1493–1500.

25. P. W. Atkins, *Atkins' Physical chemistry*, Oxford University Press, Oxford, New York, 1986.

26. P. D. E. W. Müller, Feldemission, in *Ergebnisse der exakten naturwissenschaften*, Springer, Berlin Heidelberg, 1953.

27. R. P. Morgan, *Org. Mass Spectrom.*, 1978, **13**, 490.

28. A. K. Sen, J. Darabi, D. R. Knapp and J. Liu, *J. Micromech. Microeng.*, 2006, **16**, 620.

29. *Fundamentals of Mass Spectrometry*, ed. K. Hiraoka, Springer, 2013.

30. M. K. Mandal, L. C. Chen, Y. Hashimoto, Z. Yu and K. Hiraoka, *Anal. Methods*, 2010, **2**, 1905.

31. D.-Y. Chang, C.-C. Lee and J. Shiea, *Anal. Chem.*, 2002, **74**, 2465–2469.

32. A. U. Jackson, N. Talaty, R. G. Cooks and G. J. Van Berkel, *J. Am. Soc. Mass Spectrom.*, 2007, **18**, 2218–2225.

33. H. Chen, A. Venter and R. G. Cooks, *Chem. Commun.*, 2006, 2042–2044.

34. Y. Yao, S. Hong, H. Zhou, T. Yuan, R. Zeng and K. Liao, *Cell Res.*, 2009, **19**, 497–506.

35. M. K. Mandal, L. C. Chen, Z. Yu, H. Nonami, R. Erra-Balsells and K. Hiraoka, *J. Mass Spectrom.*, 2011, **46**, 967–975.

36. L. C. Chen, Z. Yu, H. Nonami, Y. Hashimoto and K. Hiraoka, *Environ. Control Biol.*, 2009, **47**, 73–86.

37. K. Hiraoka, L. C. Chen, T. Iwama, M. K. Mandal, S. NInomiya, H. Suzuki, O. Ariyada, H. Furuya and K. Takekawa, *J. Mass Spectrom. Soc. Jpn.*, 2010, **58**, 215–220.

38. Z. Yu, L. C. Chen, H. Suzuki, O. Ariyada, R. Erra-Balsells, H. Nonami and K. Hiraoka, *J. Am. Soc. Mass Spectrom.*, 2009, **20**, 2304–2311.

39. K. Yoshimura, L. C. Chen, Z. Yu, K. Hiraoka and S. Takeda, *Anal. Biochem.*, 2011, **417**, 195–201.

40. J. M. Wiseman, D. R. Ifa, Y. Zhu, C. B. Kissinger, N. E. Manicke, P. T. Kissinger and R. G. Cooks, *Proc. Natl. Acad. Sci.*, 2008, **105**, 18120–18125.

41. S. N. Jackson, M. Ugarov, T. Egan, J. D. Post, D. Langlais, J. Albert Schultz and A. S. Woods, *J. Mass Spectrom.*, 2007, **42**, 1093–1098.

42. S. Taira, Y. Sugiura, S. Moritake, S. Shimma, Y. Ichiyanagi and M. Setou, *Anal. Chem.*, 2008, **80**, 4761–4766.
43. J. M. Wiseman, S. M. Puolitaival, Z. Takáts, R. G. Cooks and R. M. Caprioli, *Angew. Chem., Int. Ed.*, 2005, **44**, 7094–7097.
44. M. K. Mandal, K. Yoshimura, L. C. Chen, Z. Yu, T. Nakazawa, R. Katoh, H. Fujii, S. Takeda, H. Nonami and K. Hiraoka, *J. Am. Soc. Mass Spectrom.*, 2012, **23**, 2043–2047.
45. K. Yoshimura, L. C. Chen, M. K. Mandal, T. Nakazawa, Z. Yu, T. Uchiyama, H. Hori, K. Tanabe, T. Kubota, H. Fujii, R. Katoh, K. Hiraoka and S. Takeda, *J. Am. Soc. Mass Spectrom.*, 2012, **23**, 1741–1749.
46. M. K. Mandal, S. Saha, K. Yoshimura, Y. Shida, S. Takeda, H. Nonami and K. Hiraoka, *Analyst*, 2013, **138**, 1682–1688.
47. S. Gerbig, O. Golf, J. Balog, J. Denes, Z. Baranyai, A. Zarand, E. Raso, J. Timar and Z. Takats, *Anal. Bioanal. Chem.*, 2012, **403**, 2315–2325.
48. Z. Yu, L. C. Chen, R. Erra-Balsells, H. Nonami and K. Hiraoka, *Rapid Commun. Mass Spectrom.*, 2010, **24**, 1507–1513.
49. L. Konermann, J. Pan and D. J. Wilson, Biochemical Reaction Kinetics Studied by Time-Resolved Electrospray Ionization Mass Spectrometry, in *Comprehensive Analytical Chemistry*, ed. J. P. Whitelegge, Elsevier, 2008, ch. 5.
50. G. Thevenon-Emeric, J. Kozlowski, Z. Zhang and D. L. Smith, *Anal. Chem.*, 1992, **64**, 2456–2458.
51. D. A. Simmons, S. D. Dunn and L. Konermann, *Biochemistry (Moscow)*, 2003, **42**, 5896–5905.
52. J. Pan, D. J. Wilson and L. Konermann, *Biochemistry (Moscow)*, 2005, **44**, 8627–8633.
53. D. D. Weis, J. R. Engen and I. J. Kass, *J. Am. Soc. Mass Spectrom.*, 2006, **17**, 1700–1703.
54. Z. Yu, L. Chen, M. Mandal, H. Nonami, R. Erra-Balsells and K. Hiraoka, *J. Am. Soc. Mass Spectrom.*, 2012, **23**, 728–735.
55. M. N. G. James and A. R. Sielecki, *Nature*, 1986, **319**, 33–38.
56. D. W. Piper and B. H. Fenton, *Gut*, 1965, **6**, 506–508.
57. J. S. Fruton, The Specificity and Mechanism of Pepsin Action, in *Advances in Enzymology and Related Areas of Molecular Biology*, ed. F. F. Nord, John Wiley & Sons, Inc., 2006.
58. Castro, *Enzyme Microb. Technol.*, 1999, **25**, 689–694.
59. Y. Hamuro, S. J. Coales, K. S. Molnar, S. J. Tuske and J. A. Morrow, *Rapid Commun. Mass Spectrom.*, 2008, **22**, 1041–1046.
60. M. Palashoff, *Chem. Masters Theses*, MPhil, Northeastern University, 2008.
61. M. Schlamowitz, L. U. Peterson and F. C. Wissler, *Arch. Biochem. Biophys.*, 1961, **92**, 58–68.
62. M. S. Silver and M. Stoddard, *Biochemistry (Moscow)*, 1972, **11**, 191–200.
63. W. M. Belikow and T. W. Antonowa, *Nahrung*, 1976, **20**, 593–596.
64. R. Beynon, *Proteolytic Enzymes: A Practical Approach*, Oxford University Press, USA, 2nd edn, 2001.

65. B. Keil, *Specificity of Proteolysis*, Springer, 1992.
66. R. Aebersold and M. Mann, *Nature*, 2003, **422**, 198–207.
67. B. N. Pramanik, U. A. Mirza, Y. H. Ing, Y.-H. Liu, P. L. Bartner, P. C. Weber and A. K. Bose, *Protein Sci. Publ. Protein Soc.*, 2002, **11**, 2676–2687.
68. E. C. A. Stigter, G. J. de Jong and W. P. van Bennekom, *Anal. Bioanal. Chem.*, 2007, **389**, 1967–1977.
69. J. Zhou, *Rapid Commun. Mass Spectrom.*, 2010, **24**, 2236–2244.
70. Y. Noda, K. Fujiwara, K. Yamamoto, T. Fukuno and S. Segawa, *Biopolymers*, 1994, **34**, 217–226.
71. R. Blagrove, G. Lilley and A. Kortt, *Funct. Plant Biol.*, 1981, **8**, 507–513.
72. P. J. Glynn and A. L. Pulsford, *J. Mar. Biol. Assoc. U. K.*, 1993, **73**, 425–436.
73. M. K. Mandal, T. Ozawa, S. Saha, M. M. Rahman, M. Iwasa, Y. Shida, H. Nonami and K. Hiraoka, *J. Agric. Food Chem.*, 2013, **61**, 7889–7895.
74. R. B. Cody, J. A. Laramée and H. D. Durst, *Anal. Chem.*, 2005, 77, 2297–2302.
75. A. Venter, M. Nefliu and R. Graham Cooks, *TrAC, Trends Anal. Chem.*, 2008, **27**, 284–290.
76. M. O. Rahman, M. K. Mandal, Y. Shida, S. Ninomiya, L. C. Chen, H. Nonami and K. Hiraoka, *J. Mass Spectrom.*, 2013, **48**, 823–829.
77. M. K. Mandal, K. Yoshimura, S. Saha, S. Ninomiya, M. O. Rahman, Z. Yu, L. C. Chen, Y. Shida, S. Takeda and H. Nonami, *Analyst*, 2012, **137**, 4658–4661.
78. S. Saha, M. K. Mandal and K. Hiraoka, *Anal. Methods*, 2013, DOI: 10.1039/C3AY41117F.

Desorption Electrospray Mass Spectrometry

JOSHUA S. WILEY,[a] ZOLTAN TAKATS,[b] ZHENG OUYANG*[c,d] AND R. GRAHAM COOKS*[a,d]

[a] Purdue University, Dept. of Chemistry, West Lafayette, IN 47907, USA; [b] University of London Imperial College of Science and Technology and Medicine, Faculty of Medicine, Department of Surgery & Cancer, London, SW7 2AZ, UK; [c] Purdue University, Weldon School of Biomedical Engineering, West Lafayette, IN 47907, USA; [d] Purdue University, Center of Analytical Instrumentation Development, West Lafayette, IN 47907, USA
*Email: ouyang@purdue.edu; cooks@purdue.edu

12.1 Background and Concepts

The ability to manipulate ions with electric and magnetic fields is fundamental to mass spectrometry. Molecular ionization processes can conveniently be grouped into four main types, electron ionization (EI), chemical ionization (CI), desorption ionization (DI) and spray ionization (SI). Traditionally, the key MS process of ionization has been implemented internally, under vacuum. Of the four main types of ionization methods, SI in the form of electrospray ionization (ESI) is typically done at atmospheric pressure. Electrospray ionization (ESI)[1] is a solution-based ionization method that is almost always used with prior liquid chromatography separation. Most sample types are successfully ionized – and indeed give excellent analytical figures of merit – using one or other of the traditional ionization methods usually with the aid of prior chromatographic separation. The special case of rapid, direct ionization of complex mixtures is not handled

New Developments in Mass Spectrometry No. 2
Ambient Ionization Mass Spectrometry
Edited by Marek Domin and Robert Cody

well with the traditional methods although there are exceptions for selected chemical types of analytes, as in the case of alkaloids in plant materials[2] where soft ionization can be used to generate a set of intact ionized molecules followed by the use of tandem mass spectrometry to obtain structurally characteristic spectra.

Molecular ionization at atmospheric pressure was initiated with the study of vapors using a differentially pumped interface to the mass spectrometer[3] a method that proved to be very sensitive but one that was poorly suited to mixtures. This book deals with the problem of how to analyze complex mixtures directly by mass spectrometry. This task requires that condensed phase samples be ionized and to do this quickly the experiment must necessarily be done in air. ESI itself is not suited to such tasks as it requires taking up the sample in solution and its performance is degraded by salts and competition effects in complex mixture samples. Desorption electrospray ionization (DESI),[4] the subject of this chapter, is a method that extends ESI in such a way as to allow direct analysis of samples in the solid state. Complex samples can be interrogated directly because DESI has a built-in sampling step based on solvent extraction. The successes of this technique encouraged a variety of other methods of generating ions from native samples in their ordinary environment, that is, ambient ionization mass spectrometry.[5] In all of these methods, the complex sample is interrogated to provide analyte ions that are transferred into the vacuum system. This selective transfer into vacuum of a portion of sample material is a further distinction between ambient ionization and electrospray ionization. So, even though DESI is derived from and related to electrospray ionization, it is also distinct from it and with the other ambient ionization methods forms a class of methods that address the problem of rapid, direct mixture analysis.

The ambient ionization experiments can be divided into categories in several ways: one method of categorization is based on the nature of the principal agency used: solvent sprays, electrical discharges and lasers. DESI uses a single agency, charged droplets, to perform the desorption (extraction) step and also the ionization process. In other cases, two separate agents are used as in electrospray laser desorption ionization (ELDI),[6] where the desorption step is performed by a UV-laser and ionization of the resulting vapor phase neutrals by charged droplets. DESI, as depicted in the schematic in Figure 12.1, belongs to the spray-based, single agency category.

One further comment on ambient ionization is made to note that the term ambient ionization refers to a particular class of methods used to create ions from samples in the ambient environment, as just noted. It could also be defined by the objective of the experiment that is the rapid analysis of ordinary objects in their native environment. There are methods of ionization that allow ambient analysis but that might also be termed ambient ionization methods. For example, probe-based methods in which a small sample is removed from an object and ionized by spray ionization might fall into this class. Such methods include the probes used by Hiraoka *et al.*[7] To take this point further, ionization of ordinary objects in their native

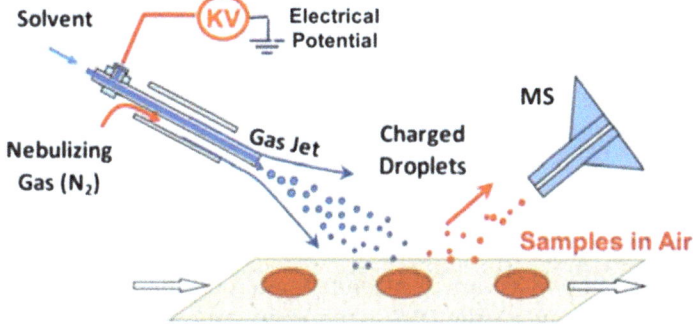

Figure 12.1 Schematic of the DESI-MS experiment.

environment can be performed in two ways: by direct methods or by sampling methods. The direct methods are by definition ambient ionization methods. Direct methods allow imaging and avoid the extra physical manipulations of the sampling methods. However, probe sampling may allow access to positions difficult to locate directly, relative to MS, although sampling using heated or unheated gas "probes" can achieve sampling by passing over any object from which sample vapor may be transferred directly into the MS. The use of physical probes followed by ESI represents an important class of ambient analysis methods.

12.2   Implementation

A DESI source utilizes a solvent flow at 0.5–5 µl min^{-1}, a nebulizing gas at linear velocities in the 100–250 m s^{-1} range and a DC potential in the range of 2–5 kV. The initial charged droplets impact the sample surface and form, by a simple splash, secondary droplets that contain dissolved analyte, and are also electrically charged. The electric charge leads to coulomb explosion to create smaller droplets that assists in the formation of dry ions, either by the ion evaporation or by the charge-residue model, both of which are familiar as processes that occur in the standard ESI experiment. Note that the DESI experiment can be done without applying a voltage and in special cases the nature of the sample might demand this. This version of the experiment is termed easy ambient sonic-spray ionization (EASI);[8] however, applying voltage increases the droplets' charge accumulation and allows for more efficient ionization.

An aspect of DESI that is relatively critical is the angle of the spray emitter with respect to the surface and MS inlet capillary. The angle between the DESI source and the sample surface is typically in the range of 40–70°, but a geometry-independent DESI has been developed in which the spray moves perpendicularly onto the sample plane and the sampling inlet is antiparallel to this.[9] This is readily implemented using a coaxial capillary to primary and secondary droplets. In addition to direct impact onto the surface by the charged droplet stream, a transmission-mode DESI may be employed, where the DESI emitter is positioned directly in front of and antiparallel to the MS inlet capillary and the sample placed on a mesh in between the DESI source and the inlet capillary.[10] Further important parameters include the spray tip-to-surface and spray tip-to-MS inlet distances.

12.3 Mechanism

The DESI experiment can be broken down into two main steps: surface wetting followed by splashing of secondary microdroplets and their transfer to the mass spectrometer (the DESI phenomenon has been observed to take place also on hot surfaces, presumably following an alternative mechanism, see below). The first step requires that the solvent dissolves the analyte from the sample surface. Hence, changing the solvent will greatly affect the overall ionization efficiency, based on the solubility and dissolution kinetics of an analyte in the spray solvent. Once the analyte is in solution on the wetted surface, new droplets impact the surface and cause a splash event in which secondary analyte-containing microdroplets are ejected. Given that the surface is either electrically insulated or kept at a nonzero potential relative to the MS inlet, the departing droplets still carry electric charge, which makes them amenable to the generation of free gas-phase ions by electrospray-like ionization mechanisms.

The combination of simulations and phase Doppler particle analysis has helped to confirm the main proposed mechanism. As seen in the histogram of droplet sizes in Figure 12.2(a),[11] the average size of the droplets in a typical experiment is 3 µm and the histogram in Figure 12.2(b)[11] reveals an average velocity of 150 m s^{-1}. In addition, the DESI-MS analysis of thermometer ions has suggested a median internal energy distribution centered at ~ 2 eV (*cf.* Figure 12.2(c)),[12] which is low and similar to ESI and corresponds well with the soft ionization that is observed with both techniques. In a less-common alternative to the described mechanism, ion formation may occur following neutral molecule desorption into the gas phase, which helps to account for DESI observed on hot and electrically conductive (and grounded) surfaces.

12.4 Analytical Performance with Small and Large Molecules in Food, Energy and Forensics

12.4.1 Wide Applicability

As a spray-based technique, DESI is able to desorb/ionize samples ranging from small molecules to salts[13] to proteins.[14,15] Furthermore, DESI is able to analyze raw biofluids deposited onto solid carrier surfaces including blood and urine with little matrix effects. A major advantage of DESI and other spray-based ambient ionization methods is the ability to derivatize an analyte in the course of ionization to increase sensitivity or specificity,

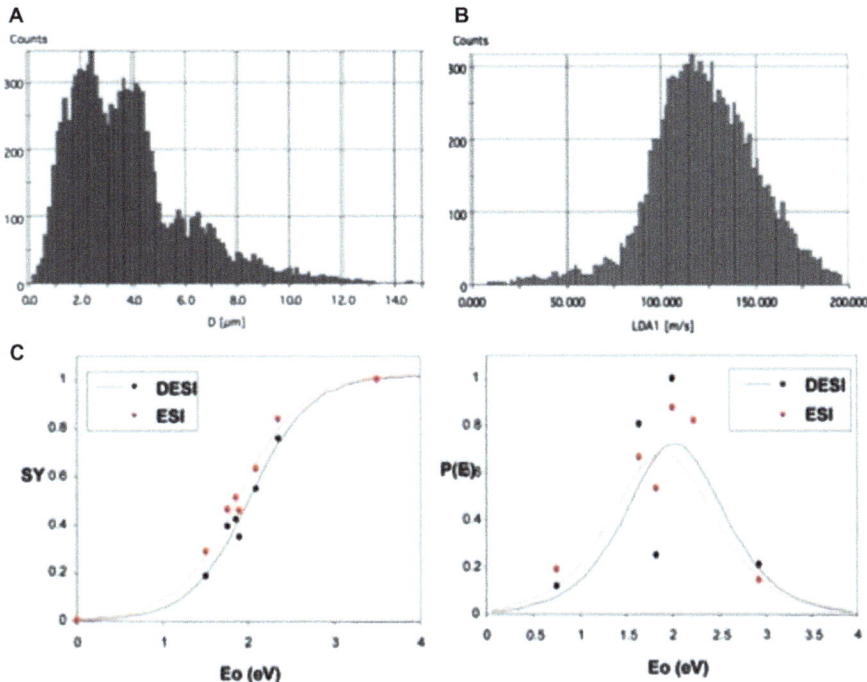

Figure 12.2 Histogram of (A) diameters of droplets at 5 mm from the sprayer tip with a 2 µL min⁻¹ solvent flow rate and (B) velocities of droplets at 2 mm from the sprayer tip with a 5 µL min⁻¹ solvent flow rate. Both histograms were obtained using phase Doppler particle analysis with 50% methanol/water solution and a gas pressure of 1130 kPa. (C) Calculated breakdown (left) and internal energy (right) curves for DESI and ESI sources using the survival yield method.

as discussed in greater detail in the next section. These reactions occur in small droplets and increase sensitivity and/or specificity, as discussed in greater detail in the next section. Reactive DESI has allowed the analysis of many analytes that are difficult to ionize, including steroids,[16] triacylglycerol species[17,18] and saturated hydrocarbons.[19] Only requiring solvents, voltage and gas flow, DESI is capable of being coupled with a miniature mass spectrometer and used for in-field analyses. A fieldable derivative of DESI called Venturi (very easy ambient sonic-spray) has also been developed; it uses no voltage, no syringe pump and an air canister to supply gas flow.[20]

Spray-based ambient ionization methods are arguably the most versatile of the three categories, combining field portability with wide applicability. Techniques based on laser desorption/ionization tend to have an excellent mass range and tend to have an excellent performance for heavier compounds, but are not well suited to in-field analysis. Plasma-based techniques rely heavily on thermal desorption as the primary desorption mechanism, and, consequently, are often limited to lower molecular weight (typically less than m/z 500) and more volatile analytes. However, plasma-based ambient ionization sources are well suited for field applicability and various associated small-molecule applications. One area where both laser- and plasma-based ambient ionization excel over spray-based methods is for the analysis of nonpolar analytes, which is difficult with DESI-MS but can be achieved, as discussed later in the section on reactive DESI.

12.4.2 Positive/Negative Ions

As is the case with ESI-MS, the ability to produce both negative and positive ions is easily achieved with DESI by applying a potential the same polarity as that of the intended ion. In the negative-ion mode, the $[M-H]^-$ is predominantly detected but electron transfer and adduct formation can also be observed, yielding for example, $M^{-\bullet}$ and $[M+Cl]^-$ ions. Analogously, proton transfer is most common in the positive-ion mode, $[M+H]^+$ ions usually predominating, however, sodium and potassium adducts are also common, with $M^{+\bullet}$ formation occurring for particular analytes.

12.4.3 Trace/Bulk Analysis

A common problem with ESI and especially nanospray ionization deals with the spray emitter clogging when a large amount of floating material is present in unprepared native samples. In a DESI-MS experiment, the sample is directly desorbed from a surface and therefore clogging of the spray emitter is not a factor. This allows the analysis of bulk samples and raw matrices, especially samples consisting of high salt content, that are not easily analyzed with ESI or nanospray MS. While the sensitivity of DESI is generally inferior to ESI methods by a small factor, the method features sufficiently low LOD values (typically subnanogram absolute) for most practical applications.

12.4.4 Necessity of Tandem MS

Any mass spectrum of ions produced under ambient conditions will be polluted with numerous peaks from contaminants in solvents, air or from surfaces. The number of peaks increases when performing an ambient MS experiment in which the unprocessed sample can be a complex mixture of molecules. Figure 12.3(A) shows the full-scan DESI-MS spectrum in negative-ion mode of diluted and dried human plasma, containing a mixture of micronutrients (including folic acid at 50 ppb). While there is a small peak for the $[M-H]^-$ of folic acid at m/z 440, it is hard to discern from chemical noise. Unless the relative abundance of the analyte of interest within the sample is unusually high, it is often necessary to use MS/MS to increase specificity and the signal-to-noise ratio to confirm the presence of an analyte peak. The MS/MS spectrum shown in Figure 12.3(B) was taken on the same human plasma sample as in Figure 12.3(A); however, the peak at m/z 440 was isolated and fragmented to unique fragments at m/z 311, 422 and 396, which

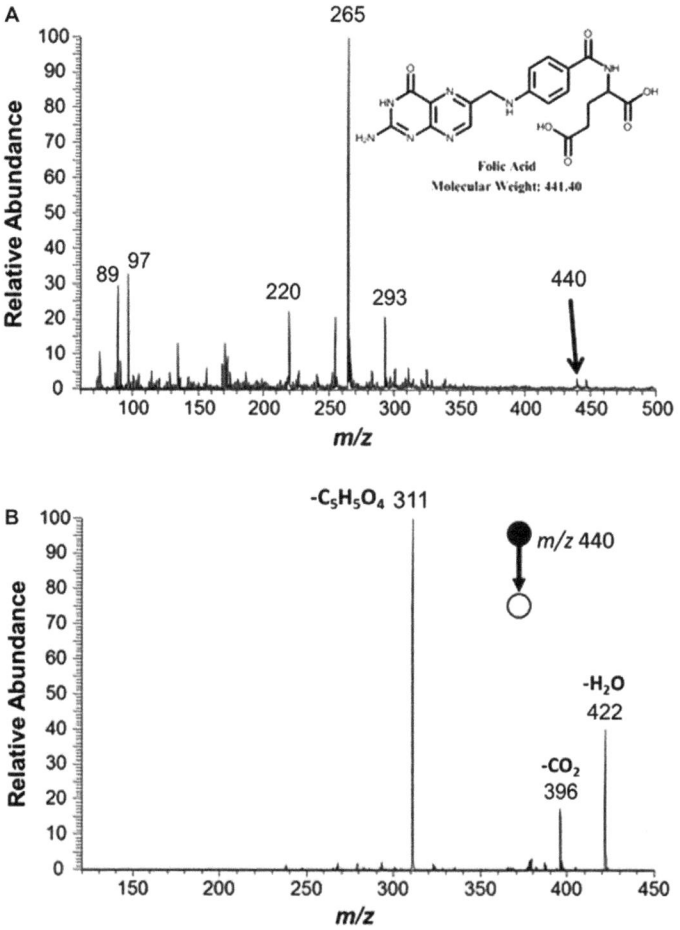

Figure 12.3 (A) Full DESI mass spectrum of some drug in blood (should not see large peak for drug). (B) DESI-MS/MS spectrum for drug in blood.

correspond to losses of $C_5H_5O_4$, H_2O and CO_2, respectively. Tandem MS not only helps confirm that the analyte of interest is present but also significantly increases the signal-to-noise ratio.

12.4.5 Solvents, Surfaces and Mass Range

As a spray-based ionization technique, DESI is capable of desorbing/ionizing analytes with a wide range of masses. Desorption of an analyte is dependent on solubility, so it is common to adapt the solvent mixture to capably dissolve the analyte. Besides analyte solubility, the solvent system has to be electrically conductive and volatile (*i.e.* it should evaporate without leaving a solid residue) under given experimental conditions. Mixtures of methanol and water are most commonly used; however, combinations of acetonitrile, ethanol and dimethylformamide are also used as well as the addition of a weak acid (*i.e.* formic or trifluoroacetic acid) to facilitate proton transfer. Selecting the appropriate solvent mixture is not always sufficient to ensure desorption of an analyte, due to unfavorable surface tension or dissolution kinetics. While the successful DESI-MS analysis of proteins has been reported from liquid samples,[14,15] attempting DESI-MS from biological tissue yields no protein signal when using typical DESI spray solvents. The non-covalent interaction of an analyte with the substrate on which it resides can decrease the probability of desorption, even when working with surfaces more inert than biological tissue. For example, many analytes yield a higher signal from a Teflon® surface as compared with a glass surface.[16]

12.5   Applications

As the first ambient ionization method developed, DESI has been used for a wide variety of applications. In this section, each of the major categories of DESI-MS applications will be discussed; however, given the extensive use of DESI-MS over the past near-decade, it will not be possible to cover each and every application.

12.5.1   Analysis of Formulated Pharmaceuticals

One application of DESI-MS is for the direct, rapid analysis of pharmaceutical drug tablets. The ability to sample from a tablet directly, without requiring destruction of the tablet, makes DESI-MS a practical technique for quality control on pharmaceutical products. Figure 12.4 (from Chen *et al.*) shows the DESI-MS spectrum of an Excedrin tablet (Figure 12.4(a)), where all three active ingredients (aspirin, acetaminophen and caffeine) were detected, and a Centrum vitamin tablet (Figure 12.4(b)) in which thirteen vitamins were successfully detected (only seven are labeled in the full MS spectrum).[21] In addition to the $[M+H]^+$ peaks for acetaminophen and caffeine, the two most abundant ions in the spectrum are the sodium adducts of acetaminophen and aspirin at m/z 174 and 203, respectively. Chen *et al.* also reported the ability to measure 2.67 Claritin tablets per second by having the tablets on a moving belt setup, demonstrating the potential of DESI-MS for high-throughput quality control of pharmaceutical tablets.[21] While pharmaceutical tablets have been heavily targeted with DESI-MS, tablets of illicit drugs have also been examined as a potential forensics application.[22,23]

12.5.2   Analysis of Biofluids

While the analysis of tablets is an example of detecting a mixture of analytes at higher abundances near the % level, the analysis of biofluids containing these drugs or other metabolites within often requires greater sensitivity. Much effort has been directed towards the analysis of biomatrices with

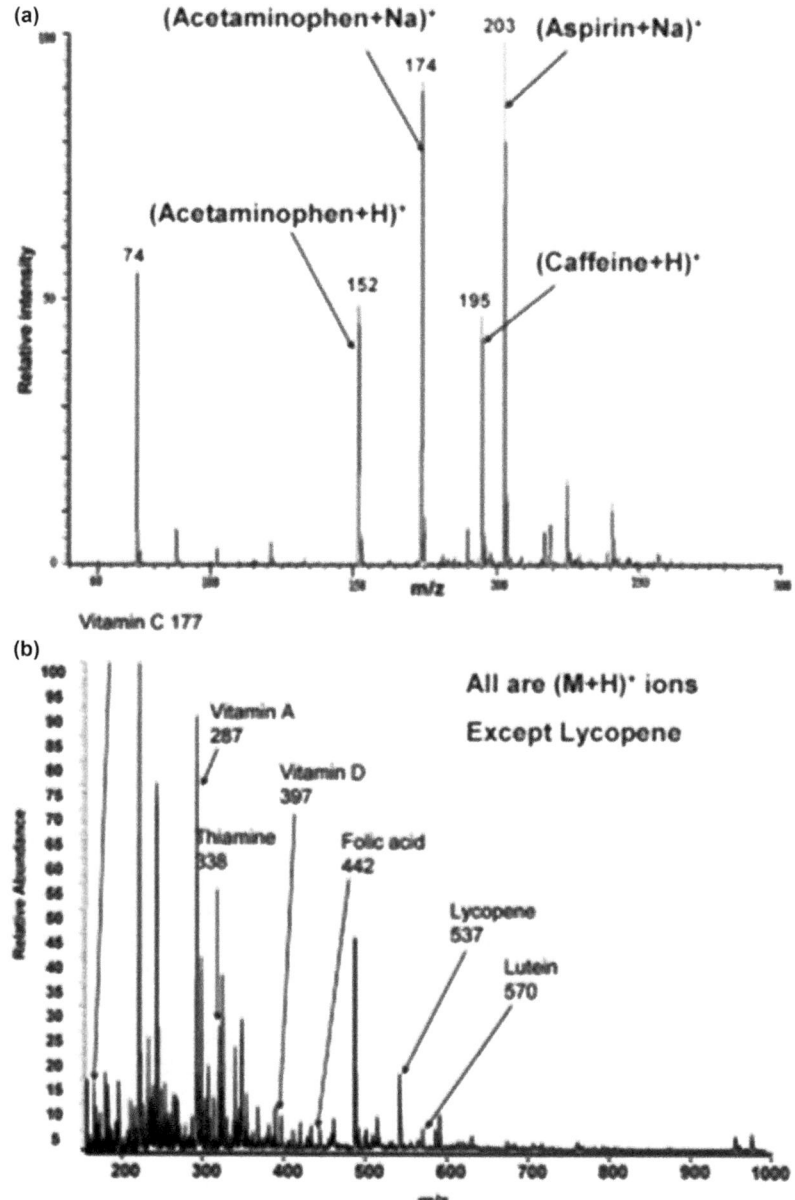

Figure 12.4 DESI-MS of an (a) Excedrin and (b) Centrum vitamin tablet with methanol:water (1:1) in positive-ion mode.[21]

DESI-MS for a wide variety of applications. Since biological fluids represent mostly liquid-phase samples, their DESI-MS analysis generally requires a simple preparation step, which results in a solid-phase sample. Sample preparation may comprise deposition of fluids onto solid surface followed by drying or the absorption of biofluid by porous medium (*e.g.* filter paper). The latter can also be used to selectively bind interfering matrix components, for instance inorganic salts by the porous medium. The DESI-MS analysis of urine has been applied to lung cancer biomarker screening in mice,[24] diagnosis of patients with inborn errors in metabolism,[25] screening for anabolic steroids (discussed further in the reactive DESI section)[16] and drugs of abuse.[26] Thanks to the nondestructive nature of DESI, skin has also been analyzed by DESI-MS for the analysis of explosives,[27] gunshot residues,[28] personal care products[29] as well as other areas of forensics and clinical chemistry.[30] The analysis of human plasma with DESI-MS has also been demonstrated for various purposes.[31–33] However, separating blood plasma from whole blood is not always a necessary step for DESI-MS analyses. The analysis of dried blood spots (DBSs) plays an important role in the medical community for applications ranging from neonatal screening to pharmaco-/toxicokinetic studies to therapeutic drug monitoring. DESI-MS has successfully been applied to the direct analysis of DBSs for model compounds (sitamaquine, terfenadine, and prazosin), achieving linearity from 10–10 000 ng mL^{-1}.[34] However, an ambient ionization method that is perhaps more suitable and has been heavily tested for DBS analysis is paper spray ionization.[35]

12.5.3 Food Safety

A nice review of the DESI-MS analysis of chemical food contaminants has been prepared by Nielen *et al.*[36] Applications of DESI-MS to food safety include: the analysis of strobilurin fungicides in wheat,[37] agrochemicals on fruits and vegetables,[38] arsenic speciation on animal feeds and plant foods,[39] Sudan dyes in food,[40] dithiocarbamate fungicides in fruit,[41] lipid profiling of bacteria,[42] triglycerides in foods[18] and melamine in milk.[43] The number of applications extend beyond these listed and should only continue to grow as demanded. As mentioned by Nielen *et al.*, the ability for DESI-MS to be performed on site to provide qualitative or semiquantitative information makes it an intriguing approach for sample prescreening in the field, limiting the number of samples that are sent back to the lab.

12.5.4 Forensic and Public Safety

As with food safety, DESI-MS has also been heavily applied to forensics, and a recent review detailing the areas of forensics that have been targeted has been written by Morelato *et al.*[44] Again, the ability for *in situ* analyses has warranted the wide application of DESI-MS to forensic studies, which demands rapid, direct and efficient methodologies for in-field analyses. Some of the forensic applications that have been targeted by DESI-MS include: the analysis of illicit drugs, gunshot residues, explosives, inks and counterfeited documents, fingerprints, and chemical warfare agents. Explosives have also been analyzed by DESI on a miniature MS.[45] Currently, many in-field forensic investigations are performed with ion-mobility or Raman spectrometers, which lack the overall combination of sensitivity, specificity and broad applicability of MS. As miniature MS instruments continue to develop and become smaller,[46] they should eventually replace these techniques, with ambient ionization as the means of directly sampling from crime scene materials, luggage at airport security, *etc.*

12.5.5 Micro-organisms (Bacteria, Algae, Embryos)

In close relation to public and food safety, micro-organisms have been another targeted area for DESI-MS analysis. As mentioned previously, various types of bacteria have been analyzed and distinguished based on their distinct lipid profiles.[42] Also relying on lipid signatures, a recent effort has been made to chemically study oocytes, blastocysts and embryos.[17,47,48] In a study where preimplanted embryos were obtained at various stages and directly analyzed with DESI-MS, two- and four-cell embryos were seemingly more synthetically active than unfertilized oocytes with a wider range of phospholipids detected as well as higher fatty acid signals.[49] While phospholipids and fatty acids have been an excellent target analyte for the characterization of various micro-organisms due to their ability to readily form anions, sometimes it is necessary to choose a target analyte that might not form an ion of either polarity very well. In a study where DESI-MS was performed directly on marine algal tissue, various anions (*e.g.* Cl^- and Br^-) were added to the spray solvent to increase the sensitivity for the detection of bromophycolide (an antimicrobial chemical defense agent) in a reactive DESI experiment.[50]

12.6 Reactive DESI

12.6.1 Concept/Overview

Reactive ionization is an emerging area of ambient mass spectrometry and is based on adding reagents that can derivatize an analyte in order to enhance desorption or ionization abilities. This is typically done by converting an analyte to a charged product or one that can be efficiently ionized. In a DESI-MS experiment, this is typically done by putting the derivatizing reagent in the spray solvent while the analyte is on a surface. After impact with primary reagent-containing droplets, secondary droplets are emitted from the surface, which contain a distribution of both compounds, allowing reactions to occur. Despite the short reaction times between dissolution of the analyte and drying of the droplets in the inlet capillary, a wide variety of reactions have been demonstrated with DESI-MS. In fact, results have indicated that reactions often occur at accelerated rates in the confined droplets. One likely reason is the large change in pH that occurs as a result of the voltage applied and the decreasing droplet size that promote acid/base-catalyzed chemistry. Two reactions that have been demonstrated are shown in

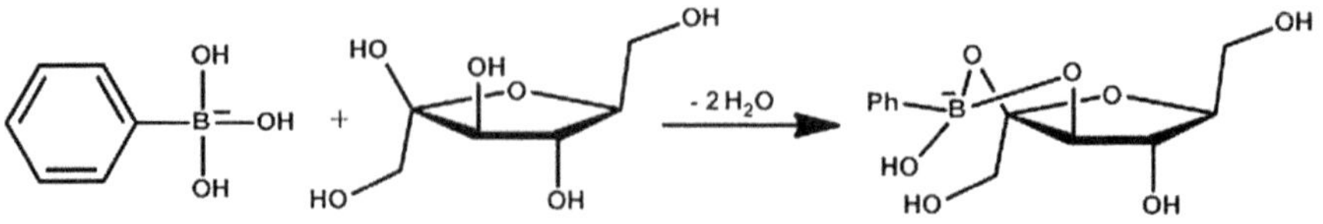

Scheme 12.1

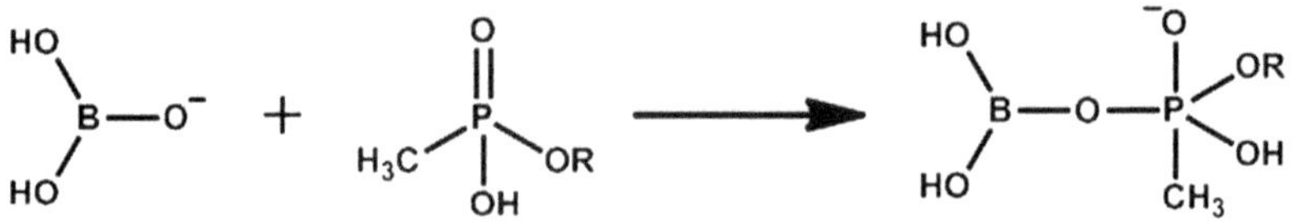

Scheme 12.2

Schemes 12.1 and 12.2. The first utilized phenylboronic acid to recognize *cis*-diol functionalities at a surface due to the selective ability to form the cyclic boronate (*cf.* Scheme 12.1).[51] In Scheme 12.2 is the reaction of boric acid with phosphonate esters, commonly used as chemical warfare agents, to form the hydrolysis product.[52] The various types of reactions that have been demonstrated with DESI-MS are discussed in this section.

12.6.2 Steroids

Two closely related reactions have been demonstrated for the DESI-MS analysis of ketosteroids, which represent an intriguing target for analytical methods due to the use of illegal anabolic steroids and testosterone in sports. The first DESI-MS reaction with ketosteroids involved the addition of hydroxylamine in the spray solvent, which becomes protonated during the electrospray process and can attack the carbonyl of the steroid.[16] As observed in Scheme 12.3, the oxime is the final product; however, the intermediate shown in the scheme just after hydroxylamine addition has been detected at a higher abundance than the oxime. Figure 12.5(a) shows the reactive DESI mass spectrum of 20 ng of epitestosterone where hydroxylamine was doped into the spray solvent. Note that while the oxime (*m/z* 304) was detected with higher abundance than the intermediate (*m/z* 322), this was not observed for all ketosteroids and DESI conditions.[16]

Another reaction that has been demonstrated with ketosteroids is shown in Scheme 12.4, and is similar to the hydroxylamine reaction except that a hydrazine, Girard's reagent T (GT), was used to attack the carbonyl.[53] The advantage of using GT is that it has a permanent charge, which remains with the product after formation, ultimately increasing sensitivity as a result of not having to ionize the product. Figure 12.5(b) shows a reactive DESI mass spectrum of 200 ng of cortisone where GT was doped into the spray solvent. The major product was the hydrazone as shown in Scheme 12.4, labeled as [MGT]$^+$ in the mass spectrum at *m/z* 474.3.

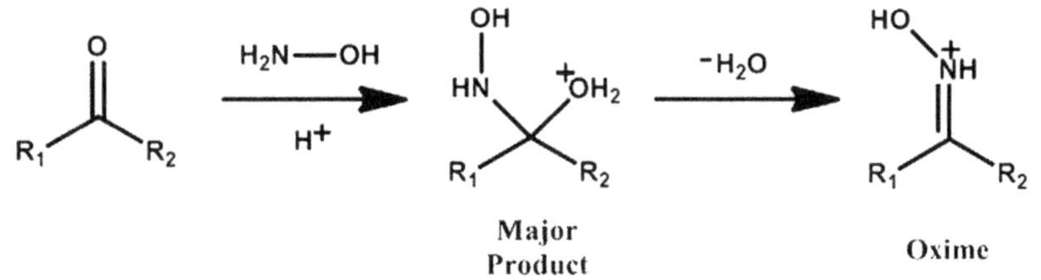

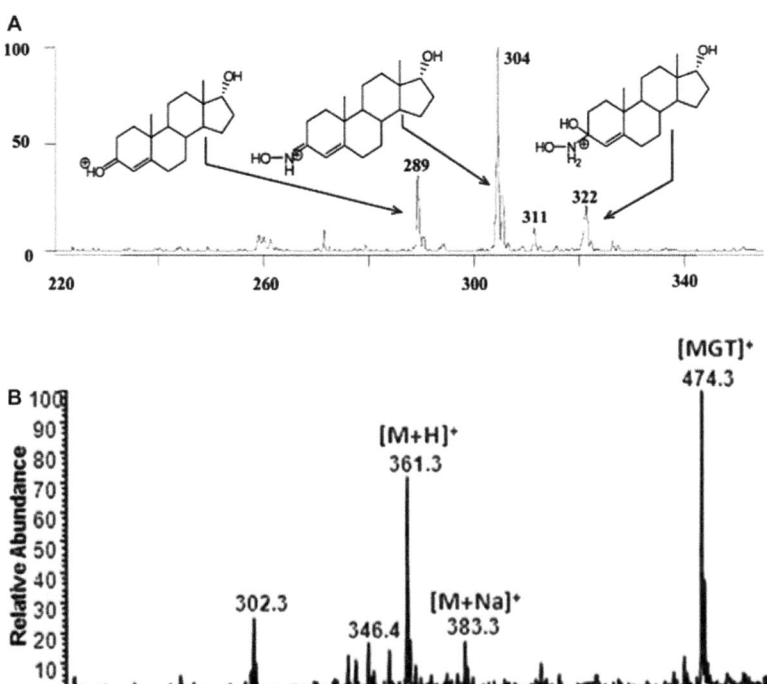

Scheme 12.3

Figure 12.5 Reactive DESI-MS of (A) 20 ng of epitestosterone with hydroxylamine[16] and (B) 200 ng of cortisone with Girard's Reagent T.[53]

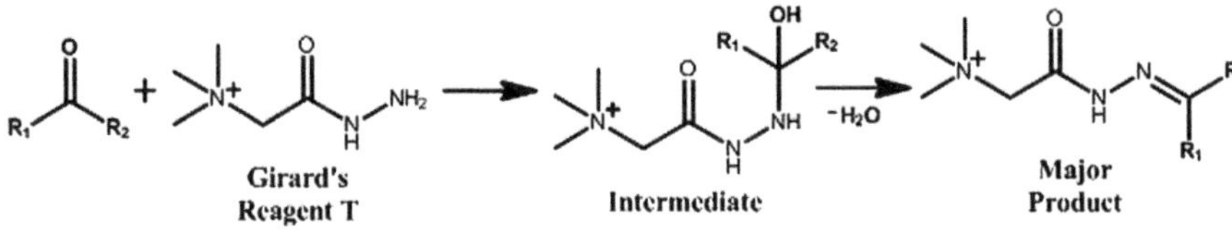

Scheme 12.4

12.6.3 Oxidation Reactions

While a reaction involving covalent-bond formation between a reagent in the spray solvent and a surface-bound analyte is typical for reactive DESI, there are other routes of reactive ionization with DESI-MS. One such alternative is based on electrochemistry and can be done with or without the addition of reagents into the spray solvent. For this to occur, special DESI spray conditions (namely higher voltages) are required. In this experiment, which has been used to oxidize carboxylic acids[54] as well as saturated hydrocarbons,[19] a discharge is formed as a result of the higher-than-normal electric field and results in the formation of oxygen radicals that are very strong oxidizing agents. In the case where reagents, such as iodine and potassium bichromate, were added to the spray solvent, signal enhancement over two orders of magnitude was achieved for the oxidation of dibutyl dithiocarbamate to form $M^{+\bullet}$.[55] Interestingly, this enhancement was only observed during the short timescale of mixing during DESI-MS; whereas, allowing the reagents to be mixed together longer during an electrosonicspray ionization (ESSI) MS experiment allowed other reactions to occur and much less enhancement of $M^{+\bullet}$ signal was observed.

12.6.4   Metal Attachment

As with electrochemistry, another reactive DESI approach that does not require covalent-bond formation is through metal adduction, which is a noncovalent association between a metal ion and a neutral analyte molecule. For some analytes, the ability to bind to a charged metal ion is highly favorable and results in increased specificity/sensitivity. This has recently been achieved by the addition of silver to the spray solvent, which is able to readily form stable adducts with olefins.[56] In this study, $AgNO_3$ was doped into the spray solvent and used for the analysis of various unsaturated lipids and fatty acids, providing greater specificity and up to 50 times signal enhancement. Further application of Ag^+ cationization has been demonstrated for the analysis of bovine oocytes, in which adduction with Ag^+ allowed detection of triacylglycerol.[17] Silver cations are also intriguing for other applications, such as chelation with aromatic compounds through pi-stacking;[57] although this has not been demonstrated with DESI-MS. While other metals may also result in adduct formation with analytes, there have been no such reports for DESI-MS besides Na^+ and K^+ adducts that often occur without doping their salts into the spray solvent. However, other adducting reagents have been doped into the DESI spray for various applications including chloride and trifluoroacetate for explosives[27,58] and bromophycolides[50] as well as ammonium for triglycerides.[18]

12.6.5   Cholesterol/Betaine Aldehyde

As a marker for monitoring heart disease such as triglycerides, cholesterol is another analyte that is difficult to ionize but has also been detected with reactive DESI-MS. The reaction involved the use of betaine aldehyde, which is susceptible to nucleophilic attack by the alcohol functional group in cholesterol (*ca.* Scheme 12.5 for the general reaction scheme).[59]

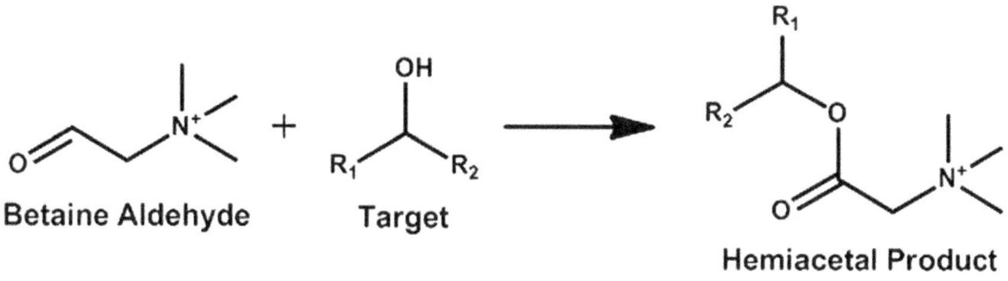

Scheme 12.5

The hemiacetal was detected at m/z 488.6 for the reactive DESI-MS analysis of cholesterol. The ability to detect cholesterol by reactive DESI was further demonstrated directly from rat brain tissue while keeping track of the spatial distribution of cholesterol along with phospholipids, an experiment known as mass spectral imaging.

12.7 DESI-MS Imaging

12.7.1 Introduction

Perhaps one of the most intriguing applications of DESI-MS is for the 2D mapping of chemical entities from a surface in an experiment known as MS imaging. Historically, secondary ion mass spectrometry (SIMS) and matrix-assisted laser-desorption ionization (MALDI) have been used extensively for MS imaging and are capable of resolving chemical features on a surface at less than ~100 nm with SIMS[60] and at just a few micrometers with MALDI-MS. While both of these techniques offer better spatial resolution than DESI-MS, which can be as low as 30 μm, the advantage of imaging with DESI-MS stems from the ability to analyze a surface with no pretreatment at atmospheric pressure. Hence, there are no interferences from matrix ions as observed with MALDI-MS and the possibility of sample contamination is greatly reduced. Another advantage of imaging with DESI-MS is due to its nondestructive nature. With no morphological damage to the sample during DESI interrogation, it can then be used for other studies without concern of any information loss.[61] Applications of DESI-MS imaging have ranged from plant, animal and human tissues to human fingerprints for forensics; each of which will be discussed further.

12.7.2 Tissue

Analysis of lipids in human/animal tissue is perhaps the most explored and fascinating application of DESI-MS imaging. The unique distribution of lipids in different types and/or disease states of tissue allows appropriate classification when relative abundances of the ionized lipids are compared in the mass spectra. Fortunately, many lipids have high ionization efficiencies and are easily detected with DESI-MS. As a result, many types of tissue have been analyzed from a wide variety of animals with DESI-MS, including brain,[62–64] bladder,[65,66] spinal cord,[67] prostate,[68] kidney,[69] eye lens[70] and adrenal glands.[71] As an example of distinguishing different types of tissue, Figure 12.6(a) and (b) reveal DESI mass spectra of gray and white matter (respectively) from a mouse brain. It is clear that the lipid profiles are very different between the gray and white matter, with m/z 834.4 (a phosphatidylserine, PS 18:0/22:6) as the base peak in the gray matter and m/z 888.8 (a sulfatide, ST 24:1) the base peak in white matter. Two-dimensional images of the relative abundances of these peaks are plotted in Figure 12.6(c) and (d). When the two ion images are overlaid, there is a clear distinction between the gray and white matter regions (Figure 12.6(e)). If the mouse brain is sliced throughout its entirety, 2D images can be acquired and

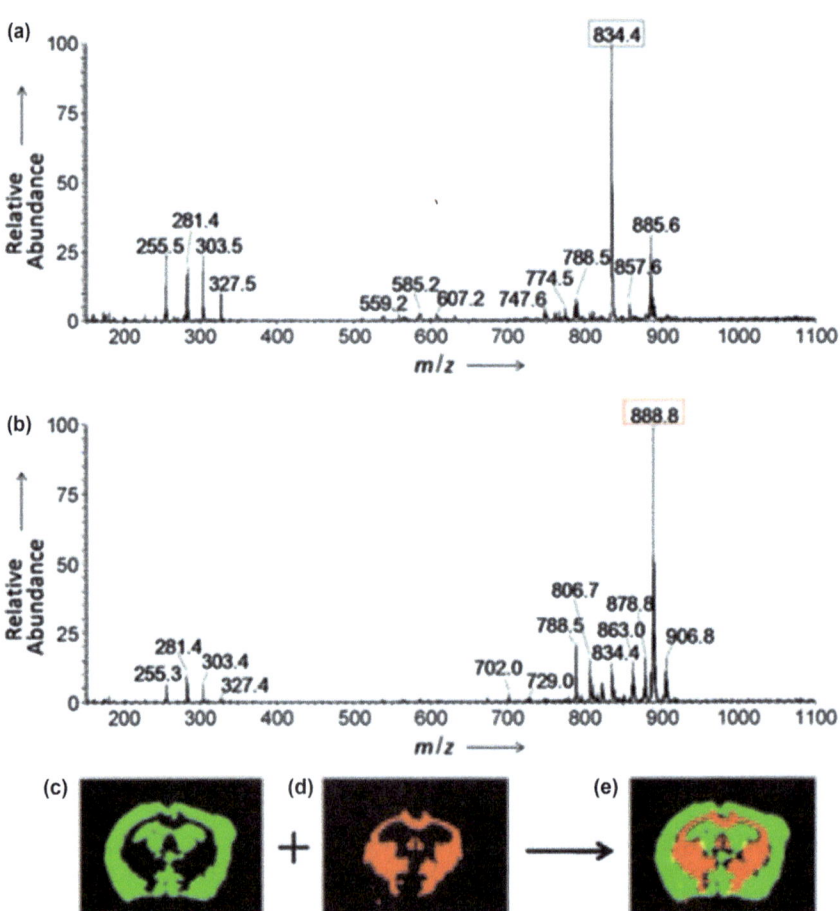

Figure 12.6   Negative-ion mode mass spectrum of (a) gray and (b) white matter of a mouse brain. The 2D plot (or ion image) of (c) m/z 834.4 and (d) m/z 888.8 as well as (e) the overlaid ion images of m/z 834.4 and 888.8.

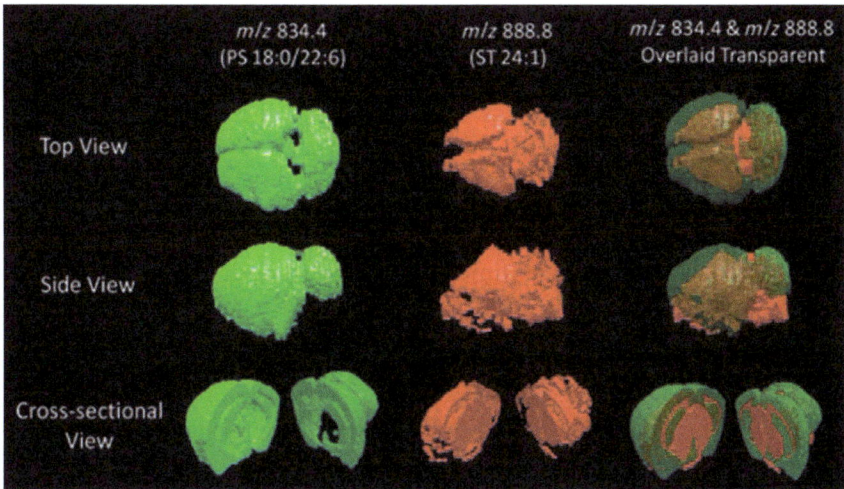

Figure 12.7 3D models of the mouse brain by DESI-MS. Top, side, and cross-sectional views are shown for the 3D construction of the distribution of (a) PS 18:0/22:6 in green (b) ST 24:1 in red, and (c) the transparent overlaid distributions of PS 18:0/22:6 and ST 24:1.

stacked for each tissue section, resulting in a three-dimensional image, as in Figure 12.7.

Not all DESI-MS imaging applications are this clear-cut in terms of being able to distinguish between different types of tissue. A prime example, and major application of DESI-MS imaging, is in the identification and classification of the disease state of tissue, especially cancerous *versus* noncancerous tissue. This has been pushed even further in some studies that have been able to determine the stage of diseased tissue, such as distinguishing astrocytoma grades,[62] using DESI-MS imaging. Classification of the disease state of tissue is often achieved by monitoring relative distributions of lipids and/or fatty acids and analyzing with a statistical classification method such as principle component analysis (PCA). Mass spectral data is then correlated with H&E stains read by a pathologist. For many samples the distribution of fatty acids and lipids is enough to allow the classification of cancerous *versus* noncancerous tissue with the aid of PCA, as observed in the DESI-MS imaging of dog and human bladders.[65,66] In the case of human prostate, classification of tissue by lipid and fatty acid profiles was difficult; however, cholesterol sulfate was found to be a cancer biomarker and allowed for accurate correlation of 64 of 68 tissue samples.[68]

12.7.3 Natural Products

Imaging of natural products is an intriguing application of DESI-MS due to the real possibility of imaging live organisms since no matrices are required and the ionization process is nondestructive and can be done with biologically friendly solvents. The use of reactive DESI-MS to allow the detection of fungal inhibitors on tropical seaweed, as discussed previously, was also performed as an imaging experiment by Lane *et al.*, revealing patches of the bromophycolides on the surface at sufficient concentrations for fungal inhibition.[72] One problem with directly imaging an organism without first cutting it into sections with smooth surfaces is the large deviations in signal that can occur as a result of spraying to/from a rough surface. One method of overcoming this has been to make an imprint of the sample to be imaged and then acquire the DESI-MS image from the smooth imprinted surface. The use of imprints for DESI-MS imaging has been applied to characterization of bacterial culture products[73] as well as various tissues including barley leaves,[74] St John's wort and Jimson weed,[75] Lam seed and mouse brain,[76] and various plant leaves for measuring nonfluorescent chlorophyll catabolites.[77] Many of these experiments were also performed directly on the sample as well, but imprints were used due to yielding better performance.

12.7.4 DESI Imaging in Forensics

One of the first applications of DESI-MS imaging was for the analysis of inks, with a forensics focus on identifying counterfeited documents by distinguishing which dyes were used on which regions of the document.[78] Another application that has gained a large amount of public interest is the DESI-MS imaging of latent fingerprints.[79] For a typical crime scene investigation, the ability to merely match a fingerprint's pattern with that of a suspect is a powerful route of incrimination. However, the ability to combine chemical information with visualization of the pattern can be very useful. One such example is when multiple fingerprints overlap on a surface. The ability to correlate certain chemicals with each fingerprint can allow the distinction and visualization of each fingerprint, which otherwise would not have been possible. In addition, searching for various drugs or their metabolites within the fingerprint could provide insight into the mental state of the suspect.

12.7.5 Spatial Resolution

Spatial resolution is defined as the shortest distance at which two distinct features may be resolved in an MS image. Since the onset of the field of MS imaging, there has been a great deal of interest in increasing spatial resolution (*i.e.* decreasing the distance at which features can be resolved). While some analyses require very high spatial resolutions, as with MS imaging of single cells, the time required must be taken into consideration. For every decrease in the pixel size for an imaging experiment, the number of pixels and, consequently, time of acquisition increases quadratically for a constant overall image area. Initial attempts to measure the spatial resolution of DESI-MS were done by Ifa *et al.* by scanning over ink lines, suggesting a spatial resolution less than 200 μm.[80] In general, a standard DESI-MS imaging experiment provides a spatial resolution of about 180 μm; however, there has been a push to go even lower with reports by Kertesz *et al.* suggesting ∼40 μm resolution for printed patterns on paper,[81] while Campbell *et al.* reported ∼35 μm resolution for DESI-MS imaging of mouse brain.[82] For each of these studies, various parameters including gas back pressure, capillary sizes, solvent selection and solvent flow rate were carefully tuned to achieve a spatial resolution below 50 μm. Further improvements in spatial resolution are possible, but will likely require added engineering in the design of the DESI source.

12.7.6 Data-analysis Strategies

Data analysis is generally aimed either at the determination of the concentration of certain species on the surface or the classification of the mass spectra obtained from spatially well-defined areas of the sample. A typical example of the former case is the quantification of drug molecules and drug metabolites in tissue sections, while pixel-by-pixel histological classification of mass spectrometric images represents a good instance for the latter case. Nevertheless, in both cases the DESI imaging data requires data preprocessing, comprising mass spectrometric calibration, normalization of the signal and optional coregistration of mass spectrometric data with any other – usually optical – image of the sample. Mass-spectrometric recalibration is critical in the case of high-resolution time-of-flight (TOF) analyzers, since in these the mass accuracy drift of these instruments can be significant within the timeframe of an imaging experiment (up to a few tens of hours). Normalization of signal is expected to compensate for the overall instrumental fluctuation of signal intensity. Various normalization strategies have been described including normalization to TIC, median intensity or certain, arbitrarily chosen peaks in the spectra, which are expected to show uniform concentration within the sample. Since DESI-MS imaging is nondestructive, the analysis is usually followed by morphological imaging, comprising histological staining and optical imaging (scanning). In order to establish correlation between the two types of data, coregistration is necessary with accuracy better than half of the MS pixel diameter. Coregistration algorithms allow translation, rotation and two dimensional stretching of images, in order to maximize the overlap.

Coarse quantification of certain detected species is easily achievable by visualizing the normalized intensity distribution of the corresponding molecular ion, however this approach provides only relative values and also does not take into account the different matrix effects taking place in different tissue types. More accurate quantification requires the determination of histology-dependent response factors by the parallel analysis of the corresponding tissue types by DESI-MS and LC-MS at different concentration levels.

DESI-MS imaging datasets can also be interpreted as a set of MS profiles associated with the biochemical/clinical status of the tissues. Since the individual spectra are treated as spectroscopic fingerprints, no identification or quantification of detected species is required. In these cases at first an authentic "training" or "learning" dataset is acquired, which serves as a basis to create a multivariate statistical model. This model is used for the identification of unknown spectra in the test datasets. As an example, DESI-MS imaging data was collected from a colon adenocarcinoma set and spectra were analyzed using a combination of principal component analysis and linear discriminant analysis. Statistical models were created to separate DESI-MS fingerprints of different histological tissue types (mucosa, adenocarcinoma, smooth muscle) and also patients with different predictive tumor marker statuses and the models were validated for test sets. Although the multivariate approach has tremendous value regarding the identification of tissue types, the training sets are expected to represent population-level variance of the spectral fingerprints which may require the analysis of several hundreds of samples for a single pathology. Further problems are associated with the experimental variance of DESI-MS data, which may exceed the biological/histological variance, making proper identifications difficult. Further standardization of DESI imaging parameters and development of more robust hardware (especially regarding spray emitters) is expected to lift these constraints and allow the construction of global, histologically specific DESI-MS spectral libraries available for all users worldwide.

12.8 Future Perspectives

12.8.1 Miniature Mass Spectrometers and *in situ* Analysis

Miniature mass spectrometers have been in development for quite some time, but until the recent develop of discontinuous atmospheric-pressure interface (DAPI),[83] miniature MS had been based on reduced-pressure ionization. The advent of DAPI, which affords the limited pumping ability of a miniature instrument by pulsing gas/ion introduction to just 10–20 ms of a ~1 s duty cycle, has allowed the use of various ionization sources that operate at atmospheric pressure. Most importantly, the coupling of ambient ionization with miniature MS is possible, which is ideal for rapid, *in situ* analyses. As previously discussed, the ability to perform MS/MS is essential for these types of analyses, so a geometrically simplified version of the linear ion trap has been employed that uses planar electrodes, the rectilinear ion trap (RIT).[84] The other advantage of using an ion trap mass analyzer is due to its ability to operate at higher pressure in comparison to most other mass analyzers.

12.8.2 Synchronized and Inductive DESI

While DESI has been coupled to miniature MS in the laboratory setting,[45,83,85,86] sampling efficiency and the need for high gas flow rates have made actual *in situ* analyses rather impractical. A method of mitigating both

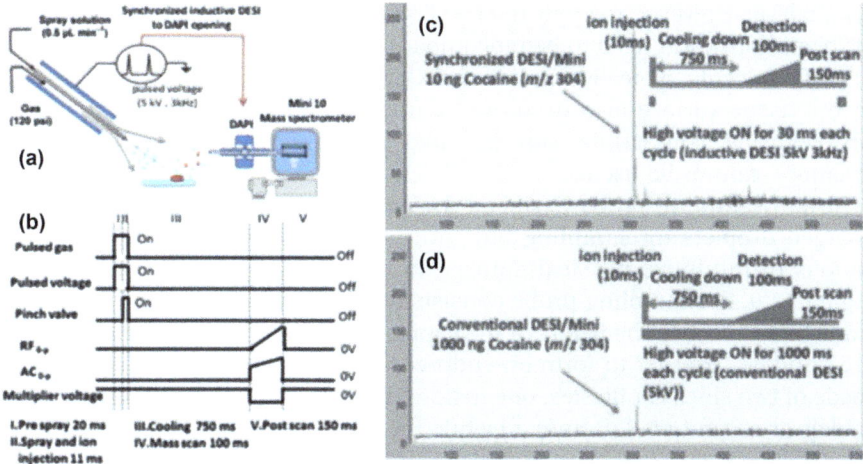

Figure 12.8 Synchronized DESI-MS (a) schematic and (b) pulse sequence. (c) Synchronized DESI-MS cocaine (10 ng) and (d) standard DESI-MS of cocaine (1 µg), both from a glass slide.

of these issues is by ion source and MS sampling synchronization. Many mass analyzers are pulsed, only introducing ion packets over small time windows, while the majority of the time is used for ion transfer and mass analysis. To better optimize sampling, a synchronized DESI-MS experiment has been demonstrated in which a voltage is applied inductively and both gas and solvent flows are pulsed in synchronization with ion introduction into the MS.[87] This has resulted in an overall reduction of N_2 flow from 2 L min^{-1} to 0.2 L min^{-1}, solvent flow from 5 µL min^{-1} to 0.5 µL min^{-1} as well as 100-fold increases in sensitivity and sampling efficiency. Figure 12.8(a) shows a schematic of this setup on a miniature MS instrument, with the overall pulse sequence of potentials used in the synchronized experiment in Figure 12.8(b). Figure 12.8(c) and (d) show the synchronized DESI-MS of 10 ng of cocaine on a glass substrate and standard DESI-MS of 1 µg of cocaine on a glass substrate, respectively. Despite analyzing 100 times less sample in with synchronized DESI-MS, the signal for cocaine at m/z 304 is still higher.

12.8.3 Endoscopic Diagnostics using DESI

DESI sampling/ionization using charged droplets allows direct analysis of endogenous compounds from biological tissues.[64,88] The characteristic lipid profiles assist in the diagnosis of pathology. To extend the method further, we are exploring the use of DESI and related methods for intrasurgical and endoscopic procedures, where *in vivo* chemical analysis would provide molecular information potentially of value for decision making. A MS system with a sampling probe that can be fitted into an endoscope would enable this capability. Analytes on the tissue are ionized and transferred to the MS for analysis. However, to apply the DESI method *in vivo*, the conditions of the ambient ionization need to be biocompatible. The high voltage (4–5 kV) and organic solvents typically applied for DESI cannot be used. The damage to the tissue surface also needs to be minimal and previously introduced histologically compatible solvents like *N,N*-dimethylformamide (DMF)/ethanol[61] cannot be used.

An MS sampling probe designed for endoscopic chemical analysis uses charged droplets for sampling but with a number of signification modifications to minimize physical damage to biological tissues.[89] As shown in Figure 12.9, the sampling probe consists of a capillary sprayer and a coaxial transmission tube. The fronts of the capillary sprayer and transmission tube are bundled together to form an endoscopic probe. The capillary sprayer is made of two silica capillaries, one inside another. The inner capillary is used to deliver solvent (4–8 μL min^{-1}) while the outer capillary is used to deliver

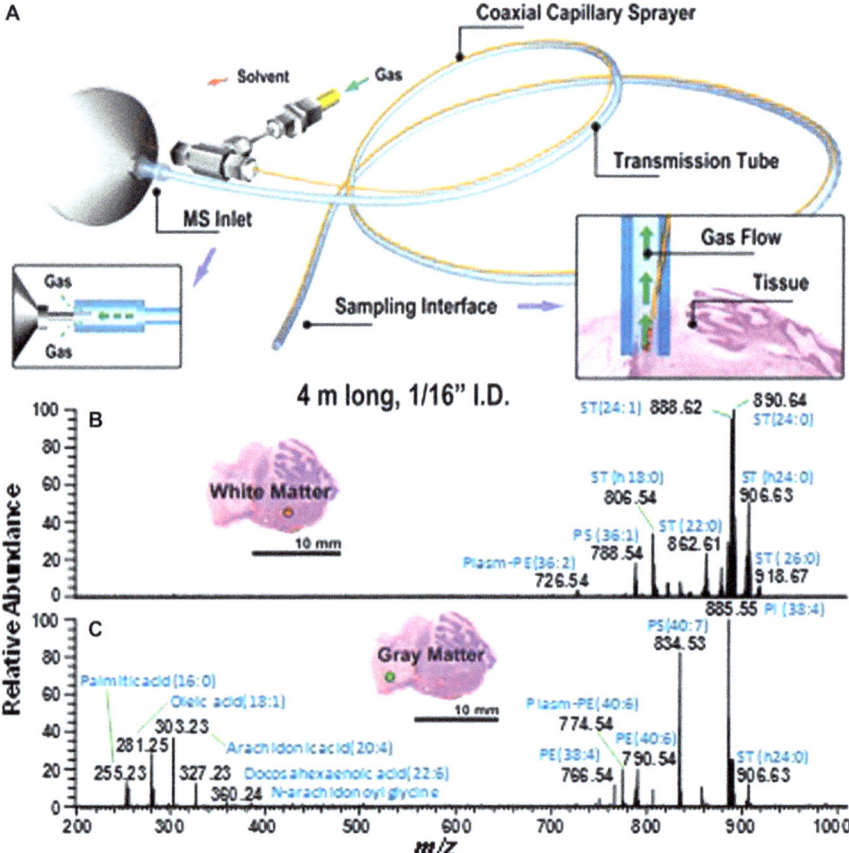

Figure 12.9 (A) Schematic setup of endoscopic sampling ionization probe. The probe is used to collect spectra from a rat brain tissue on the (B) white matter and (C) gray matter, respectively.

auxiliary nitrogen gas $(1.5–5.2 \text{ L min}^{-1})$. When the open end of the endoscopic probe is placed at the sampling spot, the spray plume generated from the capillary sprayer bounces back carrying chemicals extracted from the sampling tissue through the transmission tube. Transport of droplets through the transmission tube and into MS inlet allows spectra to be recorded. As a demonstration, a rat brain tissue section was analyzed with the sampling probe and the spectra recorded from white matter and gray matter agree with previous DESI results.

This long-distance ion transport tubing allows ions to be transported over 4 m,[90] which is adequate for endoscopic applications. The use of pure water as the DESI solvent provides higher signal intensity after the transportation, presumably due to the slower but more efficient desolation of water droplets in the tubing. The high voltage used in DESI was also abandoned just as is done in sonic-spray ionization, since the high-velocity gas flow provides enough signal intensity (Figure 12.9(b)). Typically, a tissue section would be damaged by the high-velocity gas flow, but this can be simply solved by introducing a pulling force to drag the gas backward into the MS inlet with an additional diaphragm pump (Figure 12.9(c)). With all these features, the endoscopic sampling probe is suitable for *in situ* and noninvasive biological analysis. The endoscopic sampling probe can be coupled with a portable Mini 12 mass spectrometer,[91] which provides real-time MS measurement at the point-of-care. This system could provide specific chemical information about biological tissues for biomedical studies and clinical diagnostics in the doctor's offices or in the operating room.

Acknowledgements

The authors acknowledge support of their work by National Science Foundation (CHE 13-07264) and the National Institutes of Health (EB015722-01). J.S.W. acknowledges fellowship support from the US Department of Energy.

References

1. J. B. Fenn, M. Mann, C. K. Meng, S. F. Wong and C. M. Whitehouse, *Science*, 1989, **246**, 64.
2. R. W. Kondrat and R. G. Cooks, *Anal. Chem.*, 1978, **50**, A81.
3. E. C. Horning, M. G. Horning, D. I. Carroll, I. Dzidic and R. n. Stillwel, *Anal. Chem.*, 1973, **45**, 936.
4. Z. Takats, J. M. Wiseman, B. Gologan and R. G. Cooks, *Science*, 2004, **306**, 471.
5. M. E. Monge, G. A. Harris, P. Dwivedi and F. M. Fernández, *Chem. Rev.*, 2013, **113**, 2269.
6. J. Shiea, M. Z. Huang, H. J. Hsu, C. Y. Lee, C. H. Yuan, I. Beech and J. Sunner, *Rapid Commun. Mass Spectrom.*, 2005, **19**, 3701.
7. K. Hiraoka, K. Nishidate, K. Mori, D. Asakawa and S. Suzuki, *Rapid Commun. Mass Spectrom.*, 2007, **21**, 3139.

8. R. Haddad, R. Sparrapan and M. N. Eberlin, *Rapid Commun. Mass Spectrom.*, 2006, **20**, 2901.
9. A. Venter and R. G. Cooks, *Anal. Chem.*, 2007, **79**, 6398.
10. J. E. Chipuk and J. S. Brodbelt, *J. Am. Soc. Mass Spectrom.*, 2008, **19**, 1612.
11. A. Venter, P. E. Sojka and R. G. Cooks, *Anal. Chem.*, 2006, **78**, 8549.
12. M. Nefliu, J. N. Smith, A. Venter and R. G. Cooks, *J. Am. Soc. Mass Spectrom.*, 2008, **19**, 420.
13. E. Sokol, A. U. Jackson and R. G. Cooks, *Cent. Eur. J. Chem.*, 2011, **9**, 790.
14. C. N. Ferguson, S. A. Benchaar, Z. Miao, J. A. Loo and H. Chen, *Anal. Chem.*, 2011, **83**, 6468.
15. Z. Miao, S. Wu and H. Chen, *J. Am. Soc. Mass Spectrom.*, 2010, **21**, 1730.
16. G. Huang, H. Chen, X. Zhang, R. G. Cooks and Z. Ouyang, *Anal. Chem.*, 2007, **79**, 8327.
17. A. F. Gonzalez-Serrano, C. R. Ferreira, V. Pirro, L. S. Eberlin, J. Heinzmann, A. Lucas-Hahn, H. Niemann and R. G. Cooks, *Reprod., Fertil. Dev.*, 2013, **25**, 262.
18. S. Gerbig and Z. Takats, *Rapid Commun. Mass Spectrom.*, 2010, **24**, 2186.
19. C. Wu, K. Qian, M. Nefliu and R. G. Cooks, *J. Am. Soc. Mass Spectrom.*, 2010, **21**, 261.
20. V. G. Santos, T. Regiani, F. F. G. Dias, W. Romao, J. L. P. Jara, C. F. Klitzke, F. Coelho and M. N. Eberlins, *Anal. Chem.*, 2011, **83**, 1375.
21. H. W. Chen, N. N. Talaty, Z. Takats and R. G. Cooks, *Anal. Chem.*, 2005, **77**, 6915.
22. L. A. Leuthold, J. F. Mandscheff, M. Fathi, C. Giroud, M. Augsburger, E. Varesio and G. Hopfgartner, *Rapid Commun. Mass Spectrom.*, 2006, **20**, 103.
23. S. E. Rodriguez-Cruz, *Rapid Commun. Mass Spectrom.*, 2006, **20**, 53.
24. H. W. Chen, Z. Z. Pan, N. Talaty, D. Raftery and R. G. Cooks, *Rapid Commun. Mass Spectrom.*, 2006, **20**, 1577.
25. Z. Z. Pan, H. W. Gu, N. Talaty, H. W. Chen, N. Shanaiah, B. E. Hainline, R. G. Cooks and D. Raftery, *Anal. Bioanal. Chem.*, 2007, **387**, 539.
26. T. J. Kauppila, N. Talaty, T. Kuuranne, T. Kotiaho, R. Kostiainen and R. G. Cooks, *Analyst*, 2007, **132**, 868.
27. D. R. Justes, N. Talaty, I. Cotte-Rodriguez and R. G. Cooks, *Chem. Commun.*, 2007, 2142.
28. M. Zhao, S. Zhang, C. Yang, Y. Xu, Y. Wen, L. Sun and X. Zhang, *J. Forensic Sci.*, 2008, **53**, 807.
29. T. L. Salter, F. M. Green, N. Faruqui and I. S. Gilmore, *Analyst*, 2011, **136**, 3274.
30. M. Katona, J. Denes, R. Skoumal, M. Toth and Z. Takats, *Analyst*, 2011, **136**, 835.
31. J. E. Chipuk, M. H. Gelb and J. S. Brodbelt, *Anal. Chem.*, 2010, **82**, 4130.
32. J. H. Kennedy and J. M. Wiseman, *Rapid Commun. Mass Spectrom.*, 2010, **24**, 309.
33. G. Xu, B. Chen, G. Liu and S. Yao, *Analyst*, 2010, **135**, 2415.

34. J. M. Wiseman, C. A. Evans, C. L. Bowen and J. H. Kennedy, *Analyst*, 2010, **135**, 720.
35. H. Wang, J. Liu, R. G. Cooks and Z. Ouyang, *Angew. Chem., Int. Ed.*, 2010, **49**, 877.
36. M. W. F. Nielen, H. Hooijerink, P. Zomer and J. G. J. Mol, *TrAC, Trends Anal. Chem.*, 2011, **30**, 165.
37. J. Schurek, L. Vaclavik, H. Hooijerink, O. Lacina, J. Poustka, M. Sharman, M. Caldow, M. W. F. Nielen and J. Hajslova, *Anal. Chem.*, 2008, **80**, 9567.
38. J. F. Garcia-Reyes, A. U. Jackson, A. Molina-Diaz and R. G. Cooks, *Anal. Chem.*, 2009, **81**, 820.
39. L. Ziqing, Z. Mengxia, Z. Sichun, Y. Chengdui and Z. Xinrong, *Analyst*, 2010, **135**, 1268.
40. H. W. Chen, X. Zhang and M. B. Luo, *Chin. J. Anal. Chem.*, 2006, **34**, 464.
41. T. Cajka, K. Riddellova, P. Zomer, H. Mol and J. Hajslova, *Food Addit. Contam., Part A*, 2011, **28**, 1372.
42. J. I. Zhang, N. Talaty, A. B. Costa, X. Yu, W. A. Tao, R. Bell, J. H. Callahan and R. G. Cooks, *Int. J. Mass Spectrom.*, 2011, **301**, 37.
43. Y. Shuiping, D. Jianhua, Z. Jian, H. Bin, L. Jianqiang, C. Huanwen, Z. Zhiquan and Q. Xiaolin, *Anal. Chem.*, 2009, **81**, 2426.
44. M. Morelato, A. Beavis, P. Kirkbride and C. Roux, *Forensic Sci. Int.*, 2013, **226**, 10.
45. N. L. Sanders, S. Kothari, G. Huang, G. Salazar and R. G. Cooks, *Anal. Chem.*, 2010, **82**, 5313.
46. W. Xu, N. E. Manicke, G. R. Cooks and Z. Ouyang, *JALA*, 2010, **15**, 433.
47. C. R. Ferreira, L. S. Eberlin, J. E. Hallett and R. G. Cooks, *Reprod., Fertil. Dev.*, 2012, **24**, 132.
48. J. E. Hallett, C. R. Ferreira, L. S. Eberlin and R. G. Cooks, *Reprod., Fertil. Dev.*, 2012, **24**, 163.
49. C. R. Ferreira, V. Pirro, L. S. Eberlin, J. E. Hallett and R. G. Cooks, *Anal. Bioanal. Chem.*, 2012, **404**, 2915.
50. L. Nyadong, E. G. Hohenstein, A. Galhena, A. L. Lane, J. Kubanek, C. D. Sherrill and F. M. Fernandez, *Anal. Bioanal. Chem.*, 2009, **394**, 245.
51. H. Chen, I. Cotte-Rodriguez and R. G. Cooks, *Chem. Commun.*, 2006, 597.
52. Y. Song and R. G. Cooks, *J. Mass Spectrom.*, 2007, **42**, 1086.
53. M. Girod, E. Moyano, D. I. Campbell and R. G. Cooks, *Chem. Sci.*, 2011, **2**, 501.
54. M. Benassi, C. P. Wu, M. Nefliu, D. R. Ifa, M. Volny and R. G. Cooks, *Int. J. Mass Spectrom.*, 2009, **280**, 235.
55. M. Nefliu, R. G. Cooks and C. Moore, *J. Am. Soc. Mass Spectrom.*, 2006, **17**, 1091.
56. A. U. Jackson, T. Shum, E. Sokol, A. Dill and R. G. Cooks, *Anal. Bioanal. Chem.*, 2011, **399**, 367.
57. B. Nikolova-Damyanova, *J. Chromatogr. A*, 2009, **1216**, 1815.
58. I. Cotte-Rodriguez, Z. Takats, N. Talaty, H. W. Chen and R. G. Cooks, *Anal. Chem.*, 2005, 77, 6755.

59. C. P. Wu, D. R. Ifa, N. E. Manicke and R. G. Cooks, *Anal. Chem.*, 2009, **81**, 7618.

60. A. Benninghoven, *Surf. Sci.*, 1994, **299**, 246.

61. L. S. Eberlin, C. R. Ferreira, A. L. Dill, D. R. Ifa, L. Cheng and R. G. Cooks, *ChemBioChem*, 2011, **12**, 2129.

62. L. S. Eberlin, A. L. Dill, A. J. Golby, K. L. Ligon, J. M. Wiseman, R. G. Cooks and N. Y. R. Agar, *Angew. Chem., Int. Ed.*, 2010, **49**, 5953.

63. L. S. Eberlin, D. R. Ifa, C. Wu and R. G. Cooks, *Angew. Chem., Int. Ed.*, 2010, **49**, 873.

64. J. M. Wiseman, D. R. Ifa, Q. Song and R. G. Cooks, *Angew. Chem., Int. Ed.*, 2006, **45**, 7188.

65. A. L. Dill, L. S. Eberlin, A. B. Costa, C. Zheng, D. R. Ifa, L. Cheng, T. A. Masterson, M. O. Koch, O. Vitek and R. G. Cooks, *Chem. - Eur. J.*, 2011, **17**, 2897.

66. A. L. Dill, D. R. Ifa, N. E. Manicke, A. B. Costa, J. A. Ramos-Vara, D. W. Knapp and R. G. Cooks, *Anal. Chem.*, 2009, **81**, 8758.

67. M. Girod, Y. Shi, J.-X. Cheng and R. G. Cooks, *Anal. Chem.*, 2011, **83**, 207.

68. L. S. Eberlin, A. L. Dill, A. B. Costa, D. R. Ifa, L. Cheng, T. Masterson, M. Koch, T. L. Ratliff and R. G. Cooks, *Anal. Chem.*, 2010, **82**, 3430.

69. A. L. Dill, L. S. Eberlin, C. Zheng, A. B. Costa, D. R. Ifa, L. Cheng, T. A. Masterson, M. O. Koch, O. Vitek and R. G. Cooks, *Anal. Bioanal. Chem.*, 2010, **398**, 2969.

70. S. R. Ellis, C. Wu, J. M. Deeley, X. Zhu, R. J. W. Truscott, M. I. H. Panhuis, R. G. Cooks, T. W. Mitchell and S. J. Blanksby, *J. Am. Soc. Mass Spectrom.*, 2010, **21**, 2095.

71. C. Wu, D. R. Ifa, N. E. Manicke and R. G. Cooks, *Analyst*, 2010, **135**, 28.

72. A. L. Lane, L. Nyadong, A. S. Galhena, T. L. Shearer, E. P. Stout, R. M. Parry, M. Kwasnik, M. D. Wang, M. E. Hay, F. M. Fernandez and J. Kubanek, *Proc. Natl. Acad. Sci. U. S. A.*, 2009, **106**, 7314.

73. J. Watrous, N. Hendricks, M. Meehan and P. C. Dorrestein, *Anal. Chem.*, 2010, **82**, 1598.

74. B. Li, N. Bjarnholt, S. H. Hansen and C. Janfelt, *J. Mass Spectrom.*, 2011, **46**, 1241.

75. J. Thunig, S. H. Hansen and C. Janfelt, *Anal. Chem.*, 2011, **83**, 3256.

76. D. R. Ifa, A. Srimany, L. S. Eberlin, H. R. Naik, V. Bhat, R. G. Cooks and T. Pradeep, *Anal. Methods*, 2011, **3**, 1910.

77. T. Muller, S. Oradu, D. R. Ifa, R. G. Cooks and B. Krautler, *Anal. Chem.*, 2011, **83**, 5754.

78. D. R. Ifa, L. M. Gumaelius, L. S. Eberlin, N. E. Manicke and R. G. Cooks, *Analyst*, 2007, **132**, 461.

79. D. R. Ifa, N. E. Manicke, A. L. Dill and G. Cooks, *Science*, 2008, **321**, 805.

80. D. R. Ifa, J. M. Wiseman, Q. Song and R. G. Cooks, *Int. J. Mass Spectrom.*, 2007, **259**, 8.

81. V. Kertesz and G. J. Van Berkel, *Rapid Commun. Mass Spectrom.*, 2008, **22**, 2639.

82. D. I. Campbell, C. R. Ferreira, L. S. Eberlin and R. G. Cooks, *Anal. Bioanal. Chem.*, 2012, **404**, 389.

83. L. Gao, R. G. Cooks and Z. Ouyang, *Anal. Chem.*, 2008, **80**, 4026.

84. Z. Ouyang, G. X. Wu, Y. S. Song, H. Y. Li, W. R. Plass and R. G. Cooks, *Anal. Chem.*, 2004, **76**, 4595.

85. A. Keil, N. Talaty, C. Janfelt, R. J. Noll, L. Gao, Z. Ouyang and R. G. Cooks, *Anal. Chem.*, 2007, **79**, 7734.

86. E. Sokol, R. J. Noll, R. G. Cooks, L. W. Beegle, H. I. Kim and I. Kanik, *Int. J. Mass Spectrom.*, 2011, **306**, 187.

87. G. M. Huang, G. T. Li, J. Ducan, Z. Ouyang and R. G. Cooks, *Angew. Chem., Int. Ed.*, 2011, **50**, 2503.

88. C. Wu, A. L. Dill, L. S. Eberlin, R. G. Cooks and D. R. Ifa, *Mass Spectrom. Rev.*, 2013, **32**, 218.

89. C.-H. Chen, Z. Lin, S. Garimella, L. Zheng, R. Shi, R. G. Cooks and Z. Ouyang, *Anal. Chem.*, 2013, **85**, 11843.

90. S. Garimella, W. Xu, G. M. Huang, J. D. Harper, R. G. Cooks and Z. Ouyang, *J. Mass Spectrom.*, 2012, **47**, 201.

91. L. Li, T.-C. Chen, Y. Ren, P. I. Hendricks, R. G. Cooks and Z. Ouyang, *Anal. Chem.*, 2013, **86**, 2909.

CHAPTER 13

Surface Acoustic Wave Nebulization

YUE HUANG, SCOTT HERON, SUNG HWAN YOON AND
DAVID R. GOODLETT

University of Maryland, School of Pharmacy, 20 North Pine Street,
Baltimore, MD 21201, USA
Email: dgoodlett@rx.umaryland.edu

13.1 Introduction

A surface acoustic wave (SAW) is an acoustic wave traveling along the surface of a material. This phenomenon has been studied extensively in the field of electronics where SAW devices are widely used as filters, oscillators and transformers.[1] In 2010 the first paper describing their use for producing ions for mass spectrometric detection was published, and the phrase "surface acoustic wave nebulization" (SAWN) coined to describe this phenomenon.[2] Ions produced by SAWN are done so either in a continuous electrospray ionization (ESI) like mode or an intermittent manner like matrix-assisted laser desorption ionization (MALDI). In either mode nebulization occurs from a planar surface, with the opportunity to detect either positive or negative ions. Notably, SAWN mass spectra, of chemical compounds such as peptides and proteins that are chemically basic in nature, commonly exhibit a lower average charge-state distribution than ESI produces from the same solution. Importantly, given that no DC voltage is applied to the liquid sample to produce ions, the SAWN-nebulized ions can have lower internal energy than ESI- or MALDI-generated ions.[3]

New Developments in Mass Spectrometry No. 2
Ambient Ionization Mass Spectrometry
Edited by Marek Domin and Robert Cody
© The Royal Society of Chemistry 2015
Published by the Royal Society of Chemistry, www.rsc.org

13.2 Surface Acoustic Wave Background

A surface acoustic wave (SAW) is a longitudinal wave traveling along the surface of a material (Figure 13.1).[4] Such waves are quite common in nature. For example, elephants communicate with each other using surface acoustic waves generated by stomping their feet.[5] The amplitude of a SAW typically decays exponentially with the depth into the substrate.

Surface acoustic waves, as a mode of propagation, were first explained in 1885 by Lord Rayleigh who also predicted their properties.[6] Named after their discoverer, Rayleigh waves have a longitudinal and a vertical shear component that can couple with any media in contact with the surface. This coupling strongly affects the amplitude and velocity of the wave, allowing SAW sensors to directly sense mass and mechanical properties. In electronics, SAW devices are widely used as filters, oscillators and transformers.

Going back to the late 1880s, Pierre Curie and his brother Jacques demonstrated, prior to Pierre marrying the future Nobel laureate Marie Sklodowska, that an electric potential was generated when crystals were compressed. They named this the piezoelectric effect, from the Greek word to squeeze. Important to the development of SAW actuators, they also showed the reverse property, namely that when subjected to an electric field that the crystals deformed. Today, many digital electric circuits rely on this type of crystal oscillator.

As shown in Figure 13.1, when a SAW travels through a piezoelectric substrate and reaches the liquid placed on the surface, mechanical energy will be transferred into the droplet. Usually when the sample is a liquid, SAW energy will induce a circular motion inside the liquid droplet causing it to shake or move on the chip.[7] However, as the amplitude increases, the liquid droplet may move, jet or nebulize depending on the applied power. Nebulization is also referred to as "atomization" in certain literature (especially engineering literature).[8] For semantic purposes though when the Goodlett

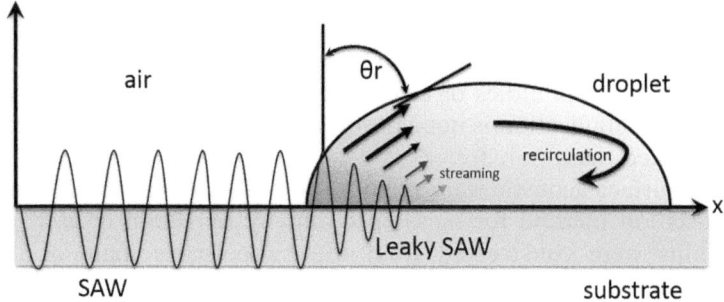

Figure 13.1 Surface acoustic wave. Surface acoustic wave (SAW) traveling on the surface of a piezoelectric material. When the SAW hits the liquid droplet sitting on the surface, mechanical energy is transferred into the liquid droplet, causing the droplet to move, or at high power, nebulize.

group developed the use of SAW action to generate ions for mass spectrometry (MS) detection, atomization was not thought to be an accurate descriptor because the generated plume is not generally composed of single atoms, but rather chemical compounds. This led to development of the phrase surface acoustic wave nebulization (SAWN) rather than surface acoustic wave atomization and thus SAWN as the preferred acronym for SAW action that leads to the detection of ions[2] by mass spectrometers.

The convenience and simplicity of energy transfer by SAW technology has made it useful within the field of microfluidics.[9,10] Importantly, and unlike other microfluidic methods for sample manipulation, SAW does not require pressure-driven pumps, interconnects, nor the integration of electrodes and microchannels.[11,12] The removal of the need for pumps and interconnects eliminates many problems commonly associated with dead volumes, which are particularly disadvantageous when using small volumes of rare, low-abundance samples.

SAW devices have already been incorporated into microfluidic workflows to enable mixing, using SAW microcentrifugation in channels; heating; liquid-droplet movement; and delivery to or from a microfluidic port.[48] SAW is also known to be able to nebulize protein samples for array writing with masks.[13,14] More recently SAW nebulization has been used to generate aerosol droplets with diameters of between 5 to 10 nm to assist with the synthesis of polymeric nanoparticles, and to generate protein aerosols and nanoparticles for drug delivery.[15,16]

13.3   Surface Acoustic Wave Nebulization

When the amplitude of the SAW is high enough, large amounts of energy will be transferred to a liquid sample on the surface of the device. As the energy absorbed by the liquid droplet on the chip increases, the surface tension of the liquid may be compromised causing the droplet to nebulize (Figure 13.2(A)). This phenomenon has been thoroughly investigated by the Yeo group and applied to develop medical devices[15] that deliver medicine to patients in the form of aerosols.[16,17] Despite this, nebulization has received little attention, compared to other SAW-driven droplet movements, especially from the microfluidics or electronics communities.[14,18] This is most likely because nebulization is not a useful way to transfer a signal or to carry out SAW-related mass transfer.

In 2010, surface acoustic wave nebulization (SAWN) was introduced as a novel ionization method for mass spectrometry by Heron *et al.*[2,19] These SAWN chips were constructed from the piezoelectric material lithium niobate ($LiNbO_3$). It was noted that when the plume of liquid droplets containing peptides generated from a SAWN device was positioned close enough to the orifice of an atmospheric-pressure ionization (API) mass spectrometer interface, the plume was pulled into the instrument under the influence of the pressure differential (Figure 13.2(B iv)). Despite the fact that no DC voltage was applied to produce the plume, the mass spectrometer

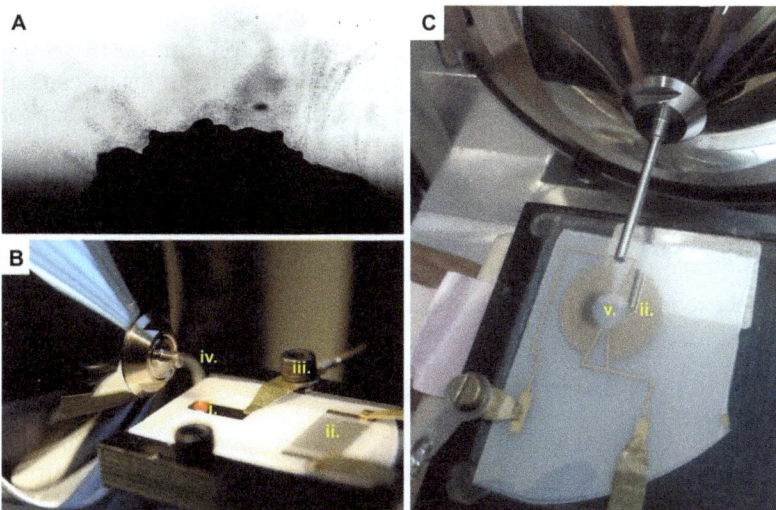

Figure 13.2 Surface acoustic wave nebulization (SAWN) device. Shown are: (A) Side view of a liquid droplet on a SAWN chip as it begins to nebulize; (B) SAWN chip in use with plume being drawn into the atmospheric-pressure ionization inlet of an LTQ ion trap mass spectrometer (ThermoFisher, San Jose, CA) where (i) is the ground pad on which a liquid droplet of red food coloring is being nebulized; (ii) is the interdigitated (IDT) electrodes that are connected to a signal generator and amplifier through electrical leads on either side of the IDT; (iii) is the ground wire connection which may also be used to add DC voltage; and (iv) is the plume being pulled into the API interface; and (C) an alternative ring electrode IDT design with heated capillary extender in place to capture plume generated from (v) without a DC voltage.

would record a mass spectrum, implying that the compounds contained within the aerosolized droplets were undergoing ionization during the desolvation stage. The results showed that SAW action on a piezoelectric substrate could be used to generate ions for MS detection, by nebulization, without application of a DC voltage as used in electrospray ionization (ESI). Tandem mass spectra were used to verify that the ions detected (in this case angiotensin I) were coming from the aerosolized plume. Since these results were published, others have shown that SAWN can produce ions of lower energy than ESI and thus should be consider a soft ionization method that imparts less internal energy during the transfer of analytes from liquid to gas phase.[3]

This finding, that SAWN generated ions had lower energy than ions produced by ESI, indicated SAWN could potentially be better for the study of fragile compounds and noncovalently bonded compounds, a topic discussed in Section 1.5 (*vide infra*). In 2011, the Go group used paper to deliver sample directly to a SAWN device, and created a paper based microfluidics system.[20] In 2012, the Goodlett group reported the use of SAWN and an hierarchical tandem mass spectrometry (HiTMS) algorithm to define lipid structures by collision-induced dissociation.[21] While numerous applications of SAWN-MS have been reported, the mechanism of SAWN remains under investigation. The Goodlett group has suggested that fast, vibration-induced shear forces may tear apart liquid droplets into both positively and negatively charged aerosolized droplets and these may be may responsible for subsequent ion formation. Additionally, in 2012, the Go group reported on the possible origin of charges based on simulations stating that it may be due to the voltage gradient generated from the potential difference between the liquid/piezoelectric material interface and liquid/gas interface.[20]

13.4   SAWN Microfabrication

To date, the Goodlett group has used standard photolithographic micro-fabrication processes to construct SAWN chips (Figure 13.3(A)). Although chip design may be different for each experiment, most fabrications have followed the same basic procedure. The most commonly used wafer material

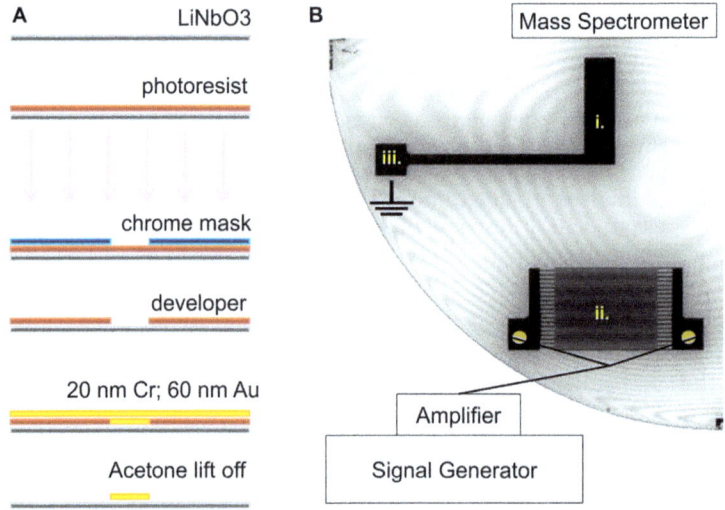

Figure 13.3   Photolithographic process for production of SAWN chips. Shown are (A) the photolithographic process for SAWN chip production (top to bottom) which is usually done on a 1.5-inch size wafer fabricated with Y-cut X-propagating LiNbO$_3$ wafer and (B) the finished SAWN chip schematically set in place with mass spectrometer (top) and key electronics (bottom) where: (i) is the ground pad on which a droplet of liquid for nebulization is placed; (ii) is the interdigitated (IDT) electrodes that are connected to a signal generator and amplifier through electrical leads on either side of the IDT; and (iii) the ground plate that may also be used to add DC voltage. Typically, the SAWN chip IDT is 10 mm in aperture with a 100 μm finger width and spacing, activated by a 9.56-MHz frequency signal. The amplitude used for nebulization varies between 70 mW to 250 mW, depending on the sample and solvent.

in microfabrication is silicon or glass. However, to generate a SAW, a piezoelectric substrate is required. There are several options for piezoelectric materials, but for historical reasons lithium niobate ($LiNbO_3$) has served as convenient material for SAWN chip construction.

The $LiNbO_3$ wafers used have been either 3- or 4-inch diameter 0.5 mm thick 128° Y-cut X-propagating of a circular design. Before the fabrication process begins wafers are cleaned according to standard published protocols followed by sonication in acetone, and then, before the acetone dries, the wafer is sequentially washed with isopropanol and deionized water. The remaining water droplets are removed with nitrogen gas.

SAWN chips have been constructed by a standard photolithographic process using chemical treatments to create the interdigitated (IDT) electrodes onto $LiNbO_3$ wafers (Figure 13.3(B ii)).[2] In experiments conducted by the Goodlett laboratory, the wafer is usually covered with an adhesion primer and then a photoresist (*e.g.* AZ 1512). After baking, the photoresist is patterned by either laser writing or mask alignment techniques. Laser writing is preferred for producing small batches, whilst mask alignment techniques are used for greater numbers of wafers.

After baking again, the wafer is metallized with a layer of 20 nm Cr following 60 nm of Au. The chip is finally washed with acetone to remove all the excess gold and photoresist, a process often referred to as acetone lift-off.

Various chip modifications may be conducted on the wafer after photolithographic production, including chemical evaporation, spin coating, plasma-enhanced chemical vapor deposition (PECVD) and dry etching.

Finally, the round $LiNbO_3$ wafers, now patterned for use as SAWN chips, are manually scored with a diamond pen and fractured along that line, or by use of a dicing saw. This produces four SAWN chips one of which is shown in Figure 13.3(B). While cutting with a dicing saw gives a cleaner edge and has better control when the target piece is small, hand cutting works best if the design is of a larger area.

As shown by the ring electrode in Figure 13.2(C), there are many different configurations of the IDT electrodes that may be produced. With the ring electrode design the liquid to be nebulized is placed at the center of the ring. It should be noted that in the photo shown in Figure 13.2(C) there is neither a ground electrode nor a means to add DC voltage.

13.5 SAWN-MS Detection of Biological Compounds

Figure 13.4 shows a comparison of a peptide, substance P, ionized be ESI (Figure 13.4(A)) and by SAWN (Figure 13.4(B)). Such mass spectra of basic peptides and proteins showing the average charge-state distribution shifted to a lower average value, are commonly produced by SAWN. In the first report of SAWN by Heron *et al.*[2] ions of angiotensin I were produced by SAWN (Figure 13.5(A)) and ESI from an identical acidic solution. Comparison of the two showed the same behavior as exhibited here by substance P. As can be seen for substance P, the SAWN-produced mass spectrum still has the 2 + ion

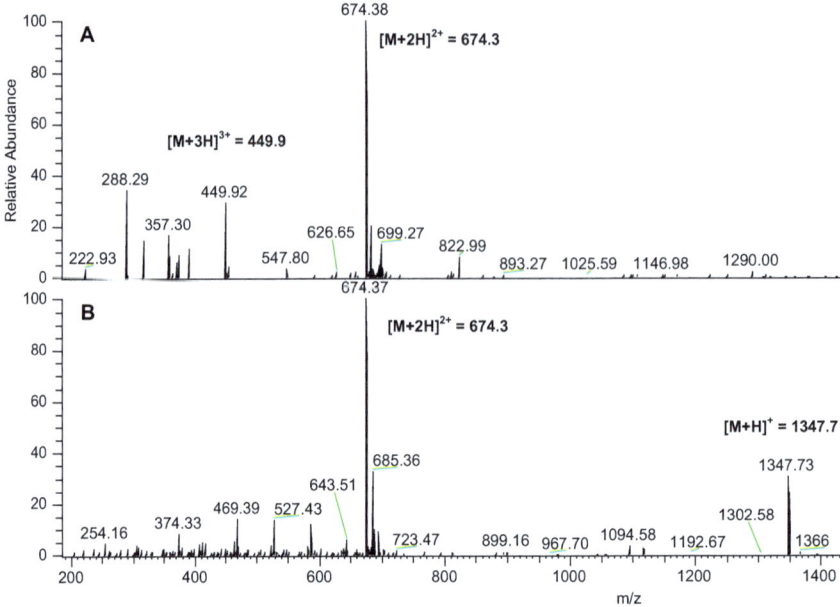

Figure 13.4 SAWN mass spectrum of substance P. Shown are the mass spectra for substance P peptide (sequence RPKPQQFFGLM) produced by: (A) ESI and (B) SAWN from the same acidified, aqueous acetonitrile solution. Note the base peak is the $[M+2H]^{2+} = 674.3$ ion in both spectra but with SAWN the $[M+H]^{+} = 1347.7$ is significantly higher than ESI.

as the base peak, but also shows a 1+ ion that is ~33% of the base peak. While this could be explained by in-source CID of the higher charge state ions, this does not appear to be the case. To date, this phenomenon has not been adequately explained, but it has been seen to be a general trend for peptides and proteins nebulized by SAWN.

Specifically, these basic polymers tend to have their average charge-state distribution shifted to lower average mass/charge when nebulized from an aqueous, acidified acetonitrile solvent, as is typically used for solvent A in a reversed phase HPLC experiment. Interestingly, the same has not been observed for acidic compounds ionized in negative-ion mode. Given that there is known to be surface charging with piezoelectric wafers used in SAWN studies, a likely explanation for this observed shift in charge-state distribution with basic peptides must be related to this affect. Lending some credibility to this hypothesis comes from unpublished studies done in the Tureček laboratory at the University of Washington. In this study, ion current from SAWN- and ESI-generated droplets of NaCl were measured against a metal collection plate. Compared to ESI, SAWN-generated droplets presented much less positive charge. However, the Go laboratory's published results on similar experiments showed the opposite result, but with different solutions (water, ethanol and glycerol) leading one to conclude more research is needed on this phenomenon.

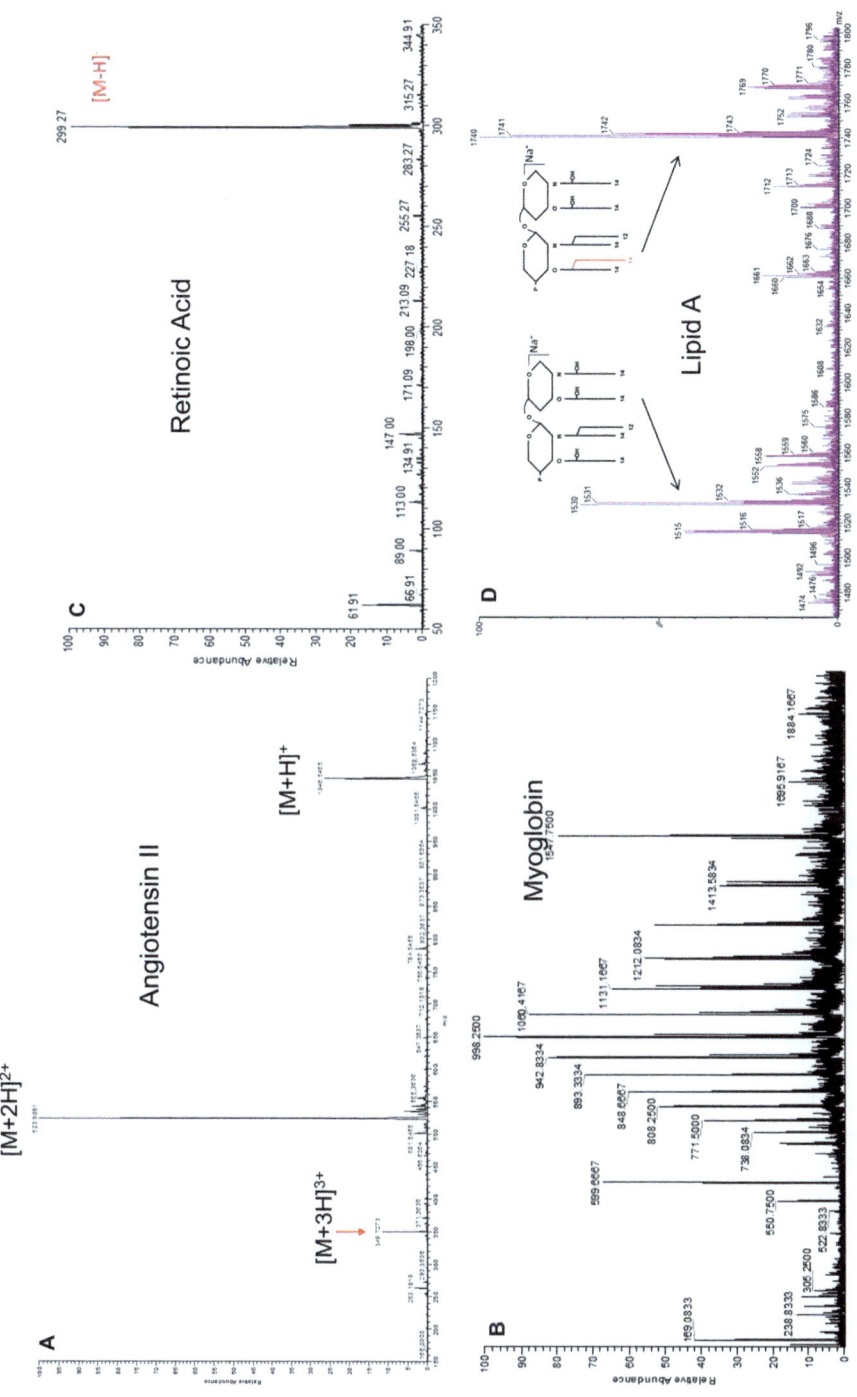

Figure 13.5 SAWN-generated mass spectra. Shown are: (A) the peptide angiotensin II (sequence DRVYIHPF) in positive-ion mode; (B) the protein myoglobin (equine, 17.2 kD) in positive-ion mode; (C) al-trans retinoic acid (MW = 300 Da) in negative mode; and (D) Purified and detoxified lipid A isolated from *Salmonella minnesota* detected in negative mode.

Proteins up to about 20 kDa have been shown to be amenable to ion generation by SAWN (Figure 13.5(B)). To date, however, detection of proteins with molecular weights above ∼ 20 kDa has required application of a voltage to the metal surface pad on which the sample rests on the SAWN wafer. This pad from which red food coloring is being nebulized in Figure 13.2(B i) normally serves as a ground, but added voltage may be easily applied. With the application of this additional voltage, ions from proteins as large as albumin have been successfully detected from SAWN (unpublished data).

As mentioned above, negative ions have also been detected by SAWN; for example, Figure 13.5(C) and (D) shows retinoic acid and lipid A mass spectra. In the case of retinoic acid, which happens to easily oxidize during ionization by ESI often resulting in loss of the retinoic acid parent ion, the molecular ion is detected as the base peak at the expected *m/z* value with minimal fragmentation or oxidation. Lipid A, a class of bacterial membrane lipids often containing phosphate groups, was also readily detected in negative-ion mode by nebulization from a solution of chloroform and methanol. As shown by Yoon *et al.*, phospholipids nebulized by SAWN produced mass spectra with twice as much molecular ion as ESI generated mass spectra from an identical solution. These and other experiments hinting at lower ion energetics led Huang *et al.* to publish a study comparing SAWN to ESI.[3]

13.6   Ion Energetics

Given that the majority of the ion detection discussed (*vide supra*) was in the absence of an applied DC voltage to the ground pad on which the liquid to be nebulized rests, it was suspected that the fundamental aspects of ion production from SAWN would be quite different from ESI. Namely, in the absence of this applied voltage one can speculate that SAWN produced droplets may not be subject to the same degree of Coulombic repulsion as ESI generated droplets.[3,22] From this idea a working hypothesis was developed that a lack of what is perceived to be the violent fission as the parent droplet splits apart by ESI, will be less so by SAWN, and that the cooling affects provided by desolvation will be different. In a series of experiments using so-called thermometer ions (*i.e.* substituted benzylpyridinium compounds of known bond-dissociation energies), Huang *et al.* showed that in most cases examined, SAWN produced ions were lower in energy than ESI-generated ions.[3] These thermometer ions gained their name by virtue of the fact that, by knowing the bond energy for specific labile bonds in the compounds, one may work out the internal energy of generated ions. In short, this is done by measuring the ratio of precursor ion to a pyridinium ion common to all of the substituted "thermometer" compounds.

Figure 13.6 shows an example of the differences detected under different conditions using the thermometer ions; *i.e.* substituted benzylpyridinium ions such as Me-, MeO-, Cl- and NO$_2$- substituted ions which were used in this research. In Figure 13.6(A), the dissociation of SAWN generated ions is shown to vary as a function of temperature. Not surprisingly as the

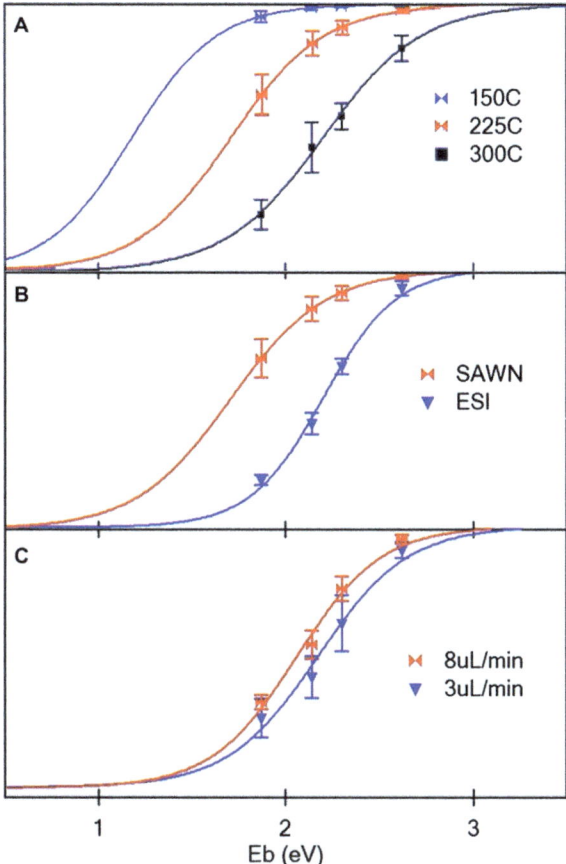

Figure 13.6 Comparison of ion internal energy between SAWN and ESI. Shown are
(A) internal ion energy distribution of SAWN under various heated ion
transfer capillary temperatures on an LTQ ion trap mass spectrometer
(Thermo Fisher, San Jose, CA); (B) comparison between SAWN and ESI
of ion internal energy distribution at a heated ion-transfer capillary
temperature of 225 °C; and (C) Effect of syringe pump flow rate delivery
of liquid to the SAWN chip surface. The compound being used to gauge
ion internal energy here was substituted benzylpyridinium ions (MeO-,
Me-, Cl-, and NO_2- substituted from lowest internal energy to highest
internal energy).

temperature of the heated capillary increases, so does the amount of ion
internal energy. In Figure 13.6(B) one can clearly see the difference between
SAWN and ESI at a single heated capillary temperature of 225 °C. Finally,
Figure 13.6(C) shows that there is a mild "cooling" effect as a result of the
delivery rate of liquid from the syringe pump used in these experiments
increases from 3 to 8 μL min^{-1}. It should be noted that this discussion regarding why SAWN-generated ions can have lower energy than ESI-generated
ions under the same conditions is speculative. Many experimental factors

that are difficult to control or even list may be affecting the measurements. Thus, why SAWN-generated ions can have lower ion internal energies than ESI generated ions remains a question open for debate.

Practically speaking though there is no question that there is some utility in these observations. In Figure 13.7 one may see the practical outcome of producing ions of lower internal energy where a mixture of 1-palmitoyl-2-oleoyl-*sn-glycero*-3-phospho-(1′-*rac*-glycerol) (PG (16:0/18:1)) and 1-palmitoyl-2-oleoyl-*sn-glycero*-3-phosphoethanolamine (PE (16:0/18:1)) are analyzed in negative-ion mode by ESI (Figure 13.7(A)) and SAWN (Figure 13.7(B)). Notice that the ion at *m/z* 748 is more than twice as abundant relative to the base peak at *m/z* 717 in the SAWN mass spectrum compared to the same ion in the ESI mass spectrum. Finally, it should be noted that the SAWN-generated mass spectra of phospholipids did not contain any detectable fatty acid

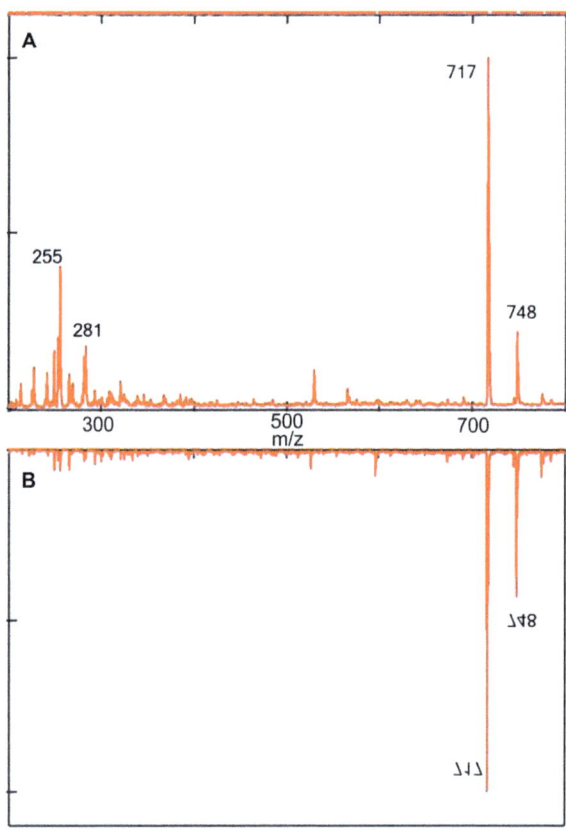

Figure 13.7   Comparison of phospholipid mass spectra between SAWN and ESI. Both mass spectra are of a mixture of 1-palmitoyl-2-oleoyl-*sn-glycero*-3-phospho-(1′-*rac*-glycerol) (PG (16:0/18:1)) and 1-palmitoyl-2-oleoyl-*sn-glycero*-3-phosphoethanolamine (PE (16:0/18:1)) recorded in negative-ion mode by (A) ESI and (B) SAWN with the lower one being inverted for comparison.

fragment ions, indicating that the ion internal energy threshold required to generate these ions was not exceeded.

13.7 Summary

In 2010 Heron *et al.* reported the first use of SAW to nebulize a chemical compound for mass-spectrometric detection and proposed the acronym SAWN to describe the process. Unlike ESI there was no need for a DC voltage to be applied for ion production. This was followed by two publications revealing that SAWN-generated thermometer ions could have lower internal energy than ESI generated ions by Huang *et al.* and that SAWN-generated phospholipid ions showed more molecular ion and no loss of fatty acids commonly observed with ESI by Yoon *et al.* Regardless of why these SAWN-generated ions had lower internal energies, the practical observation of phospholipids being detected with no loss of their fatty acids is promising for SAWN analysis of fragile molecules, such as explosives. Additionally, preliminary results in the Goodlett laboratory (Monkkonen, unpublished results) indicate that SAWN can also preserve more protein noncovalent complex than ESI. In terms of laboratory workflows, SAWN offers an advantage of simplicity and has been shown to be capable of quantitation (Yoon, unpublished results) like ESI. In order to produce ions by SAWN one need not bother with the complicated plumbing needed to connect an ESI source to a syringe pump, making SAWN much more like MALDI in terms of setup and ease of use. Sample delivery to the chip surface can be as simple as pipetting the liquid in place or liquid may be delivered continuously as was done for the thermometer ion studies by coupling to a syringe pump. Regardless, ions are subsequently produced simply by actuation of the SAW. Additionally, in terms of comparison to MALDI, SAWN-generated mass spectra have much lower chemical background due to the lack of an added chemical matrix promising greater signal/noise for some applications. This reduction in chemical noise can in part be attributed to the fact that SAWN-generated mass spectra produced without an added DC voltage applied to the liquid on chip do not contain ionized polysiloxanes pervasive in modern laboratory air. Finally, a number of investigations are under way currently in the Goodlett laboratory to refine SAWN. These include modifications to the type of IDTs being used (*e.g.* more compact ring IDTs as shown in Figure 13.2(C)) that promise more efficient ionization by producing more homogeneous nebulized droplets than those shown in Figure 13.2(A) as well as coupling SAWN to an autosampler to more rapidly dispense liquid samples to the SAWN surface.

Acknowledgments

This work was supported in part by the University of Maryland Baltimore, School of Pharmacy Mass Spectrometry Center (SOP1841-IQB2014). The authors thank Prof David P. A. Kilgour for his critical editing of the manuscript.

References

1. K. Lange, B. E. Rapp and M. Rapp, Surface acoustic wave biosensors: a review, *Anal. Bioanal. Chem.*, 2008, **391**(5), 1509–1519.
2. S. R. Heron, R. Wilson, S. A. Shaffer, D. R. Goodlett and J. M. Cooper, Surface Acoustic Wave Nebulization of Peptides As a Microfluidic Interface for Mass Spectrometry, *Anal. Chem.*, 2010, **82**(10), 3985–3989.
3. Y. Huang, S. H. Yoon, S. R. Heron, C. D. Masselon, J. S. Edgar, F. Tureček and D. R. Goodlett, Surface acoustic wave nebulization produces ions with lower internal energy than electrospray ionization, *J. Am. Soc. Mass Spectrom.*, 2012, **23**(6), 1062–70.
4. A. Qi, L. Y. Yeo and J. R. Friend, Interfacial Destabilization and Atomization Driven by Surface Acoustic Waves, *Phys. Fluids*, 2008, **20**, 074103–074117.
5. R. H. Günther, C. E. O'Connell-Rodwell and S. L. Klemperer, Seismic waves from elephant vocalizations: A possible communication mode? *Geophys. Res. Lett.*, 2004, **31**(11), L11602.
6. L. Rayleigh, On Waves Propagated along the Plane Surface of an Elastic Solid, *Proceedings of the London Mathematical Society*, 1885, **s1-17**(1), 4–11.
7. A. R. Rezk, A. Qi, J. R. Friend, W. H. Li and L. Y. Yeo, Uniform mixing in paper-based microfluidic systems using surface acoustic waves, *Lab Chip*, 2012, **12**(4), 773–9.
8. D. J. Collins, O. Manor, A. Winkler, H. Schmidt, J. R. Friend and L. Y. Yeo, Atomization off thin water films generated by high-frequency substrate wave vibrations, *Phys. Rev. E: Stat., Nonlinear, Soft Matter Phys.*, 2012, **86**(5), 056312.
9. Y. Bourquin, J. Reboud, R. Wilson, Y. Zhang and J. M. Cooper, Integrated immunoassay using tuneable surface acoustic waves and lensfree detection, *Lab Chip*, 2011, **11**(16), 2725–2730.
10. R. J. Shilton, L. Y. Yeo and J. R. Friend, Quantification of surface acoustic wave induced chaotic mixing-flows in microfluidic wells, *Sens. Actuators, B*, 2011, **160**(1), 1565–1572.
11. T. M. Tarasow, L. Penny, A. Patwardhan, S. Hamren, M. P. McKenna and M. S. Urdea, Microfluidic strategies applied to biomarker discovery and validation for multivariate diagnostics, *Bioanalysis*, 2011, **3**(19), 2233–2251.
12. X. Ding, P. Li, S.-C. S. Lin, Z. S. Stratton, N. Nama, F. Guo, D. Slotcavage, X. Mao, J. Shi, F. Costanzo and T. J. Huang, Surface acoustic wave microfluidics, *Lab Chip*, 2013, **13**(18), 3626–3649.
13. M. Alvarez, J. Friend and L. Y. Yeo, Rapid generation of protein aerosols and nanoparticles *via* surface acoustic wave atomization, *Nanotechnology*, 2008, **19**(45), 455103.
14. T. Vuong, A. Qi, M. Muradoglu, B. H.-P. Cheong, O. W. Liew, C. X. Ang, J. Fu, L. Yeo, J. Friend and T. W. Ng, Precise drop dispensation on

superhydrophobic surfaces using acoustic nebulization, *Soft Matter*, 2013, **9**(13), 3631–3639.

15. A. Qi, J. R. Friend, L. Y. Yeo, D. A. V. Morton, M. P. McIntosh and L. Spiccia, Miniature Inhalation Therapy Platform Using Surface Acoustic Wave Microfluidic Atomization, *Lab Chip*, 2009, **9**(15), 2184–2193.

16. A. Qi, L. Yeo, J. Friend and J. Ho, The extraction of liquid, protein molecules and yeast cells from paper through surface acoustic wave atomization, *Lab Chip*, 2010, **10**(4), 470–6.

17. L. Y. Yeo, J. R. Friend, M. P. McIntosh, E. N. Meeusen and D. A. Morton, Ultrasonic nebulization platforms for pulmonary drug delivery, *Expert Opin. Drug Delivery*, 2010, **7**(6), 663–79.

18. R. Raghavan, J. Friend and L. Yeo, Particle concentration *via* acoustically driven microcentrifugation: microPIV flow visualization and numerical modelling studies, *Microfluid. Nanofluid.*, 2010, **8**(1), 73–84.

19. D. Gao, H. Liu, Y. Jiang and J. M. Lin, Recent advances in microfluidics combined with mass spectrometry: technologies and applications, *Lab Chip*, 2013, **13**(17), 3309–22.

20. J. Ho, M. K. Tan, D. B. Go, L. Y. Yeo, J. R. Friend and H. C. Chang, Paper-Based Microfluidic Surface Acoustic Wave Sample Delivery and Ionization Source for Rapid and Sensitive Ambient Mass Spectrometry, *Anal. Chem.*, 2011, **83**(9), 3260–3266.

21. S. H. Yoon, Y. Huang, J. S. Edgar, Y. S. Ting, S. R. Heron, Y. Kao, Y. Li, C. D. Masselon, R. K. Ernst and D. R. Goodlett, Surface acoustic wave nebulization facilitating lipid mass spectrometric analysis, *Anal. Chem.*, 2012, **84**(15), 6530–7.

22. P. Kebarle and L. Tang, From ions in solution to ions in the gas phase – the mechanism of electrospray mass spectrometry, *Anal. Chem.*, 1993, **65**(22), 972A–986A.

Laser Ablation Electrospray Ionization Mass Spectrometry: Mechanisms, Configurations and Imaging Applications

PETER NEMES* AND AKOS VERTES*

Department of Chemistry, W. M. Keck Institute for Proteomics Technology and Applications, The George Washington University, Washington, DC, USA
*Email: vertes@gwu.edu; petern@gwu.edu

14.1 Introduction

In recent years, the combination of midinfrared (mid-IR) laser ablation sampling with efficient ionization methods for direct analysis by mass spectrometry (MS) has been successfully implemented in a growing number of ambient ionization techniques.[1–3] The primary driving force behind this approach is to combine the microsampling capabilities of focused laser beams with the high ion yields achievable by electrospray and photoionization sources. Laser-ablation sampling at 2.94 μm wavelength is especially advantageous in biomedical analysis, where the native water content of the sample acts as a matrix due to the strong absorption by the OH stretching mode. Since the first implementation of a mid-IR laser-ablation electrospray ionization (LAESI) source,[4] several closely related techniques have followed,[5–9] and the corresponding literature has expanded to include numerous journal publications (see Table 14.1) and three book chapters.[10–12]

New Developments in Mass Spectrometry No. 2
Ambient Ionization Mass Spectrometry
Edited by Marek Domin and Robert Cody
© The Royal Society of Chemistry 2015
Published by the Royal Society of Chemistry, www.rsc.org

Table 14.1 Milestones in LAESI development and applications.

Operation modality	Geometry	Figures of Merits	Specimen	Detected Compounds	Ref.
In situ/ in vivo	Conventional	Quantitation (4 orders of magnitude); LOD = 8 fmol, up to 66.5 kDa; *in vivo* analysis	Urine; whole blood; plant organs (*T. patula*)	Drug standards, lipids, metabolites, proteins	4
		Reactive LAESI for structural elucidation	*M. musculus* brain	Metabolites, lipids	24
		Animal tissue, *in vitro*	*T. californica* electric organ	Metabolites	23
		Soft ion generation	Chemical standards	Thermometer ions	37
	Heat-assisted	Polar-apolar compounds; quantitation (3 orders of magnitude)	Avocado mesocarp and *M. musculus* sections	Metabolites, lipids	57
Biomarker discovery	Conventional	Human cell lines transfected with Human T-lymphotropic virus type 1	T lymphocyte cells (CEM and H9), kidney epithelial cells (293T)	Metabolites, lipids, proteins	25
		Oncovirus-infected cell lines	B lymphocytes infected with Kaposi's sarcoma-associated herpesvirus (KSHV)	Metabolites, lipids, proteins	27
	Plume collimation	~600 cyanobacterial cells	*Anabaena sp.* PCC7120	Metabolites, proteins	26
		Small cell populations: 6 cells; quantitation (6 orders of magnitude); LOD = 600 amol	Cheek cells (human), rat insulinoma β-cells (RIN5mF), sea urchin eggs	Metabolites, drug standards	22
	Fiber optic	Single cells	*A. cepa, N. pseudonarcissus, L. pictus*	Metabolites	18–20
	Fiber optic	Subcellular organelles	*A. cepa*	Metabolites	21,22
		50 μm depth resolution	*S. lynise* leaf	Metabolites, ink dyes	13
Depth profiling	Conventional	350 μm lateral	*S. lynise* and *A. squarrosa* leaves;	Metabolites (primary, secondary)	10,13,15
2D MSI	Conventional	250 μm lateral	*R. norvegicus* section	Metabolites, osmolites, lipids	10,15,16
	Fiber optic	Cell-by-cell imaging	*A. cepa*	Sugars, polysaccharides, metabolites	17
3D MSI	Conventional	350 μm lateral by 50 μm depth resolution	*A. squarrosa*	Metabolites	10,14,15

Despite the growing number of analytical applications of LAESI MS in tissue imaging,[13–17] single-cell analysis[18–22] and metabolomics,[11,14,16,18,21–27] the understanding of mechanisms underlying the LAESI process is incomplete. There are four major factors that contribute to ion production in LAESI. In the sampling step, mid-IR laser ablation of the water-containing target takes place, resulting in an expanding plume of vapor and fine particulates representing the sample. Simultaneously, an electrospray is produced that generates a charged dynamic droplet plume that travels toward the mass spectrometer.[28] The third process takes place in the region where the laser-ablation plume merges with the electrospray. In this phase, the particulates from the sample coalesce with the charged droplets from the electrospray. This is followed by ion production that unfolds similarly to the conventional electrospray process, except that in LAESI the charged droplets are seeded with the ablated sample particles. As this last step has been extensively discussed in the literature, in the second segment of this chapter we focus our attention on the three first steps, laser ablation, electrospray formation and coalescence in the merging plumes. Each of these processes is complex in itself, and understanding their combination is essential to gain an insight into the mechanisms of LAESI.

In order to improve the efficiency of the individual processes, various LAESI configurations have been developed. Two of these variants of the technique rely on changes in the delivery of the laser energy (optical fiber-based focusing and ablation in transmission geometry), one interferes with the plume dynamics (plume collimation), and another facilitates droplet desolvation (heat-assisted LAESI). In the third segment of this chapter we explore how these altered modalities perform compared to the conventional LAESI source, and what additional information they provide on the mechanistic steps.

In the final segment of the chapter we describe the different approaches pursued in molecular imaging applications. We demonstrate the advantages of MS imaging (MSI) with LAESI under ambient conditions. Combining lateral imaging and depth profiling, enables three-dimensional imaging on rectangular grids. We show the utility of ion-mobility separation (IMS) in reducing chemical interferences in MSI of complex samples. To retain more of the biological information in MSI of tissues, cell coordinates are extracted from microscope images for a cell-by-cell imaging strategy.

14.2 Sampling and Ionization Mechanisms

14.2.1 Midinfrared Laser Ablation of Biomedical Samples

LAESI measurements capitalize on atmospheric-pressure (AP) ablation of the specimen of interest. The experimental setup for a traditional LAESI system is depicted in Figure 14.1(A). Instructions on assembling the setup and developing measurement methods have been provided elsewhere.[10,15] Sampling is initiated by focusing a midinfrared (mid-IR) laser beam of

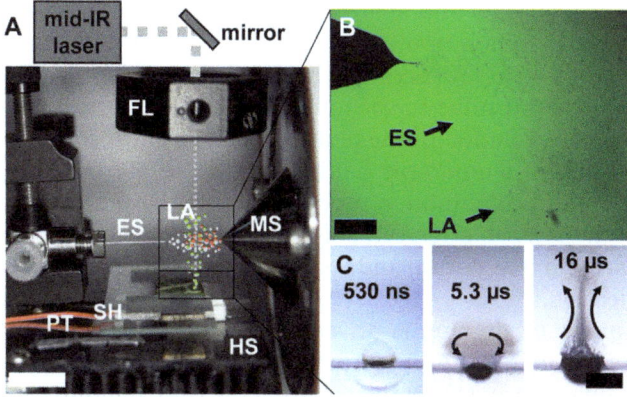

Figure 14.1 Schematic of laser-ablation electrospray ionization (LAESI) and relevant processes. (A) Resonant excitation of water by a midinfrared laser beam (dashed lines) ablates the sample. The generated plume (LA, *green dots*) is intercepted by charged droplets from an electrospray (ES, *grey dots*). Ions produced from charged droplets seeded by sample molecules (*red dots*) are analyzed by a mass spectrometer (MS). The sample holder (SH) is equipped with Peltier stage (PT) for cooling and a heat sink (HS). Scale bar = 25 mm. (B) Fast imaging captures interception of LA particles by ES plume. Scale bar = 100 μm. (C) Steps of LA include a hemispherical plume produced by phase explosion (left panel), collapse of the primary ablation plume (shown at 5.3 μs in the middle), and ejection of particulate matter (shown at 16 μs in the right panel). Scale bar = 1 mm.
(Parts of this figure were adapted from ref. 4 and 30 with permission.)

2.94 μm wavelength and 1–100 ns duration to the surface of the sample.[4] This wavelength is strategically selected because it allows for efficient coupling of the light energy into the specimen material through the adjacent absorption maximum related to the symmetric stretching mode of OH in water that is natively present or added to the sample. As the temperature rises water exhibits nonlinear absorption behavior and becomes increasingly transparent.[29] At sufficiently high fluences, usually at 0.1–5 J cm^{-2} for tissues of high water content, this sudden deposition of energy causes ablation of the sample (Figures 14.1(B) and (C)) on a microscopic scale.

The mechanistic details of mid-IR ablation of water and water-containing biological tissues can be discussed based on results from fast-imaging experiments.[30] Ablation of water occurs through three partially overlapping phases at AP (Figure 14.1(C)). During the first ~100 ns, fast surface evaporation takes place that creates hemispherical vapor plume that exhibits relatively slow expansion.[31] When the temperature in the irradiated surface layer reaches ~0.9T_c, where T_c is the critical temperature of water, phase explosion occurs that generates a rapidly expanding dense plume of mostly neutral nanosized droplets and particles.[32] This primary ablation plume also contains a small portion of ions, with a yield inferred to be below ~0.01%,[33] that are directly collected and analyzed in another technique, AP IR matrix-assisted laser desorption ionization (MALDI).[34,35] The AP environment imparts important consequences on the dynamics of the plume created by the phase explosion. Unlike in vacuum, where plume expansion continues indefinitely, at AP evolution of the plume is slowed down by the presence of the background gas, temporarily halting it ~1 mm above the sample surface. Due to rapid recombination processes at AP, the primary ions directly generated by the laser ablation can only be collected in close proximity of the sample surface (up to ~2 mm) and require a high voltage for extraction and efficient collection.[36]

The next phase of ablation is driven by the recoil pressure of the phase explosion exerted on the sample surface. Shock-wave velocity measurements in the liquid phase suggest that the resulting pressure buildup can reach 500 to 900 MPa that relaxes through the expulsion of sample material in the next ~500 μs (Figure 14.1(C)).[30] The momentum carries the droplets and particles of this secondary ablation plume to several millimeters above the sample surface (Figure 14.1(C)). Systematic studies employing fast-imaging experiments, and chemical analysis of the secondary ablation plume suggested that ejected particles of ~5 μm average diameter travel as far as 45 mm above the sample surface before being stopped by the drag force.[36,37] Although this process casts sample material to large distances, the secondary ablation plume lacks the ions necessary for MS analysis.

14.2.2 Electrosprays for Ionization in LAESI

In LAESI, material in the secondary ablation plume is intercepted by an electrospray and ionized to enable mass spectrometric determination of its chemical composition. Fine charged droplets (less than ~5 μm in diameter) are generated by supplying a suitable spray solution (*e.g.*, acidified or basified 50% methanol) through an electrified capillary (*e.g.*, metal or fused silica) (Figure 14.1). Operating parameters of the electrospray such as solution composition, supply rate, and voltage are judiciously chosen to produce a stable spray, preferably in the pulsating or cone-jet regime[28] (Figure 14.1(B)) to continuously supply small droplets for efficient LAESI MS.[4] Incorrectly chosen operating parameters result in spraying modes with low ion yields (dripping and burst regimes)[38,39] or low stability (a stable regime).[40] Nonaxial spraying regimes (rim emission) are detrimental to the LAESI signal as the density of the droplets close to the spray axis is low and the droplet velocities point away from the mass-spectrometer inlet. The pulsating regime provides relatively high ion yields with current oscillations in the 1 to 7 kHz range.[38,41] These pulsation frequencies provide a sufficiently high duty cycle, so the temporal overlap between the laser-ablation plume and the electrospray is acceptable. Although the ion yield with the pulsating regime does not reach the values seen for the cone jet regime, it is substantially easier to establish and maintain without spray-current monitoring.[38]

14.2.3 Merging Ablation Plume and Electrospray

Control experiments confirmed that normal operating conditions in LAESI do not generate ions by corona discharge at the emitter, *i.e.* ionization by an electrospray is pivotal. The electrospray plume, with its axis typically 15 to 30 mm above the sample surface, intercepts the secondary laser-ablation plume, thus promoting coalescence between the ablated particulate matter and the electrospray droplets, seeding the latter with molecules of the sample.

Phase Doppler anemometry measurements indicate that methanol–water mixtures electrosprayed in the pulsating regime, produce droplets of 2 to 10 µm in diameter with 0.5 to 5 m s^{-1} velocities gradually slowing as they move away from the emitter.[28,42,43] Fast-imaging experiments of water ablation by an Er:YAG laser at 2.94 µm wavelength and 5.4 J cm^{-2} fluence indicate that the initial velocity of the ejected material at the sample surface, v_0, is 150 m s^{-1}.[30] As the ablation plume rises, the particle velocity diminishes due to the drag force in the ambient environment. Assuming that Stokes law holds for the drag force, the stopping distance for these particles can be expressed as:

$$x_{stop} = 2\rho R^2 v_0 / 9\mu \qquad (14.1)$$

where ρ is the density of the sample, R is the radius of the particle, and μ is the dynamic viscosity of air.[37] Substituting a 15 mm and 30 mm stopping distances, 150 m s^{-1} initial velocity, and room-temperature material properties for the environment, particle diameters of 5.7 µm and 8.1 µm are obtained. This means that the laser-ablated particles, which slow down close to stopping at the axis of the electrospray, have diameters similar to those of the charged droplets. Indeed, slow ablated particles are desired as in typical LAESI configurations they travel perpendicular to the trajectories of the charged droplets. By reducing the particle velocities close to zero, they can be entrained by the velocity field induced by the electrospray droplets and move together with them.

Several models exist to describe the scavenging of microscopic aerosol particles by small charged raindrops[44,45] but their validity is limited to situations when the droplet diameter is at least ten times larger than that of the particle. Under these circumstances the trajectory of the droplet is not affected by a collision with the particle. Collisions are also facilitated by the attractive polarization forces induced by the charged droplets.[46] Another consequence of the large size difference is that every collision leads to particle capture by the droplet.

As the particle and droplet sizes in LAESI are similar to each other, the collision efficiencies and coalescence efficiencies can be significantly different. The governing parameter in these collisions is the Weber number, a relative measure of inertia and surface tension, We $= \rho_1 (v_L - v_S)^2 D_S / \sigma_1$, where D_S is the diameter of the smaller colliding partner, v_L and v_S are the velocities of the larger and the smaller colliding partners, and ρ_1 and σ_1 are the density and the surface tension of the liquid, respectively. Depending on the value of the Weber number, the collision events follow different scenarios.[47] For We < 2, there is no coalescence, as the collision results in either bouncing or grazing. For 2 < We < 5, the outcome of the collision depends on the impact parameter. For small impact parameters, coalescence occurs but for large impact parameter values after transient fusing separation takes place. In the 5 < We <7 range, the coalescence efficiency is close to one, whereas at higher We values, 7 < We, it gradually declines to zero. Thus, as long as the relative velocity of the particle and the droplet are in a range that results in 5 < We <7, the coalescence is very efficient.

The resulting seeded droplets are entrained by fluid dynamic and electrostatic forces into the mass spectrometer through the orifice of a sampling cone located ∼10 mm away from the electrospray emitter tip (Figures 14.1 and 14.2). Throughout this process, ions are produced by the solvated ion evaporation[48] or the charge residue mechanism.[49] The chemical composition of generated gas-phase ions is determined by the mass spectrometer.

14.2.4   Ion Internal Energies and Conformations

Based on accumulated evidence, the LAESI ionization is efficient and nondestructive to most biomolecules. The ion yield has been found to be ∼10–100-fold higher than in typical AP IR-MALDI.[36] This is because ∼99.99% of the ablation plume is in the form of neutrals that are efficiently ionized by the electrospray. Early observations indicated that LAESI was capable of producing intact ions from molecules as large as human serum albumin with a molecular mass of 66 556 Da.[4] The presence of multiply charged ions in LAESI mass spectra of proteins and peptides supported the notion of mechanistic similarity with electrospray ionization.

To gauge the internal energy of ions generated by LAESI, survival yields of benzyl-substituted benzylpyridinium thermometer ions were measured using solution and tissue samples.[37] The results showed that the ions generated by LAESI and electrospray had indistinguishable internal energies. Furthermore, LAESI mass spectra of labile biomolecules, such as peptides and vitamin B_{12}, showed no signs of fragmentation. In contrast, MALDI mass spectra of the latter from 2,5-dihydroxybenzoic acid (DHB) matrix exhibited two major fragment ions. Similarly to LAESI, low internal energies are observed in other ambient ionization sources utilizing electrosprays, including desorption electrospray ionization (DESI), electrosonic spray ionization, and paper spray ionization.[50,51] A qualitative explanation of the observation that the mid-IR laser ablation in LAESI does not increase the internal energy of the ions relies on a mechanistic argument. As the ionized neutrals are mostly produced by the recoil pressure-induced material ejection, the produced particulates are dislodged by a pressure pulse and are not heated significantly by the laser. Therefore, the ions generated from the ejected particulates only reflect the internal energy contributed by the electrospray process.

Another property associated with the internal energy of ions is their conformations. Higher internal energy makes more extreme conformations accessible for the ion. To assess the conformation of LAESI-generated ions, mobility measurements were performed in a traveling-wave IMS system on protein standards, cell pellets and tissue samples.[52,53] Using multiply charged myoglobin ions for calibration,[54] the collision cross sections of lysozyme and ubiquitin ions are found to be within 4% of the values established by electrospray ionization.[55] Such close similarity in collision cross section is an indication of analogous conformations of ions produced by LAESI and electrospray.

14.3   LAESI Geometries and Configurations

Because ionization is decoupled both spatially and temporally from ablation sampling, LAESI has been afforded various configurations with many having spawned specific applications. In the original reflection arrangement (Figure 14.2), the focused laser beam impinging on the sample surface and the ensuing ablation plume sampled by the electrospray are on the same side of the sample (front side). In the less-explored inverted geometry (Figure 14.3), the back side of the sample is illuminated and the ablation plume evolving on the front side is intercepted by the electrospray (transmission geometry). Although these configurations share analogy with those developed for MALDI,[56] the relative positions of various components (*e.g.*, the electrospray source and inlet orifice of the mass spectrometer) enable additional variants in LAESI. The selected configurations influence the interception efficiency between the laser-ablation plume and electrospray, thus affecting the ion signal intensities.

Both of these basic LAESI geometries give rise to additional variants based on the relative configuration of specific components. Examples of the reflection geometry include the conventional, heat-assisted, and fiber-optic arrangements, as presented in Figure 14.2. Likewise, the transmission geometry has found applications in conventional and fiber-optic (plume collimation) arrangements (Figure 14.3). The different configurations support specific application niches. A brief summary of developmental milestones and applications is reviewed in Table 14.1. Below, the fundamental considerations and leading performance metrics are described for the various arrangements.

14.3.1   Conventional LAESI in Reflection Geometry

The original or *conventional* LAESI system developed in 2007 demonstrated broad utility in the direct analysis of biological fluids, tissues and cells. The experimental setup, depicted in Figure 14.1, utilized reflection geometry (see left panel of Figure 14.2(A)). A laser beam of 2.94 µm wavelength was generated by an Er:YAG laser with less than 100-ns pulse length and steered using gold-coated mirrors through a calcium fluoride (CaF_2) lens to yield

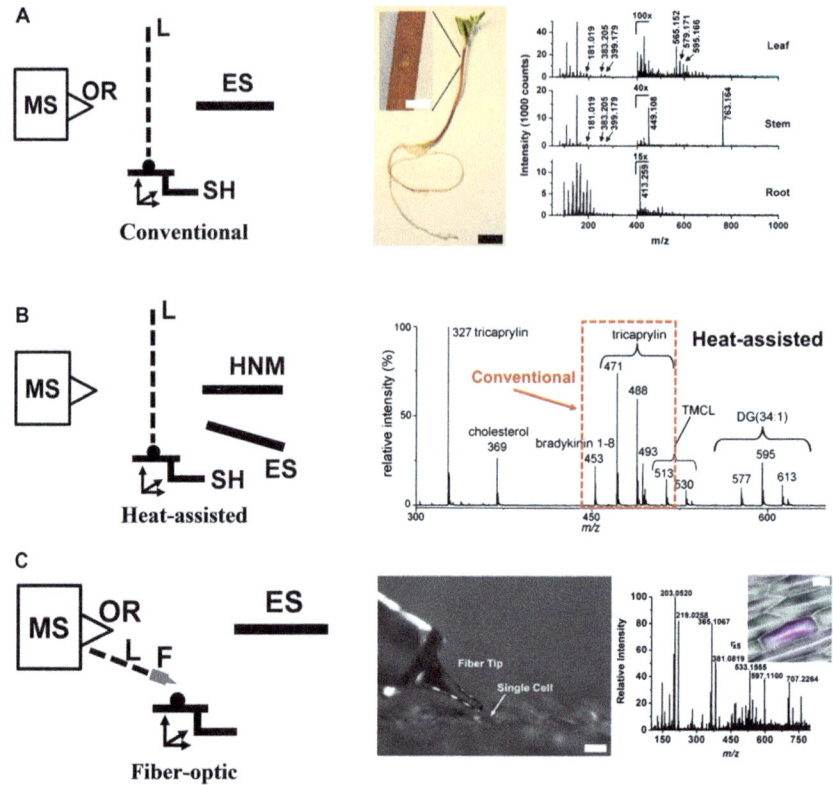

Figure 14.2 Reflection geometry LAESI in representative configurations. (A) Conventionally, the laser beam (L) is focused using lenses or mirrors to enable *in situ* and *in vivo* sampling. Root, stem, and leaf organs of a live plant were profiled with ~250 μm resolution. Scale bars = 10 mm (black), 1 mm (white). (B) In heat-assisted LAESI, adding a gas jet heated by a microchip (HNM) extends the detectable compound types to less-polar molecules. The ions detected by conventional LAESI are in the red rectangle. (C) In fiber-LAESI, a sharpened optical fiber (F) produced significantly smaller ablation areas. By holding the fiber tip 30 μm above the specimen surface, single epidermal cells were ablated with 50 μm resolution. Scale bar = 100 μm. Key: mass spectrometer (MS), MS sampling cone orifice (OR), sample holder (SH) and electrospray (ES).
(Parts of this figure were adapted from ref. 4, 18 and 57 with permission.)

circular ablation craters of ~350–400 μm diameter. As shown in the right panel of Figure 14.2(A), these ablation dimensions proved sufficiently small to profile various organs of a French marigold seedling (*Tagetes patula*) without compromising its viability. The leaf, stem and root were measured *in vivo*, revealing organ-specific chemical composition. Apart from small analysis areas, rapid operation was equally important to enable deciphering the undisturbed metabolic composition of the specimen. The time requirement of the analysis was less than ~1 s/sample, enabling the study of metabolic changes that occur on the time scale of seconds or slower. As a consequence, conventional LAESI can be used to take a snapshot of the chemical composition of biological systems or follow temporal changes on a timescale of seconds and longer.

Analysis by LAESI is chemically specific and quantitative. Detected ions range from singly charged small compounds (drugs, metabolites, lipids, *etc.*) to multiply charged species (midsized proteins) (see Table 14.1). The high-mass capabilities were demonstrated for 66.5 kDa human serum albumin detected with a charge state distribution between ~39 and ~52 protons in the mass spectra. The LAESI MS signal was quantitative for solutions of drug standards for over 4 orders of magnitude in concentration. For conventional LAESI, the lower limit of detection was 8 fmol.[4] These figures of merit have recently been extended by plume collimation and a high-performance mass spectrometer to 6 orders of magnitude dynamic range and 600 amol limit of detection.[22]

In the years following the original design, technological advances were pivotal to achieving improved analytical performance in conventional LAESI. Short laser pulses (5–10 ns full width at half-maximum) based on a Nd:YAG laser-driven optical parametric oscillator (OPO), high repetition rates (up to 100 Hz), and laser pulse energy stability (< 7% variation) promoted improved signal-to-noise ratios and analytical reproducibility. A planoconvex ZnSe lens or a reflective microscope objective (50105-02, Newport, Irvine, CA) ensured improved focusing of the incident mid-IR beam, yielding circular sampling spots with diameters reduced to ~100 μm.

As smaller ablation spots, required for improved local analysis, contain less material for detection, the efficiency of ionization has to be improved. A systematic study found that LAESI ion yield increased with a decreasing flow rate of the electrospray solution (see Figure 9.3(b) in Reference 10). For example, a change from 300 nL min^{-1} to ~50 nL min^{-1} resulted in a ~10-fold improvement in signal intensity.

In turn, higher ion yields facilitated measurements on various animal tissues, making, *e.g.*, the detection of 200–300 endogenous metabolites possible in rat brain tissue sections.[16] To avoid the evaporative loss of water and chemical degradation in these samples, a Peltier cooling element was incorporated into the sample holder that kept the sample frozen during analyses (Figure 14.1). Likewise, control of the ambient temperature and humidity *via* an enclosure prevented condensation of water on the cooled sample surfaces (*e.g.*, frozen brain tissue section). As a result, LAESI MS has evolved to achieve nontargeted analysis of multiple compounds in diverse samples types (see Table 14.1).

14.3.2 Heat-assisted LAESI

The conventional setup has recently been modified to enhance the ionization of less polar species. Maintaining the reflection geometry, the electrospray plume was intercepted at a 20–30° angle by a jet of nitrogen gas (180–360 mL min^{-1}) heated to ~170–220 °C using a microchip.[57] Schematics of this heat-assisted setup are shown in the left panel of Figure 14.2(B). A systematic study found the ionization efficiency to increase with gas temperature and flow rate for compounds of low polarity (*e.g.*, estrone), whereas ionization of polar species (*e.g.*, bradykinin 1-8) appeared independent of these experimental variables. Heat-assisted LAESI successfully ionized a number of apolar specimens that were not detected using conventional LAESI, including polyaromatic hydrocarbons (*e.g.*, perylene and naphtho[2,3-α]pyrene) and neutral lipids (cholecalciferol, cholesterol, DG(34:1), and tricaprylin) (see right panel of Figure 14.2(B)). This LAESI modality was also quantitative, spanning over a 3 orders of magnitude concentration range.

Complementary performance by heat-assisted LAESI proved particularly useful for the interrogation of complex biological samples. Measurements on avocado (*Persea americana*) mesocarp, mouse brain, and pansy (*Viola*) petals yielded mass spectra enhanced for compounds of low polarity. Polar lipids such as phospholipids and monogalactosyl diglycerides gave generally strong response with conventional LAESI, whereas detection of apolar lipids such as triglycerides and cholesterol benefited from the heated gas stream.

14.3.3   Microsampling using Fiber Optics

Further development of LAESI to enable microsampling of specimens required changes in laser pulse delivery. Due to the limitations of coherent mid-IR sources, primarily the large beam divergence (*e.g.*, 5 to 10 mrad for OPOs) and spherical aberrations associated with conventional methods of focusing based on lenses and mirrors, the produced spot diameters are many times larger (~100 to 200 μm) than the diffraction-limited value (~1.5 μm). To alleviate these difficulties, the sharpened end of an optical fiber can be used to deliver laser light to much smaller spots. In this arrangement, similar to scanning near-field optical microscopy, the illuminated area is determined by the sharpness of the tip, not by the diffraction limit defined by the wavelength of the light. Although the spot sizes used in fiber-based LAESI (~30 μm) are still significantly above the diffraction limit, they are much smaller than in the case of conventional focusing. Schematics of the setup are shown in Figure 14.2(C).

A typical setup used a germanium oxide-based (GeO$_2$) fiber (outlet) whose end was etched to create a tip with a radius of curvature, $R = $~15 μm, following protocols discussed elsewhere.[20] The tip was positioned ~2R distance at ~45° incidence angle from the specimen surface. Mid-IR light of appropriate energy (*e.g.*, <600 μJ) and repetition rate (*e.g.*, 100 Hz) was coupled through the inlet. As a result, the material was ablated from under the tip across a circular area of ~2.5R average diameter, usually within 1 s or 100 laser pulses.[18] Ablation craters on thin plan tissues usually measured 30–40 μm in diameter, affording an opportunity to access the chemical composition of area- or volume-limited samples such as single cells under ambient conditions.

Ablation of single cells followed a remarkable mechanism. Unlike free solutions ablated in the conventional setup (on a sample holder such as a glass slide), plant cells have rigid walls that act as a natural boundaries, confining the ablation in space. Time-lapse imaging and study of the LAESI signal revealed that at least 2 consecutive laser pulses delivered 10–100 ms apart were necessary to rupture the cell wall, thus providing access to the intracellular content.[19] Ejection or evaporation of the cytosol completed the process, beyond which no mass spectrometric signal was detected.

Interrogations requiring single-cell resolution have particularly benefited from fiber-optic LAESI. Representative examples include metabolic analysis of onion (*Allium cepa*) and daffodil (*Narcissus pseudonarcissus*) single plant cells as well as individual eggs of sea urchin (*Lytechinus pictus*) with ∼25–50 µm probe diameters. More than 300 different ions were registered below *m/z* 800 with acceptable signal-to-noise ratios. Of the detected ions, 35 were assigned to primary (*e.g.*, oligosaccharides) and secondary metabolites (*e.g.*, glycosidic flavonoids).[18] Furthermore, analysis at single-cell resolution permitted profiling cells of different pigmentation as well as age, and finding differences in metabolic composition (*e.g.*, arginine *vs.* alliin concentrations). Likewise, the small analysis area made chemical differentiation of oil gland and neighboring cells possible in leaves of bitter orange (*Citrus aurantium*).[17]

When combined with advanced protocols in sample preparation, *e.g.*, microdissection, fiber optics in LAESI has also opened the door to subcellular investigations. In a recent example, microsurgical needles of approximately 1 µm tip diameter helped to cut away the cell wall, exposing the cytoplasm and organelles of single cells.[21] Because the sharpened tip of the optical fiber was dimensionally comparable to the nucleus, this LAESI modality was able to attain the cell compartment-specific chemistries *in situ*. Nuclear and cytoplasmic small metabolites were differentiated with a multivariate data interpretation tool, orthogonal projections to latent structures discriminant analysis (OPLS-DA). The results showed higher abundance for the hexose and alliin in the cytoplasm and arginine and glutamine in the nucleus. By extension, increasingly sophisticated microdissection can help to extend subcellular analysis to progressively smaller cells and subcellular organelles, or other volume-limited samples.

14.3.4 LAESI in Transmission Geometry

In transmission geometry, a thin sample, *e.g.*, tissue section, is illuminated by the laser beam from the back side, and the ablation plume develops on the front side. The advantage of this approach is that the electrospray that intercepts the ablation plume and the focusing lens are on different sides of the sample (see the left panel in Figure 14.3(A)). Thus, high-performance, short working distance, lenses can be used to achieve ablation with

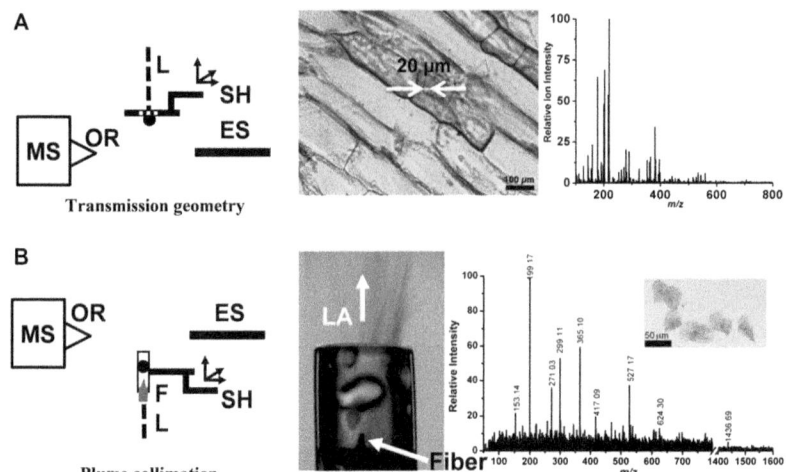

Figure 14.3 Transmission geometry LAESI with representative configurations. The laser beam (L) was focused on the back side of the sample in the transmission geometry, allowing (A) to ablate thin specimens (*e.g.*, dried residues, tissue sections, and single cells). (B) Confinement of the ablation plume (LA) in a capillary enhances the interaction with the electrospray for more efficient ionization. The sharpened optical fiber (F) is visible in the capillary. Primary metabolites were measured for 6 cheek cells (inset). Key: mass spectrometer (MS), MS sampling cone orifice (OR), sample holder (SH), and electrospray (ES). (Parts of this figure were adapted from ref. 22 with permission.)

significantly reduced spot sizes. For example, using a short working distance aspherical lens (BD-2, Thorlabs, Newton, NJ), 20 μm ablation spot and LAESI signal were achieved on a monolayer of *A. cepa* epidermal cells (see the right panel in Figure 14.3(A)).[58] This arrangement also enables the *in situ* monitoring of the sample during ablation by a long distance microscope from the front side. In a proof-of-principle application, transmission geometry LAESI was used to compare the flavonoids in the tulip leaf and bulb epidermis.[59]

14.3.5   Plume Collimation

In this variant of the transmission geometry arrangement, the sample was ablated under spatial confinement. The specimen was loaded into a fused silica capillary, and a sharpened optical fiber was inserted coaxially through the back end of the capillary to achieve ablation.[22] A systematic study of this configuration using a high-performance mass spectrometer found that the analytical figures of merit were significantly better than those reported for the conventional LAESI. The 600 amol limit of detection and the 6 orders of magnitude dynamic range for quantitation achieved by plume collimation compared favorably to the 8 fmol and 4 orders of magnitude values established for conventional LAESI, respectively.[4]

These improvements were ascribed to fundamental differences in the dynamics of the ablation plume as well as a higher ion collection and transfer efficiency associated with the high-resolution mass spectrometer. Unlike during the laser pulse exposure of a free-standing specimen, such as a droplet or liquid film, ablation of a sample in an enclosure results in the confinement of the generated plume. Results inferred from optical imaging of the ablation plume corroborate this point. In conventional LAESI, significant radial expansion of the plume is observed (see Figure 14.1(B)), whereas in the case of plume collimation by confinement in a capillary the divergence of the ablation plume is significantly reduced (see right panel in Figure 14.3(B)). Thus, a greater portion of the ejected material reaches the electrospray plume resulting in a stronger signal.

The lower limit of detection and wider dynamic range with plume collimation extended ambient MS to smaller amounts and higher complexity of samples. For example, less than 100 rat insulinoma β-cells (RIN5mF) with an average diameter of 10 μm were successfully analyzed, yielding significant ion intensities for 35 different ions (primary metabolites and lipids). The studied cell populations were progressively reduced for larger cell types: mass spectra of 20 megakaryoblast cells (50 μm diameter), six cheek cells (30–50 μm diameter) (see right panel in Figure 14.3(B)), and a single sea urchin egg (~ 100 μm diameter) were measured using plume collimation.

14.4 Spatial Profiling and Imaging Applications

14.4.1 Lateral Molecular Imaging

Local analysis in conventional LAESI enabled MS imaging (MSI) under ambient conditions. In this modality, the chemical composition of the tissue is determined pixel by pixel according to the workflow shown in Figure 14.4(A). The tissue section is mounted on a sample holder in the focal plane of the mid-IR laser beam (Figure 14.1) and translated in two dimensions (X and Y) using a computer-controlled two-axis translation stage, while the location (X–Y coordinates of the pixel) and the corresponding LAESI mass spectra are coregistered. A distribution of a chemical species in the interrogated area can be reconstructed into a false-color image by selecting the corresponding ion and plotting its signal intensities against pixel locations.

Successful LAESI MSI requires the consideration of various experimental factors. Imaging in two dimensions calls for pixel-by-pixel ablation over the domains of interest and the entire depth of the tissue section. Reducing the imaging time requirement is important to improve the throughput, and minimize the data file size to facilitate subsequent data analysis. The time allocated for two-dimensional (2D) MSI, $t_{\text{MSI,2D}}$, is given by:

$$t_{\text{MSI,2D}} = \frac{d_x \, d_y}{R_x \, R_y} \left(t_{\text{dwell}} + t_{\text{pos}} \right) \quad (14.2)$$

where a rectangular area of the sample with d_x and d_y dimensions is interrogated at R_x and R_y step sizes in the X and Y dimensions, respectively, t_{dwell} is the time of analysis at each pixel (dwell time), and t_{pos} is the average time required to translate between consecutive pixels/rows. The dwell time is judiciously determined prior to an imaging experiment by considering specimen thickness, repetition rate of the laser, and the material removal rate by the focused beam, with the latter mainly dependent on laser fluence and the tensile strength of the sample. For soft tissues of plants and animals of 50–500 μm thickness, dwell times of 1–5 s/pixel were successful under ~ 1 J cm^{-2} radiant exposes.[13,15,16] At 10 Hz laser pulse repetition rate and 1 s spectrum integration time, the mass spectrometric signal was found to decay in less than 3 s, permitting to complete imaging of a 12.5×10.5 mm^2 sample at 200 μm resolution in ~ 3 h.

On plant and animal tissue sections, conventional 2D LAESI MSI provided molecular distributions for various polar compounds, including primary and secondary metabolites, lipids, and proteins. Imaging of frozen rat brain sections (Figure 14.4(B)) by conventional LAESI coupled to a high mass-resolution mass spectrometer deciphered different distribution patterns for the isobars γ-aminobutyric acid (GABA) and choline (Figure 14.4(B)). To help discover the colocalization of selected ions, Pearson crosscorrelation analysis was implemented.[16] In one example, the cholesterol and PC(O-33:3) and/or PE(O-36:3) plasmalogens were observed to colocalize in the corpus callosum of the brain. Unlike tissues of animal origin, plant tissue surfaces are directly amenable to imaging because their waxy cuticle acts as a natural barrier to dehydration rendering freezing of the specimen unnecessary during MSI. For example, the variegated leaves of the zebra plant contained high amounts of methoxykaempferol glucoronide in the yellow sectors, whereas the kaempferol-(diacetylcoumaryl-rhamnoside) was distributed evenly across the yellow and green areas (Figure 14.4(B)).[13] These examples illustrate that 2D LAESI MSI can determine the distribution of diverse compound types in tissues without requiring special treatment.

More recently, the heat-assisted LAESI configuration has also been implemented in MSI. The imaging protocol was identical to conventional LAESI. Molecular images determined for polar anthocyanins showed good agreement with results obtained by the conventional approach (Figure 14.4(D)), with the added potential to extend MSI to apolar compounds.[57] The performance metrics for conventional and heat-assisted LAESI imaging are comparable to other ambient MSI techniques, such as DESI, AP MALDI, surface desorption AP chemical ionization, and probe electrospray ionization.[1,3]

14.4.2 Depth Profiling

Tissue ablation by consecutive laser pulses at the same location results in the deepening of the crater and serves as the basis of molecular depth profiling. Scanning electron microscopy (SEM) and Z-stack imaging by optical microscopy indicated that a single laser pulse of ~ 2 J cm^{-2} fluence impinging on the upper surface (adaxial) of a leaf ablated the waxy cuticle and portions of the epidermal and palisade cell layers.[13] The SEM image of the ablation mark is shown in Figure 14.5(A) (inset). The cylindrical voxel assigned to the crater measured ~ 350 µm in diameter and ~ 50 µm in height corresponding to ~ 2 nL in volume.[13] In control experiments consecutive laser pulses of 0.2 J cm^{-2} in fluence were delivered to a peace lily leaf whose lower (abaxial) surface had been marked with rhodamine 6G dye. By reducing the pulse-to-pulse variation of the laser energy to less than 7% (*e.g.*, by attenuation *via* a neutral density filter or a Brewster-angle attenuator), a ~ 50-µm average depth resolution was achieved for routine profiling.[13]

LAESI MS analysis provided depth-resolved metabolite profiles in plant leaves. The first layer, the top ~ 50 µm, exhibited abundant signal for hexose and disaccharide ions as well as kaempferol or cyanidin glucoside rhamnoside.[13] The latter results agreed with the known accumulation of glycosides in the epidermal layer of leaves to protect the underlying tissue from ultraviolet radiation. In a separate study, the mesophyll layer, between ~ 50 and 100 µm from the top surface, revealed various ions derived from chlorophyll *a*.[14] Chlorophyll is known to localize in the chloroplasts concentrated in chlorenchyma cells of the mesophyll layer, supporting the findings from LAESI depth profiling.

14.4.3 Imaging in Three Dimensions

The *in situ* combination of 2D LAESI MSI with depth profiling enabled three-dimensional (3D) reconstruction of molecular distributions. In this modality, the depth profile is determined voxel by voxel (Z coordinate) for each X–Y position until a preselected area is rastered. The alternative approach to *in situ* 3D MSI by LAESI would be to perform multiple 2D MSI experiments using only one laser pulse for a given layer. This method, however, is inferior due to positioning errors in returning to a desired pixel for consecutive imaging layers as well as potential drying/contamination of the tissue layer that is exposed to the environment. The imaging resolution is dependent on the same factors that underpin 2D imaging and depth profiling in LAESI: tight focusing of the laser beam (*e.g.*, using reflective microscope objectives), appropriate laser energy with high pulse-to-pulse stability, and accurate positioning elements are needed for a reliable performance. The duration of a 3D MSI experiment, $t_{\mathrm{MSI,3D}}$, can be estimated based on:

$$t_{\mathrm{MSI,3D}} = \frac{d_x\, d_y\, d_z}{R_x R_y R_z} \left(t_{\mathrm{dwell}} + t_{\mathrm{pos}} \right) \qquad (14.3)$$

where a volume of d_z thickness is depth-profiled with R_z resolution, and other variables are the same as in eqn (14.2). Similar to lateral imaging, 3D MSI by LAESI requires that ions generated during the ablation of

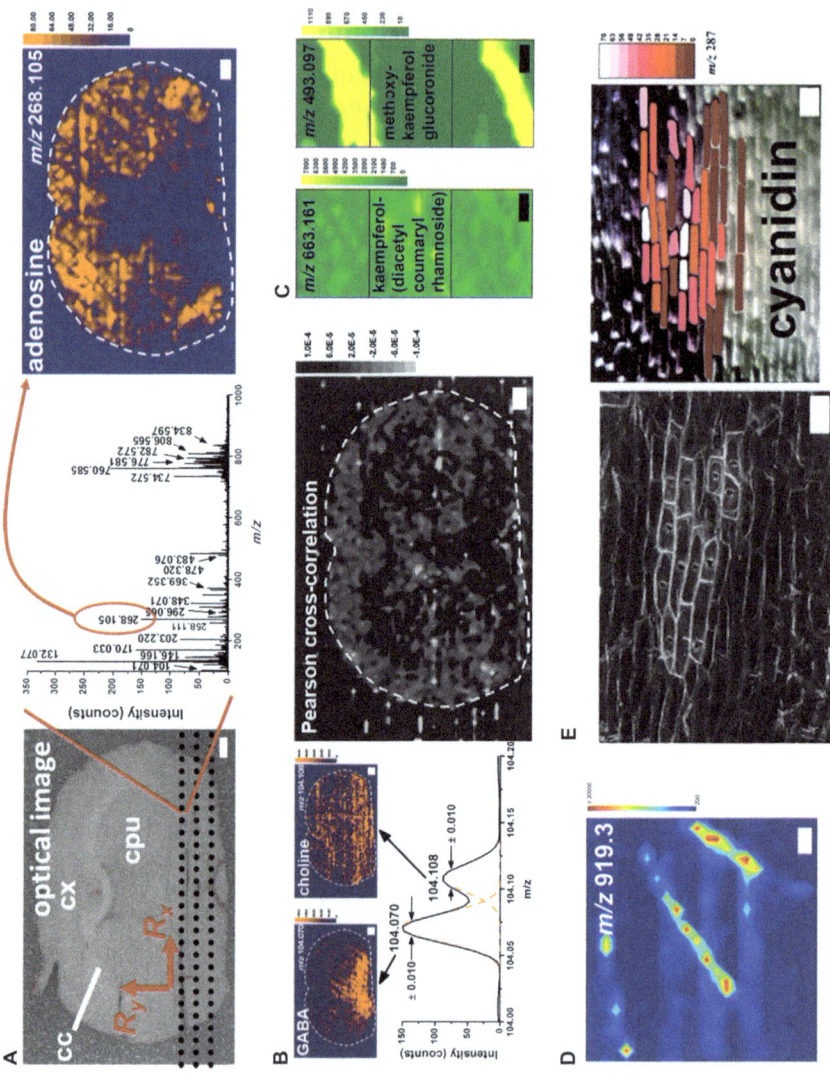

individual voxels be distinguished upon detection. For a mass spectrometer operating at 1 s spectrum integration time, a 0.2 Hz laser repetition rate offers a compromise between lateral/depth resolution and the total time required for imaging, $t_{MSI,3D}$, permitting to explore a $12.5 \times 10.5 \times 0.45$ mm^3 volume with ~ 350 μm lateral ($R_x = R_y$) and ~ 40 μm depth resolutions in ~ 5 h.[14]

3D LAESI MSI has uncovered complex distributions for endogenous metabolites in plant tissue.[14] For example, the molecular ion of cyanidin/kaempferol rhamnoside glucoside was detected only in the top 40 μm of peace lily leaf (Figure 14.5(B)). Flavonoids such as kaempferol glycosides often populate in the upper epidermal layers, presumably to protect the underlying tissue from UV-A and UV-B radiation.[60] Particularly diverse distributions were noted in the variegated leaf of the zebra plant. Acacetin accumulated in the yellow sectors of the leaf in the second and third layers but was present homogeneously in others (Figure 14.5(C)). Kaempferol-(diacetyl coumarylrhamnoside) exhibited elevated levels with homogeneous distributions in the third and fourth layers (Figure 14.5(D)).

Alternative 3D volume reconstruction methods for *in situ* MSI rely on computational assembly of 2D images obtained on serial tissue sections. The procedure originally applied for vacuum MALDI MS images[61] has recently been extended to DESI imaging in the ambient environment.[62] Cryogenic microtomes provide high depth resolution (typically 0.5 to 100 μm) for 3D volume reconstruction, making the technique reproducible, albeit demanding on user experience and analytical throughput. In contrast, 3D LAESI MSI, based on depth profiling, eliminates the need for multiple sectioning of the tissue but the depth resolution, *e.g.*, ~ 30 μm, is limited

---

Figure 14.4 Molecular imaging in 2D by LAESI. (A) Imaging workflow: a LAESI mass spectrum is registered for each pixel of a selected area (left panel, rat brain section shown). The false-color image of an ion (*e.g.*, *m/z* 268.105 for adenosine, right panel) is constructed from the spectra (middle panel) by plotting the ion signal intensity against the pixel coordinates. (B) High mass-resolution imaging by conventional LAESI differentiates the distributions of isobars γ-aminobutyric acid (GABA) and choline in rat brain (left panel). Pearson crosscorrelation analysis shows the colocalization of cholesterol, and PC(O-33:3) and/or PE(O-36:3) plasmalogens (right panel). (C) In the variegated leaves of the zebra plant, the kaempferol-(diacetyl coumaryl-rhamnoside) distributes homogeneously in the sample (left panel), whereas methoxykaempferol glucuronide accumulates in the yellow sectors (right panel). (D) Heat-assisted LAESI is used to image the *m/z* 919.3 ion, putatively identified as delphinidin-3-p-coumaroylrhamnosyl-glucoside-5-glucoside, in pansy (*Viola*) petals. (E) Single-cell analysis and cell-by-cell imaging in a heterogeneous cell population by fiber-LAESI (left panel) shows the accumulation of cyanidin in the pigmented cells (right panel). Key: cc, corpus callosum; cx, cerebral cortex; cpu, caudate putamen. For A to D the scale bars $= 1$ mm, and for E the scale bar $= 200$ μm.
(Parts of this figure were adapted from ref. 1, 13, 16, 17, 19 and 57 with permissions.)

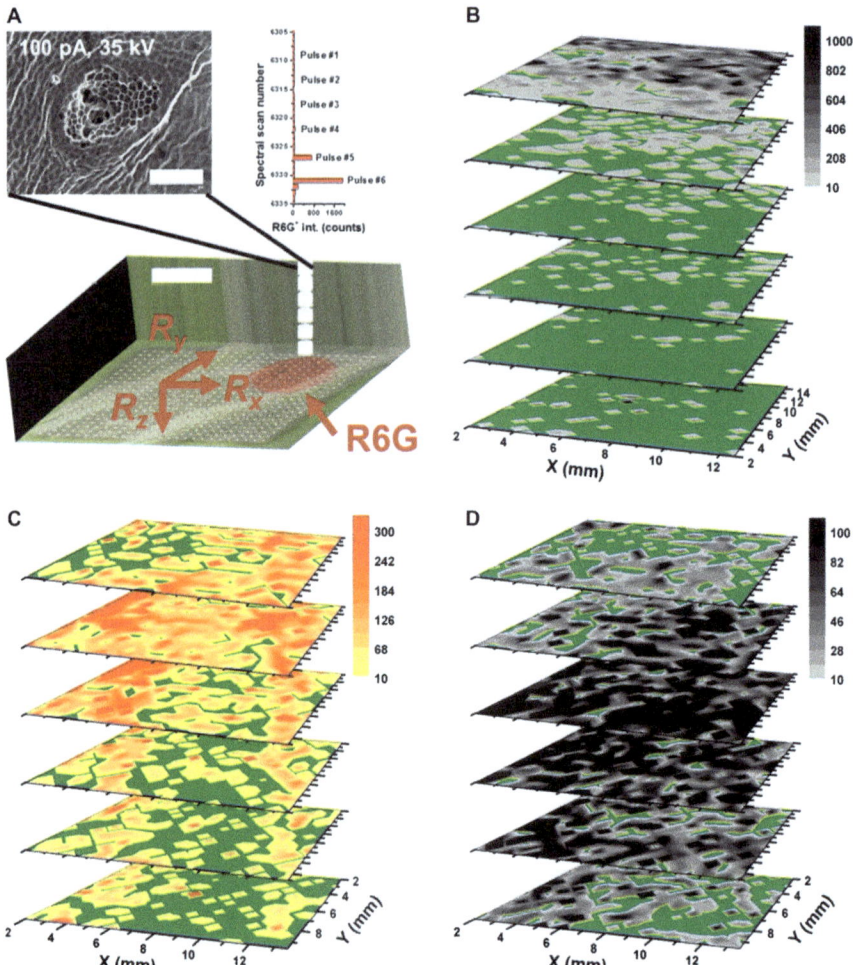

Figure 14.5 Depth profiling and ambient molecular imaging in three dimensions by LAESI. (A) Ablation of a peace lily leaf by consecutive single laser pulses enables depth profiling by LAESI with ~50 μm depth resolution. (B) A combination of depth profiling and 2D MSI results in *in situ* 3D imaging. The epidermal and palisade mesophyll layers of peace lily leaves (*S. lynise*) indicated high abundances of cyanidin/kaempferol rhamnoside glucoside. (C) In the variegated leaves of the zebra plant (*A. squarrosa*), acacetin accumulates in the yellow sectors in the second and third layers, whereas it shows homogeneous distributions in the others. (D) The mesophyll layers are rich in kaempferol-(diacetyl coumarylrhamnoside).
(Parts of this figure were adapted from ref. 1 and 14 with permissions.)

by the tissue removal depth achieved by a single laser pulse. Further improvement of this technique requires enhanced depth-resolution capabilities.

14.5   Cell-by-cell Imaging

The common approach to MSI is to lay a rectangular grid over the tissue surface and collect mass spectra in the grid points. The concern with this method is that the focal spot of the laser in a grid point, *i.e.* a pixel in the image, might overlap with several cells resulting in spectra that reflect the composition of potentially different cells. To overcome this difficulty we introduced the concept of cell-by-cell imaging, where the sampling points are allocated to the centroids of individual cells, thereby assigning the related spectral information as the natural pixels of the image. The obvious advantage of this approach is the greater correlation between cellular composition and pixels in the image.

The tighter focusing and smaller ablation dimensions produced by fiber optics opened the door to imaging with single-cell resolution. As shown in Figure 14.4(E), epidermal cells of *A. cepa* were ablated one cell at a time using a manually positioned, sharpened optical fiber.[19] Multivariate analysis of LAESI mass spectra collected for 36 neighboring cells indicated characteristic differences among the adjacent cells. Cyanidin was detected in significantly higher signal intensity in the pigmented cells, whereas sucrose was uniformly distributed in the cell population.[17]

Manual positioning of the cells for ablation renders this technique slow and cumbersome. To overcome these deficiencies, an automated version has recently been introduced based on object recognition in microscope images of the tissue.[63] Through image-processing algorithms, individual cells were recognized and the coordinates of the corresponding centroids were calculated. This dataset was used to drive the motorized translation stage and bring the cells one-by-one under the tip of the optical fiber for laser ablation. This approach enabled the automated collection of LAESI spectra for cell-by-cell imaging or the selective analysis of a particular cell type based on its morphology, *e.g.*, guard cells in a leaf epidermis. As in the resulting molecular images chemical information from adjacent cells is not mingled, this method can be used to explore cell-to-cell transport and signaling processes.

14.6   Conclusions

The development of LEASI instrumentation and methodology, described in this chapter, has reached the point when the focus is shifting to the ever-growing array of applications. Based on the demonstrated ability of this technique to rapidly identify a wide array of metabolites and lipids in tissues and cells, numerous applications are expected in nontargeted metabolomics and lipidomics. The method is also ripe for comparative studies and biomarker discovery in health, disease and treatment. LAESI has the demonstrated ability to follow biochemical transformations in cells and tissues induced by viral (*e.g.*, Kaposi's sarcoma-associated herpesvirus and human T-lymphotropic virus type 1 and type 3) and bacterial infection. At this point, the coverage for proteins is insufficient for large-scale proteomics, but selected proteins with elevated local concentrations can be discovered and compared by LAESI MS. Examples include the determination of the subunit composition of phycobilisomal antenna proteins in cyanobacteria, and the discovery of gender-specific protein composition in the gill glands of certain fish.

In conventional LAESI MS experiments, typically ~ 300 different ions are detected. This number can be indicative of the metabolite and lipid coverage of the LAESI technique. However, significant improvement is demonstrated, with over 1100 different ions detected, when IMS is performed before MS. The introduction of IMS helps to differentiate isobaric species, structural isomers, and different conformations of the same molecule. Further improvement in molecular coverage is expected by combining IMS with reactive LAESI, a technique based on introducing reactants in the electrospray that can enhance the ionization and/or the fragmentation of otherwise non-responsive chemical species.

An important attribute of LAESI MS is its high-throughput capability. Typical measurements of unprocessed samples take on the order of a few seconds. The elimination of sample preparation and lengthy separation steps for metabolomics combined with the reduced analysis time can radically accelerate measurements in this developing field. This is further supported by the recently introduced commercialized version of the LAESI system (DP-1000, Protea Biosciences, Morgantown, WV). With automated sample handling, this instrument offers dramatically improved throughput for metabolomics, reaching hundreds of samples per hour. Commercial availability of the LAESI source will have an increasing impact on the number and variety of applications for this emerging technique.

Acknowledgements

Support is gratefully acknowledged from the U.S. Department of Energy, Office of Basic Energy Sciences, Chemical Sciences, Geosciences, and Biosciences Division under Award DE-FG02-01ER15129 and from the U.S. National Science Foundation under Grant no. CHE-1152302 to AV, and from the US Medical Countermeasure Initiative under Award MCM2J277MW to PN.

References

1. P. Nemes and A. Vertes, *TrAC, Trends Anal. Chem.*, 2012, **34**, 22–34.
2. M. E. Monge, G. A. Harris, P. Dwivedi and F. M. Fernandez, *Chem. Rev.*, 2013, **113**, 2269–2308.
3. C. Wu, A. L. Dill, L. S. Eberlin, G. Cooks and D. R. Ifa, *Mass Spectrom. Rev.*, 2013, **32**, 218–243.
4. P. Nemes and A. Vertes, *Anal. Chem.*, 2007, **79**, 8098–8106.
5. Y. H. Rezenom, J. Dong and K. K. Murray, *Analyst*, 2008, **133**, 226–232.
6. J. S. Sampson, K. K. Murray and D. C. Muddiman, *J. Am. Soc. Mass. Spectrom.*, 2009, **20**, 667–673.

7. I. X. Peng, R. R. O. Loo, E. Margalith, M. W. Little and J. A. Loo, *Analyst*, 2010, **135**, 767–772.

8. G. A. Harris, S. Graf, R. Knochenmuss and F. M. Fernandez, *Analyst*, 2012, **137**, 3039–3044.

9. G. Robichaud, J. A. Barry, K. P. Garrard and D. C. Muddiman, *J. Am. Soc. Mass. Spectrom.*, 2013, **24**, 92–100.

10. P. Nemes and A. Vertes, in *Mass Spectrometry Imaging: Principles and Protocols*, ed. S. S. Rubakhin and J. V. Sweedler, Springer, Berlin, 2010, pp. 159–171.

11. A. Vertes, B. Shrestha and P. Nemes, in *Methodologies for Metabolomics: Experimental Strategies and Techniques*, ed. N. W. Lutz, J. V. Sweedler and R. A. Wevers, Cambridge University Press, Cambridge, 2013, pp. 140–158.

12. B. Shrestha and A. Vertes, in *Plant Metabolism*, ed. G. Sriram, Springer, Berlin, 2014, pp. 31–39.

13. P. Nemes, A. A. Barton, Y. Li and A. Vertes, *Anal. Chem.*, 2008, **80**, 4575–4582.

14. P. Nemes, A. A. Barton and A. Vertes, *Anal. Chem.*, 2009, **81**, 6668–6675.

15. P. Nemes and A. Vertes, *J. Visualized Exp.*, 2010, **43**, e2097.

16. P. Nemes, A. S. Woods and A. Vertes, *Anal. Chem.*, 2010, **82**, 982–988.

17. B. Shrestha, J. M. Patt and A. Vertes, *Anal. Chem.*, 2011, **83**, 2947–2955.

18. B. Shrestha and A. Vertes, *Anal. Chem.*, 2009, **81**, 8265–8271.

19. B. Shrestha, P. Nemes and A. Vertes, *Appl. Phys. A*, 2010, **101**, 121–126.

20. B. Shrestha and A. Vertes, *J. Visualized Exp.*, 2010, **43**, e2144.

21. J. A. Stolee, B. Shrestha, G. Mengistu and A. Vertes, *Angew. Chem., Int. Ed.*, 2012, **51**, 10386–10389.

22. J. A. Stolee and A. Vertes, *Anal. Chem.*, 2013, **85**, 3592–3598.

23. P. Sripadi, J. Nazarian, Y. Hathout, E. P. Hoffman and A. Vertes, *Metabolomics*, 2009, **5**, 263–276.

24. B. Shrestha, P. Nemes, J. Nazarian, Y. Hathout, E. P. Hoffman and A. Vertes, *Analyst*, 2010, **135**, 751–758.

25. P. Sripadi, B. Shrestha, R. L. Easley, L. Carpio, K. Kehn-Hall, S. Chevalier, R. Mahieux, F. Kashanchi and A. Vertes, *PLoS One*, 2010, **5**, e12590.

26. G. Parsiegla, B. Shrestha, F. Carriere and A. Vertes, *Anal. Chem.*, 2012, **84**, 34–38.

27. B. Shrestha, P. Sripadi, C. M. Walsh, T. T. Razunguzwa, M. J. Powell, K. Kehn-Hall, F. Kashanchi and A. Vertes, *Chem. Commun.*, 2012, **48**, 3700–3702.

28. P. Nemes, I. Marginean and A. Vertes, *Anal. Chem.*, 2007, **79**, 3105–3116.

29. R. K. Shori, A. A. Walston, O. M. Stafsudd, D. Fried and J. T. Walsh, *IEEE J. Sel. Top. Quantum Electron.*, 2001, **7**, 959–970.

30. I. Apitz and A. Vogel, *Appl. Phys. A*, 2005, **81**, 329–338.

31. Z. Chen, A. Bogaerts and A. Vertes, *Appl. Phys. Lett.*, 2006, **89**, 041503.

32. Z. Y. Chen and A. Vertes, *Phys. Rev. E*, 2008, 77, 036316.

33. C. D. Mowry and M. V. Johnston, *Rapid Commun. Mass Spectrom.*, 1993, **7**, 569–575.
34. P. V. Tan, N. I. Taranenko, V. V. Laiko, M. A. Yakshin, C. R. Prasad and V. M. Doroshenko, *J. Mass Spectrom.*, 2004, **39**, 913–921.
35. Y. Li, B. Shrestha and A. Vertes, *Anal. Chem.*, 2007, **79**, 523–532.
36. A. Vertes, P. Nemes, B. Shrestha, A. A. Barton, Z. Y. Chen and Y. Li, *Appl. Phys. A*, 2008, **93**, 885–891.
37. P. Nemes, H. H. Huang and A. Vertes, *Phys. Chem. Chem. Phys.*, 2012, **14**, 2501–2507.
38. L. Parvin, M. C. Galicia, J. M. Gauntt, L. M. Carney, A. B. Nguyen, F. Park, L. Heffernan and A. Vertes, *Anal. Chem.*, 2005, 77, 3908–3915.
39. I. Marginean, P. Nemes and A. Vertes, *Phys. Rev. Lett.*, 2006, **97**, 064502.
40. I. Marginean, P. Nemes and A. Vertes, *Phys. Rev. E*, 2007, **76**, 026320.
41. I. Marginean, L. Parvin, L. Heffernan and A. Vertes, *Anal. Chem.*, 2004, **76**, 4202–4207.
42. Z. Olumee, J. H. Callahan and A. Vertes, *J. Phys. Chem. A*, 1998, **102**, 9154–9160.
43. Z. Olumee, J. H. Callahan and A. Vertes, *Anal. Chem.*, 1999, **71**, 4111–4113.
44. S. N. Grover and K. V. Beard, *J. Atmos. Sci.*, 1975, **32**, 2156–2165.
45. K. V. Beard and S. N. Grover, *J. Atmos. Sci.*, 1974, **31**, 543–550.
46. B. A. Tinsley, R. P. Rohrbaugh, M. Hei and K. V. Beard, *J. Atmos. Sci.*, 2000, **57**, 2118–2134.
47. N. Ashgriz and P. Givi, *Int. Commun. Heat Mass Transfer*, 1989, **16**, 11–20.
48. V. Znamenskiy, I. Marginean and A. Vertes, *J. Phys. Chem. A*, 2003, **107**, 7406–7412.
49. J. F. de la Mora, *Anal. Chim. Acta*, 2000, **406**, 93–104.
50. M. Nefliu, J. N. Smith, A. Venter and R. G. Cooks, *J. Am. Soc. Mass. Spectrom.*, 2008, **19**, 420–427.
51. H. Wang, J. J. Liu, R. G. Cooks and Z. Ouyang, *Angew. Chem., Int. Ed.*, 2010, **49**, 877–880.
52. B. Shrestha and A. Vertes, Proceedings of the 61st ASMS Conference on Mass Spectrometry and Allied Topics, Minneapolis, MN, 2013.
53. B. Shrestha and A. Vertes, *Anal. Chem.*, 2014, **86**, 4308–4315.
54. B. T. Ruotolo, J. L. P. Benesch, A. M. Sandercock, S.-J. Hyung and C. V. Robinson, *Nat. Protoc.*, 2008, **3**, 1139–1152.
55. Clemmer Group Cross Section Database, http://www.indiana.edu/~clemmer/Research/Cross%20Section%20Database/cs_database.php, Last accessed: 12/12/2013.
56. K. Dreisewerd, *Chem. Rev.*, 2003, **103**, 395–425.
57. A. Vaikkinen, B. Shrestha, J. Nazarian, R. Kostiainen, A. Vertes and T. J. Kauppila, *Anal. Chem.*, 2013, **85**, 177–184.
58. B. Shrestha and A. Vertes, Proceedings of the 60th ASMS Conference on Mass Spectrometry and Allied Topics, Vancouver, Canada, 2012.

59. R. S. Jacobson, B. Shrestha and A. Vertes, Proceedings of the 61st ASMS Conference on Mass Spectrometry and Allied Topics, Minneapolis, MN, 2013.

60. G. Agati, C. Brunetti, M. Di Ferdinando, F. Ferrini, S. Pollastri and M. Tattini, *Plant Physiol. Biochem.*, 2013, **72**, 35–45.

61. M. Andersson, M. R. Groseclose, A. Y. Deutch and R. M. Caprioli, *Nat. Methods*, 2008, **5**, 101–108.

62. L. S. Eberlin, D. R. Ifa, C. Wu and R. G. Cooks, *Angew. Chem., Int. Ed.*, 2010, **49**, 873–876.

63. H. Li, B. K. Smith, B. Shrestha, L. Márk and A. Vertes, in *Mass Spectrometry Imaging of Small Molecules*, ed. L. He, Springer, Berlin, 2014, DOI: 10.1016/j.ijms.2014.06.025.

CHAPTER 15

Electrospray Laser Desorption Ionization Mass Spectrometry

MIN-ZONG HUANG, SIOU-SIAN JHANG, YA-TING CHAN,
SY-CHI CHENG, CHUN-NIAN CHENG AND JENTAIE SHIEA*

Department of Chemistry, National Sun Yat-Sen University, Kaohsiung,
Taiwan
*Email: jetea@mail.nsysu.edu.tw

15.1   Introduction

The development of ambient ionization mass spectrometry (AMS) arose
from the need for rapid, real-time, and sensitive analyses of various substances under ambient conditions with little to no sample pretreatment.[1–7]
With a wide selection of sampling, desorption, and ionization methods, AMS
is able to characterize analytes regardless of polarity or volatility. Existing
AMS techniques are differentiated based on their desorption, ionization, or
sampling processes. For most AMS techniques, analytes are desorbed
through laser desorption, thermal desorption, or impact by ions or charged
droplets, and are predominantly ionized *via* electrospray ionization (ESI) or
atmospheric-pressure chemical ionization (APCI). Because the sampling
processes of most AMS techniques are essentially nondestructive, they are
ideal tools for investigating organic surface compositions of various
samples. A number of techniques have been developed to address such requirements. Examples of these techniques include desorption electrospray
ionization (DESI),[8] direct analysis in real time (DART),[9] and electrospray
laser desorption ionization (ELDI).[10] These techniques use a variety of de-
sorption/ionization processes or sampling/ionization processes (*e.g.* sample

New Developments in Mass Spectrometry No. 2
Ambient Ionization Mass Spectrometry
Edited by Marek Domin and Robert Cody
© The Royal Society of Chemistry 2015
Published by the Royal Society of Chemistry, www.rsc.org

bombardment by charged droplets and metastable atoms, thermal desorption, postionization, *etc.*) and have been applied to areas such as food safety screening, characterization of pharmaceuticals and drugs of abuse, monitoring of environmental pollutants, antiterrorism and forensic detection of explosives, characterization of protein and metabolic biomarkers, molecular imaging analysis, and chemical and biochemical reaction monitoring.[11–24]

Among the AMS techniques, electrospray laser desorption ionization (ELDI) has several unique analytical features. Analytes are desorbed through irradiation by a pulsed laser without the assistance of organic matrices, unlike matrix-assisted laser desorption ionization (MALDI), which requires such matrices for analysis. Desorbed molecules subsequently enter an electrospray plume and are postionized through reactions with charged solvent species (species produced from a methanol-based electrospray solution). Since the desorption energy from the pulsed laser is high, multiply charged ions can be formed through these ESI processes, which makes ELDI/MS not only useful for detecting small biological compounds, but large ones as well. This chapter introduces the ELDI technique and details its development, underlying principles of operation, ionization processes, unique features, related techniques and representative applications.

15.2   Development of ELDI

The principle of ELDI is based on fused-droplet electrospray ionization (FD-ESI, or two-step electrospray ionization) developed in 1998.[25] In an FD-ESI source, gaseous analytes or neutral droplets containing analytes are generated from a pyrolyzer, an ultrasonic nebulizer, or a pneumatic nebulizer, and are conducted to the tip of an electrospray capillary where they are postionized through fusion or reactions with charged solvent species present in the electrospray plume.[26–28] Interferences from small organic and inorganic salts (*e.g.*, NaCl, NH$_4$Cl, Tris, and SDS) in the samples were eliminated by interacting charged methanol species (generated by electrospraying a pure methanol solution) with protein droplets;[29] since these salts have much lower solubilities in methanol than in water, they are separated from the methanol droplets during the drop fusing processes, so that protein signals were still obtained from a saturated NaCl solution.[28] Another example of a two-step electrospray ionization technique includes the analysis of fatty acid mixtures, which was achieved through GC-ESI/MS (gas chromatography-electrospray ionization mass spectrometry), in which gaseous molecules exiting the gas chromatograph were directed into an ESI plume for ionization.[25,30] A third example involves the ionization and characterization of highly reactive ketenes and polar pyrolysates of the synthetic polymers, which were generated in a flow pyrolyzer.[31]

An advantage of using FD-ESI combined with chromatography and pyrolysis for chemical analyses is that polar components in the gaseous pyrolysates can be selectively ionized. However, information regarding the spatial distribution of the compounds on the sample is lost. This is because the sample is heated rapidly at high temperatures during pyrolysis; therefore, the detected ion signals are an average of data from the sample. In addition, the technique cannot be used for protein analysis because proteins decompose at high temperatures during pyrolysis. Solving these problems requires an energy source that is capable of providing spatial resolution and rapid energy input, *i.e.* one that will generate gaseous intact protein molecules from a defined area on the biological sample's surface. An added bonus of developing such a technique is its potential ability to perform direct protein profiling and molecular imaging of tissue.

Laser desorption (LD) allows sampling over a small and defined sample area and guarantees high spatial resolution for solid samples. However, only relatively small biological and chemical compounds (*e.g.*, those with masses smaller than 3000 Da) can be detected using LD/MS.[32–34] Failure to detect larger molecules such as proteins after laser desorption may be due to their low ionization efficiencies, decomposition induced by the laser energy, or rapid neutralization during the desorption/ionization processes. However, it has been reported that a larger number of neutral molecules are produced during a laser desorption event.[35–38] Consequently, several different approaches have been applied to postionize these neutral organic species to increase detection sensitivies. The use of a number of different ionization techniques for postionization of laser-desorbed analytes, such as electron ionization (EI), chemical ionization (CI), corona discharge atmospheric-pressure chemical ionization (APCI), Ni-63 APCI, inductively coupled plasma (ICP), and vacuum-UV photoionization (PI), have been explored in the past.[35,36,38–45] However, in the absence of a matrix these techniques are still limited to the study of analytes with relatively low molecular weights where attempts to analyze proteins have all failed. Since ESI has been successfully used to postionize the droplet proteins in FD-ESI, the same approach was then applied to ionize the protein desorbed by laser desorption. The technique that combines laser irradiation for analyte desorption and electrospray for postionization is known as electrospray laser desorption ionization (ELDI).

15.3   Concept of ELDI

In ELDI, a pulsed laser is used to desorb analytes from liquid or solid samples, after which the desorbed analytes subsequently enter an ESI plume for ionization (Figure 15.1(a)). The main ionization mechanisms appear to be electrospray-based processes, since the formation of multiply charged ion species from proteins have been observed to produce ESI-like mass spectra. The principle of ELDI was first proved in its detection of bovine cytochrome c.[10] The protein standard was applied on the sample plate located a few millimeters below an ESI capillary (which continuously generated charged methanol species) and the MS inlet. Laser desorption alone did not yield signals for the cytochrome c ion. However, when laser

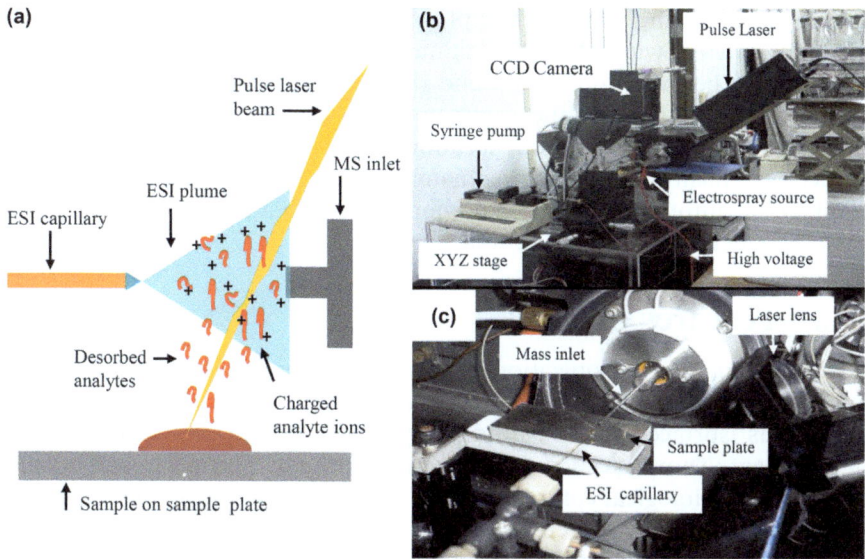

Figure 15.1 (a) Schematic representation of the desorption and ionization of molecules from the sample surface. (b) Photograph of the ELDI experimental setup. (c) Close-up of the ELDI source showing the relative positions of the ESI capillary, MS inlet and sample. Note that the figure is not to scale.

desorption and ESI were simultaneously operated, a typical ESI mass spectrum of cytochrome c was obtained, which showed extensive multiply charged protein ion signals. The experimental results indicated that even without the assistance of an organic matrix, intact protein molecules were still produced by irradiating the protein sample with a pulsed laser, regardless of the solid or liquid state of the sample.

15.4 ELDI Instrumental Setup

The ELDI experimental set-up is shown in Figure 15.1(b). The typical operating parameters were as such: an ESI emitter continuously sprayed an acidic methanol solution (50% MeOH with 0.1% acidic acid) through a fused silica capillary (100 μm i.d.) at a flow rate of 150 μL h^{-1}. A nebulizing gas, commonly used in conventional ESI, was not used during ELDI. The ESI plume was directed toward an ion-sampling orifice parallel to the sample plate (Figure 15.1(c)). The electrospray needle, the sample plate, and the sampling tube voltages were maintained at +4.5, 0, and −0.5 kV, respectively. A 266 nm pulsed Nd:YAG laser operating at a frequency of 10 Hz, a pulse energy of approximately 250 μJ, a pulse duration of 4 ns, and a spot size of approximately 200 μm was used. The sample plate was positioned on an *XYZ* stage in front of a sampling capillary of a mass analyzer. The strongest ion signals were obtained at an incident laser angle of 45 degrees.

The geometry of the source was optimized to achieve efficient mixing of ablated analytes with the ESI plume for maximum signal strength. The distance between the ESI tip and the MS inlet tube was set at 8 mm, while the distance between the electrospray capillary and sample surface was set at 3 mm and the optimum location of the laser spot on the sample surface was positioned at approximately 1 mm below the tip of the ESI capillary. An automated triaxial precision stage with a travel range of 10 cm was used when scanning the sample surface.

15.5 Features of ELDI

ELDI/MS is an ambient ionization technique that combines laser desorption and electrospray ionization. It can rapidly characterize chemical compounds on various surfaces under ambient conditions with minimal sample pretreatment. Samples such as proteins, synthetic polymers, and small chemical and biological compounds (including phosphatidylcholines, dyes, and drugs) are readily detected.[10,20–24,46–48]

ELDI has several unique features that allow it to perform its analysis: (1) chemicals on soft or hard solid surfaces are directly desorbed and ionized, avoiding many of the inconveniences from using a vacuum required for the conventional operation of a MALDI or LD system; (2) since the operation of the technique is under ambient conditions, sample switching is simple and fast, which allows for high-throughput sample analysis; (3) organic and inorganic matrices are unnecessary in ELDI analysis, simplifying sample-preparation procedures and avoiding interference from matrix ions in the lower m/z regions; (4) the energy produced by pulsed laser irradiation onto the sample surface is very high, which allows for the desorption of chemical compounds on hard surfaces; (5) different types of chemical compounds can be selectively detected by simply varying the polarity of the electrospray solution; (6) ELDI desorption and ionization processes are performed at atmospheric pressure and are thus compatible with most mass analyzers such as ion traps and quadrupoles; and (7) due to the extremely high spatial resolution of the laser beam for desorption, ELDI/MS can perform chemical profiling and imaging of biological tissues.

15.6 Techniques Related to ELDI

In addition to ELDI, several hybrid ambient ionization techniques that combine ESI with laser desorption or ablation have been reported. The main differences between these techniques are the presence or absence of an organic matrix and the type of laser used for desorption. The technique known as matrix-assisted laser desorption electrospray ionization (MALDESI) is similar to the MALDI technique in that an organic matrix is used to prepare the sample prior to laser desorption, with the desorbed analytes postionized in the ESI plume. High-resolution mass spectra were obtained by combining MALDESI with Fourier transform ion cyclotron resonance mass spectrometry (FT-ICR-MS) for the analysis of biological samples. Applications of MALDESI-MS in several biochemical studies have been reported, including "top-down" proteomics analyses and the direct characterization of intact proteins, polypeptides, carbohydrates, and other biomolecules on tissue samples.[49,50]

Laser-ablation electrospray ionization (LAESI) employs a pulsed IR laser (2.94 μm) for ablation and an electrospray for ionization. The water-rich samples are particularly useful for LAESI-MS analysis as they may absorb more IR than UV laser energy. LAESI has been applied to characterize proteins, lipids, and metabolites in biological fluids (*e.g.*, urine, blood, and serum), tissue sections, and cells for *in vivo* spatial molecular profiling and imaging analysis.[51–53] Infrared laser-assisted desorption electrospray ionization (IR-LADESI) that employs an IR laser (10.6 μm) for sample desorption, has been used to characterize pharmaceutical and biological compounds in various biological fluids.[54] Laser electrospray mass spectrometry (LEMS or fs-ELDI) uses a nonresonant, femtosecond-duration (70 fs) laser pulse for sample desorption, and has been applied to rapidly characterize explosives and several pharmaceutical compounds on various surfaces including glass, fabric, steel, and wood.[55–57] Laser desorption spray postionization mass spectrometry (LDSPI-MS) has been used for the mass spectrometric analyses of yogurt and dairy-related products.[58]

Laser-induced acoustic desorption electrospray ionization mass spectrometry (LIAD-ESI-MS) is an ambient ionization technique that couples LIAD with ESI mass spectrometry, and has been developed to characterize analytes in solid or liquid samples.[59,60] LIAD-ESI differs from ELDI (and its related techniques described above) in that it uses a high-power laser pulse (*e.g.*, 10 mJ) to irradiate the rear side of a thin metal foil (10–20 μm). Shockwaves and heat are then induced, resulting in desorption of the analyte from the sample applied on the other side of the foil, after which the desorbed analyte species enter an ESI plume for further ionization. Again, both small organic and large biological compounds, including amino acids, peptides, and proteins (as large as albumin at 66 000 kDa) can be ionized and detected using LIAD-ESI-MS.[59] this technique has also been applied to characterize small organic compounds separated on a TLC plate.[60] Because the sample is not directly irradiated by the laser beam, the technique is useful for the analyses of light-sensitive compounds.

15.7 Applications of ELDI

15.7.1 Characterizing Peptides and Proteins for Biological Sample Analysis

ELDI/MS is one of a few ambient mass spectrometric techniques that can be used to detect large biological compounds. It is also the first time that mass spectra of intact proteins in the sample are obtained using laser desorption/ ESI without the addition of a matrix. Figure 15.2 shows the mass spectrum

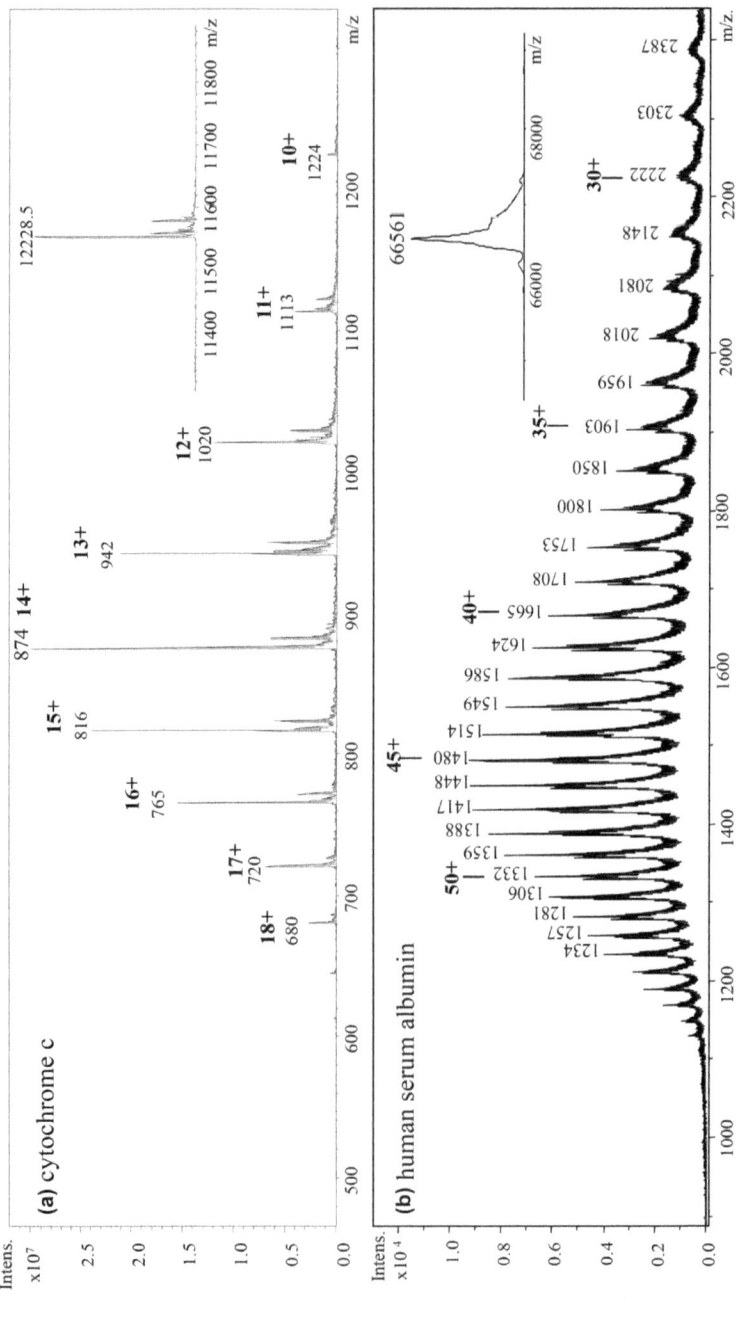

Figure 15.2 ELDI mass spectra of samples of (a) dry cytochrome c and (b) wet human serum albumin proteins. Deconvoluted mass spectra are shown in the insert.

obtained from dry cytochrome c (10^{-4} M) and wet human serum albumin (10^{-3} M) proteins using ELDI/MS without a matrix. The results demonstrate that the ELDI technique is amenable to proteins up to 66 kDa without any fragmentation. ELDI has also been applied to detect hemoglobin α- and β-chains in dried whole blood, lysozymes in dry human tears and saliva, lipids in brain tissue, and proteins in porcine heart and liver tissues.[20,21] Native protein ions such as those of myoglobin, cytochrome c, and hemoglobin can also be obtained using ELDI/MS.[47] The molecular weights of intact proteins can be measured more accurately for top-down proteomics analyses, and it is possible to sequence peptides and proteins directly from biological media by using ELDI coupled with a high-resolution mass spectrometer.

15.7.2 Detecting Organic Chemical Compounds and Chemical Reaction Monitoring

In addition to analysis of large proteins, ELDI/MS can be used to detect chemical compounds dissolved in either polar or nonpolar solvents (such solvents include water, methanol, methylene chloride, tetrahydrofuran, toluene, and hexane).[22] Fine carbon powders were suspended in the sample solution to absorb the laser energy so that the energy would be subsequently transferred to the analyte for desorption. ELDI/MS has been applied to monitor the progress of ongoing chemical reactions such as the epoxidation of chalcone in ethanol solution, the chelation of ethylenediaminetetraacetic acid (EDTA) with copper and nickel ions in aqueous solutions, the chelation of 1,10-phenanthroline with iron (II) in a methanol solution, and the tryptic digestion of cytochrome c in aqueous solutions.[22] The so-called reactive-ELDI was used to perform rapid biochemical reactions during these processes with the addition of reactive agents in the ESI solution to induce the aforementioned reactions.[61] The results for online disulfide bond reduction using dithiothreitol on oxidized glutathione and insulin show the effectiveness of reactive-ELDI in exploring chemical and enzymatic reactions, in which the technique is able to analyze.

15.7.3 Forensic Applications

It is generally known that irradiation of a sample surface by a focused pulsed laser produces very high energy on the sample surface. This makes it possible to directly characterize chemical compounds on hard sample surfaces using ELDI/MS, which is particularly useful for trace amounts of drugs, explosives, firearm residues, and chemical and biological evidence. ELDI/MS can also be used to distinguish authentic from counterfeit compounds through differences in chemical composition between the two. For example, different ink compounds on regular and thermal papers, paintings, and compact discs have been rapidly distinguished using ELDI/MS without any sample pretreatment.[21,48] In addition, ELDI/MS has been used to

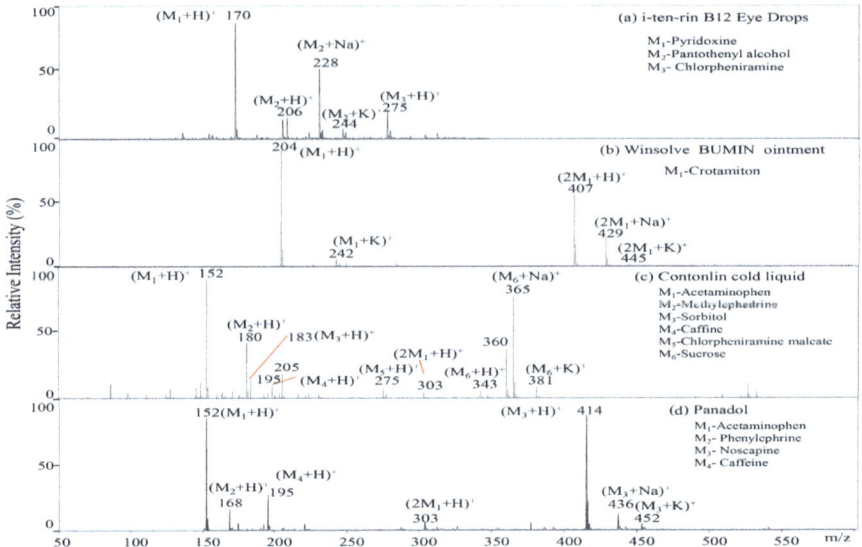

Figure 15.3 ELDI mass spectra of pharmaceutical drugs. (a) *i-ten-rin* B12 eye drops, (b) Winsolve BUMIN ointment, (c) Contonlin cold liquid, and (d) Panadol.

characterize the active ingredients in a wide variety of over-the-counter drugs formulated as liquids (eye drops), ointments, syrups and tablets (Figure 15.3). Since the laser used to desorb the molecules from the sample surface is focused over a very small area, the desorbed area is always invisible to the naked eye; as a result, a small amount of sample is needed for analysis. Given the rapidity of ELDI/MS analysis and the minimal sample volume and preparation requirements, analysis of forensic evidence can be performed in a short time after collection with negligible damage to samples.

15.7.4 Molecular Profiling and Chemical Imaging

Due to the high spatial resolution of the laser beam for desorption, ELDI/MS is an efficient molecular imaging technique that is performed under ambient conditions.[24] The spatial resolution of laser irradiation in ELDI can be as low as several micrometers. Unlike MALDI, a sample can be imaged by ELDI/MS without adding a matrix. This is particularly important when imaging biological tissues, since analysis of such tissues under a vacuum in conventional MALDI requires further time-consuming pretreatment steps. Furthermore, ELDI/MS can be used to characterize analytes on hard surfaces like dry plants and fungi since the high energy produced by pulsed laser irradiation. Since the operation is performed under ambient conditions, both volatile and nonvolatile compounds can be simultaneously characterized.[24] Figure 15.4 shows the results of directly profiling an

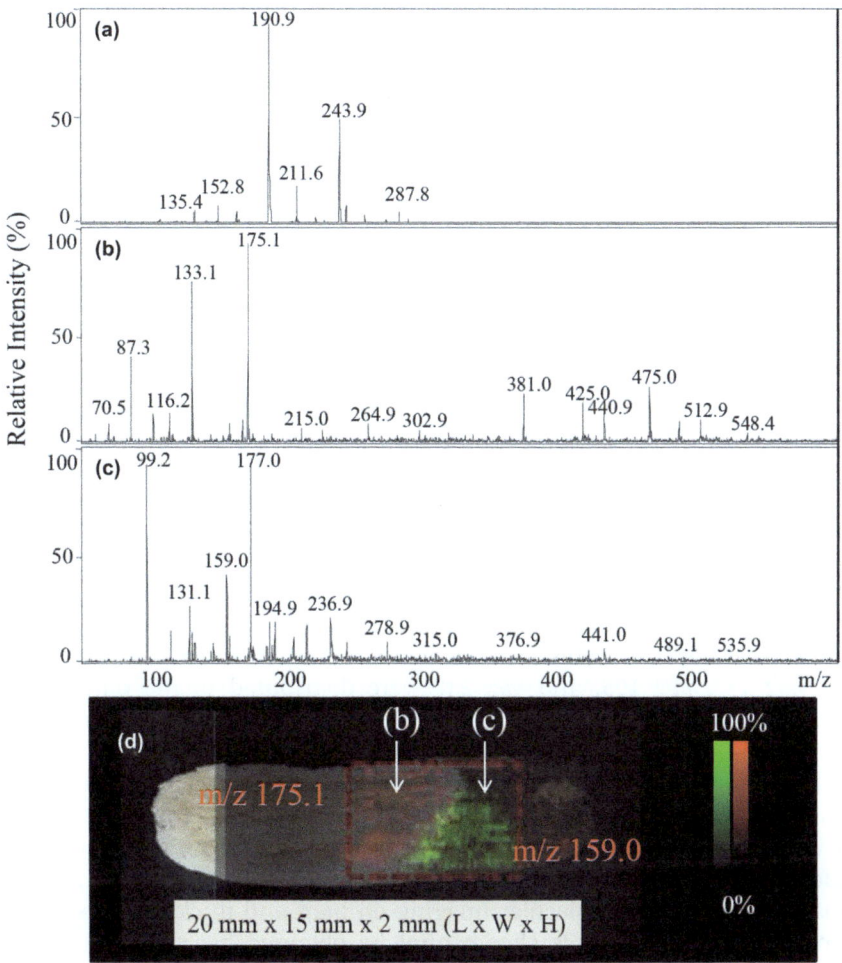

Figure 15.4 ELDI mass spectra of the surface of a dried *Astragalus membranaceus* tissue slice. Both volatile (a) and nonvolatile (b)–(e) organic components from the tissue surface were detected using IR-ELDI/MS under ambient conditions. (d) An *Astragalus membranaceus* slice with a size of 20 mm × 15 mm × 2 mm (L × W × H) was analyzed by IR-ELDI/MS under ambient conditions to obtain molecular images, where these images are overlaid by arrows showing the localizations of ions at *m/z* 175.1 (red) and *m/z* 159.0 (green).

anhydrous *Astragalus membranaceus* tissue slice using ELDI/MS with a 1064 nm pulsed laser beam. Smaller, volatile, and odorous compounds spontaneously evaporated from the sample slice without laser irradiation, after which they then fused and reacted with charged solvent species in an electrospray plume in a manner similar to that in the fused droplet electrospray (FD-ESI) technique.[28] Several ion signals at *m/z* 135.4, *m/z* 152.8, *m/z* 190.9, *m/z* 211.6, *m/z* 243.9 and *m/z* 287.8 were successfully detected in the fusion step (Figure 15.4(a)). Nonvolatile compounds from the inner and outer layers of the slice were also characterized. Ion signals at m/z 71, m/z 87, m/z 116, m/z 133, m/z 175, m/z 381, m/z 425, m/z 475, and m/z 513 were obtained from the inner layer of the anhydrous *Astragalus membranaceus* tissue (Figure 15.4(b)). In contrast, analysis of the outer tissue layer revealed ion signals at m/z 99, m/z 117, m/z 131, m/z 159, m/z 177, m/z 195, and m/z 237 (Figure 15.4(c)). Ion signals at m/z 133 and m/z 441 were found in both layers, albeit with differences in ion intensities. From the results described above, the direct ELDI profiling of small organic compounds in plant tissue shows different ion distributions between the inner and outer layers. The anhydrous plant tissue of *Astragalus membranaceus* was then sliced into thin sections that measured approximately 66 mm in length, 15 mm in width, and 2 mm in thickness. An area of size 20 mm $\times$ 15 mm within the *Astragalus membranaceus* tissue slice was selected for further ELDI imaging analysis. The distribution of different natural product compounds in the plant tissue of *Astragalus membranaceus* is clearly divided into two groups as seen in Figure 15.4(d). According to the molecular profiling and images of the *Astragalus membranaceus* tissue slice, analyte with ion signals at m/z 71, m/z 87, m/z 116, m/z 133, m/z 175, m/z 381, m/z 425, m/z 475, and m/z 513 were mainly distributed in the inner layer of the *Astragalus membranaceus* tissue; on the other hand, analytes with ion signals at m/z 99, m/z 117, m/z 131, m/z 159, m/z 177, m/z 195, and m/z 237 were distributed in the outer layer. Figure 15.4(d) shows a photograph of the anhydrous *Astragalus membranaceus* tissue overlaid with colored ion intensities indicating the presence of analyte ions at m/z 175 (red) and m/z 159 (green), which shows the differential localizations of the two molecules. The results show that analytes from tissues can be successfully desorbed *via* thermal desorption processes from irradiation by an IR pulsed laser. The ELDI/MS technique demonstrated its capability to perform molecular imaging on hard and thick surfaces where a micrometer-scale sample slice cannot be obtained because of the high degree of texture and fragile dry fungi or plants.

15.7.5   Depth Profiling

One of the unique advantages of a pulsed laser system is its ability to produce information on the depth composition around sample surfaces. This capacity was also demonstrated in the ELDI system since a pulse laser was involved. For example, the depth profile of a STONA tablet sample was obtained by laser pulses that removed an average of $\sim$ 50 μm each laser shot (Figure 15.5(a)). The active ingredients, acetaminophen (m/z 152) and methylephedrine (m/z 180), were found to exist in different layers of the STONA tablet. With this technique, a small amount of sample was ablated from the tablet using a focused laser with a wavelength of 1064 nm and a spot size of approximately 200 μm has been achieved. As illustrated in Figure 15.5(b), the first and second layers of the tablet mainly consist of acetaminophen and methylephedrine, respectively. Figures 15.5(c)–(e) shows

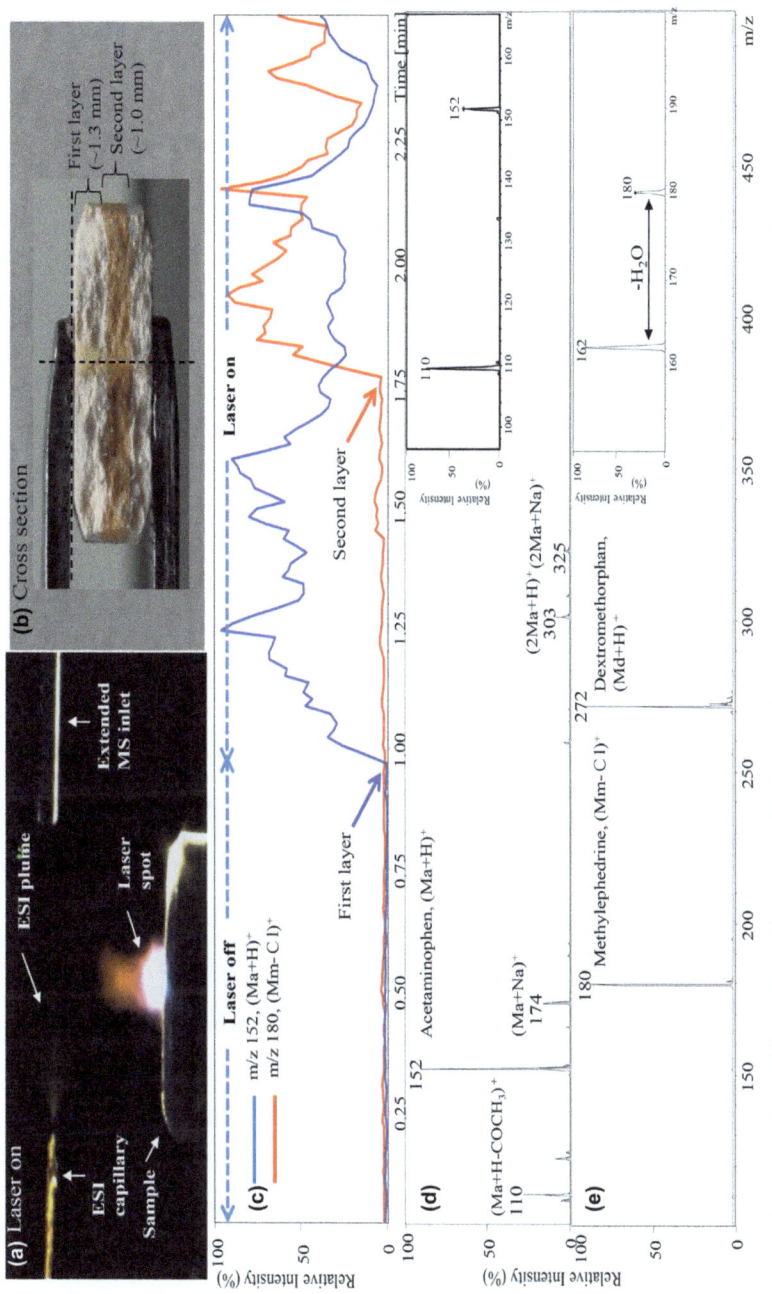

Figure 15.5 Photographs of (a) the ELDI/MS setup displaying the relative positions of the extended MS inlet, electrospray capillary, laser spot, and the STONA tablet sample during ELDI analysis, and (b) the cross-sectional and depth profiling of the ablated STONA tablet. Mass spectra indicate the (c) ion distribution and mass spectra of (d) acetaminophen (m/z 152, blue) and (e) methylephedrine (m/z 180, red) ions from the STONA tablet.

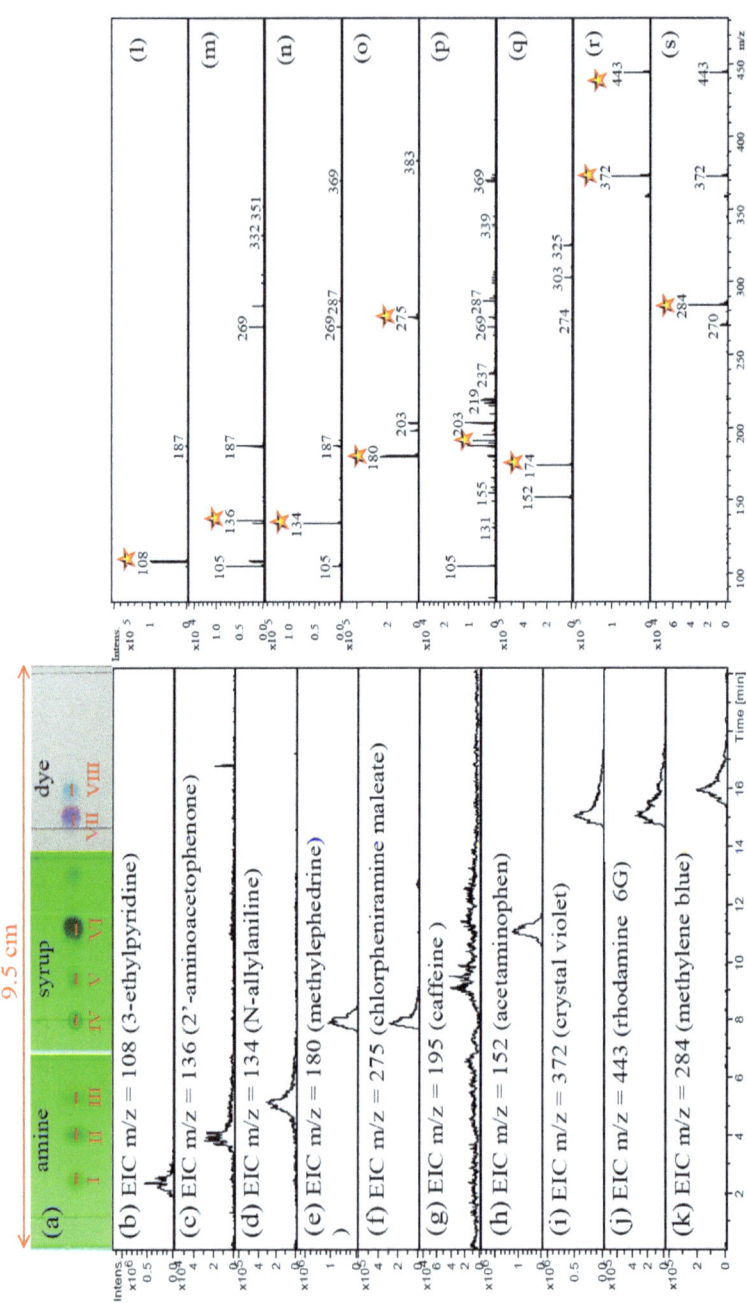

Figure 15.6 Extracted ion chromatograms and mass spectra of different analytes from three separated TLC plates (*i.e.* an amine mixture and a syrup and dye mixture) detected by ELDI/MS.

the extracted ion chromatograms and related mass spectra during depth analysis by ELDI/MS. Using a laser energy of 20 mJ, ELDI/MS was able to obtain depth information of over 3 mm with a resolution of ~ 50 μm on the surface of a drug tablet. Of course, the attainable depth resolution depended on the material being analyzed, the laser power and spot size.

15.7.6　TLC High-throughput Analysis

The ELDI technique is able to efficiently desorb sample analytes on solid surfaces and in liquid solutions because a large amount of energy is introduced to analytes by a pulsed laser. Furthermore, the innately high spatial resolution and scanning capability of the laser beam makes ELDI a useful technique for the rapid and continuous characterization of chemical entities directly from the surfaces of thin-layer chromatography (TLC) plates.[46] Various organic compounds such as FD&C dyes, amines, and extracts of a drug tablet were directly characterized from TLC plates using ELDI/MS. In addition, a LEGO-made TLC plate dealing and delivery system was successfully combined with ELDI/MS to detect chemical compounds separated on TLC plates.[62] Samples including a mixture of synthetic dyes and extracts of pharmaceutical drugs were analyzed to demonstrate the capability of this TLC-ELDI/MS system. Figure 15.6 shows the extracted ion chromatograms and mass spectra from three separated TLC plates detected by ELDI/MS. Mixtures of chemical compounds, such as amines (3-ethylpyridine, 2′-aminoacetophenone and *N*-allylaniline), syrups (methylephedrine, chlorpheniramine maleate, and caffeine), and dyes (crystal violet, rhodamine 6G and methylene blue) were all detected by ELDI/MS. In addition, the ambient ionization feature of ELDI meant that little or no sample pretreatment was required, thus reducing time, labor, and material costs for analyses. This green methodology shows the capacity of high-throughput analysis of TLC plates using ELDI/MS.

15.8　Conclusions

Electrospray laser desorption ionization mass spectrometry (ELDI/MS) is useful for the rapid characterization of major chemical ingredients in various solids under ambient conditions. Particularly exciting is the use of ELDI in the study of proteins without the necessity of sample preparation and in the absence of a matrix. The combined desorption and ionization processes in ELDI are shown to be very soft, as evidenced by the mass spectra of proteins shown to be in their native conformations. Furthermore, both volatile and nonvolatile molecules can be detected in samples in liquid, ointment, syrup and solid forms. This technique also shows its ability for high-throughput analysis of complex samples separated on TLC plates. Employing a focused laser beam for analyte ablation offers 2D imaging of solid materials in their untreated states. The alteration of parameters such as laser wavelengths would permit imaging of different materials, while increasing laser power, and the small size of the laser focal spot, drastically lowers analyte detection limits. ELDI/MS can greatly extend the analytical capabilities of ambient mass spectrometry since it allows independent control of ionization and desorption processes.

References

1. R. G. Cooks, Z. Ouyang, Z. Takats and J. M. Wiseman, *Science*, 2006, **311**, 1566.
2. G. J. Van Berkel, S. P. Pasilis and O. Ovchinnikova, *J. Mass Spectrom.*, 2008, **43**, 1161.
3. A. Venter, M. Nefliu and R. G. Cooks, *TrAC, Trends Anal. Chem.*, 2008, **27**, 284.
4. H. W. Chen, G. Gamez and R. Zenobi, *J. Am. Soc. Mass Spectrom.*, 2009, **20**, 1947.
5. R. M. Alberici, R. C. Simas, G. B. W. Sanvido, W. Romao, P. M. Lalli, M. Benassi, I. B. S. Cunha and M. N. Eberlin, *Anal. Bioanal. Chem.*, 2010, **398**, 265.
6. M. Z. Huang, C. H. Yuan, S. C. Cheng, Y. T. Cho and J. Shiea, *Annu. Rev. Anal. Chem.*, 2010, **3**, 43.
7. M. Z. Huang, S. C. Cheng, Y. T. Cho and J. Shiea, *Anal. Chim. Acta*, 2011, **702**, 1.
8. Z. Takats, J. M. Wiseman, B. Gologan and R. G. Cooks, *Science*, 2004, **306**, 471.
9. R. B. Cody, J. A. Laramee and H. D. Durst, *Anal. Chem.*, 2005, 77, 2297.
10. J. Shiea, M. Z. Huang, H. J. HSu, C. Y. Lee, C. H. Yuan, I. Beech and J. Sunner, *Rapid Commun. Mass Spectrom.*, 2005, **19**, 3701.
11. J. M. Wiseman, C. A. Evans, C. L. Bowen and J. H. Kennedy, *Analyst*, 2010, **135**, 720.
12. J. M. Wiseman, S. M. Puolitaival, Z. Takats, R. G. Cooks and R. M. Caprioli, *Angew. Chem., Int. Ed.*, 2005, **44**, 7094.
13. Z. Takats, J. M. Wiseman and R. G. Cooks, *J. Mass Spectrom.*, 2005, **40**, 1261.
14. Z. Takats, I. Cotte-Rodriguez, N. Talaty, H. W. Chen and R. G. Cooks, *Chem. Commun.*, 2005, 1950.
15. H. W. Chen, N. N. Talaty, Z. Takats and R. G. Cooks, *Anal. Chem.*, 2005, 77, 6915.
16. J. M. Nilles, T. R. Connell and H. D. Durst, *Anal. Chem.*, 2009, **81**, 6744.
17. C. Petucci, J. Diffendal, D. Kaufman, B. Mekonnen, G. Terefenko and B. Musselman, *Anal. Chem.*, 2007, **79**, 5064.
18. T. Rothenbacher and W. Schwack, *Rapid Commun. Mass Spectrom.*, 2009, **23**, 2829.
19. R. R. Steiner and R. L. Larson, *J. Forensic Sci.*, 2009, **54**, 617.
20. M. Z. Huang, H. J. Hsu, L. Y. Lee, J. Y. Jeng and L. T. Shiea, *J. Proteome Res.*, 2006, **5**, 1107.

21. M. Z. Huang, H. J. Hsu, C. I. Wu, S. Y. Lin, Y. L. Ma, T. L. Cheng and J. Shiea, *Rapid Commun. Mass Spectrom.*, 2007, **21**, 1767.
22. C. Y. Cheng, C. H. Yuan, S. C. Cheng, M. Z. Huang, H. C. Chang, T. L. Cheng, C. S. Yeh and J. Shiea, *Anal. Chem.*, 2008, **80**, 7699.
23. M. Z. Huang, S. S. Jhang, C. N. Cheng, S. C. Cheng and J. Shiea, *Analyst*, 2010, **135**, 759.
24. M. Z. Huang, S. C. Cheng, S. S. Jhang, C. C. Chou, C. N. Cheng, J. Shiea, I. A. Popov and E. N. Nikolaev, *Int. J. Mass Spectrom.*, 2012, **325**, 172.
25. C. Y. Lee and J. Shiea, *Anal. Chem.*, 1998, **70**, 2757.
26. H. J. Hsu, T. L. Kuo, S. H. Wu, J. H. Oung and J. Shiea, *Anal. Chem.*, 2005, 77, 7744.
27. J. Shiea, D. Y. Chang, C. H. Lin and S. J. Jiang, *Anal. Chem.*, 2001, **73**, 4983.
28. D. Y. Chang, C. C. Lee and J. Shiea, *Anal. Chem.*, 2002, **74**, 2465.
29. I. F. Shieh, C. Y. Lee and J. Shiea, *J. Proteome Res.*, 2005, **4**, 606.
30. N. Brenner, M. Haapala, K. Vuorensola and R. Kostiainen, *Anal. Chem.*, 2008, **80**, 8334.
31. C. M. Hong, F. C. Tsai and J. Shiea, *Anal. Chem.*, 2000, **72**, 1175.
32. M. A. Posthumus, P. G. Kistemaker, H. L. C. Meuzelaar and M. C. Tennoeverdebrauw, *Anal. Chem.*, 1978, **50**, 985.
33. J. C. Tabet and R. J. Cotter, *Anal. Chem.*, 1984, **56**, 1662.
34. M. Benazouz, B. Hakim, J. L. Debrun, D. Strivay and G. Weber, *Rapid Commun. Mass Spectrom.*, 1999, **13**, 2302.
35. R. J. Cotter, *Anal. Chem.*, 1980, **52**, 1767.
36. R. J. Perchalski, R. A. Yost and B. J. Wilder, *Anal. Chem.*, 1983, **55**, 2002.
37. R. Zenobi and R. Knochenmuss, *Mass Spectrom. Rev.*, 1998, **17**, 337.
38. J. J. Coon and W. W. Harrison, *Anal. Chem.*, 2002, 74, 5600.
39. F. Drewnick and P. H. Wieser, *Rev. Sci. Instrum.*, 2002, **73**, 3003.
40. S. T. Fountain and D. M. Lubman, *Anal. Chem.*, 1993, **65**, 1257.
41. J. J. Coon, K. J. McHale and W. W. Harrison, *Rapid Commun. Mass Spectrom.*, 2002, **16**, 681.
42. L. Kolaitis and D. M. Lubman, *Anal. Chem.*, 1986, **58**, 2137.
43. J. Kosler, M. Wiedenbeck, R. Wirth, J. Hovorka, P. Sylvester and J. Mikova, *J. Anal. At. Spectrom.*, 2005, **20**, 402.
44. L. Yang, R. E. Sturgeon and Z. Mester, *J. Anal. At. Spectrom.*, 2005, **20**, 431.
45. E. Woods, G. D. Smith, Y. Dessiaterik, T. Baer and R. E. Miller, *Anal. Chem.*, 2001, **73**, 2317.
46. S. Y. Lin, M. Z. Huang, H. C. Chang and J. Shiea, *Anal. Chem.*, 2007, **79**, 8789.
47. J. Shiea, C. H. Yuan, M. Z. Huang, S. C. Cheng, Y. L. Ma, W. L. Tseng, H. C. Chang and W. C. Hung, *Anal. Chem.*, 2008, **80**, 4845.
48. S. C. Cheng, Y. S. Lin, M. Z. Huang and J. Shiea, *Rapid Commun. Mass Spectrom.*, 2010, **24**, 203.
49. J. S. Sampson, A. M. Hawkridge and D. C. Muddiman, *J. Am. Soc. Mass Spectrom.*, 2006, **17**, 1712.

50. J. S. Sampson, K. K. Murray and D. C. Muddiman, *J. Am. Soc. Mass Spectrom.*, 2009, **20**, 667.
51. P. Nemes and A. Vertes, *Anal. Chem.*, 2007, **79**, 8098.
52. P. Nemes, A. A. Barton, Y. Li and A. Vertes, *Anal. Chem.*, 2008, **80**, 4575.
53. B. Shrestha and A. Vertes, *Anal. Chem.*, 2009, **81**, 8265.
54. J. S. Sampson and D. C. Muddiman, *Rapid Commun. Mass Spectrom.*, 2009, **23**, 1989.
55. J. J. Brady, E. J. Judge and R. J. Levis, *Rapid Commun. Mass Spectrom.*, 2009, **23**, 3151.
56. E. J. Judge, J. J. Brady, D. Dalton and R. J. Levis, *Anal. Chem.*, 2010, **82**, 3231.
57. J. J. Brady, E. J. Judge and R. J. Levis, *Rapid Commun. Mass Spectrom.*, 2010, **24**, 1659.
58. J. Liu, B. Qiu and H. Luo, *Rapid Commun. Mass Spectrom.*, 2010, **24**, 1365.
59. S. C. Cheng, T. L. Cheng, H. C. Chang and J. Shiea, *Anal. Chem.*, 2009, **81**, 868.
60. S. C. Cheng, M. Z. Huang and J. Shiea, *Anal. Chem.*, 2009, **81**, 9274.
61. I. X. Peng, R. R. O. Loo, J. Shiea and J. A. Loo, *Anal. Chem.*, 2008, **80**, 6995.
62. S. C. Cheng, M. Z. Huang, L. C. Wu, C. C. Chou, C. N. Cheng, S. S. Jhang and J. Shiea, *Anal. Chem.*, 2012, **84**, 5864.

CHAPTER 16

Paper Spray

JIANGJIANG LIU,[a] NICHOLAS E. MANICKE,[b] XIAOYU ZHOU,[a]
R. GRAHAM COOKS*[c] AND ZHENG OUYANG*[a,d]

[a] Weldon School of Biomedical Engineering, Purdue University, West
Lafayette, Indiana 47907, USA; [b] Departement of Chemistry and Chemical
Biology, Indiana University-Purdue University Indianapolis, Indianapolis,
Indiana 46202, USA; [c] Departement of Chemistry, Purdue University, West
Lafayette, Indiana 47907, USA; [d] Departement of Electrical and Computer
Engineering, Purdue University, West Lafayette, Indiana 47907, USA
*Email: cooks@purdue.edu; ouyang@purdue.edu

16.1 Introduction

Paper spray was first introduced in 2009[1] and has been used for development of a wide range of quantitative and qualitative applications.[2] A paper
substrate serves for multiple functions, including sample loading, analyte
extraction and spray ionization. The studies of the processes involved in the
paper spray have been carried out and the performance of PS-MS could be
optimized by selecting proper spray solvent, paper substrate, spray voltage
and other parameters of operation. A wide range of compounds have been
successfully analyzed by PS-MS directly from complex samples with a variety
of matrices. PS-MS has been demonstrated for direct chemical analysis in
application fields including food safety, public safety, forensics and bio-
medicine, *etc.* Among all the ambient ionization methods, paper spray has
been particularly developed for quantitation analysis, as extensively demonstrated with analysis of drugs in blood samples. A series of efforts in the
development has led to solutions for quantitation with small sample
amounts and simple operation procedures, which assumes its potential in

New Developments in Mass Spectrometry No. 2
Ambient Ionization Mass Spectrometry
Edited by Marek Domin and Robert Cody
© The Royal Society of Chemistry 2015
Published by the Royal Society of Chemistry, www.rsc.org

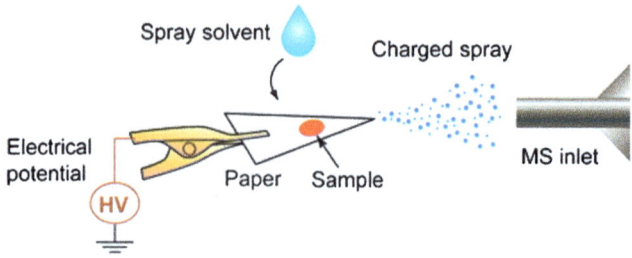

Figure 16.1   Experimental setup for performing PS-MS.
Figure taken from ref. 1 with permission.

clinical diagnosis. In addition, point-of-care (POC) analysis using paper
spray has been explored with miniature ion trap mass spectrometers.

The original setup of PS-MS is extremely simple (Figure 16.1).[1] A triangular
paper substrate is held by a metal clip and placed in the front of the inlet of a
mass spectrometer. The metal clip is connected to a DC high-voltage power
supply. To perform an analysis, the sample is deposited on the paper triangle to form a sample spot. The paper substrate is then wetted by a spray
solvent. When the DC high voltage is applied, charged droplets are generated at the tip of the paper triangle carrying the analytes extracted from the
deposited sample toward the MS inlet. A representative set of parameters for
the working condition of PS-MS include 3–5 kV for high voltage, 0.4–20 µL of
liquid sample or 1–10 mg of solid sample, and 10–30 µL of spray solvent
(*e.g.* methanol or methanol/water, 1 : 1, v:v).

A wide range of chemicals can be ionized by paper spray, from small
molecules to large biomolecules (Figure 16.2).[2] Typically, protonated ions
are generated in positive-ion mode, while deprotonated ions are generated
in negative-ion mode. The spray duration is greatly affected by the volume of
spray solvent applied on paper, since the solvent is consumed through both
spray and evaporation. Under a typical experimental condition with 10 µL
methanol as spray solvent, signal may last more than 1 min, which is long
enough to perform a number of MS/MS scans for several different target
analytes of interest. When the angle of the spray tip is relatively small, the
spray plume is strong and adequately covers the vicinity of the MS inlet. As a
characterization of the ease of implementation, the signal intensity was
monitored with different offsets between the spray tip and the MS inlet. A 2D
contour map of the measured signal intensity was obtained quantitatively
indicating that a stable signal can be obtained as long as the paper triangle
stays within the 5×10 mm area in front of the inlet of mass spectrometer.[2]

16.2   Fundamentals

16.2.1   Spray from a Paper Tip

The paper spray can be described as a three-step process.[1] The first step is
the extraction of chemicals from the deposited sample, which occurs

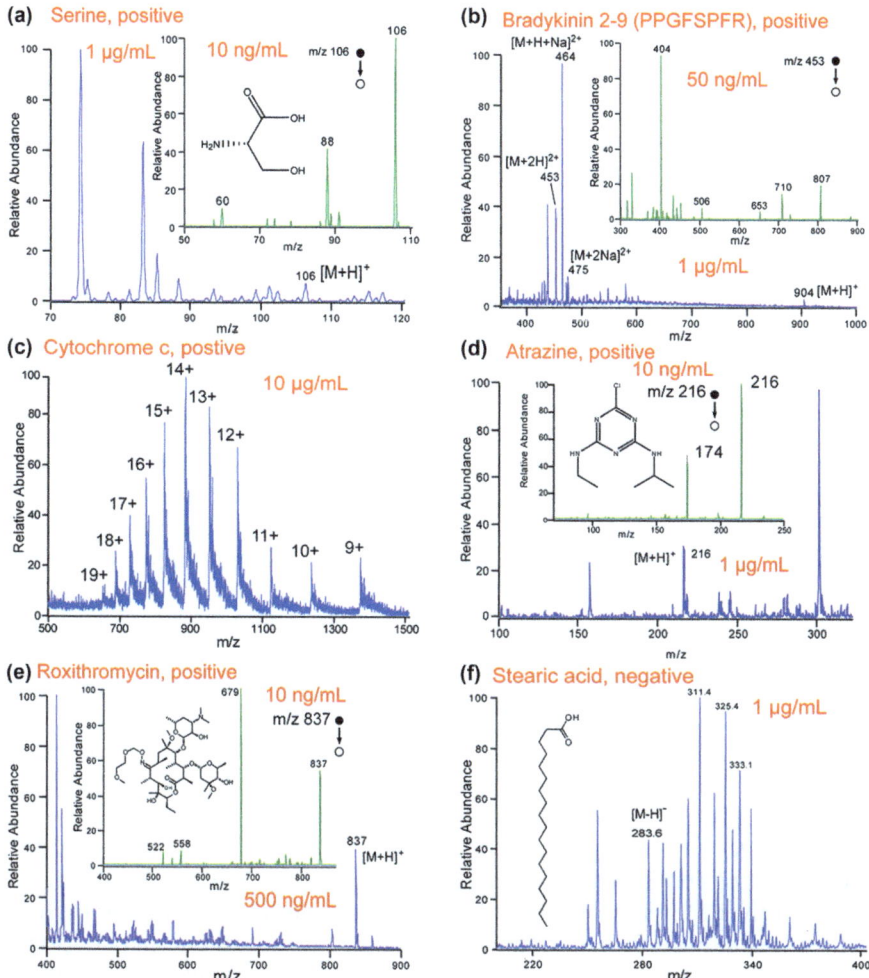

Figure 16.2 Mass spectra collected from pure chemical solutions with PSI-MS, including (a) serine, (b) bradykinin 2–9, (c) cytochrome C, (d) atrazine, (e) roxithromycin in positive-ion mode and (f) stearic acid in negative-ion mode.
Figure taken from ref. 2 with permission.

immediately with the application of spray solvent on the paper substrate. The second step is the transport of the extracted analytes by the solvent wicking through the paper and the third step is the generation of the charged droplets by a process similar to the electrospray ionization (ESI). The spray solvent carries the compounds extracted from the sample and delivers them to the tip of paper triangle, where the ionization occurs with the charged droplets formed with the high voltage applied.[3] The paper substrate plays a key role in the solvent–substrate and substrate–sample interactions for the chemical extraction, which allows a very simple but fast and effective sample purification in real time[3,4] prior to the spray ionization for MS analysis.

Studies for comparison between paper spray and nano-ESI were performed with the results obtained shown in Figure 16.3.[1] Both ionization methods generate very similar spray plumes (Figure 16.3(a)) and provide similar spectra in terms of peak patterns and absolute ion intensities (Figure 16.3(b)). However, PS-MS requires a much higher spray voltage, which is presumably due to the relatively large substrate tip for the spray. The size and velocity of the charged droplets during the ionization process were measured with a phase Doppler particle ancmometer.[5] The droplet size for paper spray ranges from submicrometer to over ten micrometers and the velocity of the droplets varies from 3 m s^{-1} to 10 m s^{-1}. This experiment also indicates that corona discharge ionization may occur under a condition with relatively less wetted paper substrate.

Paper spray is a soft ionization method that allows the generation of gas-phase ions without extensive fragmentation. The "survival yield" method[6] was deployed to measure the internal energy of the ions generated by paper spray and nano-ESI.[1] The fragmentation fractions were measured for a set of thermometer ions that are partially fragmented at different degrees during ionization. The breakdown curves (Figure 16.3(c)) reflected the internal energy distribution of ions generated with different ionization methods. The internal energies is slightly lower (~ 0.4 eV) for ions generated by paper spray (Figure 16.3(d)).

The relationship between the voltage applied for spray and the spray current is an important characteristic for any spray ionization. Figure 16.4(a) and (b) show the spray currents increase with the increasing spray voltage for both paper spray and nano-ESI in positive-ion mode. The trends of ion intensity as a function of spray voltage, however, is significantly different for the two ionization methods. A much higher voltage is preferred in paper spray. In the negative-ion mode paper spray, spray plume could also be generated using polar spray solvents. However, no spray plume was observed with solvents of relatively low polarity and high surface tension. Instead, an electron emission from the paper tip occurs and was proved by the formation of radical anion $[M]^{-\bullet}$ (Figure 16.4(c) and (d)).

As a porous and hydrophilic material, cellulose paper allows liquid transport over the surface and through the inside microchannels by capillary action. When connected to a high voltage, the charged droplets are generated only at the sharp corners of a paper substrate (Figure 16.5(a)). This is presumably due to the high local electric fields around these corners. The effect of macroscopic structure of a corner was characterized by altering the angle of the corner for paper spray. Sharp corners were shown to generate stronger spray plume at relatively lower voltage (Figure 16.5(b)–(f)).[8] Paper triangles with 30–45° corners are typically used for PS-MS.

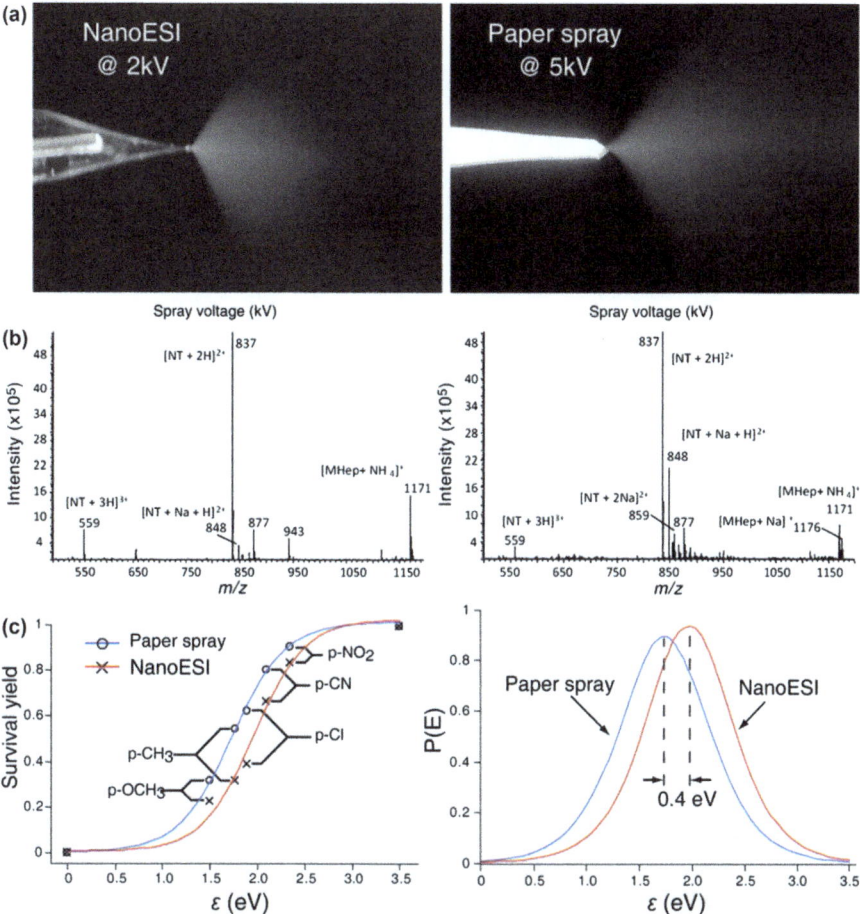

Figure 16.3 Comparison of nano-ESI and PSI-MS in positive-ion mode. (a) Optical photograph of the spray plumes generated with nano-ESI and PSI-MS, respectively. (b) Full-scan mass spectra of 10 μM neurotensin (NT) and maltoheptaose (MHep) obtained in positive-ion mode with nano-ESI and PSI-MS, respectively. Sample prepared in methanol/water (1 : 1, v:v) with 10 mM ammonium acetate. (c) Survival ion yields for PSI-MS (blue curve) and nano-ESI (red curve). (d) Internal energy distributions for PSI-MS (blue curve) and nano-ESI (red curve). Tested chemicals: *p*-OCH₃, *p*-methoxybenzylpyridinium tetrafluoroborate, *p*-CH₃, *p*-methylbenzylpyridinium bromide, *p*-Cl, *p*-chlorobenzylpyridinium chloride, *p*-CN, *p*-cyanobenzylpyridinium chloride and *p*-NO₂, *p*-nitrobenzylpyridinium bromide.
Figure taken from ref. 1 with permission.

16.2.2 Analyte Elution

The elution of analytes from complex mixture is not only affected by the interactions among solvent, paper and analytes, but also affected by how the

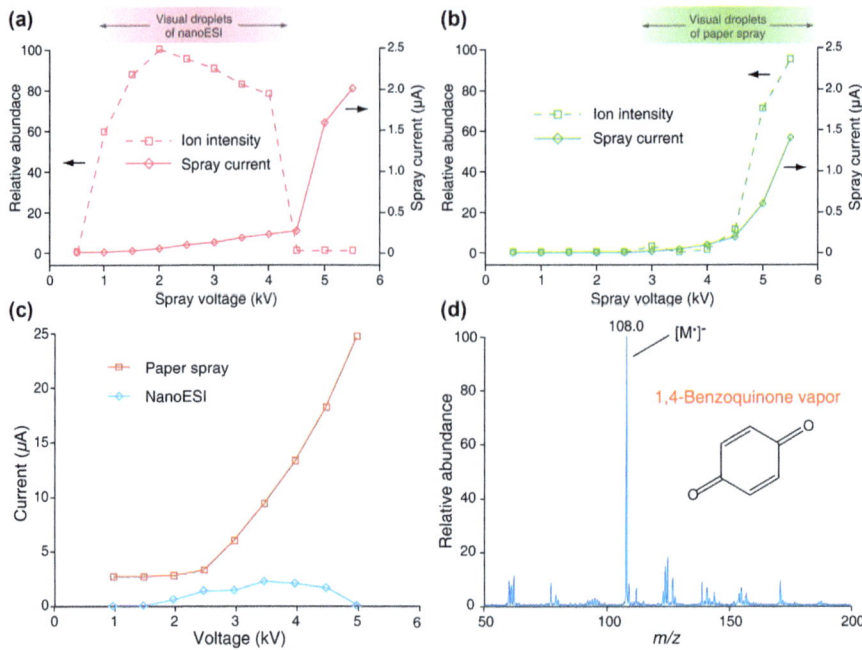

Figure 16.4 Comparison of signal intensity in positive-ion mode and spray current as a function of increasing spray voltage from 0.5 to 5.5 kV for (a) Nano-ESI and (b) PS-MS. Sample: 200 ng mL^{-1} cocaine in methanol. (c) Spray current of nano-ESI and PSI-MS as a function of increasing spray voltage in negative-ion mode. (d) Spectrum obtained from 1,4-benzoquinone vapor with PSI-MS in negative-ion mode. Spray voltage, –4 kV; solvent, methanol/water (1:1, v:v). 1,4-benzoquinone solid was placed close to the paper tip to produce 1,4-benzoquinone vapor. Figure taken from ref. 1 with permission.

spray solvent is applied on the paper. Spray solvents can be applied on paper *via* two distinct ways, in a dumping mode or a wicking mode. As a demonstration, propranolol was deposited on paper and analyzed by PS-MS with solvent applied in these two modes.[7] In the dumping mode, 10 μL of solvent was applied all at once and a relatively constant ion intensity for propranolol was obtained (Figure 16.6(a)). In the wicking mode, the solvent was continuously applied to the paper *via* capillary action. The ion intensity of propranolol went up quickly to a maximum and then decreased gradually and then stayed at a relatively constant value for minutes (Figure 16.6(b)). The variation of the signal intensity does not affect significantly the precision of the quantitation since internal standard was used with propranolol-d7 spiked in the samples before the elution (Figure 16.6(c)). The ion intensities of analyte and internal standard varied in the same pattern, leading to a constant ratio between the two compounds. When a mixture of multiple analytes of different properties is analyzed, the elution patterns may be different for each analyte in the

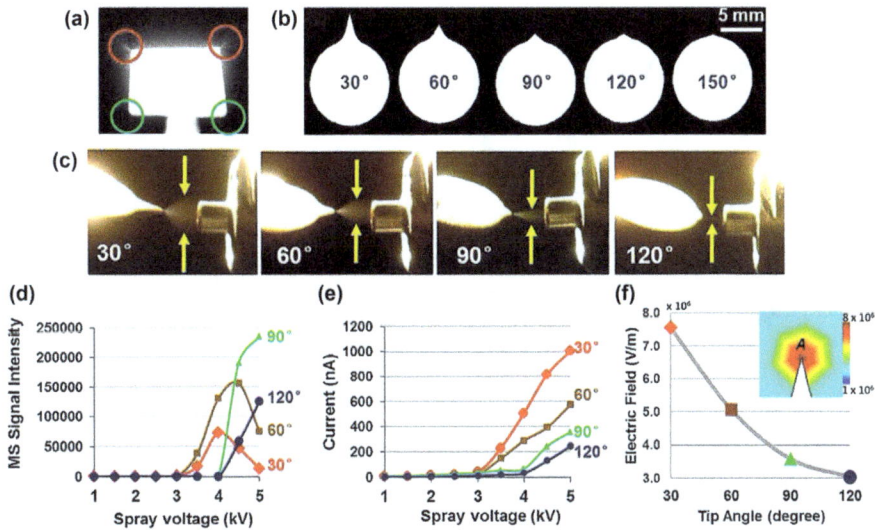

Figure 16.5   (a) A paper substrate used to generate spray plumes with PSI-MS setup. The spray plumes were only observed at the sharp corners (red circles), rather than round corners (green circles). (b) Paper substrates were fabricated with different tip angles of 30°, 60°, 90°, 120° and 150°. (c) Photographs of spray plumes generated on different paper substrates with different tip angles. (d) Signal intensity (cocaine fragment ion m/z 182) and (e) spray current as a function of the increasing spray voltage from 1 kV to 5 kV. (f) Simulated electric-field strength at the tip of paper substrate as a function of the tip angle from 30° to 120°. Inset: zoomed-in view of simulated electric field distribution at a paper substrate tip angle of 30°.
Figure adapted from ref. 8 with permission.

wicking mode, just like for a paper chromatography. Figure 16.6(d)–(f) show the separation of methyl violet 2B and methylene blue, which were eluted and detected one after another.

16.2.3   Paper Substrate and Solvent

The performance of paper spray has been systematically characterized and the working condition has been optimized to improve its sensitivity, stability and versatility for direct MS analysis. The analyte extraction and the subsequent transport toward the tip of the substrate are two of the essential processes during the paper spray, in addition to the spray ionization. The properties of the paper substrate and spray solvent are expected to have a significant impact on the efficiency of the PS-MS. Both the physical and chemical properties of a paper substrate may affect the interactions among analyte, paper and solvent, resulting in an ultimate impact on the analysis result. A large number of different types of paper are commercially available, many of which were tested for PS-MS. Analysis of cocaine (200 ng mL^{-1} in methanol/water, 1 : 1, v:v) by paper spray was performed with glass fiber

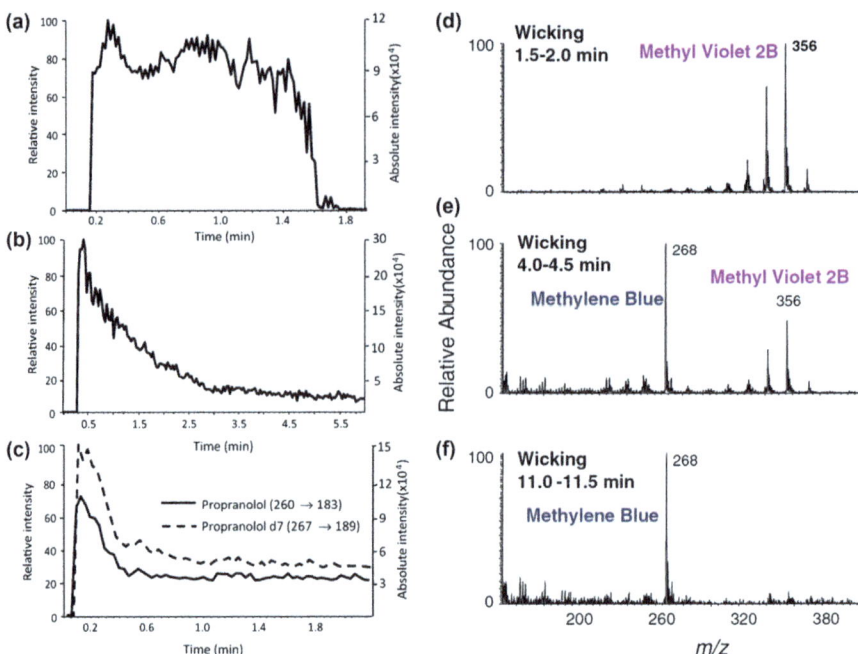

Figure 16.6   Ion chronograms for propranolol (500 ng) tested with PSI-MS. The propranolol was deposited on paper and eluted by (a) applying 10 µL of solvent all at once or (b) and (c) continuous solvent wicking. A mixture of methyl violet 2B and methylene blue was tested with paper spray using continuous solvent wicking. Spectra are collected over different periods (d) 1.5–2.0 min, (e) 4.0–4.5 min, and (f) 11.0–11.5 min.
Panel (a) and (b) taken from ref. 7 and panel (d)–(f) taken from ref. 3 with permissions.

paper, chromatography paper and filter papers of different pore sizes. The best MS spectra, in terms of chemical noise and analyte signal intensity, were obtained with chromatography paper, which thus has been used for most of the subsequent PS-MS studies.[2]

Chemical treatments can modify both the physical and chemical properties of the paper substrates, which thus have been applied for the performance improvement of PS-MS. A silica-coated paper was used for analysis of drug compounds in blood with paper spray and a comparison was made with chromatography paper.[4] The silica coating changes the microstructure of the paper, preventing the diffusion of blood into the paper matrix (Figure 16.7(a)–(i)) as well as altering the affinity of the substrate to the extracted drug molecules. A better extraction and elution of analytes from blood was observed using the silica-coated paper, resulting a 5–50-fold enhancement of sensitivity for analyzing the drugs in blood samples using PS-MS (Figure 16.7(j) and (k)).

In further studies[3,9] of the properties of the paper substrate and its impact on the PS-MS analysis, silanized paper and print paper were used in addition

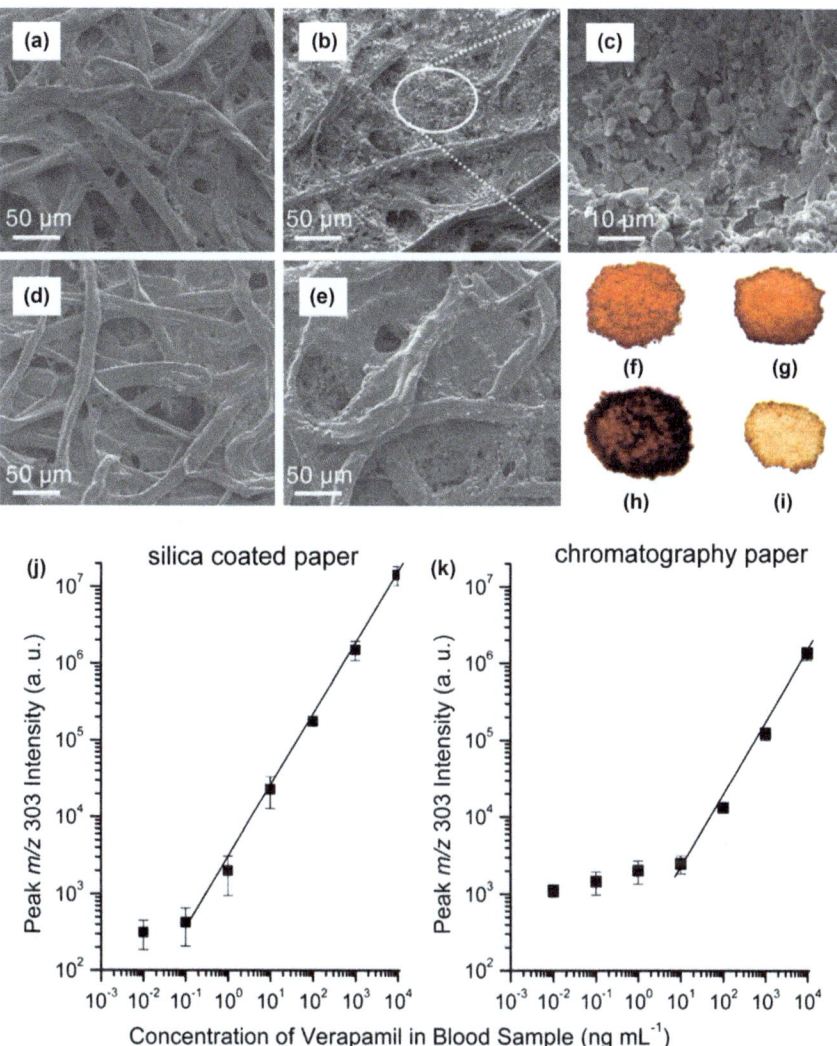

Figure 16.7 SEM images of blank (a) chromatography paper and (b) and (c) silica coated paper. SEM images of (d) chromatography paper and (e) silica-coated paper loaded with dried blood spots. Blood spots formed on chromatography paper, (f) top and (g) back view; on silica-coated paper, (h) top and (i) back view. Comparison of the analysis of verapamil in blood using (j) silica coated paper and (k) chromatography paper. Figure adapted from ref. 4 with permission.

to chromatography and silica-coated paper for paper spray. The cellulose paper is normally hydrophilic but can be changed to hydrophobic *via* a treatment with silanizing reagent, such as (Tridecafluoro-1,1,2,2-tetrahydrooctyl) trichlorosilane. The silanized paper has been shown to be a good substrate for analyzing protein hemoglobin in whole blood using paper

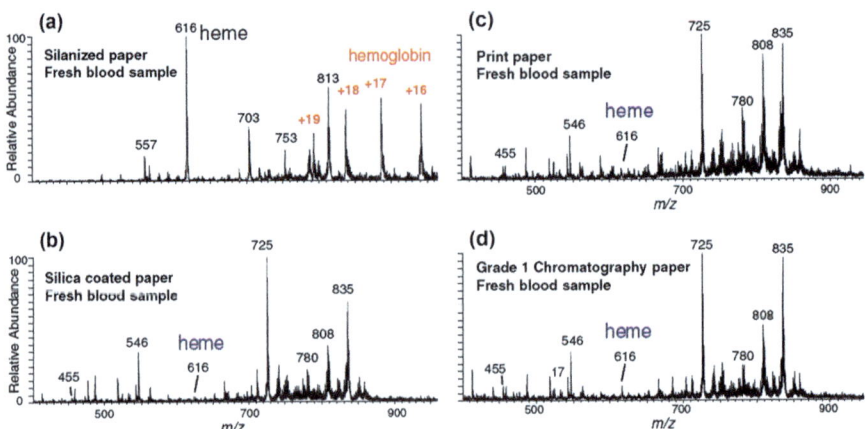

Figure 16.8   Analysis of fresh blood samples using (a) silanized paper, (b) silica-coated paper, (c) print paper, and (d) grade 1 chromatography paper. Dried blood spots formed with 2 μL blood containing 1 μg mL^{-1} verapamil (*m/z* 455), 20 μL methanol used as spray solvent. Figure taken from ref. 3 with permission.

spray (Figure 16.8).[3] Presumably the salinization of the paper weakened the binding interaction between the hydrophilic proteins and the cellulose paper, which resulted in a better elution of proteins during the paper spray. The print paper is of much smaller pore sizes (*e.g.* 0.1 μm), in comparison with the chromatography paper (*e.g.* 11 μm for Whatman Grade 1), and can have various additional pigment coatings. In the development of direct quantitation of nicotine and its metabolites in blood (to be further described later),[9] the print paper was demonstrated to have much lower background signals thus significantly improving the limit of quantitation (LOQ).

The effect by spray solvent and its optimization for PS-MS have also been studied. The solvent is used for on paper liquid extraction and the subsequent spray ionization. In most cases for chemical analysis using paper spray, methanol or methanol/water provides an adequate extraction and ionization efficiency. Methanol-based spray solvent has moderate polarity and low surface tension, which helps to produce a spray ionization at high efficiency and good stability. Therapeutic drugs, lipids, and proteins in whole blood could all be observed using methanol in paper spray.[3] An ultimately optimal solvent condition is dependent on the properties of the target analytes, sample matrices, and the paper substrates. In a test of paper spray with methanol/water solvents each containing a single compound, the optimal percentage of methanol was 40% for PC (16:0–05:0), 50% for epinephrine, and 70% for imatinib.[10] For analysis of acylcarnitines in serum or blood samples, the solvents of methanol/water/formic acid (80 : 20 : 0.1) and acetonitrile/water (90 : 10) were found to be optimal for the dried serum and blood spots, respectively, prepared on the chromatography paper.[11] Different organic solvent and their combination were tested for the analysis of

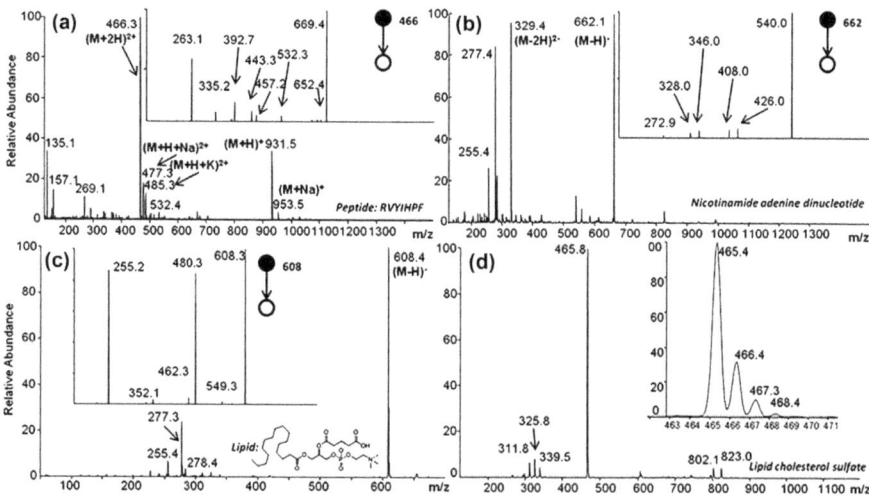

Figure 16.9 Using n-hexane as spray solvent for PSI-MS. Spectra collected from (a) angiotensin III (500 ng), (b) nicotinamide adenine dinucleotide (500 ng), (c) 1-palmitoyl-2-glutaryl-*sn-glycero*-3-phosphocholine (500 ng) and (d) Cholesterol sulfate (3 ng).
Figure taken from ref. 12 with permission.

therapeutic drugs in whole blood using paper spray with silica-coated paper. The dichloromethane/isopropanol solvent was identified as the best spray solvent, which improved the sensitively 3–5-fold in comparison with methanol, ethanol and butyl alcohol.[4]

An interesting phenomenon was observed for using nonpolar solvents in paper spray ionization of insoluble analytes (Figure 16.9).[12] Drugs, peptides, nucleotides and phospholipids were deposited on paper substrate as dried solids. Nonpolar organic solvents, such as hexane, toluene and dioxane, were applied as the spray solvents for paper spray, which occurred at a relatively low positive or negative voltage (0.8–2 kV). The ions $(M+H)^+$ and $(M+Na)^+$ in the positive mode and $(M–H)^-$ in the negative mode were observed. The spray was possibly due to a field-assisted evaporation. The transport of insoluble analyte was likely to be the result of hoping motions between adjacent cellulose platelets, analogous to those observed for chromatography.[13] Paper spray with nonpolar solvents is potentially useful for direct, online analysis of synthetic mixtures.

16.2.4 On-paper Derivatization

In addition to the condition optimization for the extraction, elution and spray ionization, online derivatization has also be demonstrated for improving the sensitivity and the selectivity of the analysis for target analytes using paper spray. Due to the nature of the specific molecular structure, cholesterol is typically difficult to form protonated ions or salt adducts by

electrospray or similar methods. In order to effectively analyze the cholesterol directly from human serum, betaine aldehyde chloride was applied
on the paper substrate, prior to the deposition of the serum sample.
The chemical reaction occurs when liquid serum sample and spray solvent
were applied on the paper. Charge labeling of cholesterol with betaine
aldehyde through its reaction with hydroxyl groups significantly improved
the ionization efficiency of the cholesterol and $[Chol + BA]^+$ at m/z 488.6
were observed with paper spray using a solvent acetonitrile/CHCl$_3$ (1 : 1 v : v)
(Figure 16.10).[2] In applications of paper spray with online reactions, sample
cartridges can potentially be manufactured with different reagents pre-
deposited on the paper substrates to enhance the analysis sensitivity for
target analytes.

16.3   Applications

Paper spray has been used for developing a wide variety of applications for
direct analysis, including therapeutic drug monitoring, tissue analysis,
foodstuff analysis and many others. A major effort has been put on the development of high-precision quantitation with simple operational procedures for the analysis, which is mandatory for applications such as clinical
and especially point-of-care (POC) diagnosis.

16.3.1   Analysis for Paper Chromatographic Separation

Paper-based chromatographic separation has been studied and developed
for decades. The analyte extraction and elution in paper spray, especially in
the wicking mode, is very similar to the paper chromatography. The paper
substrate behaves as a stationary phase while the spray solvent behaves as a
mobile phase. Chromatographic separation and subsequent direct analysis
using mass spectrometry can now be performed with a simple procedure as
shown in Figure 16.11.[1] A mixture containing two dyes on chromatography
paper, methylene blue and methyl violet, was first separated on paper with
methanol by paper chromatographic separation. Instead of extracting the
separated samples of the paper for analysis, two paper triangles were cut off
from the chromatography paper and analyzed directly using PS-MS. The
spectra obtained from each paper triangle show a dramatic difference in
terms of the ion intensities of the two dyes, reflecting the chemical separation on the chromatography paper.

16.3.2   Blood Analysis

Paper spray is an easy and simple method for direct analysis of small
amounts of blood samples, and is highly suitable for the analysis of dried
blood spots (DBSs) on paper. The DBSs are formed by collecting small
amounts (~ 30 µL) of blood samples on paper and letting them dry. DBSs are
easy for storing and transport and are used as a standard format for neonatal

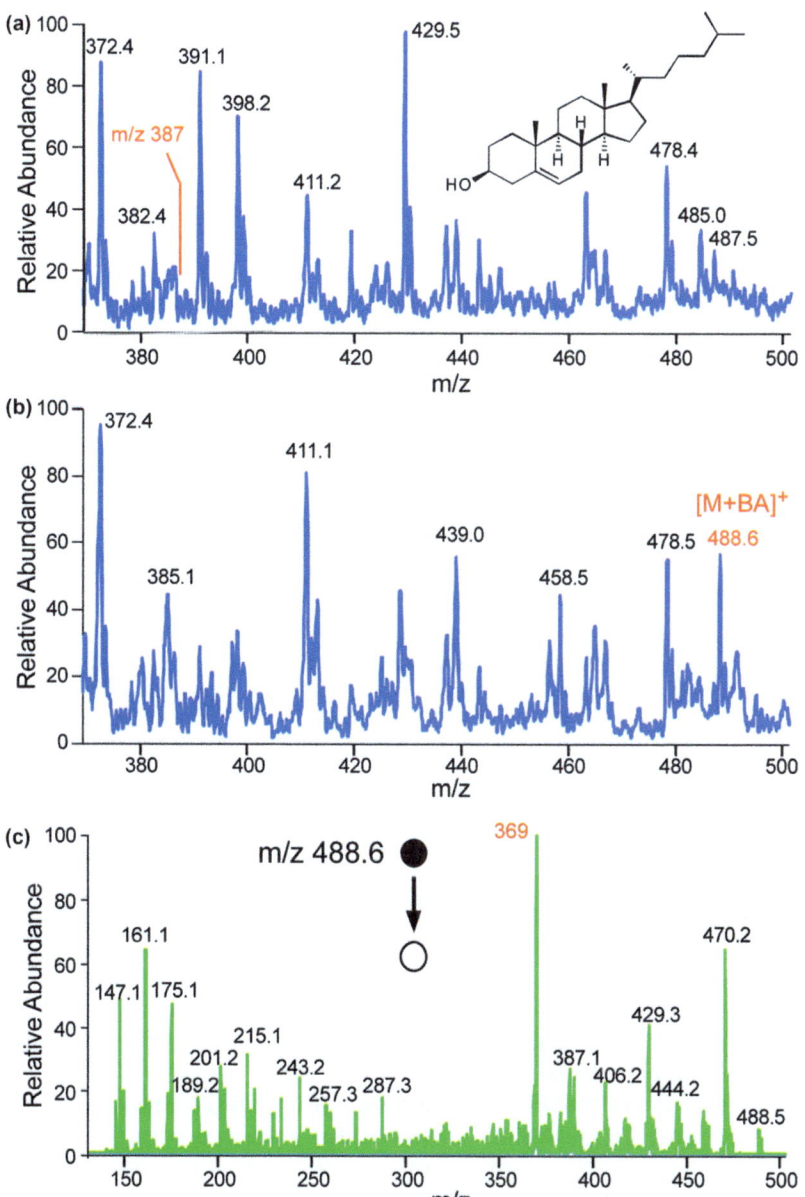

Figure 16.10 Cholesterol analysis with PSI-MS in human serum using (a) blank paper and (b, c) paper preloaded with betaine aldehyde chloride (5 µL, 100 µg mL^{-1}). 2 µL of human serum spotted onto the paper and 10 µL of ACN/CHCl$_3$ (1 : 1 v:v) used as spray solvent.
Figure taken from ref. 2 with permission.

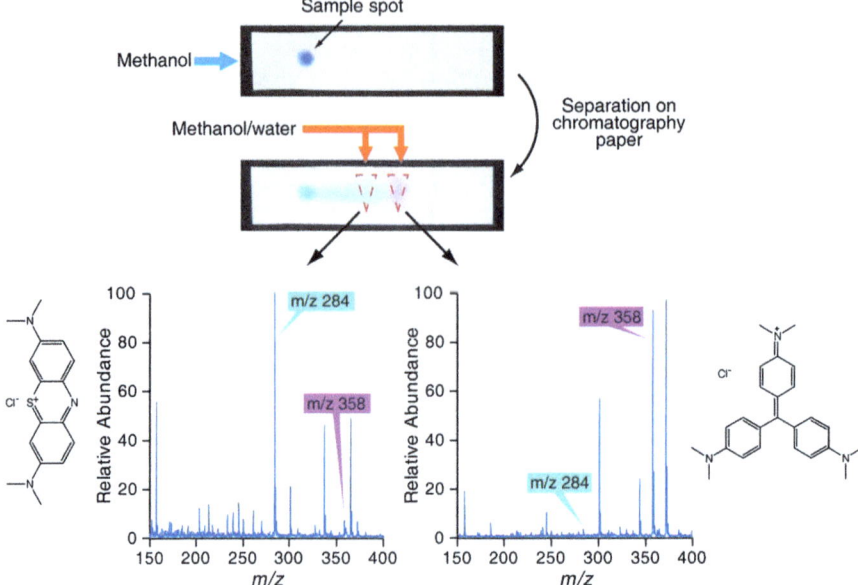

Figure 16.11 Separation and direct analysis of two dyes on chromatography paper
with PSI-MS. 100 μL of the mixture solution (1 mg L^{-1} methylene blue
and 1 mg mL^{-1} methyl violet in water) was deposited onto one end of
the chromatography paper (4 cm×0.5 cm) to form a spot. Dyes were
separated on the paper strip by dipping the left end of the paper strip
into bulk methanol. After 90 s elution on the dyes, two triangles
(indicated by dashed lines) were cut out from the paper strip and used
for PSI-MS test, respectively. The spectra obtained from the two paper
triangles show significant difference on the distribution of the methylene blue (*m/z* 284) and methyl violet (*m/z* 358) after.
Figure taken from ref. 1 with permission.

screening samples. Paper spray is highly compatible with DBS, providing a
fast and simple means for blood analysis at high sensitivity with no sample
preparation. With internal standards incorporated, high-precision quantitation has also been achieved.

The application of PS-MS for quantitation of imatinib (Gleevec) in blood is
shown in Figure 16.12. A DBS on paper was prepared with 0.4 μL blood
containing imatinib (62.5 ngmL^{-1} to 4 μgmL^{-1}, *m/z* 494) and the internal
standard imatinib-d8 (1 μgmL^{-1}, *m/z* 502). Spray solvent of 10 μL methanol/
water (1:1 v/v) was applied and MRM (multiple reaction monitoring) was
performed with transitions *m/z* 494 → *m/z* 394 and *m/z* 502 → *m/z* 394 to
measure the relative abundances of the imatinib and imatinib-d8 for
quantitation. Without delicate sample purification or chromatographic
separation, quantitation with good linearity and RSD was achieved with
paper spray. A matrix effect is typically a major concern for quantitative
analysis in blood, since salts, amino acids, peptides, proteins and many
other molecules would interfere the analysis and suppress the signal

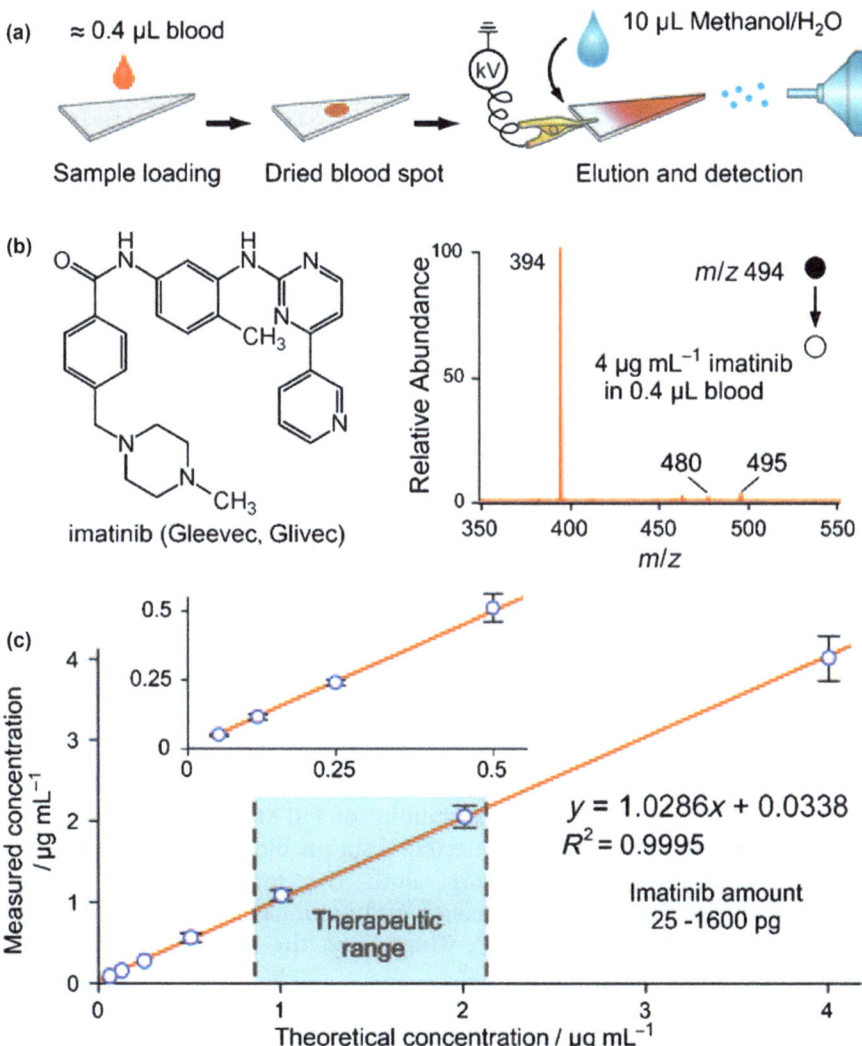

Figure 16.12 (a) Dried blood spot analysis using PS-MS. 0.4 μL whole blood was applied directly to a triangular chromatography paper. 4.5 kV DC voltage was applied to the paper triangle wetted with 10 μL methanol/water (1:1 v/v) to generate charged droplet for ionization. (b) Molecular structure of imatinib and tandem mass spectrum collected from 0.4 μL blood (spiked with 4 μg mL^{-1} imatinib). (c) Quantitative analysis of imatinib in whole blood using PS-MS. Imatinib-d8 was used as internal standard.
Figure taken from ref. 1 with permission.

intensity of the analytes of interest. This initial demonstration showed that matrix effect was sufficiently suppressed with paper spray and adequate analytical performance was obtained for therapeutic drug monitoring (TDM).

Table 16.1 Intraday and interday tests for quantitation of imatinib in dried blood spots using PS-MS.[14]

Imatinib concentration	Intraday ($n = 6$ per day)			Interday ($n = 18$)
	Day 1	Day 2	Day 3	
4 ng mL^{-1} (LOQ)				
Mean accuracy (%)	117	102	108	109
Standard deviation (%)	9.6	10.6	6.2	10.2
32 ng mL^{-1}				
Mean accuracy (%)	106	107	104	105
Standard deviation (%)	4.4	4.6	4.3	4.3
320 ng mL^{-1}				
Mean accuracy (%)	102	105	101	103
Standard deviation (%)	2.4	3.3	1.8	3.2
3200 ng mL^{-1}				
Mean accuracy (%)	104	106	101	104
Standard deviation (%)	1.8	1.8	2.8	2.8
8000 ng mL^{-1}				
Mean accuracy (%)	103	103	100	102
Standard deviation (%)	2.5	1.1	2.0	2.2

The relative matrix effects in the quantitative analysis of drugs in blood using paper spray were assessed with blood samples from five donors. Each blood sample was spiked with internal standard and tested with PS-MS.[14] The calibration curves obtained for blood samples from each donor were plotted and compared. The variation in slope among the five curves was only 1.3%, indicating a minimal impact by the relative matrix effects on the drug quantitation by PS-MS. The reproducibility of PSI-MS was also evaluated by conducting the intraday and interday tests on blood samples. The test results are shown in Table 16.1. Both the intraday and interday assessments give the adequate precision and accuracy over the tested range from 4 ng mL^{-1} to 8000 ng mL^{-1}, which meet the requirement for therapeutic drug monitoring.

Paper spray has also been used for direct quantitative analysis of tobacco alkaloids from biofluid samples,[9] which is of great importance for testing of tobacco use, tobacco-cessation treatment and studies on exposure to secondhand smoke. The half-life of nicotine in the body is relatively short and the major metabolite of nicotine, cotinine, is typically used as the biomarker for tests of cigarette smoking or exposure to tobacco smoke. The cotinine is further metabolized to trans-3′-hydroxycotinine and their ratio can be used as an activity measure of CYP2A6, an enzyme responsible for nicotine metabolism. In the case of differentiation of taking nicotine medications from cigarette smoking, anabasine is used as marker, as it only exits in tobacco but not in nicotine-medication products. Paper spray was applied for direct analysis of these tobacco alkaloids in blood, urine, and saliva samples in liquid and dried forms. LOQs of several nanograms per milliliter were obtained. Direct analysis of fresh blood was also developed using print paper. The fresh blood sample of 5 μL was deposited onto the print paper and a thin blood film formed on the surface of the paper without immediate diffusion into the paper. Acetonitrile (20 μL) was then added to facilitate the diffusion of the blood into the paper and a high voltage of 3.5 kV was applied to produce the paper spray. Improvement in LOQ of 1 order of magnitude was achieved using this method and an LOQ of 0.1 ng mL^{-1} was obtained for nicotine. The developed paper spray method was validated by quantitation of cotinine in the blood of a rat, with a comparison analysis using liquid chromatography MS.

DBS has been used as a blood-sampling and -storage method for neonatal screening for decades. The direct analysis of DBS using paper spray has a strong implication on the potential of PS-MS in the field of neonatal screening. As a demonstration, direct quantitation of acylcarnitines in blood and serum was carried out using paper spray.[11] The levels of acylcarnitines in blood are used for diagnosis of fatty acid oxidation disorder. Dried spots on paper prepared with serum and blood samples containing a panel of 10 acylcarnitines (C2–C18) were analyzed in the precursor ion scan mode with product ion *m/z* 85 for the common loss of the acyl side chain. Five internal standards (C2-d3, C3-d3, C5-d3, C8-d3 and C16-d13) were added for quantitation (Figure 16.13(a) and (b)). The LODs are shown in Figure 16.13(d), in relevance to the concentration ranges for diagnosis. Linearity with $R^2 > 0.95$ and reproducibility with RSD of ∼10% were achieved in the concentration ranges from 100 nM to 5 μM for the C2 acylcarnitine, and from 10 to 500 nM for other acylcarnitines tested.[11] This study not only established the feasibility of using PS-MS for direct quantitation of acylcarnitines, but also explored a method for simultaneously quantifying a number of biomarkers using paper spray with a single DBS, which are highly relevant to many other clinical applications.

High-throughput DBS analysis using paper spray has also been developed and demonstrated (Figure 16.14).[15] A paper substrate with an array of triangles was fabricated and used to hold multiple DBS samples. An automated platform was used to move the array of paper triangles across the inlet of the mass spectrometer, while the spray solvent and a high voltage was applied to generate the paper spray for analysis. The speed of the analysis was varied and its impact on quantitation and carryover between each two samples were characterized. Good LODs, eg. 1 ng mL^{-1} for sunitnib in blood, were obtained at an analysis speed of 7 s per sample with no crosscontamination problem observed.

16.3.3 Biological Tissue Analysis

Besides solution-phase samples, solid-phase samples like biological tissues can also be analyzed using paper spray. The tissue samples obtained by needle-aspiration biopsy or punch biopsy were deposited on paper and directly analyzed using PS-MS by applying the spray solvent, which extracts the analytes out of the tissue for subsequent spray ionization at the tip of the paper substrate (Figure 16.15).[10] The tissue debris was fixed on the paper without causing contamination to the mass spectrometer while the

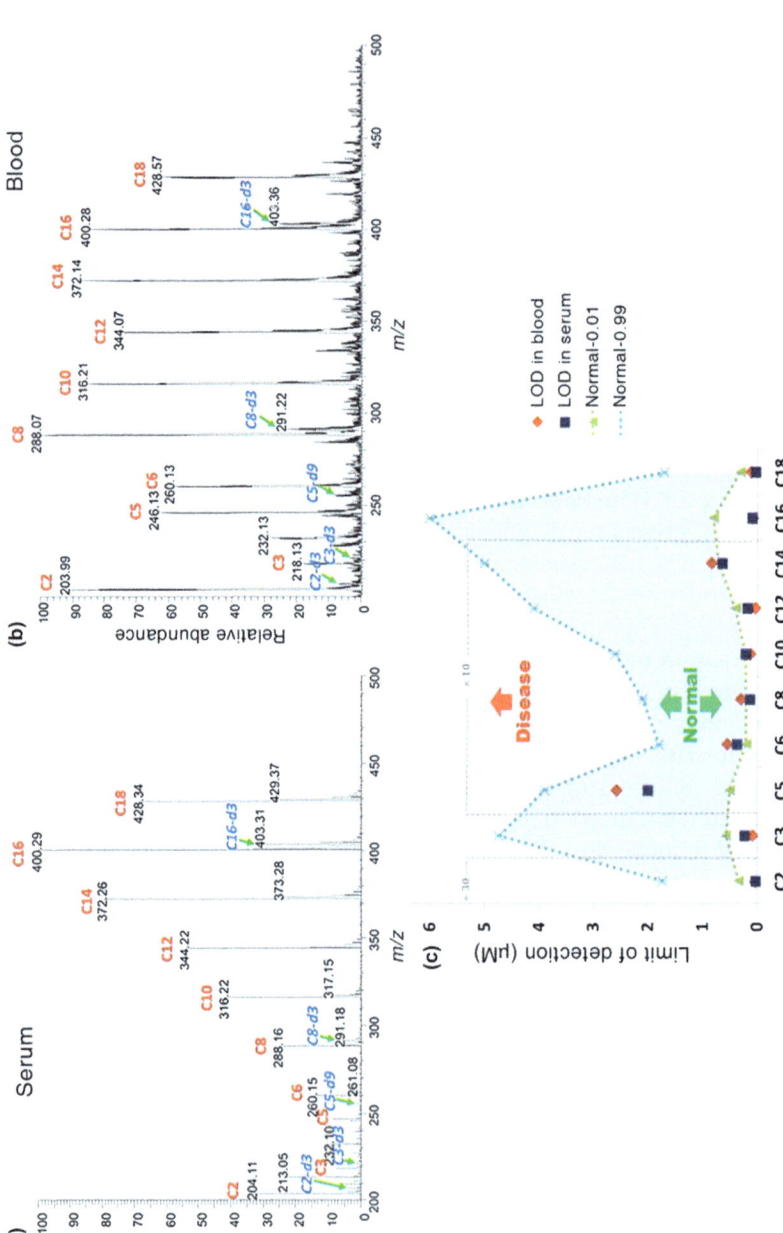

Figure 16.13 Spectra recorded in the precursor ion MS/MS mode for acylcarnitine in (a) dried serum spot and (b) dried blood spot, respectively. Acylcarnitine calibrators: C2, 5 μM; other acylcarnitines, 500 nM. Internal standards: C2-d3: 1 μM, C16-d3: 400 nM, others: 200 nM. (c) Limit of detection obtained with PSI-MS for acylcarnitines in serum and in blood, respectively. The shaded area between the two dotted lines is the normal range of acylcarnitines. Blue squares: the limit of detection for acylcarnitines in serum. Red diamonds: the limit of detection for acylcarnitines in blood. Figure adapted from ref. 11 with permission.

(a)

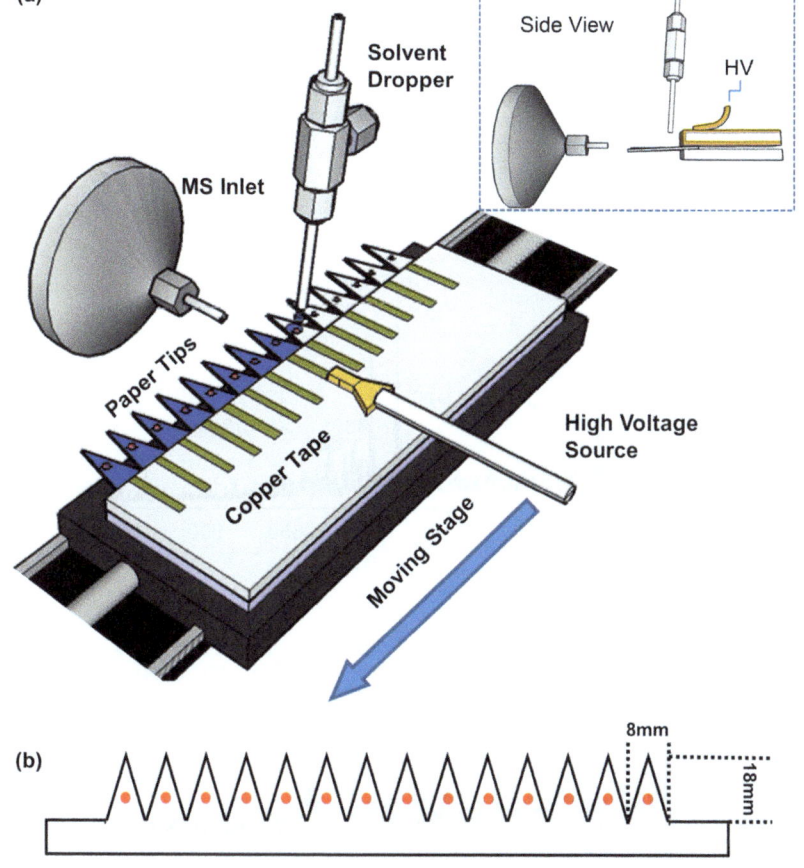

Figure 16.14 (a) High-throughput PS–MS platform. Inset shows a side view of the device. (b) The paper substrate with multiple triangles for loading samples.
Figure taken from ref. 15 with permission.

endogenous chemicals could be effectively extracted and ionized for MS analysis. A wide range of chemicals were observed from different tissue samples, including lipids, hormones, and ingested drugs. The lipid profiles obtained for tumor and normal tissues from human prostate are shown in Figure 16.15(b) and (c). A clear distinction could be made based on the difference in the relative intensities, which could be used for developing biomarkers in clinical diagnosis of cancer.

In another study with analysis of plat tissues, the polar lipids of green microalgae were characterized using paper spray and multiple MS experiments including MS/MS, exact mass measurements and gas-phase ozone reactions (Figure 16.16).[16] The lipid species were extracted and ionized by paper spray. Product ion scan experiments were used to determine the lipid head groups and fatty acid compositions; the precursor ion scan

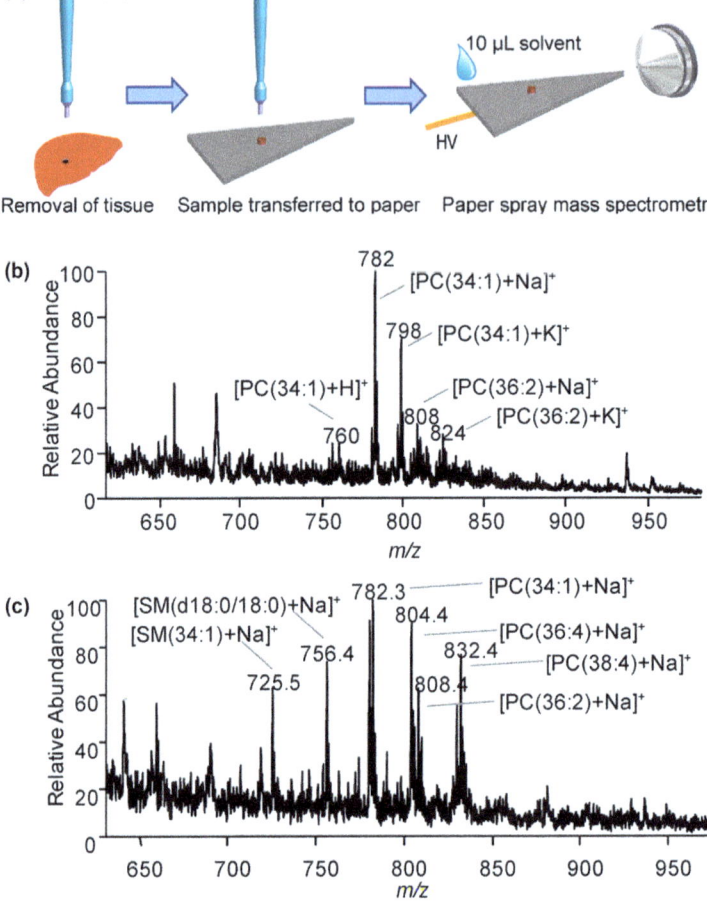

Figure 16.15   (a) Procedure of using tissue biopsy to collect tissue for PSI-MS analysis. Spectra of (b) tumor and (c) normal sections of human prostate tissue tested with PSI-MS.
Figure adapted from ref. 10 with permission.

experiments, using fragment ions such as *m/z* 184 for common phosphocholine headgroup, were used to confirm the lipid classification; exact mass measurements were used to determine the elemental compositions; the cleavage of double bonds by ozone reactions was finally used to determine the position of the unsaturation. By applying this experimental protocol, the molecular formulas and key aspects of the structures of glycerophosphocholines (PCs) such as 9Z-16:1/9Z,12Z-16:2 PC and 6Z,9Z-18:2/6Z,9Z,12Z-18:3PC and monogalactosyldiacylglycerols (MGDGs) such as 18:3/16:3MGDG were identified in the positive-ion mode, while glycerophosphoglycerols (PGs) such as 18:3/16:0 PG and sulfoquinovosyldiacylglycerols (SQDGs) such as 18:3/16:0 SQDG were identified in the negative-ion mode.

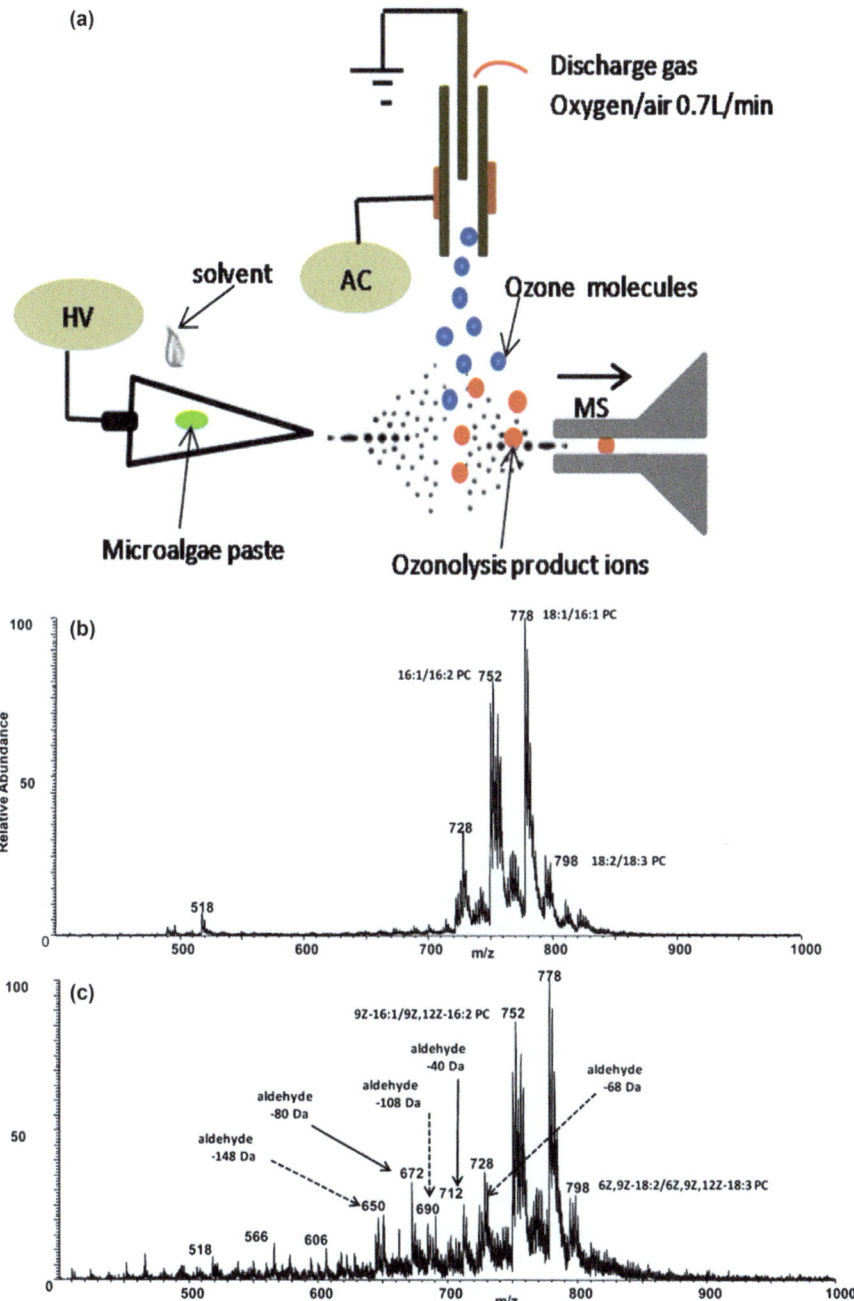

Figure 16.16   (a) Configuration of reactive PSI-MS using ozone generated by LTP as
the reagent. A kyo-Chlorella microalga was tested with this setup in
positive-ion mode. Spectra were collected (b) before and (c) after the
introduction of ozone.
Figure adapted from ref. 16 with permission.

16.3.4   Food Safety

Food safety gains increasing attention in many countries and can certainly benefit from the potentially fast and possibly on-site screening by ambient ionization mass spectrometry. Paper spray MS was demonstrated as a simple and fast method for direct analysis of contaminants or illicit additives in foodstuffs, including many chemicals involved in recent incidents of food safety, such as melamine, clenbuterol, plasticizers, sudan red and other chemical dyes.[17] These chemicals were tested with a variety of food samples, including meat, milk, chili powder and juice. Limited sample preparation was adopted, such as the homogenization of meat or sonication of powder samples in water. Tandem MS was applied for the analysis to minimize the interferences from complex matrices. The quantitative results and working conditions for each sample are shown in Table 16.2.

16.3.5   Checkpoint Chemical Detection

The paper used for paper spray is also a good material for collecting the samples through a "Swiffer" method. The samples in a large surface or on multiple subjects can be easily collected onto the paper by wiping the surfaces, which is highly compatible with the operation for checkpoints. The samples over an area larger than 100 cm^2 can be easily collected on a small area (<1 cm^2) of a piece of paper and subsequently analyzed by paper spray MS. The sample-collection process acts as a preconcentration of the analytes thus the signal intensity can be greatly improved. Figure 16.17 shows the analysis of heroin on a benchtop and thiabendazole on a lemon peel with the "Swiffer" sampling method and PS-MS.[2] The same sampling method was also used for cleaning validation of pharmaceutical equipment, which is critical for quality control in pharmaceutical industry.[18]

16.4   Ambient Ionization Methods Related to Paper Spray

16.4.1   Leaf Spray

Leaf spray[19] is an ambient ionization method sharing similar principles with paper spray. The spray solvent is applied on a plant material cut with a sharp tip and the endogenous compounds are extracted and ionized when a high voltage is applied directly on the plant material to generate the spray at the tip (Figure 16.18). Different from paper spray, the plant material serves as both the substrate and sample. For juicy plant parts such as a green onion leaf, the spray ionization can occur with their natural sap and the spray solvent is not required. Leaf spray ionization is a simple and fast method for phytochemical analysis, applicable to a wide variety of species from various

Table 16.2 Quantitative analysis of contaminants in various food stuffs using PSI-MS.[17]

Analyte	Matrix	Linear range	LOD	Solvent
Clenbuterol	Pork	1.0–1000 ng g^{-1}	1.0 ng g^{-1}	9 : 1 methanol–water
	Beef	5.0–1000 ng g^{-1}	5.0 ng g^{-1}	
Terbutaline	Pork	1.0–1000 ng g^{-1}	1.0 ng g^{-1}	9 : 1 methanol–water
	Beef	5.0–1000 ng g^{-1}	5.0 ng g^{-1}	
Salbutamol	Pork	5.0–1000 ng g^{-1}	5.0 ng g^{-1}	9 : 1 methanol–water
	Beef	5.0–1000 ng g^{-1}	5.0 ng g^{-1}	
Ractopamine	Pork	1.0–1000 ng g^{-1}	1.0 ng g^{-1}	9 : 1 methanol–water
	Beef	1.0–1000 ng g^{-1}	1.0 ng g^{-1}	
Melamine	Milk	20–1000 ng mL^{-1}	20 ng mL^{-1}	99 : 1 acetonitrile–water containing 26 μM citric acid and 5.0 mM ammonium acetate
	Formula powder	50–1000 ng g^{-1}	50 ng g^{-1}	
Sudan Red II	Chili powder	50–1000 ng g^{-1}	50 ng g^{-1}	9 : 1 dichloromethane–isopropanol containing 0.5% water
Sudan Red III	Chili powder	50–1000 ng g^{-1}	50 ng g^{-1}	9 : 1 dichloromethane–isopropanol containing 0.5% water
Sudan Red IV	Chili powder	50–1000 ng g^{-1}	50 ng g^{-1}	9 : 1 dichloromethane–isopropanol containing 0.5% water
Sudan Red G	Chili powder	100–1000 ng g^{-1}	100 ng g^{-1}	9 : 1 dichloromethane–isopropanol containing 0.5% water
Bis(2-ethylhexyl) phthalate (DEHP)	Sports juice	200–10 000 ng mL^{-1}	200 ng mL^{-1}	9 : 1 methanol–water containing 0.2% formic acid
Bis(2-ethylhexyl) adipate (DEHA)	Sports juice	50–10 000 ng mL^{-1}	50 ng mL^{-1}	9 : 1 methanol–water containing 0.2% formic acid

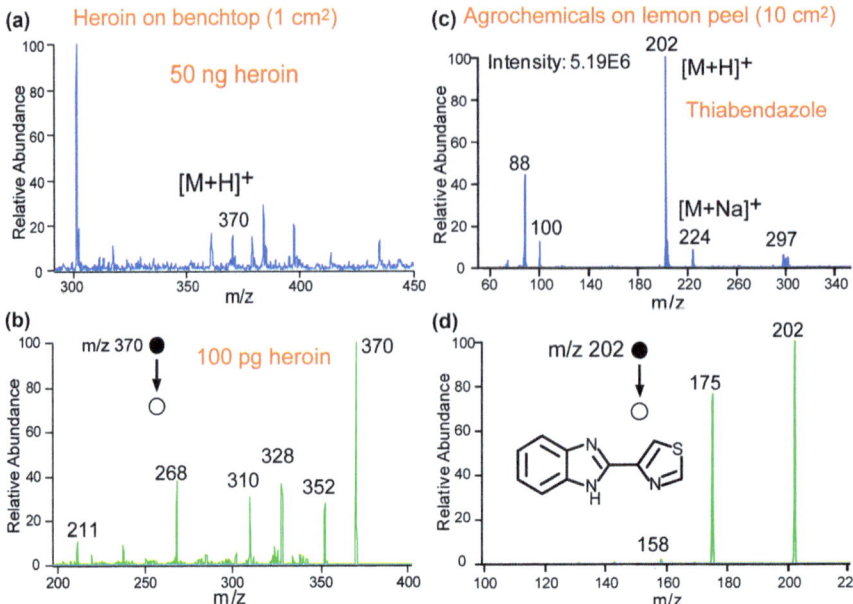

Figure 16.17 Direct analysis of chemicals from different surfaces using paper wiping and PSI-MS. (a) Full-scan spectrum and (b) tandem mass spectrum of heroin detected from a dirty desktop. (c) Full-scan spectrum and (d) tandem mass spectrum of thiabendazole detected from a lemon peel.
Figure taken from ref. 2 with permission.

plant parts. Information becomes readily available for small compounds in plants, including sugars, amino acids, fatty acids, lipids, and alkaloids. A comparison study has been made for direct analysis of steviol glycosides in *Stevia* leaves,[20] using various methods including leaf spray, paper spray, desorption electrospray ionization and low-temperature plasma probe. Leaf spray has been shown to produce the highest quality spectra. The allergenic urushiols from poison ivy (*T. radicans*) leaves could be directly analyzed with leaf spray without any sample preparation.[21] The pesticide residues (acetamiprid, diphenylamine, imazalil, linuron, thiabendazole) on fruits and vegetables (apple, pear, lemon, orange, carrot, cucumber, eggplant, potato) could also be direct quantified using leaf spray.[22]

The response by the plants to external stimulations and its reflection in chemical composition can now be monitored in real time by leaf spray MS analysis, which potentially could serve as an important research tool for plant science. As a demonstration, the stress-induced cleavage of glucosinolates in leaves of cabbage, Brussels sprout and cauliflower and Arabidopsis thaliana was monitored in real time. The abundances of intact glucosinolates were observed to quickly decrease after the leaves were cut, due to the hydrolysis reactions with myrosinase as a response to herbivore attack.[19]

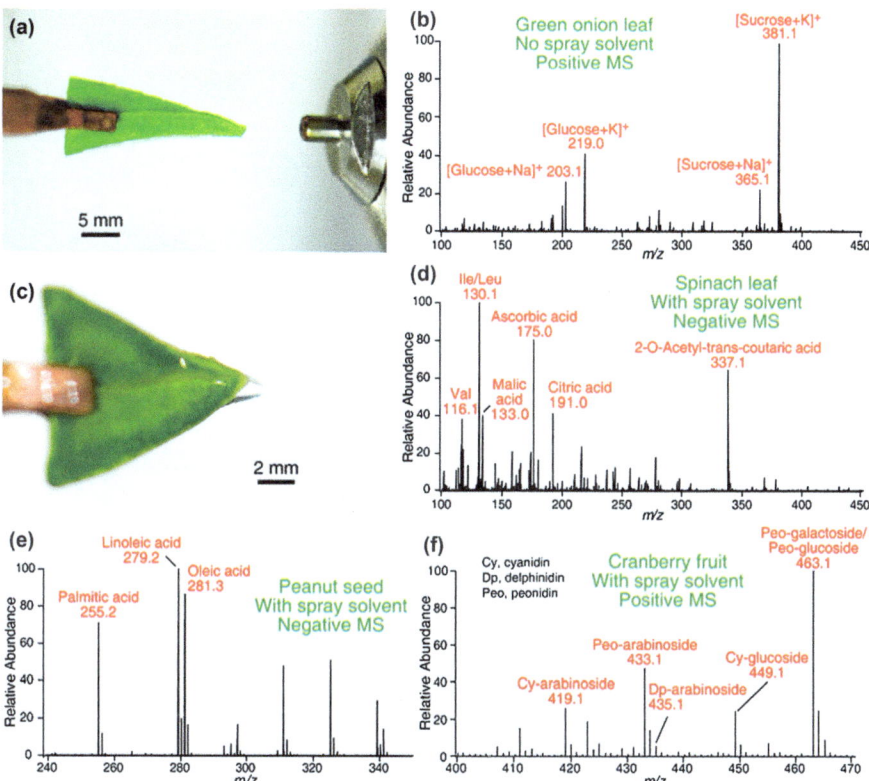

Figure 16.18 (a) Photograph of leaf spray ionization of green onion leaf without the application of spray solvent and (b) the corresponding spectrum in positive-ion mode, showing sucrose and glucose ions. (c) Photograph of leaf spray ionization of spinach leaf with the application of methanol and (d) the corresponding spectrum, showing amino acids and organic acids. (e) Spectrum acquired from peanut seed in negative-ion mode, showing three fatty acids. (f) Spectrum acquired from cranberry fruit in positive-ion mode, showing a series of phytochemicals. Figure taken from ref. 19 with permission.

16.4.2 Tissue Spray

The direct analysis of tissue using paper spray was described above, by applying the tissue samples onto the paper substrate prior to analysis. However, a direct spray off the needle biopsy tissue samples has also been explored for MS analysis.[23] Needle biopsy is a less-invasive method with a small amount of tissue extracted from a patient for pathologic test. The tissue biopsy itself can be used to generate charged droplets as shown in Figure 16.19. The tissue biopsy held in the needle tip was wet by spray solvent (2–5 µL) and a high voltage was directly applied on the metal needle to generate the emission of droplets containing chemicals extracted from the tissue.[23] This method was performed for tissues biopsies obtained from

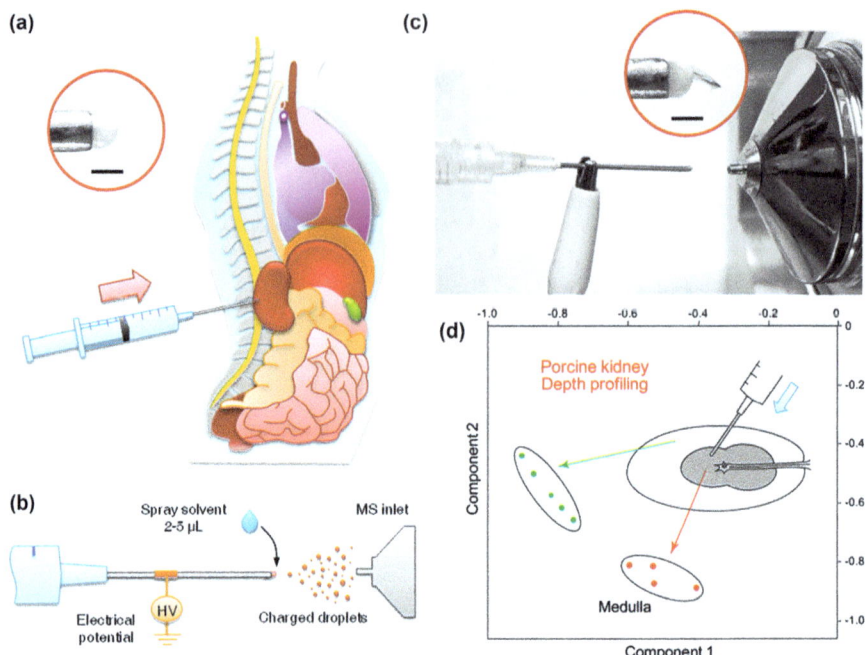

Figure 16.19 Direct spray ionization from a biopsy needle for biological tissue analysis. (a) Extraction of tissue biopsy from the organ by inserting a biopsy needle into the organ. The extracted tissue was partially released from the needle by pushing the plunger along the barrel and held by the needle tip. (b) Schematic and (c) photograph of the direct spray ionization of needle biopsy tissue. Spray voltage (4.5 kV) and spray solvent (25 μL methanol) were applied to form a Taylor cone on the tissue to generate charged droplets. Scale bar is 1 mm. Figure adapted from ref. 23 with permission

different organs including brain, liver, kidney, adrenal gland, stomach, and spinal cord. Distinct patterns for lipids were obtained for normal and cancerous tissues from a human kidney. Needle biopsy can be used to collect tissues from the internal part of an organ, providing a feasible way to perform a depth profiling of the chemical distribution in organ tissues within a human or animal body (Figure 16.19(d)). This was demonstrated with the direct analysis of tissues samples taken at different depths from a porcine kidney.

16.4.3   Extraction Spray Ionization

Paper spray represents the simplest format of sampling ionization and is highly suitable for development of disposable sample cartridges. The same paper substrate is used for separation of the analyte from the sample matrix, transport of the extract and spray ionization. Another method, extraction spray ionization,[24] was developed with an initial intention to improve the stability of the spray and has been proved to be highly sensitive for analysis of ultralow amounts of samples. As shown in Figure 16.20(a), a paper strip (1 cm length, 0.5 mm width, 0.18 mm thickness) was loaded with sample (<1 μL) and inserted into a nano-ESI capillary tube. A solvent of 10 μL methanol, was added into the tube to soak the paper strip (Figure 16.19). A metal wire was then inserted into the tube, in contact with the solvent, and a high voltage (2.0 kV) was applied to generate a nano-ESI from the tip of the glass tube. During this process, the analytes in the sample fixed on the paper strip were extracted into the solvent and ionized by the nano-ESI for MS analysis. The entire process takes less than 1 min.

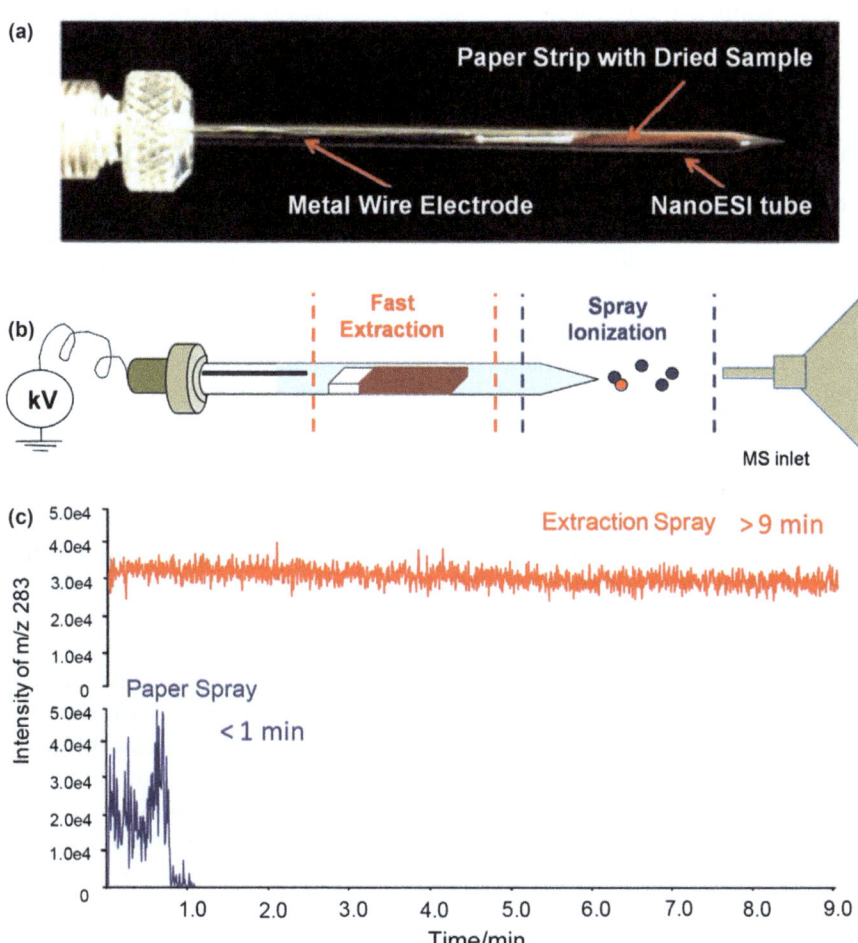

Figure 16.20 (a) Photography and (b) schematic setup of extraction spray ionization source used for direct analysis of whole blood sample. (c) Total ion current of sunitinib (200 ng mL^{-1}, *m/z* 399 → 283) tested in blood using extraction spray and PSI-MS, respectively.
Figure adapted from ref. 24 with permission.

As the same for paper spray, the suppression of the matrix effect has been achieved using extraction spray, with LODs as good as 0.1–1 ng mL^{-1} obtained for analysis of methamphetamine, nicotine, imatinib, verapamil, or sunitinib in blood, melamine in milk, clenbuterol in pork, atrazine in river water, and thiabendazole in orange.[24] In contrast with the short spray time of about 1 min for paper spray, spray longer than 20 min spray was achieved with only 10 μL methanol. The stability of the spray current and the analyte signals are also significantly improved, especially for the instruments like a QTrap 4000, which has an atmospheric-pressure interface with a curtain gas, drying the solvent very fast when paper spray was used. The desolvation of the spray was also found to be improved, which is less significant for lab-scale instruments but could be critical for quantitation using miniature mass spectrometers with an extremely simplified atmospheric-pressure interface.[25]

16.5   Development Toward Point-of-care Analysis

Paper spray has been demonstrated for direct quantitation of organic bio-markers in blood and other biofluid samples. The consumable for each analysis includes only a piece of paper substrate and a tiny amount of solvent, which makes it high suitable for point-of-care (POC) analysis. A disposable sample cartridge has been developed as shown in Figure 16.21(a)–(c), with a paper substrate and a stainless steel ball as-sembled in a polymer case.[8,25] A high-throughput paper spray source has been developed (Figure 16.21(d)) to allow a stack of these sample cartridges to be analyzed automatically using triple quadrupole instruments such as a TSQ (ThermoFisher Scientific, San Jose, CA). An ultimate goal is to have complete, compact analytical systems developed for nurses and physicians to use in clinical settings, which requires the development of miniature mass spectrometers (Figure 16.21(e)) and extremely simplified operation procedures.

As an important system component for chemical analysis outside tradi-tional analytical laboratories, miniature mass spectrometer have been under development for many years. A critical technical development with the dis-continuous atmospheric-pressure interface (DAPI)[26] allowed the coupling of miniature ion trap mass spectrometers, supported by small pumping sys-tems, with ambient ionization methods such as paper spray or a low-temperature plasma probe.[27] The instruments using ion traps retain the MS/MS capability, which is important for the direct analysis using ambient ionization methods. Without delicate sample purification and chromato-graphic separation, the chemical noise is likely to be high in MS spectra (Figure 16.22(a)) but the adverse impact can be minimized with the MS/MS scans (inset in Figure 16.22(a)).[1] Adequate specificity can also be provided at the same time based on the characteristic fragment pathways. The results for quantitation of lidocaine and verapamil[9] in blood using paper spray with silica-coated paper and a handheld Mini 10 ion trap mass spectrometer[26,28]

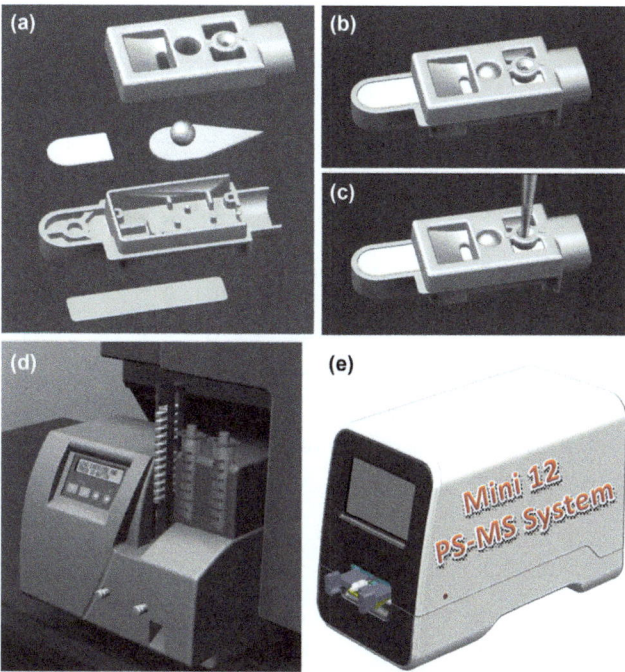

Figure 16.21 (a) Configuration of the disposable cartridge used for PSI-MS, (b) assembled cartridge, (c) blood sample loading onto the cartridge, (d) the automatic paper spray ion source mounted on a mass spectrometer. (e) Mini 12 miniature paper spray MS system for POC analysis of DBS in the disposable sample cartridge.

are shown in Figure 16.22(b) and (c). Recently the Mini 12 system (Figure 16.21(e)) was developed as a prototype for a complete POC solution using disposable paper spray cartridges. It reads a barcode on the cartridge, loads and executes the MS/MS scan functions, and calculates the concentrations of the analytes based on presaved calibration curves automatically. The entire analysis process requires minimal human intervention. An LOD of 7.5 ng mL^{-1} and an RSD better than 10% over the entire monitoring range have been achieved for quantitation of amitriptyline in blood.[25] The Mini 12 has also been demonstrated with applications for food safety and agrochemical monitoring.

The operational procedures for POC systems need to be simple, while the mandatory criteria of quantitation for clinical diagnosis need to be met. It has been extensively demonstrated with paper spray that the use of internal standards assures the highest precision of quantitation with direct MS analysis using ambient ionization methods. The standard means for IS incorporation is the premixing method widely used in analytical laboratories. However, the premixing method requires pipetting and vortex mixing that are not compatible with POC operation. The amount of blood needed is also much more than what is necessary for analysis using paper spray, where

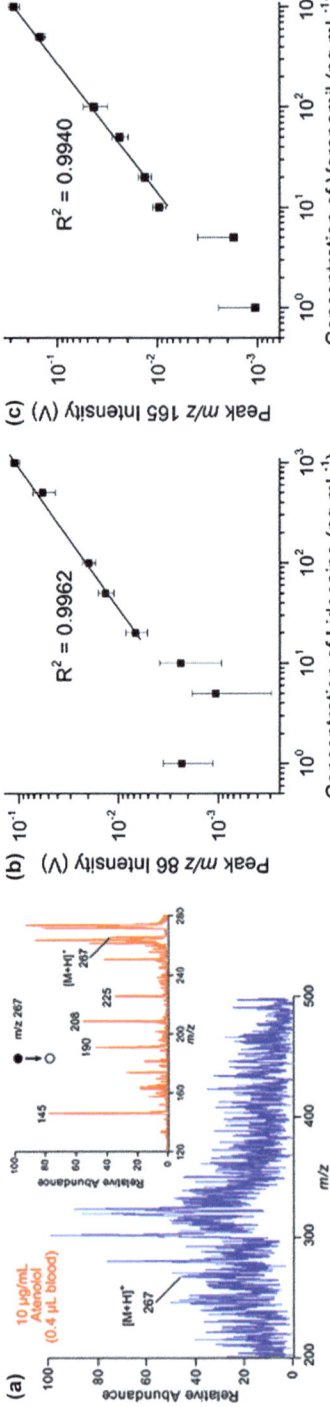

Figure 16.22 (a) Analysis of atenolol in blood (10 μg mL^{-1} in 0.4 μL blood) with PSI-MS using a handheld mass spectrometer (Mini 10). Paper substrate: chromatography paper. Calibration curves for (b) lidocaine and (c) verapamil in blood using paper spray and a handheld mass spectrometer (Mini 10). Paper substrate: silica coated paper. Panel a taken from ref. 29, panels (b) and (c) taken from ref. 30 with permissions.

Table 16.3 Recovery and precision for drug analysis with PS-MS using various methods of internal standard addition.[7]

Method of IS addition	Mean propranolol/ propranolol-d7 (AUC, $N = 7$)[a]	Standard deviation	Relative standard deviation
Mixed in liquid blood	1.05	0.03	3%
Dried on paper prior to blood spotting	0.73	0.06	8%
Added to dried blood punch	0.72	0.04	6%
Present in spray solvent	0.26	0.04	16%

[a]AUC, area under the curve.

blood samples of tens of microliters are sufficient and can be obtained by a finger prick. A series of alternative methods have been investigated for blood sampling using paper spray MS, including (i) prespotting the internal standard on the paper before the deposition of samples, (ii) spiking the internal standard into the spray solvent, and (iii) pouching out the blood-sampling card out to another paper substrate and adding the spray solvent spiked with the internal standard. Their performance was compared with the premixing method as shown in Table 16.3.[7]

Recently, a sampling method using IS-coated capillaries was developed, which is an extremely simple procedure while retaining high quantitation precision with small amount ($\sim$1 µL) of blood samples.[31] A glass capillary (0.4 mm I.D., $\sim$8 mm long) for the sampling was prepared by filling it with an IS solution through capillary action and letting it dry to form a IS coating on its inner surface. The capillary was then used to collect finger-prick blood containing the analyte, also through capillary action, and to transfer the blood sample to a paper substrate for subsequent PS-MS analysis. The internal standard was automatically mixed into the blood during the sampling and dispensing process. Analysis of imatinib in 1 µL blood was performed with the capillary transfer method and PS-MS (Figure 16.23(a)–(d)). RSDs <5% were achieved over the entire tested range from 10 ng mL^{-1} to 4000 ng mL^{-1}. As a versatile method for IS incorporation during sampling, the IS-coated capillaries also worked well with other ambient ionization methods, as demonstrated with a low-temperature plasma (LTP, Figure 16.23(e)) and desorption electrospray ionization (DESI, Figure 16.23(f)). The capability of performing highly precise quantitation with small amounts of blood benefits significantly the neonatal screening for infants and preclinical studies with small animals. The capillaries can be easily mass produced without requiring accurate control in length or volume, since the volumes of the internal standard solution and blood sample are both regulated by the volume of the capillary.

POC analysis can also benefit from a short analysis time. With paper spray already demonstrated for analysis of dried blood spots on paper, methods have also been explored for fast drying the blood samples on paper or direct analysis of fresh blood in liquid form. Alum as a coagulant was deposited on

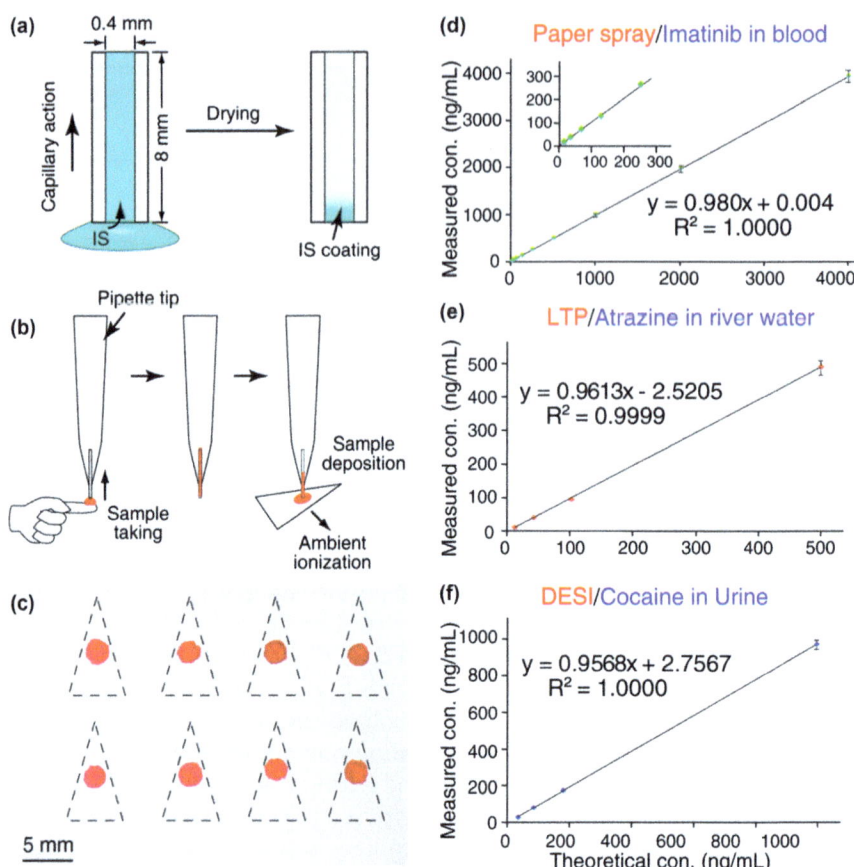

Figure 16.23   (a) Preparation of a capillary sampler (0.4 mm I.D., ~8 mm long) by filling it with an internal standard solution through capillary action and drying in air to form a coating of internal standard on the inner surface. (b) Blood sampling and deposition using a capillary sampler with a pipette tip as holder. The paper substrate loaded with blood was used for dried blood spot analysis. (c) Array of eight dried blood spots prepared on chromatography paper using eight capillary samplers. Paper triangles were cut out along the dashed lines for PSI-MS. Capillary sampler was used for quantitative analysis of (d) imatinib in blood with PSI-MS, (e) atrazine in river water with low-temperature plasma (LTP) and (f) cocaine in urine with desorption electrospray ionization (DESI).

Figure adapted from ref. 31 with permission.

paper to clot the blood in a short time such that a direct analysis can be immediately performed using PS-MS.[32] The entire process takes only 45 s for analysis of oncology drugs in blood samples. The method described above for direct analysis of nicotine alkaloids in liquid blood using print paper,[9] also remains interesting and promising for POC analysis.

Acknowledgements

The work reported was supported by the National Science Foundation (Grant CHE 0847205 and CHE 1307264) and the National Institutes of Health (1R01GM106016-01, 8R21GM103454, 1R21EB015722 and 1R21EB009459).

References

1. H. Wang, J. Liu, R. G. Cooks and Z. Ouyang, *Angew. Chem., Int. Ed.*, 2010, **49**, 877–880.
2. J. Liu, H. Wang, N. E. Manicke, J. M. Lin, R. G. Cooks and Z. Ouyang, *Anal. Chem.*, 2010, **82**, 2463–2471.
3. Y. Ren, H. Wang, J. Liu, Z. Zhang, M. N. McLuckey and Z. Ouyang, *Chromatographia*, 2013.
4. Z. Zhang, W. Xu, N. E. Manicke, R. G. Cooks and Z. Ouyang, *Anal. Chem.*, 2012, **84**, 931–938.
5. R. D. Espy, A. R. Muliadi, Z. Ouyang and R. G. Cooks, *Int. J. Mass Spectrom.*, 2012, **325**, 167–171.
6. V. Gabelica and E. D. Pauw, *Mass Spectrom. Rev.*, 2005, **24**, 566–587.
7. N. E. Manicke, Q. Yang, H. Wang, S. Oradu, Z. Ouyang and R. G. Cooks, *Int. J. Mass Spectrom.*, 2011, **300**, 123–129.
8. Q. Yang, H. Wang, J. D. Maas, W. J. Chappell, N. E. Manicke, R. G. Cooks and Z. Ouyang, *Int. J. Mass Spectrom.*, 2012, **312**, 201–207.
9. H. Wang, Y. Ren, M. N. McLuckey, N. E. Manicke, J. Park, L. Zheng, R. Shi, R. G. Cooks and Z. Ouyang, *Anal. Chem.*, 2013, **85**, 11540–11544.
10. H. Wang, N. E. Manicke, Q. Yang, L. X. Zheng, R. Y. Shi, R. G. Cooks and O. Y. Zheng, *Anal. Chem.*, 2011, **83**, 1197–1201.
11. Q. Yang, N. E. Manicke, H. Wang, C. Petucci, R. G. Cooks and Z. Ouyang, *Anal. Bioanal. Chem.*, 2012, **404**, 1389–1397.
12. A. Li, H. Wang, Z. Ouyang and R. G. Cooks, *Chem. Commun.*, 2011, **47**, 2811–2813.
13. A. W. Adamson and V. Slawson, *J. Phys. Chem.*, 1981, **85**, 116–119.
14. N. E. Manicke, P. Abu-Rabie, N. Spooner, Z. Ouyang and R. G. Cooks, *J. Am. Soc. Mass Spectrom.*, 2011, **22**, 1501–1507.
15. L. Shen, J. Zhang, Q. Yang, N. E. Manicke and Z. Ouyang, *Clin. Chim. Acta*, 2013, **420**, 28–33.
16. S. Oradu and R. G. Cooks, *Anal. Chem.*, 2012, **84**, 10576–10585.
17. Z. Zhang, R. G. Cooks and Z. Ouyang, *Analyst*, 2012, **137**, 2556–2558.
18. S. Jain, A. Heiser and A. R. Venter, *Analyst*, 2011, **136**, 1298–1301.
19. J. Liu, H. Wang, R. G. Cooks and Z. Ouyang, *Anal. Chem.*, 2011, **83**, 7608–7613.
20. J. I. Zhang, X. Li, Z. Ouyang and R. G. Cooks, *Analyst*, 2012, **137**, 3091–3098.
21. F. K. Tadjimukhamedov, G. Huang, Z. Ouyang and R. G. Cooks, *Analyst*, 2012, **137**, 1082–1084.

22. N. Malaj, Z. Ouyang, G. Sindona and R. G. Cooks, *Anal. Methods*, 2012, **4**, 1913–1919.
23. J. Liu, R. G. Cooks and Z. Ouyang, *Anal. Chem.*, 2011, **83**, 9221–9225.
24. Y. Ren, J. Liu, L. Li, M. N. McLuckey and Z. Ouyang, *Anal. Methods*, 2013.
25. L. Li, T.-C. Chen, Y. Ren, P. I. Hendricks, R. G. Cooks and Z. Ouyang, *Anal. Chem.*, 2014, **86**, 2909–2916.
26. L. Gao, R. G. Cooks and Z. Ouyang, *Anal. Chem.*, 2008, **80**, 4026–4032.
27. L. Gao, A. Sugiarto, J. D. Harper, R. G. Cooks and Z. Ouyang, *Anal. Chem.*, 2008, **80**, 7198–7205.
28. L. Gao, Q. Song, G. E. Patterson, R. G. Cooks and Z. Ouyang, *Anal. Chem.*, 2006, **78**, 5994–6002.
29. H. Wang, J. J. Liu, R. G. Cooks and Z. Ouyang, *Angew. Chem., Int. Ed.*, 2010, **49**, 877–880.
30. Z. Zhang, W. Xu, N. E. Manicke, R. G. Cooks and Z. Ouyang, *Anal. Chem.*, 2012, **84**, 931–938.
31. J. Liu, R. G. Cooks and Z. Ouyang, *Anal. Chem.*, 2013, **85**, 5632–5636.
32. R. D. Espy, N. E. Manicke, Z. Ouyang and R. G. Cooks, *Analyst*, 2012, **137**, 2344–2349.

Inlet and Vacuum Ionization from Ambient Conditions

SARAH TRIMPIN* AND BEIXI WANG

Wayne State University, Department of Chemistry, Detroit, MI, USA
*Email: strimpin@chem.wayne.edu

17.1 Introduction

Mass spectrometry (MS) is a powerful analytical technology for characterization of soluble materials at a molecular level with high sensitivity. However, many diseases and pathological pathways transform to intermediates and deposits that cannot or should not be dissolved under ionization-friendly conditions. These materials are problematic with MS as well as other analytical methods and technologies. Our, as well as other research groups, goal is to develop new analytical technologies capable of unraveling problems associated with complexity and solubility from ever-decreasing sample sizes while providing increasing speed and convenience of the measurement. High-resolution mass spectrometry (HRMS) in association with both gas- and solution-phase separation methods, and in conjunction with state-of-the-art ion-fragmentation methods, is a powerful approach to solving difficult analytical problems associated with complexity and dynamic range. Ion mobility spectrometry (IMS) provides effective separations of ion structures differing in charge and cross section (size and shape), increases dynamic range, and is applicable without the necessity of solvents. Thus, this separation approach is applicable with direct ionization methods. Solvent-free sample preparation combined with high resolution IMS-MS provides powerful 2-dimensional (2D) separation of drift

New Developments in Mass Spectrometry No. 2
Ambient Ionization Mass Spectrometry
Edited by Marek Domin and Robert Cody
© The Royal Society of Chemistry 2015
Published by the Royal Society of Chemistry, www.rsc.org

time *vs.* mass-to-charge (*m/z*). This approach requires ionization methods that can operate from the solid state, and where necessary with solvent-free sample preparation methods, ionize both volatile and nonvolatile compounds, and preferably produce multiply charged ions so that powerful high-performance mass spectrometers designed for electrospray ionization (ESI) can be employed.

New ionization technologies for MS have been recently introduced and are based on an unprecedented ionization process that is assisted by heat and vacuum. These soft and highly sensitive *inlet* ionization methods are termed laserspray ionization *inlet* (LSII),[1–15] matrix assisted ionization *inlet* (MAII),[13–17] and solvent assisted ionization *inlet* (SAII),[18–24] and, in analogy, *vacuum* ionization methods are termed laserspray ionization *vacuum* (LSIV),[13,14,25–27] and matrix assisted ionization *vacuum* (MAIV).[28–31] These newly discovered, so-called, *inlet* and *vacuum* ionization methods are operational from atmospheric pressure conditions. While neat small molecules and nonvolatile proteins can be ionized with *inlet* ionization,[14] the presence of a matrix compound, solid or solution, extends the method to *vacuum* ionization approaches and enhances the method's sensitivity and analytical utility. These ionization developments are widely applicable ranging from drug applications to top-down protein characterization.[3,15,31–33] Fundamentals and application areas are discussed with emphasis on high-performance and high-throughput measurements performed with an open source at atmospheric pressure.

17.2 Classification of Novel Ionization Processes into Ionization Methods

17.2.1 Fundamentals

Fundamental studies intended to understand the mechanism of the initial LSII invention that uses a laser to ablate a matrix-assisted laser desorption/ionization (MALDI) matrix, but produces ESI-like multiply charged ions (Scheme 17.1(A1))[2,3] led to the discovery of MAII (Scheme 17.1(A2))[13–17] for which no voltage or laser is required to produce equivalent results and SAII (Scheme 17.1(A3))[18–24] for which solvents replace the matrix. The family of *inlet* ionization methods (Scheme 17.1(A)) encompasses therefore samples in solid matrix common to MALDI and in solution common to ESI and in all cases produces ions having charge states nearly identical to those produced with ESI. The recent innovation of *vacuum* ionization (Scheme 17.1(B)) is operationally similar to LSII, but uses a laser to ablate the matrix:analyte from the surface under vacuum conditions (Scheme 17.1(B1)). In LSI (*inlet* or *vacuum*), contrary to MALDI (atmospheric pressure conditions or vacuum), molecules are not desorbed/ionized to give gas-phase ions.[9,13,14] For peptides, singly or multiply charged ions can be formed at will in LSI through proper acquisition conditions (*e.g.*, voltages and laser fluence) at atmospheric and intermediate pressure.[2,5,14] As noted above, MAII uses the same matrices as LSII and produces identical mass spectra by introducing the

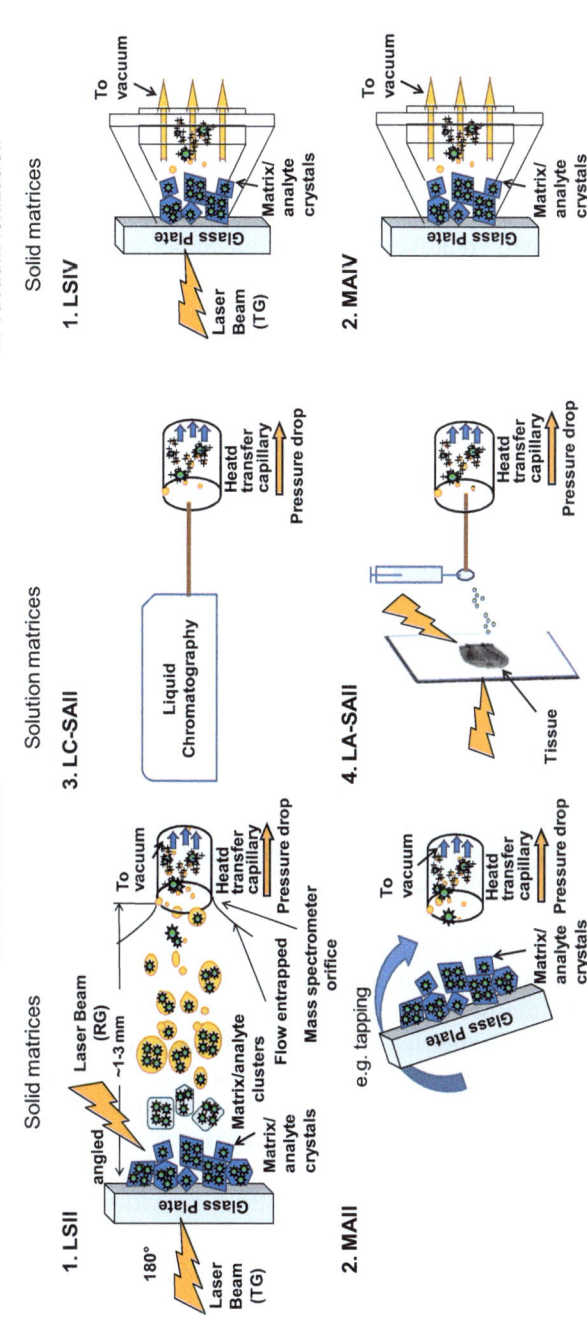

Scheme 17.1 A: Novel ionization process produces ESI-like charge states and ion abundances from small and large nonvolatile compounds: A. *Inlet* ionization: (1) Laserspray ionization *inlet* (LSII): laser ablation of a matrix in transmission geometry or reflection geometry in which the ablated matrix particles or droplets enter the inlet of a mass spectrometer where ionization occurs. (2) Matrix assisted ionization *inlet* (MAII): a solid organic matrix being physically introduced into the inlet for ionization. (3) Solvent assisted ionization *inlet* (SAII): introduction of a solution, here, from a liquid chromatograph into the inlet for ionization. (4) Laser ablation (LA)–SAII: solvent droplet held in front of a surface and collects materials ablated by a laser from the surface. The collected material is then transferred into the heated inlet of the mass spectrometer. B. *Vacuum* ionization: (1) Laserspray ionization *vacuum* (LSIV): the matrix:analyte is exposed to vacuum and a laser is used to ablate the matrix:analyte into the vacuum of a mass spectrometer. (2) Matrix assisted ionization *vacuum* (MAIV): spontaneous ionization of analyte in a matrix when introduced to the vacuum of the mass spectrometer without need of a laser. *Inlet* ionization can be performed on atmospheric pressure sources and *vacuum* ionization on vacuum and atmospheric pressure sources of commercially available mass spectrometers with minor or no instrument modifications[adapted from ref. 32].

matrix:analyte sample into a heated inlet tube by means other than a laser. MAIV is similar to MAII in that ionization occurs under vacuum conditions without use of a laser, but MAIV does not require a heated inlet tube. In MAIV, no external energy source is available except the energy in the system (matrix:analyte and the sample plate). This latent energy is believed to be released when the sample is exposed to vacuum conditions (Scheme 17.1(B2)).

The *vacuum* ionization methods produce multiply charged ions at low pressure on a MALDI time-of-flight (TOF) using LSIV[13,14,26] or at inter-mediate pressure with either LSIV or MAIV.[13,14,25–33] *Vacuum* ionization has the potential for higher sensitivity relative to ionization at atmospheric pressure, but is restricted relative to sample introduction. Initially, the sample was introduced to the vacuum using commercially available MALDI systems that requires minutes to place the plate into the position for ions to be analyzed by the mass spectrometer. Because MAIV relies on sublimation of the matrix for ionization,[28] initially only a single MAIV sample can be introduced at a time. In order to improve the speed and simplicity of an-alysis, *vacuum* ionization has been adapted for use on atmospheric-pressure ionization sources. In this configuration the sample is introduced to vacuum conditions directly from atmospheric pressure within seconds[30,34] instead of requiring minutes on vacuum MALDI sources.[28,29,31,35] Lasers can also be used to potentially enable a fast high spatial resolution means of *vacuum* ionization directly from atmospheric pressure conditions on mass spectrometers absent a heated inlet tube[36] or simply a stream of warm air.[37]

Common to the ionization methods discovered over the past four years is that ESI-like mass spectra are produced from solution or solid surfaces. In the *inlet* ionization, the sample in a matrix or solution is ionized in a region, preferably a tube, linking atmospheric pressure and the first vacuum stage of the mass spectrometer and in *vacuum* ionization by exposure to vacuum with or without laser ablation of the matrix. There exist relationships between applied heat, vacuum, and the use of a solid matrix or solvent. As noted above, in *inlet* ionization, a matrix is not needed (although encapsulated solvent cannot be ruled out) to produce ions from neat samples,[14] provided the material is exposed to appropriate heat and vacuum conditions. With matrix assistance, however, the temperature or vacuum requirement can be adjusted depending on the matrix properties and the sample can be greatly diluted in the matrix similar to MALDI, or for solvents, as in ESI. A large number of solid matrices have been shown to produce multiply charged ions from peptides and proteins at inlet temperatures in excess of 400 °C.[13] However, a few matrix compounds were found to produce ions by exposure to vacuum at room temperature[28] or by exposure to mild heat without ex-tensive vacuum.[37] Thus, neither a high voltage nor a laser is required and, with *inlet* ionization, the natural gas flow through the inlet aperture elim-inates the necessity of a nebulizing gas used in ESI.

Multiply charged ions produced by these methods extend the mass range of high performance mass spectrometers,[2–9,11,13–16,18–21,23,25–37] enhances separations in IMS,[4,6,11,14,19,27–31,33,36] and allows advanced fragmentation using electron-transfer dissociation (ETD).[3,8,13,28,30] Unlike ESI or MALDI, multiply charged ions are observed from surfaces, opening up an entire field of novel applications. Potentially, several modes of imaging a surface are available providing high spatial resolution or high-speed imaging.[10,12,27,32,36] *Inlet* ionization methods are also compatible with liquid chromatography (LC)/MS analyses.[19,21,22] These ionization methods are highly sensitive even in their early development.[2,19–22,27–37]

17.2.2 *Inlet* Ionization

17.2.2.1 *SAII*

Inlet ionization operates from atmospheric pressure through a heated inlet tube encompassing a pressure drop.[1–24] As can be seen in Figure 17.1(I),

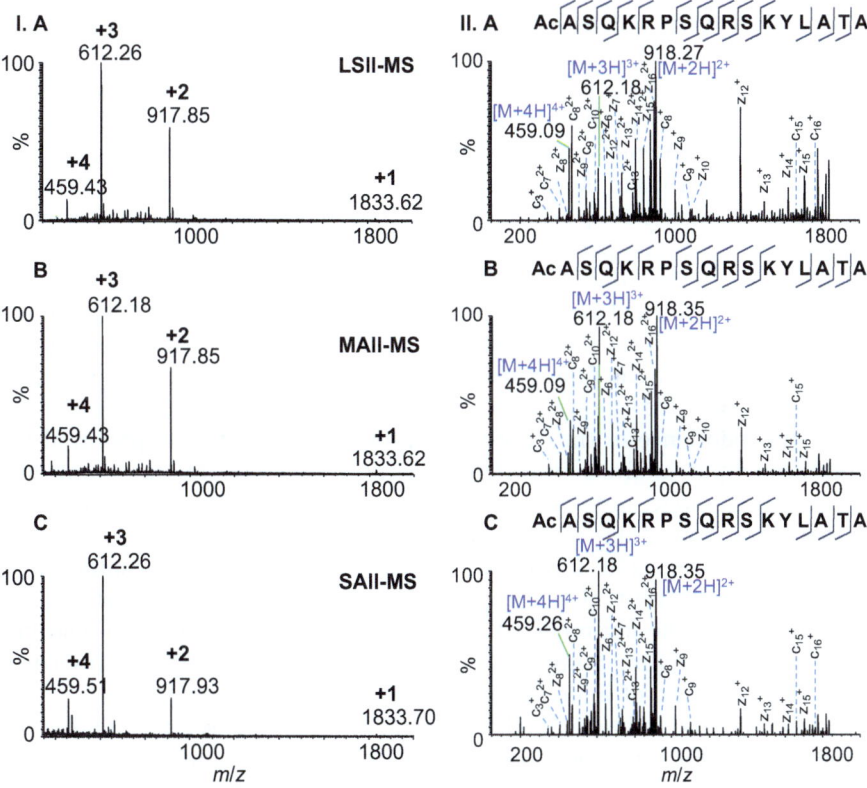

Figure 17.1 (I) MS and (II) MS/MS spectra of myelin basic protein fragment using (A) LSII, (B) MAII, and (C) SAII. The inlet temperature was 250 °C for all methods using the same analyte solution of water with 0.1% formic acid. 2,5-Dihydroxyacetophenon was used as matrix for LSII and MAII; water with 0.1% formic acid was used as solvent matrix for SAII. The +4 ions were selected and fragmented by electron-transfer dissociation (ETD). Similar MS/MS sequence coverage was obtained from all the three sample introduction methods.

using LSII, MAII, and SAII, similar mass spectra are obtained relative to charge states and ion abundance using the same sample and inlet temperature of 250 °C, but different matrices. Characterization of the +4 ions using ETD shows similar sequence coverage despite differences in how the sample was introduced to the mass spectrometer (Figure 17.1(II)).

SAII and LSII are subcategories of MAII. In other words, in LSII the laser ablates matrix particles or droplets into the inlet for ionization and in MAII other physical means are used to introduce the matrix to the inlet, and in SAII the matrix is a solvent. Therefore, SAII operates without a laser or a voltage and is capable of coupling to LC and nano-LC/MS/MS with high sensitivity.[19,21,22] While the "hot spot"[18] within the inlet capillary may gain some sensitivity, excellent results are obtained especially at low flow rates by allowing the natural gas flow from atmospheric pressure to vacuum to nebulize the solution into the inlet.[22,24] This approach introduces a great deal of flexibility. We demonstrated that nano-LC coupled with SAII-MS produces excellent quality chromatographic separation and mass spectra from injection of just 7 ng of a bovine serum albumin (BSA) tryptic digest.[22] This was achieved simply by positioning the exit end of the fused silica extension of the capillary column very near the instrument inlet as shown in Figure 17.2(A). No connections, special emitters, or nebulizing gas are required. A typical example is shown for a BSA tryptic digest analyzed by LC-SAII in Figure 17.2(B) providing good chromatographic resolution and ion abundance. Mass spectra of multiply charged ions of individual peptide can be extracted from the LC chromatogram as is shown in Figure 17.2(C) for the peptide extracted at 18.0 min. SAII has sensitivity comparable to, and in most cases studied to date better than, ESI, while displaying a higher degree of simplicity at flow rates ranging from 400 nL min^{-1} to at least 100 μL min^{-1}.[19,21,22]

The independence of the "hot spot" provides SAII with the potential of high-throughput analysis by driving multichannel pipette tips using an automated x,y-stage. In proof of principle experiments, 42 samples ranging in structures from drugs to proteins, were analyzed within 5 min using an inlet temperature of 250 °C without crosscontamination between samples.[24] Speeds of 1 second per sample were achieved with minor crosscontamination. The automated SAII high-throughput method can also be coupled with mapping software to show not only which pipette tips contained a compound of interest, but also its relative amounts.[24]

Further understanding gained from MAII and SAII allowed postulation of a mechanism for ionization that involves charge separation in matrix or solvent droplets related to the surface having an excess of one charge and the bulk the opposite charge. By rapidly removing surface, as in bubble formation, nonstatistical charging can impart hundreds of charges to a matrix or solvent droplet.[14] Desolvation of these charged droplets by evaporation or sublimation of solvent or matrix would produce the observed multiply charged ions by mechanisms related to the charge residue or ion evaporation models proposed for ESI.[14,38,39] Particle formation, postulated to be a transition stage for this new ionization process, have been verified by

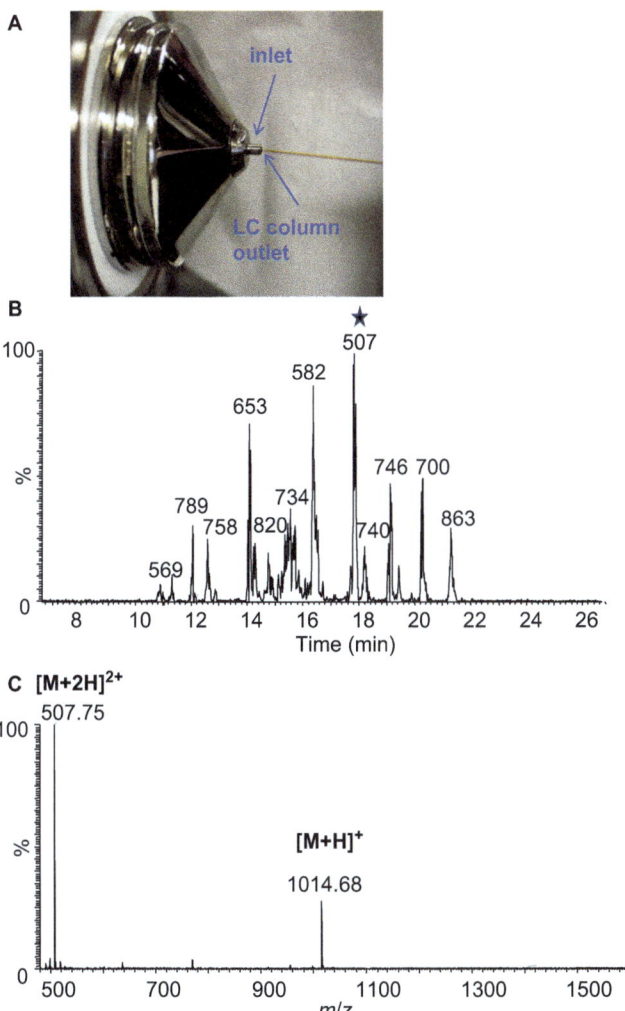

Figure 17.2 (A) Picture of the setup of LC–SAII where the outlet of the LC column was directly placed near the orifice of the MS inlet. The temperature of the inlet was held at 300 °C. (B) LC chromatogram of 7 ng BSA digest at the flow rate of 800 nL min^{-1}. Numbers labeled on the LC chromatogram are the *m/z* of the most abundant peaks observed in the mass spectra. (C) Mass spectrum extracted from the peak detected at 18.0 min and labeled 507. Singly and doubly charged ions of the peptide QTALVELLK were produced [adapted from ref. 22].

Zenobi and coworkers and Murray and coworker for LSII and MAII.[40–42] High-throughput methods based on SAII are being explored that are simple, fast, and improves on current high-throughput approaches.[24] SAII liquid-junction technology (Scheme 17.1(A.4))[32] is also being explored as another means of rapid analysis or imaging similar to reports using nano-DESI by other groups.[43–45]

17.2.2.2   LSII

LSII is a high spatial resolution method from the family of *inlet* ionization.[1,8,10,12] As noted above, LSII uses laser ablation to transfer the matrix:analyte sample into the heated inlet of a mass spectrometer where ionization occurs.[3,9,14] LSII is as sensitive as atmospheric pressure MALDI with the added benefit of producing abundant highly charged ions.[2,9] The energy for ionization is provided by the heat and pressure drop within the heated mass spectrometer inlet as demonstrated by the ability to produce nearly identical results without use of the laser.[16,17] LSII combines the attributes of MALDI, including speed of analysis[7] and high spatial resolution for imaging,[1,8,10,12] and those of ESI such as improved structural information through enhanced fragmentation (*e.g.* ETD)[3,8,13,28,30] and collision-induced dissociation (CID),[12] and better IMS separation.[4,6,11,14,19,27–31,33,36] Additionally, the range of compounds that can be studied using atmospheric pressure laser ablation methodology is extended to high-performance mass spectrometers that have limited *m/z* range. Thus, using LSII, a mass spectrum of ubiquitin (molecular weight (MW) 8561) was obtained directly from the solid state using 100 000 mass resolution and <5 ppm mass accuracy, and ETD fragmentation produced full sequence coverage (Figure 17.3).[3] LSII-ETD was employed in conjunction with high resolution MS for the first time to identify a peptide, the endogenous *N*-acetylated myelin basic protein fragment (MBP), directly from mouse brain tissue.[8]

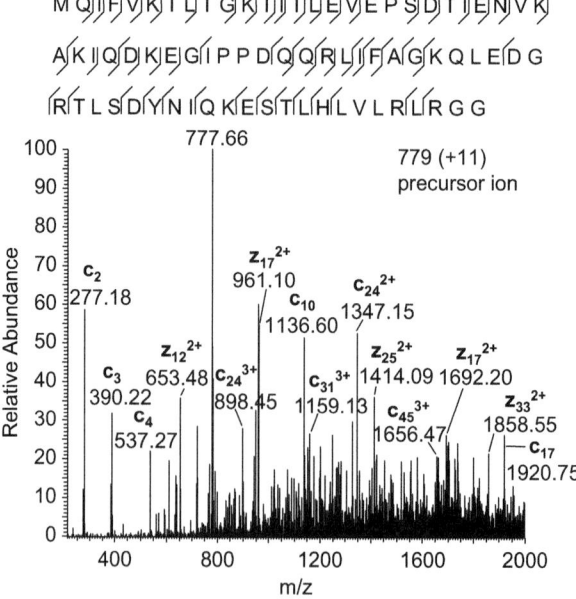

Figure 17.3   The ETD spectrum after proton-transfer reaction obtained from the +11 charge state of ubiquitin ionized by LSII on an LTQ mass spectrometer [adapted from ref. 3].

An obvious advantage of laser ablation is the high spatial resolution that can be achieved using transmission geometry,[1,2] which lends itself to imaging.[10,12] A great deal of research has gone into developing imaging using MALDI and because of its sensitivity and ability to image quasimolecular ions directly from biological tissue, vacuum MALDI has become the gold standard for molecular imaging, especially of nonvolatile biological compounds.[46,47] Nevertheless, imaging at atmospheric pressure is very attractive and has received considerable attention. For example, Spengler and coworkers have introduced a nearly zero angle laser-impact method for imaging with a spatial resolution at ~5 μm using at atmospheric pressure MALDI.[48] LSII addresses a number of shortcomings of MALDI including imaging at atmospheric pressure, higher mass resolution and mass accuracy afforded by the applicable mass analyzers, extended mass range, advanced fragmentation and improved IMS separations aided by multiple charging, faster acquisitions, and a softer ionization process. The imaging capability of LSII provides good spatial resolution (*ca.* 20 micrometer diameter laser-ablated spots (Figure 17.4),[8] fast imaging with inexpensive low repetition lasers (Figure 17.5(A)),[10] using a single laser shot per pixel, and imaging of labile compounds such as ganglioside lipids (Figure 17.5(B)).[12] The efficiency of LSII ionization also allows MS/MS to look specifically for a compound of interest (*e.g.* clozapine from the tissue section of a drug-treated mouse (Figure 17.5(C))[32] by LSII-MS/MS using CID for imaging the major fragment ion at *m/z* 270). The imaging of different charge states of the same

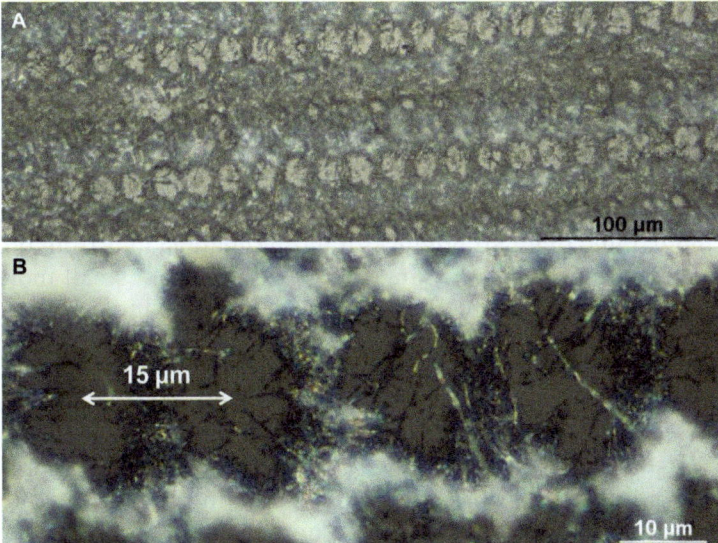

Figure 17.4 The microscopy images of mouse brain tissue section covered by matrix after LSII analysis using a xy-stage to systematically ablate the tissue section by the focused laser aligned in transmission geometry. The distance between the centers of two adjunct ablation areas is about 15 μm [adapted from ref. 8].

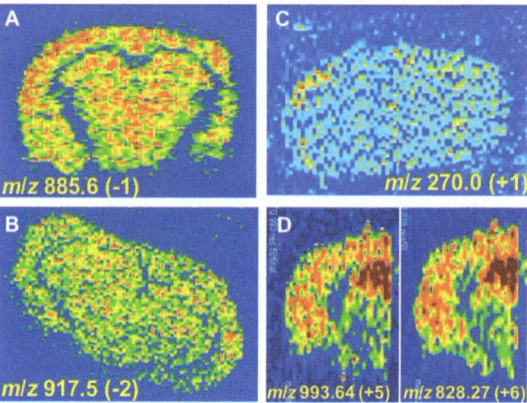

Figure 17.5 LSII images using MS of 10 μm thick mouse brain tissue sections displaying: (A) the −1 ions of an endogenous lipid at *m/z* 885.6; (B) the −2 ions of fragile ganglioside at *m/z* 917.5; (C) the +1 of the major fragment from clozapine at *m/z* 270 with a MW of 326. The mouse was treated by clozapine for 90 min before being sacrificed; (D) the +5 ions (*m/z* 993.64) and the +6 ions (*m/z* 828.27) of the same endogenous peptide [adapted from ref. 10, 12, 32].

protein ensures the accuracy of the technique (Figure 17.5(D)).[32] At its current stage of development, LSII-MS achieves attomole detection limits for peptides such as angiotensin I and II and low femtomole detection limits for proteins such as insulin and lysozyme.[9]

Initially, advanced solvent-free sample preparation methods were developed for MALDI-MS analyses of solubility restricted materials.[50] Combining LSII with solvent-free sample preparation holds promise for analyzing solubility restricted materials directly from surfaces with high mass resolution and mass accuracy. The solvent-free approach combines the analyte and matrix in the solid state without solvent.[51] Multisample solvent-free approaches[52] were developed and more recent solvent-free work demonstrated that LSII produces multiply charged ions from proteins and other macromolecules and singly charged ions from small molecules so long as the matrix : analyte sample is thoroughly ground together.[1,10,11] Solvent-free relative to solvent-based sample preparation approaches for MALDI and LSII analyses provide reliable ionization of hydrophobic and solubility restricted materials. The first MS imaging applications of the solvent-free approach applying matrix crystals of xxx micron using a mesh were shown for mouse brain tissue surfaces.[10,53] This approach is expected to minimize dislocation of compounds in the tissue, something that is inherent with solvent-based approaches.[1,10,53] IMS can be applied with solvent-free sample preparation to achieve total solvent-free analysis (TSA) encompassing solvent-free sample preparation, ionization, separation, and mass measurement.[11,50] The TSA approach is being studied as a possible means of addressing complexity involving solubility issues.

17.2.3 *Vacuum* Ionization

17.2.3.1 *LSIV*

The goal of producing highly charged ions similar to ESI and *inlet* ionization methods directly from surfaces using commercially available vacuum MALDI sources led to the discovery of what we have termed *vacuum ionization*. Ionization in vacuum eliminates ion losses inherent in ion transmission from atmospheric pressure to the mass analyzer and thus offers the potential for higher sensitivity. In this research, LSII matrices that required low inlet temperatures for ion formation using *inlet* ionization were tested to see if they formed multiply charged ions by laser ablation in the intermediate pressure MALDI source of a Waters SYNAPT G2 mass spectrometer.[13,14,25–27] The first observation of highly charged ions produced from the solid state in a vacuum MALDI-like procedure was discovered[25] using the matrix 2,5-dihydroxyacetophenon (2,5-DHAP) by tuning the vacuum MALDI SYNAPT G2 for ESI-like settings and using a low laser fluence. 2-Nitrophloroglucinol (2-NPG) was subsequently found to be a more volatile matrix that absorbed the laser energy, but did not evaporate or sublime during the process of inserting the sample into the vacuum MALDI source. This matrix not only produces abundant highly charged ions under intermediate pressure vacuum conditions, but the highest charge state ions observed to date under low-pressure conditions in a MALDI-TOF mass spectrometer.[26] Unlike the multiply charged ions produced with the matrix α-cyano-4-hydroxycinnamic acid (CHCA) under MALDI conditions, the ions produced with 2-NPG were stable and passed through the TOF reflectron intact allowing improved mass accuracy. For LSIV, without a heated inlet, the only means of supplying external energy is through absorption of the laser photon energy and this in the presence of vacuum conditions was sufficient to produce highly charged ions similar to LSII.

On the SYNAPT G2, multiply charged ions produced by this method enhances separation of ions by charge and shape (size) using IMS as well as fragmentation with CID or, potentially, ETD. The IMS dimension allowed imaging of endogenous peptides, including the MBP peptides previously characterized using LSII-ETD,[8] directly from mouse brain tissue ionized by LSIV and using a commercially available intermediate pressure source of the SYNAPT G2.[27] The extension of LSIV to Fourier transform (FT) mass spectrometers for protein characterization,[35] imaging[54] and/or electron-transfer dissociation (ECD) are anticipated from these results.

The results described above in combination with other fundamental studies, especially related to matrices,[13] suggests a common mechanism for ionization for all *inlet* and *vacuum* ionization methods.[14] The similarity of the mass spectra to ESI and the use of matrices, especially those used with MALDI, reinforce this conclusion. The ability to use MAII to remove the necessity of absorption at the laser wavelength allowed a wide variety of chemical structures to be discovered that act as MAII matrices producing multiply charged ions of peptides and proteins so long as the inlet provides sufficient heat presumably to produce charged droplets/particles and desolvate the droplets/particles.[13,14] Compounds such as anthracene and 2,2′-azobis(2-methylpropionitrile) produced multiply charged ions in MAII with the inlet heated to 450 °C, but only if the peptide was first dissolved in an acidic solution.[13] This work suggests that solid matrices act as solvents linking the method to SAII.

17.2.3.2   MAIV

Because 2-NPG is an especially good matrix compound for producing multiply charged ions and because its properties fit well the critical need to desolvate charged particles or droplets to produce multiply charged ions, similar aromatic nitro-containing compounds that are either volatile or known to sublime under vacuum conditions were tested for their ability to produce ions using low-temperature *inlet* ionization. 3-Nitrobenzonitrile (3-NBN) was discovered to produce abundant multiply charged ions of proteins and peptides using an inlet temperature as low as 50 °C,[29] and in a separate study even room temperature using a vacuum MALDI source inlet.[28] This compound is known to sublime in vacuum, and nanoparticles have been observed to dislodge from the surface,[55] possibly due to the pressure from sublimation[56] or expansion of trapped solvent. 3-NBN is also known to triboluminescence,[57] a process in which the opposite faces of a cracked crystal are charged and at atmospheric pressure a discharge between the faces produces the observed luminescence. In MAIV, the initial dislodging is hypothesized to create charged gas-phase particles from which the observed bare ions are produced by an ion evaporation (or field desorption) process occurring when matrix evaporation or fracturing produces sharp edges or points with high electric fields, or alternatively, by sublimation producing the bare ions from charged particles by the charge residue mechanism.[28,29,32]

Astonishingly, the 3-NBN matrix spontaneously transfers small molecules as well as proteins, at least as large as BSA (66 kDa), from the solid state to gas-phase ions in vacuum without a laser or other means of additional energy input.[28–31] The mass spectra are nearly identical to those observed using ESI, LSII, MAII, SAII, or LSIV. The MAIV method is extremely simple, only requiring a matrix and vacuum, available by default on any mass spectrometer. Simply placing a 1 μL drop of matrix on mouse brain tissue and inserting it into the intermediate pressure MALDI source of the SYNAPT G2 provides ions of proteins, peptides, lipids, and metabolites present in the tissue.[28,31] The solvent selected for dissolving the matrix can be used to select certain compound types as incorporation of the analyte in the matrix seems to be a necessity. MAIV-IMS-MS and MAIV-MS/MS are applicable to, for example, identify cocaine from a *specific* area on a dollar bill.[29] This indicates sufficient sensitivity for spatially resolved measurements.

In spite of the fundamental importance of this discovery, initial attempts to publish this work met with rejections. One consistent comment from the reviewers was that the method was not analytically useful because only one sample could be introduced at a time to the vacuum.[28,29] While this is true using the vacuum MALDI source, exposing the MAI sample to the vacuum at the inlet aperture of an atmospheric pressure ion source allows ionization under vacuum condition but sample manipulation from ambient conditions.[30] Initially, the atmospheric pressure inlet of the ESI source on the SYNAPT G2 mass spectrometer was modified by opening the skimmer cone inlet to provide the vacuum necessary for ion formation. The isolation valve was closed to prevent venting the instrument when the sample was not applied to the inlet using, *e.g.*, glass plates, metal foil, paper, or well plates (Scheme 17.2).[29] Thus, samples could be exposed to vacuum while being manipulated from an ambient environment.[28–31] Interestingly, even this modification was not necessary, as exposure of the sample to the commercial inlet cone or to the inlet of any mass spectrometer without heat applied to the inlet produced ions with excellent sensitivity. With 3-NBN as matrix, matrix:analyte sample dry or in solution can be introduced to any mass-spectrometer inlet on, for example, paper (Kimwipes or filter paper), from well plates, or from pipette tips (Scheme 17.2).[30] The simple operation ensures the potential of rapid progress for high-throughput analyses.[30] In a recent paper, it was demonstrated that 3-NBN as well as 2,5-DHAP, a common MALDI matrix, lifts small molecules and proteins into the gas phase as ions simply by the application of a warmed gas flow to the matrix:analyte solid placed on a glass capillary (Chapter 4).[37] Frozen and cold solutions of water or methanol were also demonstrated to produce ions of small and large nonvolatile compounds.[58]

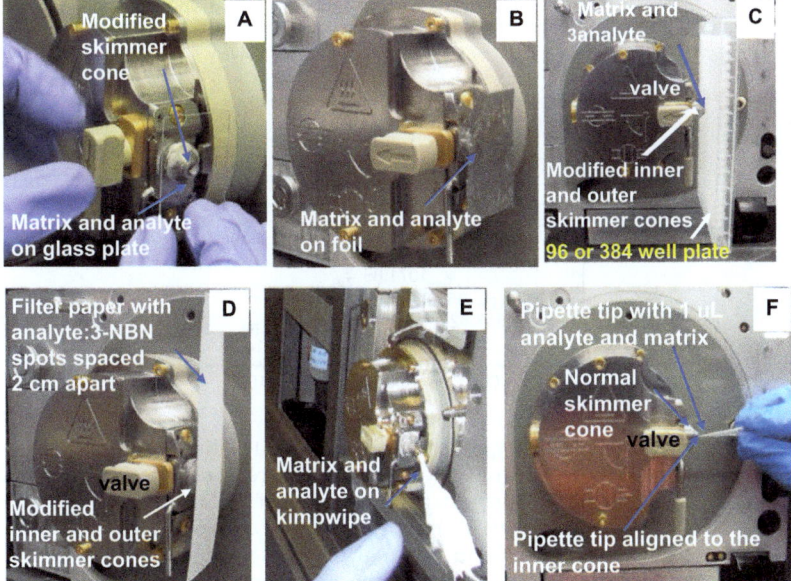

Scheme 17.2 Photos of different sample introduction methods using the atmospheric pressure source of the SYNAPT G2 and various sample holders including: (A) glass plate, (B) foil, (C) 384 well plate; (D) filter paper strip, (E) Kimwipes, (F) pipette tip [adapted from ref. 28 and 30].

MAIV was later made operational from ESI sources with minor or no instrument modifications other than overriding the source housing.[33-35] Four samples in less than one minute were analyzed from paper strips[30] and we expect this can be extended to high throughput using paper rolls (Scheme 17.2(D), Figure 17.6).[30] Multisample plates using a nozzle to

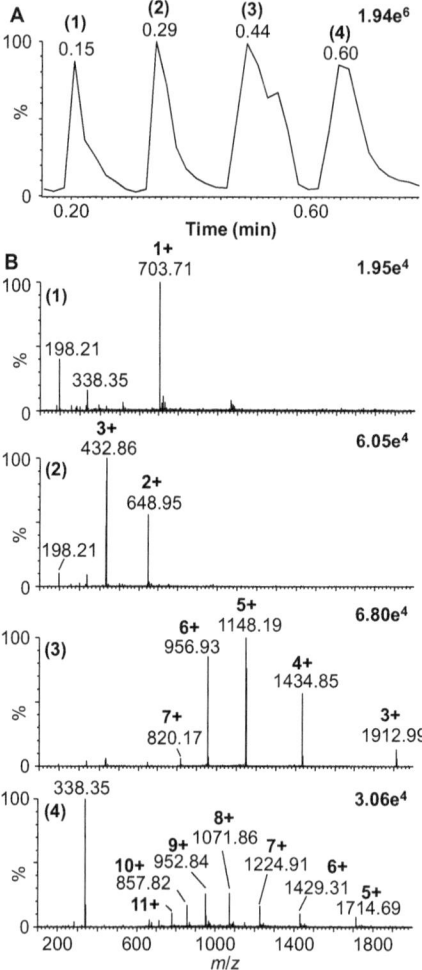

Figure 17.6 Toward high-throughput analysis by MAIV directly at atmospheric pressure. The matrix:analyte mixtures were spotted on a filter paper strip. The paper strip adheres to the inlet of the widened skimmer cone when the vacuum valve is opened, and is pulled across the inlet to introduce individual matrix:analyte spots sequentially as shown in Scheme 17.2(D). (A) The total ion chronogram of the analysis of the four analytes. (B) Mass spectra extracted from the total ion chronogram, displaying the production of singly and multiply charged ions, respectively, from (1) sphingomyelin, (2) angiotensin I, (3) bovine insulin, and (4) ubiquitin [adapted from ref. 30].

connect the skimmer opening with, *e.g.*, a 96-well plate (Scheme 17.2(C)), also produces ions by MAIV. MAIV is applicable with ETD,[28,30] and both positive and negative ions are produced.[28,29] All that is necessary, other than the proper matrix, is the vacuum of the mass spectrometer, by default available at the inlet of all atmospheric-pressure ionization mass spectrometers.

Fundamental studies on MAIV indicate that bare ions are produced without the direct exposure of the matrix:analyte to the vacuum. Means of dislodging include a gas flow,[37] or a laser.[35,36,59] The matrix:analyte sample can be delivered to the vacuum in solution using a pumping system and a Tee connector for mixing the analyte and matrix solution before the atmospheric pressure inlet,[60] offering a means of LC and high charge states without the need for the addition of heat, as in *inlet* ionization, voltages, or a laser. An example is shown in Figure 17.7. Analyte solution and matrix were introduced separately by a dual-syringe pump, and were mixed by a TEE connector before being delivered into the inlet of the mass spectrometer (Figure 17.7(A)). The silica tube carrying the mixture of analyte and matrix solution was placed in front of the inlet to allow the ionization of the analyte (Figure 17.7(B)).

Some matrices, *e.g.* 2,5-DHAP,[60] do not sublime as efficiently as 3-NBN when exposed to vacuum, and a laser can be employed to ablate the matrix:analyte using transmission geometry to transport materials from the sample holder to a higher-vacuum region, using principles from LSIV, but operating directly from atmospheric pressure.[36] Additionally, modification of the skimmer cone that allows less air flow, minimizes or overcomes the depletion of the matrix:analyte from the surface (Figure 17.8(A)) even using the rather volatile 3-NBN matrix and mouse brain tissue sections. Due to the insufficiency of spontaneous sublimation, multiply charged ions (Figure 17.8(IIA–C)) were only produced when the laser was ablating the mixture:analyte. Upon laser ablation, a sharp signal is observed (Figure 17.8(III)) and mass spectral signal in good ion abundance with little

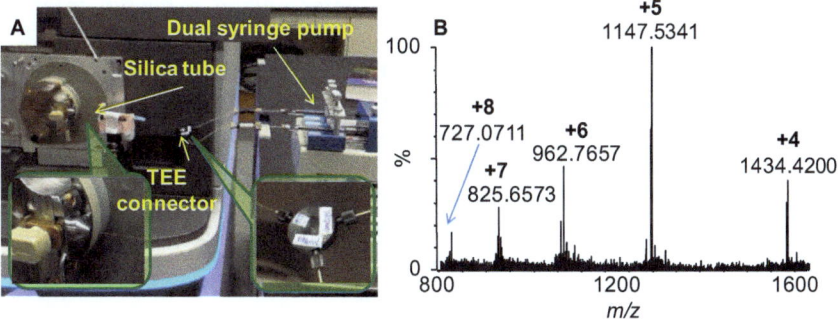

Figure 17.7 MAIV analysis of bovine insulin by introducing the analyte and matrix solutions by a dual syringe pump and mixing them in a TEE connector. (A) The silica tube carrying the mixed matrix:analyte solution was positioned in front of the inlet. (B) Mass spectrum of multiply charged bovine insulin ions.[60]

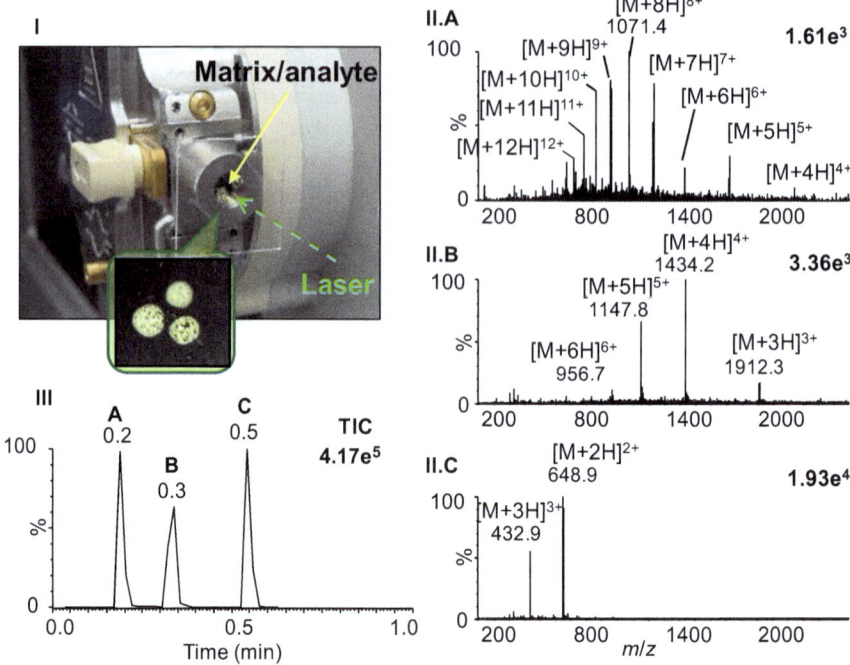

Figure 17.8 MAIV analysis with the assistance of a laser to ablate the matrix : analyte
into the vacuum of the mass spectrometer. (A) Photo of the source
setup. The inset of (I) shows the modified skimmer cone for improved
vacuum assistance and 2,5-DHAP:analyte spots on a glass slide.
(II) Mass spectra and (III) total ion chronogram and (A) ubiquitin, (B)
bovine insulin, and (C) angiotensin I [adapted from ref. 36].

chemical noise is obtained (Figure 17.8(II)). Additionally, 3-NBN can be used
in binary matrix mixtures and employed for imaging studies even of proteins
on a high mass resolution mass spectrometer   using a heated inlet for
ionization.[49]

Currently, MAIV technology is consuming considerable research time in
our group as we strive to understand the fundamentals of this astonishing
process, discover new matrix compounds that can be used with water-based
solutions, and determine its utility for observing labile compounds and
complexes from their native environment.[36] One such new compound is
2-methyl-2-nitro-1,3-propanediol, which contrary to 3-NBN attaches sodium
cations to for example, synthetic polymers, and by utilizing IMS-MS can be
used to differentiate polymer architectures.[59]

17.3   Novel Characterization Strategies

The discovery of new ionization methods provides opportunities to making
measurement inroads into intractable problems. One such area is analysis of

complex biological systems, having components of widely divergent concentrations, with temporal and spatial resolution irrespective of solubility, as can be expected in direct analysis of cells under molecular crowding conditions. Such a goal involves near-term developments in ionization technology. Advancements that need to be achieved include further improvements in sensitivity and dynamic range, universal ionization or the ability to achieve selective ionization of the desired compound types, and developments related to insolubility, labile compounds and complexes, speed of analysis, and spatial resolution. Addressing these issues will involve continued fundamental research, modifying instrumentation, and demonstrating applicability to real-world problems. While we apply a methodical approach, we remain vigilant as "Chance favors the prepared mind".[61]

The wide variety of ionization methods available for atmospheric-pressure and *vacuum* ionization that produce similar charge states from proteins provides the opportunity to study structural conformations in a way not before possible. In a fundamental research project, IMS separation on the SYNAPT G2 is being used to look at structural conformations of the various charge states of the well-studied ubiquitin protein, not only with ESI, but also LSII, LSIV, and MAIV.[60] The objective is to determine any differences in conformations observed with the various ionization methods under native and denaturing conditions with positive and negative ionization, and determine the effect of charge repulsion on structures. Success will enable a view of molecules and their shapes *directly* in tissue, not achievable by any current technology.

Development of ionization technology is only a first step to observe and identify chemicals directly from complex biological systems with high sensitivity and dynamic range with single-cell spatial resolution and observe temporal changes. A recent review summarizes past developments of traditional and newly discovered ionization processes and outlines our view on how this unprecedented ionization process can be utilized to potentially solve long-standing biological and chemical problems.[32,50] IMS is a key component to this strategy because it not only provides nearly instantaneous gas-phase separation and conformation (shape) analyses, but also provides improved dynamic range of the mass spectrometer by separating high- and low-abundance ions. Isomers are often separated and cross section measurements aid in assigning the 3-dimensional structure of, for example, enzymes.[62] Instantaneous ionization and instantaneous separation by adding IMS and MS/MS characterization can be achieved.[31,63] Shown in Figure 17.9 is an example of a small molecule mixture.[31] The ions are fragmented by CID without precursor ion selection so that ionization, separation, and characterization of individual components of a mixture are obtained instantaneously. These new ionization processes are applicable to other mass analyzer technologies such as FT mass spectrometers[1–3,8,14–16,18–21,23,35] as well as low-resolution bench-top mass spectrometers such as the Waters QDa.[64]

MAIV-IMS-MS is applicable for a variety of different compounds, ranging from clinical[31] and proteomics[33] to synthetic polymers,[59] as well as some polymer additives.[63] In the latter study, it has been reported that a polymer modifier, 6-(dibutylamino)-1,8-diazabicyclo[5.4.0]undec-7-ene (DADBU), can be detected below femtomole. 100 fmol DADBU was readily detected from a mixture with 20 pmol of poly(ethylene glycol) and 10 pmol of poly(ethylene glycol) bis(2-ethylhexanoate) (Figure 17.10).[63] A real Sample was analyzed and identified nylon fibers obtained from a new car's carpet,[63] producing doubly charged ions of an oligomer when the fiber was simply acidified and covered with a MAIV matrix.

Another necessary component is improved fragmentation technology to provide direct structural characterization or identification of compounds dislodged from biological tissue. In this regard ETD technology exploiting the new ionization methods, and when desired IMS-MS, has been implemented.[33] While back-end ETD technology (available, *e.g.*, on Thermo instruments) is readily applicable with these new ionization methods there is room for improvements for front-end ETD technology (available, *e.g.*, on Waters instruments), because the ETD reagent gas and new ion source, the mass spectrometer inlet hypothesized to form matrix:analyte clusters, interfere because of their close proximity, at least for protein ions. Such developments in combination with highly sensitive soft ionization approaches will allow, for example, identifying drug interactions in a spatial and temporal context. In the future, success will be measured by application of these technological developments to better understanding complex biological systems that are currently inaccessible, especially directly with spatial and temporal resolution. While all the above considerations and

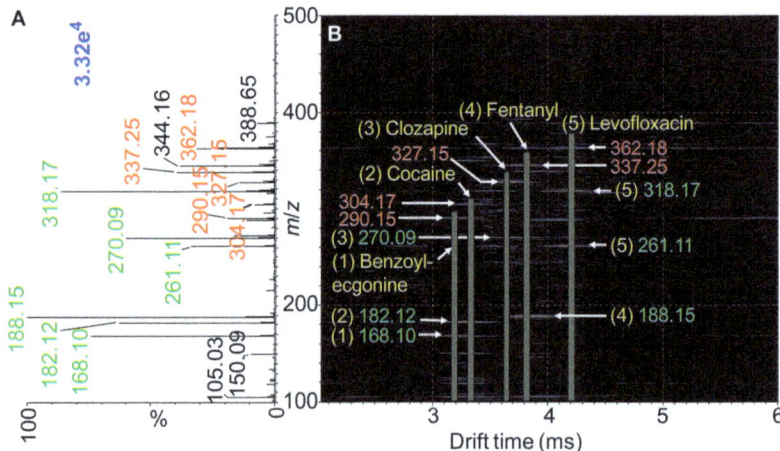

Figure 17.9 Instantaneous separation and characterization of a mixture of small molecules ionized by MAIV. The ions are fragmented by MS/MS using CID without precursor ion selection. (A) Total MS/MS spectrum and (B) two-dimensional plots of drift time *versus* m/z of the IMS-MS/MS experiment of the entire mixture. The fragment ions belonging to the same compound fall in the same drift time enabling rapid MS/MS characterization directly of the mixture [adapted from ref. 31].

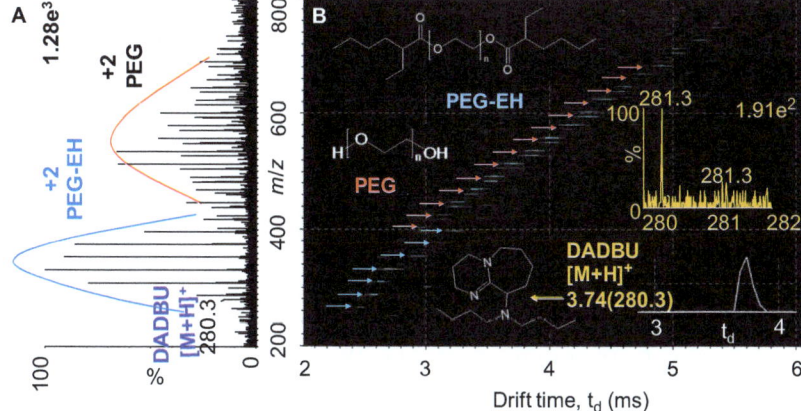

Figure 17.10   A mixture containing 100 fmol polymer additive DADBU, 20 pmol of poly(ethylene glycol) (PEG) and 10 pmol of poly(ethylene glycol) bis(2-ethylhexanoate) (PEG-EH) was analyzed by MAIV-IMS-MS. (A) The total mass spectrum and (B) two-dimensional plot of drift time *versus* m/z. Good dynamic range was achieved for the detection of the polymer additive [adapted from ref. 63].

developments are critical, perhaps most critical of all are simplicity, ease of use, and safety of operation. Towards this end, in proof of principle, a retired nurse has been invited to test MAIV and obtained outstanding results of bovine insulin at the first trial.

17.4   Future Outlook

Novel ambient ionization and sample preparation technology combined with IMS-MS and ETD has the potential to elevate MS materials character-ization, especially regarding complexity and insolubility including those of isomeric compositions. Potential application areas are diverse. One specific interest lies with Alzheimer's disease and membrane proteins in general. There is a great risk in this research, but a huge payoff when successful. About 75% of drugs target membrane proteins. Protein/ligand drug devel-opment directly from tissue is another area. There is also interest in utilizing MS for rapid analysis of antibodies. In addition to developing technology to unravel the molecular makeup of complex, soluble, and solubility restricted materials on the molecular level, the novel technology will provide the sci-entific community with a versatile simple-to-use ion source for high-throughput screening and potentially clinical use to discover disease-related pathways, drug targets, and chemical probes relevant to a variety of diseases. Field portable mass spectrometers could benefit from the simple, low-energy ionization methods described here. Fundamental knowledge regarding ionization and IMS separation for use in MS is also obtained, providing insights to traditional ionization methods such as ESI and MALDI. The innovations described in this chapter are applicable to many other areas including energy research and polymer characterization.[59,60,63]

References

1. S. Trimpin, T. N. Herath, E. D. Inutan, S. A. Cernat, J. Wager-Miller, K. Mackie and J. M. Walker, *Rapid Commun. Mass Spectrom.*, 2009, **23**, 3023.
2. S. Trimpin, E. D. Inutan, T. N. Herath and C. N. McEwen, *Anal. Chem.*, 2010, **82**, 11.
3. S. Trimpin, E. D. Inutan, T. N. Herath and C. N. McEwen, *Mol. Cell. Proteomics*, 2010, **9**, 362.
4. E. D. Inutan and S. Trimpin, *J. Am. Soc. Mass Spectrom.*, 2010, **21**, 1260.
5. C. N. McEwen, B. S. Larsen and S. Trimpin, *Anal. Chem.*, 2010, **82**, 4998.
6. E. D. Inutan and S. Trimpin, *J. Proteome Res.*, 2010, **9**, 6077.
7. A. L. Richards, D. D. Marshall, E. D. Inutan, C. N. McEwen and S. Trimpin, *Rapid Commun. Mass Spectrom.*, 2011, **25**, 247.
8. E. D. Inutan, A. L. Richards, J. Wager-Miller, K. Mackie, C. N. McEwen and S. Trimpin, *Mol. Cell. Proteomics*, 2011, **10**, 1.
9. C. N. McEwen and S. Trimpin, *Int. J. Mass Spectrom.*, 2011, **300**, 167.
10. A. L. Richards, C. B. Lietz, J. Wager-Miller, K. Mackie and S. Trimpin, *Rapid Commun. Mass Spectrom.*, 2011, **25**, 815.
11. B. Wang, C. B. Lietz, E. D. Inutan, S. M. Leach and S. Trimpin, *Anal. Chem.*, 2011, **83**, 4076.
12. A. L. Richards, C. B. Lietz, J. Wager-Miller, K. Mackie and S. Trimpin, *J. Lipid Res.*, 2012, **53**, 1390.
13. J. Li, E. D. Inutan, B. Wang, C. B. Lietz, D. R. Green, C. D. Manly, A. L. Richards, D. D. Marshall, S. Lingenfelter, Y. Ren and S. Trimpin, *J. Am. Soc. Mass Spectrom.*, 2012, **23**, 1625.
14. S. Trimpin, B. Wang, E. D. Inutan, J. Li, C. B. Lietz, V. S. Pagnotti, A. F. Harron, D. Sardelis and C. N. McEwen, *J. Am. Soc. Mass Spectrom.*, 2012, **23**, 1644.
15. L. Nyadong, E. D. Inutan, X. Wang, C. L. Hendrickson, S. Trimpin and A. G. Marshall, *J. Am. Soc. Mass Spectrom.*, 2013, **24**, 320.
16. C. N. McEwen, V. Pagnotti, E. D. Inutan and S. Trimpin, *Anal. Chem.*, 2010, **82**, 9164.
17. C. B. Lietz, A. L. Richards, Y. Ren and S. Trimpin, *Rapid Commun. Mass Spectrom.*, 2011, **25**, 3453.
18. V. S. Pagnotti, N. D. Chubatyi and C. N. McEwen, *Anal. Chem.*, 2011, **83**, 3981.
19. V. S. Pagnotti, E. D. Inutan, D. D. Marshall, C. N. McEwen and S. Trimpin, *Anal. Chem.*, 2011, **83**, 7591.
20. V. S. Pagnotti, S. Chakrabarty, A. F. Harron and C. N. McEwen, *Anal. Chem.*, 2012, **84**, 6828.

21. N. D. Chubatyi, V. S. Pagnotti, C. M. Bentzley and C. N. McEwen, *Rapid Commun. Mass Spectrom.*, 2012, **26**, 887.
22. B. Wang, E. D. Inutan and S. Trimpin, *J. Am. Soc. Mass Spectrom.*, 2012, **23**, 442.
23. V. S. Pagnotti, S. Chakrabarty and C. N. McEwen, *J. Am. Soc. Mass Spectrom.*, 2013, **24**, 186.
24. B. Wang and S. Trimpin, *Anal. Chem.*, 2014, **86**, 1000.
25. E. D. Inutan, B. Wang and S. Trimpin, *Anal. Chem.*, 2010, **83**, 678.
26. S. Trimpin, Y. Ren, B. Wang, C. B. Lietz, A. L. Richards, D. D. Marshall and E. D. Inutan, *Anal. Chem.*, 2011, **83**, 5469.
27. E. D. Inutan, J. Wager-Miller, K. Mackie and S. Trimpin, *Anal. Chem.*, 2012, **84**, 9079.
28. E. D. Inutan and S. Trimpin, *Mol. Cell Protemics*, 2013, **12**, 792.
29. S. Trimpin and E. D. Inutan, *J. Am. Soc. Mass Spectrom.*, 2013, **24**, 722.
30. S. Trimpin and E. D. Inutan, *Anal. Chem.*, 2013, **85**, 2005.
31. E. D. Inutan, J. Wager-Miller, S. B. Narayan, K. Mackie and S. Trimpin, *Int. J. Ion Mobility Spectrom.*, 2013, **16**, 145.
32. S. Trimpin, B. Wang, C. B. Lietz, D. D. Marshall, A. L. Richards and E. D. Inutan, *Crit. Rev. Biochem. Mol. Biol.*, 2013, **48**, 409.
33. C.-W. Liu, B. Wang, R. Kumar, J. Wager-Miller, K. Mackie and S. Trimpin, Proteomics Applications Using Matrix Assisted Ionization Vacuum-Ion Mobility Spectrometry/Mass Spectrometry, 62nd ASMS Conference on Mass Spectrometry and Allied Topics, Baltimore, Maryland, June 15–19, 2014.
34. D. Woodall, B. Wang, T. El-Baba, E. Inutan and S. Trimpin, High-Throughput Analysis and Characterization of Small and Large Molecules by Matrix Assisted Ionization Vacuum Ion Mobility Spectrometry Mass Spectrometry, 62nd ASMS Conference on Mass Spectrometry and Allied Topics, Baltimore, Maryland, June 15–19, 2014.
35. B. Wang, E. Tisdale, S. Trimpin and C. L. Wilkins, *Anal. Chem.*, 2014, **86**, 6792.
36. C. A. Lutomski, T. J. El-Baba, E. D. Inutan, C. D. Manly and S. Trimpin, *Anal. Chem.*, 2014, **86**, 6208.
37. S. Chakrabarty, V. S. Pagnotti, E. D. Inutan, S. Trimpin and C. N. McEwen, *J. Am. Soc. Mass Spectrom.*, 2013, **24**, 1102.
38. J. V. Iribarne and B. A. Thomson, *J. Chem. Phys.*, 1976, **64**, 2287.
39. M. Dole, L. L. Mack, R. C. Mobley, L. D. Ferguson and M. B. Alice, *J. Chem. Phys.*, 1968, **49**, 2240.
40. V. Frankevich, R. J. Nieckarz, P. N. Sagulenko, K. Barylyuk, R. Zenobi, L. I. Levitsky, A. Y. Agapov, T, Y. Perlova, M. V. Gorshkov and I. A. Tarasova, *Rapid Commun. Mass Spectrom.*, 2012, **26**, 1567.
41. T. Musapelo and K. K. Murray, *J. Am. Soc. Mass Spectrom.*, 2013, **24**, 1108.
42. T. Musapelo and K. K. Murray, *Rapid Commun. Mass Spectrom.*, 2013, **27**, 1283.
43. L. Gao, R. G. Cooks and Z. Ouyang, *Anal. Chem.*, 2008, **80**, 4026.

44. P. J. Roach, J. Laskin and A. Laskin, *Analyst*, 2010, **135**, 2233.
45. J. Laskin, B. S. Heath, P. J. Roach, L. Cazares and O. J. Semmes, *Anal. Chem.*, 2012, **84**, 141.
46. P. Chaurand, S. A. Schwartz and R. M. Caprioli, *Curr. Opin. Chem. Biol.*, 2002, **6**, 676.
47. L. A. McDonnell and R. M. A. Heeren, *Mass Spectrom. Rev.*, 2004, **165**, 1057.
48. Y. Schober, S. Guenther, B. Spengler and A. Römpp, *Anal. Chem.*, 2012, **84**, 6293.
49. A. F. Harron, K. Hoang and C. N. McEwen, *Int. J. Mass Spectrom.*, 2013, **352**, 65.
50. S. Trimpin, *J. Mass Spectrom.*, 2010, **45**, 471.
51. S. Trimpin, in *Encyclopedia of Mass Spectrometry: Molecular Ionization*, ed. M. L. Gross and R. M. Caprioli, Elsevier, p. 683.
52. S. Trimpin, K. Wijerathne and C. N. McEwen, *Anal. Chim. Acta*, 2009, **654**, 20.
53. S. Trimpin, T. N. Herath, E. D. Inutan, J. Wager-Miller, P. Kowalski, E. Claude, J. Walker and K. Mackie, *Anal. Chem.*, 2010, **82**, 359.
54. D. G. Rizzo, J. M. Spraggins, K. I. Rose, R. M. Caprioli, 61th ASMS Conference and Allied Topics on Mass Spectrometry, WP155, 2013, June, MN.
55. S. Chakrabarty, V. S. Pagnotti, C. N. McEwen, 61th ASMS Conference on Mass Spectrometry and Allied Topics, ThOA pm 2:50, 2013, June, MN.
56. J. Mitchel and H. D. Deveraux, *Anal. Chim. Acta*, 1992, **52**, 45.
57. L. M. Sweeting, *Chem. Mater.*, 2001, **13**, 854.
58. S. Chakrabarty, V. Pagnotti, B. Wang, S. Trimpin and C. N. McEwen, *Anal. Chem.*, 2014, DOI: 10.1021/ac500132j.
59. C. Foley, T. El-Baba, B. Zhang, S. Grayson, B. S. Larsen and S. Trimpin, Characterization of Homo-arm and Mikto-arm Poly(ethylene Glycol) Stars using Vacuum Ionization-Ion Mobility Spectrometry-Mass Spectrometry, 62nd ASMS Conference on Mass Spectrometry and Allied Topics, Baltimore, Maryland, June 15–19, 2014.
60. S. Trimpin, C. A. Lutomski, T. J. El-Baba, D. W. Woodall, C. D. Foley, R. Kumar, C. D. Manly, C. W. Liu, B. Wang, B. M. Harless, L. F. Imperial and E. D. Inutan, *Int. J. Mass Spectrom.*, 2014, DOI: 10.1016/j.ijms.2014.07.033.
61. Dans les champs de l'observation le hasard ne favorise que les esprits prepares, Lecture, University of Lille; 7 December 1854.
62. M. F. Bush, Z. Hall, K. Giles, J. Hoyes, C. V. Robinson and B. T. Ruotolo, *Anal. Chem.*, 2010, **82**, 9557.
63. T. J. El-Baba, C. A. Lutomski, B. Wang and S. Trimpin, *Rapid Commun. Mass Spectrom.*, 2014, **28**, 1175.
64. S. Trimpin, C. Lutomski, T. El-Baba, B. Wang, L. Imperial, D. Woodall, R. Kumar, B. Harless, C. Foley, C.-W. Liu and E. Inutan, Surprising New Ionization Methods for Mass Spectrometry, Mechanistic Insights and Potential Practical Utility, 62nd ASMS Conference on Mass Spectrometry and Allied Topics, Baltimore, Maryland, June 15–19, 2014.

CHAPTER 18

Enabling Automated Sample Analysis by Direct Analysis in Real Time (DART) Mass Spectrometry

BRIAN MUSSELMAN,* JOSEPH TICE AND
ELIZABETH CRAWFORD

Ion Sense, 999 Broadway, Suite 404, Saugus, MA 01906-4510, USA
*Email: musselman@IonSense.com

18.1 Introduction

With the development of direct ionization in real time (DART®)[1] and desorption electrospray ionization (DESI)[2] the promise of being able to use ambient ionization methods to streamline chemical analysis proved very attractive. Who could resist the idea of eliminating the often tedious sample preparation from their everyday routine? Initial interest in the ambient technology, indeed many of the first hundred publications, focused on description of direct analysis of plant materials,[3] tablets,[4] foodstuffs such as edible oils[5] and pesticides on the surface of foods.[6,7] Cody et al.[1] and later Jagerdeo at the FBI laboratory[8] successfully demonstrated the quantitative capability of the DART-MS method using the addition of stable isotope-labeled standards to samples prior to analysis. Using a TOF-MS the group reported detection of several date-rape drugs in alcoholic solutions by comparing the relative abundance of the protonated molecule of rohypnol to

New Developments in Mass Spectrometry No. 2
Ambient Ionization Mass Spectrometry
Edited by Marek Domin and Robert Cody
© The Royal Society of Chemistry 2015
Published by the Royal Society of Chemistry, www.rsc.org

a known volume of the deuterated standard added to a known volume of the liquid. This method permitted accurate determination for the presence of the drug at concentrations in excess of that expected in normal human metabolism. The test solution in this case was ideal since it was a clear liquid that was easily sampled onto a glass melting-point capillary and the isotopic standard was commercially available. The work demonstrated that ambient technology could be used to generate both rapid qualitative and quantitative results.

Shortly after the first publication of the DART and DESI methods IonSense, Inc. was founded as an independent spin-off from JEOL USA. The original focus of the company was to enable the integration of DART technology with a wider range of LC/MS and LC/MS/MS instruments in order to facilitate uptake of the technology in the analytical community. Coming shortly after disclosure of the existence of DART the source was originally available only in combination with the JEOL AccuTOF LC/MS but within a year interface to Thermo and Agilent LC/MS were available. In the expansion to other instrumentation it became obvious that other types of mass analyzers such as quadrupole and ion-trap devices were not as suitable for ambient ionization since they lacked the advantages of rapid scanning afforded by time-of-flight (TOF) -based mass spectrometers. The act of having the human hand positioning samples in the ionization region led to generation of total ion chromatograms with very inconsistent peak shapes, for example, and the requirement for an isotopically labeled material in order to quantitate was not practical for many samples. It became clear that some form of automated sample handling was necessary to allow DART users to generate higher-quality data.

18.2 Considerations for Automating DART

Sample handling in DART-MS involved positioning a sample in the desorption ionization region between the exit of the DART source where heated gas could be directed at it and close to the atmospheric-pressure inlet (API) so that the ions generated could enter the mass spectrometer. The human hand is perfect for the job, unfortunately few investigators wanted to be robots, so we had to devise a strategy for collecting a sample, moving it into and through the ionization region, and then repeating this process without human intervention over and over. The first task was to design a sample collector that could be manipulated by using a robot. Glass capillary melting-point tubes were preferred for the ambient experiment as they were inexpensive, readily available, and thermally stable. The question then became how to pick up and move the narrow, fragile glass tube without breaking it. Working with Orochem Technologies of Naperville, IL we devised a combination glass and plastic consumable, the DIP-it® sampler, which was mass produced and packaged in 96-well format boxes for safe transport, and more significantly serve to position the samplers on the CTC-PAL autosampler deck. As many laboratories had access to CTC PAL units we entered discussions with LEAP Technologies that enabled the software and custom hardware to support automated sampling of liquids.

Leveraging the CTC-PAL we commercialized the AutoDART® providing an automation platform which could pick up and move the DIP-it® sampler from its initial position dipping it into a liquid and then moving its tip through the ionization region at a controller speed to complete the experiment. After use the DIP-it was ejected from its position on the end of the PAL arm and the cycle repeated until samples were all analyzed. The use of the glass tip insured low background at all normal DART gas temperatures and the closed end of the glass capillary facilitated collection of only a limited volume of material when the tip was inserted into a volume of liquid. Results were reported in early 2008 by multiple investigators.[9,10] The AutoDART with its DIP-it® devices permitted sampling from 96-well microtiter plates thus facilitating high-throughput operation for an entire plate in one experiment.

Several problems were encountered in development of the AutoDART. The first was the problem of transferring ions over a distance from the ionization region to the API inlet and the second was the need to physically position the robot in close proximity to the LC/MS.

The atmospheric-pressure inlet of LC/MS instruments serve two purposes; they are the opening through which ions and gases can enter the mass spectrometer and they have a potential applied to their outer surface to permit electrostatic focusing of ions, thus improving sensitivity. In our case, the DART source was able to produce a rich mix of ions and gas in close proximity to the API, however, since the sampling point for the experiment is an uncharged surface at ground the ions were free to drift away from the ionization region in a random motion. In order to improve sensitivity we developed a device capable of drawing the gas containing ions to the API region for more efficient sampling. This device, the gas ion separator (GIST) was introduced, as the VAPUR interface in 2008 as described by Zhao *et al.*[10] and utilized by Yu *et al.*[11] In its initial design the VAPUR used a 6 mm ceramic tube with a 4.75 mm internal diameter and positioned 2–3 mm from the glass surface of the DIP-it in close proximity to the desorption ionization region. The objective of the VAPUR was to facilitate collection of both ions and neutral gas and transport them to the API inlet region.[11] An interface, designed specifically to enclose the API-4000 ionization region, replaced the standard APCI source thus forming a vacuum chamber around the API inlet. A rubber tube connected to a small membrane pump (Vacuubrand Model MZ 2) provided the means to draw the gases through the chamber. From an operational point of view the VAPUR interface increased sensitivity by drawing gases close enough to the API inlet through the ceramic tube when the pressure in the enclosure approximately 200 Torr as measured at the exhaust of the membrane pump. At this pressure most of the gas was being removed from the enclosure by the action of the membrane pump, however, the vacuum of the mass-spectrometer inlet was sufficient to draw a significant volume of gas containing ions into the LC-MS/MS for analysis. As first reported[12] the device improved the detection limit of the DART-MS/MS by a factor of 1000 over the same system without the VAPUR. Quantitative analysis by LC-MS/MS utilizing this method was demonstrated in several publications by our group[13] and others.[10] The AutoDART device has shown more practical value in the analysis of edible oils,[5] soft drinks[14] and beers[15-17] where the sample matrix is less challenging. In the case of blood analysis, proteins, complex lipids and salts in the sample matrix were too significant to permit adoption of the method.

The AutoDART demonstrated that automation of the open-air experiment would permit more quantitative analysis. At the same time, engineers at JEOL had investigated direct analysis of chemicals from TLC plates directly with a nonautomated device. Their experiments showed that cutting the glass plate into a 1 cm wide strip prior to DART analysis increased the efficiency of the desorption. The company subsequently incorporated a computer-controlled motorized stage into its design to permit pushing of the narrow TLC plate through the desorption ionization region, for the collection of mass spectra *vs.* time, creating a thin layer mass chromatogram. The AutoDART and JEOL's automated TLC device demonstrated that automation of the open-air experiment would permit more reliable qualitative and quantitative analysis as demonstrated by using TLC of herbal drug preparations by Kim *et al.*[18] Knowing this, IonSense undertook to automation of the DART experiment in order to expand its use as a high-throughput screening device.

The first step taken was a redesign of the DART controller to improve lab-to-lab reproducibility by fixing the gas flow rate and pressure at 3.5 L min^{-1} at 20 psi for He and 2.5 L min^{-1} at 20 psi for nitrogen. Fixing the flow rate enabled a more consistent gas-temperature assignment between different DART users. In effect, because early DART users with the original DART-100 source could vary both the gas flow and the temperature of the heater, the temperature of the gas being reported by the software could vary considerably from one lab to another. For example, using a lower gas flow where the gas remains in the heater chamber longer, could heat the gas to a higher temperature. Characterization of the dynamics of the DART-SVP source were published[19] for analysis of a sample from a glass tube by imaging the heat profile using a thermal imaging camera. The results of this study were incorporated in part into the design of successive experiment modules described below.

18.3   DART-SVP and Use of the Linear Rail

Along with the development of the new DART source with standardized voltage and pressure controls, the so-called DART SVP, a simple motor-driven linear-rail-based sampler was introduced to facilitate automated sample handling for more types of experiments. With the DIP-it liquid-sampling devices already available the first experiment module commercialized was the 12 DIP-it module. This device enabled serial analysis of up to 12 samples in under 2 min for a fixed gas temperature experiment. Application of a sample to the individual DIP-it was completed by dipping the glass tip end into the sample in separate wells of a standard microtiter plate. The collection of 12 DIP-it samplers was then presented sequentially for DART-MS and DART-MS/MS analysis as one continuous run. The DIP-it experiment module was designed to mount on the linear rail and the controller software program provided the means to adjust the speed of sample presentation into the ionization region. The module made it possible to standardize the distance and height of the sample from the DART source exit resulting in an improvement in reproducibility. A picture of the DIP-it module is shown in Figure 18.1.

The second important module introduced with the DART SVP provided the capability to operate the DART source at an angle relative to the sampling surface. Since there was little data suggesting an optimum angle for operation of the source a unit with capability to operate at angles up to 45 degrees was commercialized in 2009. This source, the DART-SVPA proved efficient at desorption ionization of molecules from flat surfaces. Operating at 45 degrees relative to the target the heated gas from the DART source would bathe the surface and with the VAPUR inlet inline transfer gas containing ions to the API inlet region of the mass spectrometer. The angled DART source also enabled direct analysis of large flat surfaces. The design of this module incorporated a second linear rail to which the sampling stage was mounted thus permitting sample movement in two dimensions. This experiment module, the 3 + D Scanner, utilized a VAPUR interface having a much longer ceramic tube for gas collection at approximately 10 cm from the API inlet of the mass spectrometer. Movement of the surface was completed under digital control to permit scanning of the entire surface. The use of the 3 + D Scanner module for determination of plant mycotoxins on large-format HPTLC plates was published by Chernetsova *et al.*[20]

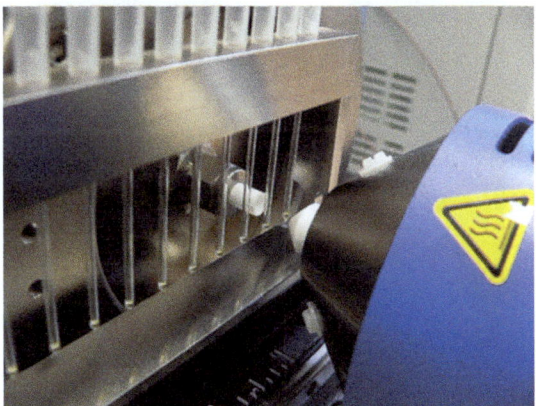

Figure 18.1   The 12 DIP-it module mounted on the linear-rail automation system of the DART SVP. The module permits reproducible positioning of liquid samples for rapid analysis.

Operation at an angle enabled direct analysis of solid samples and because of intense interest from the user community for a device to analyze the surface of tablets an experiment module capable of holding up to 10 tablets in position just under the VAPUR inlet was produced. This device was called the Tablet module, since it facilitated determination of the chemical constituents in individual tablets. Significantly, the module isolated tablets from one another during the desorption ionization process experiment by using one of three different cover plates to limit the ionization region to the center region of the tablet. The individual cover plates were designed to hold tablets with three different shapes; circular, narrow oblong, and large oblong. Very large tablets could be analyzed by removing the cover plate and simply resting the tablet in the V-shaped grove in the holder plate. With the cover plate in place multiple tablets of the same size were pushed up against the bottom of that cover plate by using a spring-mounted plate to ensure sample position relative to the desorbing gas (Figure 18.2). The Tablet module was mounted on the linear rail ensuring both accurate positioning and control of sample presentation speed. The use of this module allowed analysis of multiple tablets facilitating the use of DART for comparison of good *versus* bad product and product of known origin *versus* potential counterfeit tablets. The successive analysis of the same tablets at

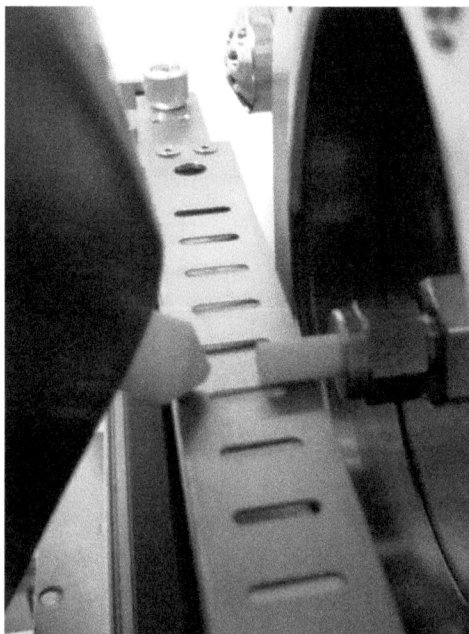

Figure 18.2   DART SVP mounted on 45 degree angle base with the gas exiting the DART source directed at the top surface of the Tablet module. The module holds ten tablets at equidistant spacing and a cover plate to further isolate the desorption site.

different gas temperatures comprises the temperature profile for that tablet. Results from the temperature profile of two types of Tylenol tablets, day and night, were presented at the Pittcon Conference in 2009,[21] Tylenol Daytime tablet data are shown in Figure 18.3. In the experiment, identical tablets from a single bottle were loaded into positions 1 and 2, 4 and 5, and 7 and 8 with every other position left empty. The use of an empty spot between the tablets served as a place where the linear rail motor could be turned off while the DART gas temperature was increased to the next higher target temperature without vaporizing material from the next tablet. The intention of this study was to show how a more complete characterization of a solid sample might bc undertaken by analyzing the same sample at ever-increasing gas temperature. Results of the analysis of four Tylenol daytime tablets show remarkably different mass spectra at each desorption gas temperature. Analysis of the first tablet clearly permits detection of the acetaminophen, along with a variety of other molecules not listed on the product packaging. Utilizing a high-resolution EXACTIVE MS later we confirmed that many of these were either compounds related to flavor or sugars from the tablet coating. As expected, at increasing temperature those compounds were vaporized quickly and not detected by the slow-scanning Finnigan LCQ DECA we utilized for this work. Remarkably, at higher temperatures the acetaminophen, which makes up nearly 75% of the tablet

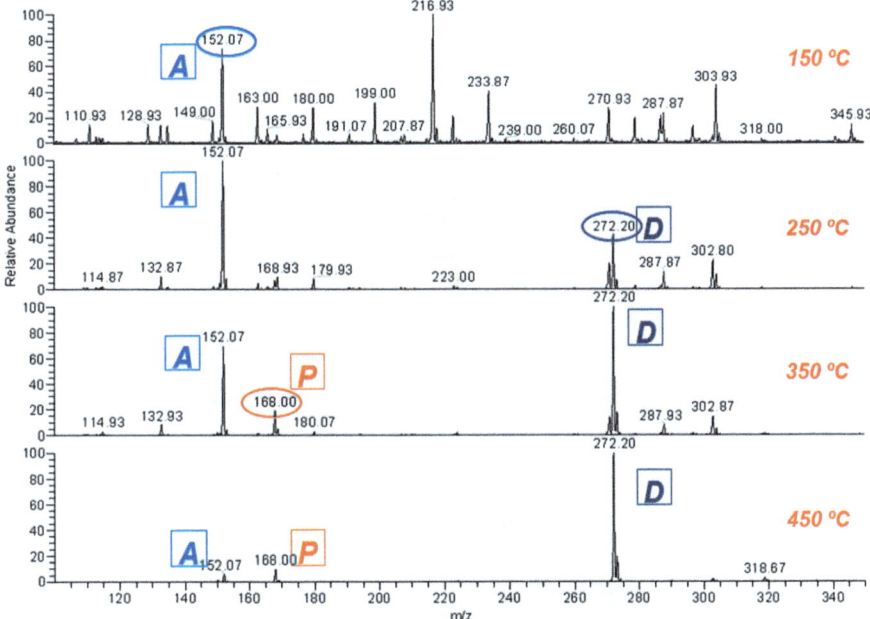

Figure 18.3   DART-MS thermal profile of Tylenol tablets. Desorption ionization mass spectra at different temperatures enable detection of Acetaminophen (A), Dextromethorphan (D), and Phenylephrine (P) in one analytical method.

material is suppressed as the other active ingredients desorb. Characterization of the tablet was completed with the analysis of the same tablets by negative ion DART-MS. The tablets were repositioned in the modules so that a fresh surface is exposed to the heated gas exiting the source. In the case of the Tylenol analyzed here detection of stearic acid, a common excipient, completed the analysis.

18.4   DART using Porous Surfaces

Collection of samples on porous samples and their direct analysis was identified as Transmission mode DART in experiments by Perez *et al.*[22] where samples of pesticide coated mosquito bed-netting were inspected by positioning them between the source exit and mass-spectrometer inlet. Use of the method to desorb materials from other porous materials including cotton tip swabs,[23] and polyurethane swabs.[6,24] In an interesting experimental scheme investigators at the FDA Forensic Chemistry Center in Cincinnati devised a scheme for collection of pesticide residues from the surface of fruit and vegetables by using foam swabs for sampling followed by DART-based desorption for ionization direct from the swab. The experiment enabled by the FDA investigators required greater stability and determination of multiple samples over a wide temperature range so we undertook to design an experiment module specifically for this purpose. The Transmission module (Figure 18.4) devised held ten swabs in the vertical plane between the exit of the DART source and in close proximity to the

Figure 18.4   Transmission module configured for the desorption ionization of pesticides from foam swabs that have been used to wipe residues from the surface of fruit as in ref. 6.

VAPUR inlet. Once mounted on the linear rail this module required only a few minutes per sample set of ten swabs to desorb analyte at a series of increasing DART gas-temperature values. This resulted in a method capable of determining a wide range of pesticides with little sample prep required in the basic experiment.[6,7]

In our work we expanded on the use of transmission-mode DART by coupling it to tandem mass-spectrometry instrumentation in order to complete quantitative analysis of a select fungicide, carbendazim, in a variety of orange juice products. Pesticide residue analysis is typically carried out using LC-MS/MS[25,26] with limits of detection, RSDs, linear dynamic ranges that exceed the current capabilities of ambient mass spectrometry. The LC/MS/MS experiment requires a significant time period for completion, as well as the use of considerable quantities of solvent. As our initial investigations[27] we had utilized wire mesh as the desorption ionization surface in order to improve reproducibility of the ambient ionization experiment. Use of the wire mesh enabled more rapid desorption of ions generating more reliable peak shapes in the mass chromatogram. More reliable peak shapes, resulted in better peak detection and quantitation. Sample analysis times averaged 10 s per run *versus* a 10 min run per each LC-MS/MS sample. Two types of tandem mass spectrometers, a high-resolution accurate mass and a triple quadrupole mass spectrometer were used for these experiments. All of the sampling for the liquid sample analysis was performed by spotting the solutions onto a cleaned stainless steel mesh that is held in the transmission module. For all experiments 3 µL aliquots of juice were deposited onto the mesh using a micropipette and the sample spots were allowed to fully dry under ambient conditions before being subjected to the DART source for analysis. A saved method in the DART SVP software with the following parameters was used to run the heater-temperature-profile experiments for the carbendazim standard at 5 µg mL^{-1}. The automated linear rail on the DART source was set to a speed of 0.5 mm s^{-1} for all experiments and the DART heater was operated at three different temperatures to determine the optimal heater settings: 150, 250, 350 °C. From the DART heater optimization, 250 °C was used for all of the targeted Carbendazim experiments. A stock solution containing the target compound Carbendazim was diluted 1:20 with methanol yielding a 25 µg mL^{-1} spiking solution. The spiking solution was used to generate the highest concentration spiked orange juice sample concentration at 5000 ng mL^{-1} by taking 300 µL of the spiking solution and mixing into 1.2 mL of orange juice. A standard curve was generated by serial dilution from the 5000 ng mL^{-1} sample yielding concentrations at 2000, 1000, 500, 200, 100, 50, 20, 10, 5, 2 and 1 ng mL^{-1}.

18.5   DART-MS/MS Setup 1: Q EXACTIVE

A DART SVP was coupled to a bench-top quadrupole Orbitrap mass spectrometer (Q EXACTIVE, Thermo Fisher Scientific, Bremen, Germany) *via* a Vapur® interface (IonSense, Inc.). The Q EXACTIVE was operated in targeted

MS/MS (tMS^2) positive acquisition mode and the $[M+H]^+$ parent ion at m/z 192.07675 was subjected to higher-energy collisional dissociation (HCD) normalized collision energy (NCE) of 50, yielding the exact mass product ion fragment at m/z 160. 05032. The universal tMS^2 instrument method parameters were as follows: S-lens RF level 50, capillary temperature 200 °C, spray voltage and all source gas values were set to zero, 1 microscan, resolution setting of 140 000 FWHM at m/z 200 at 1.5 Hz, automated gain control (AGC) target 5e^4, 80 ms maximum inject time, multiplexing count 1, isolation window 3.5 m/z. The prepared positive-ion mode calibration solution was obtained from Thermo Fisher Scientific (Pierce ESI Positive Ion Calibration Solution, P/N 88323) and contained caffeine, L-methionyl-arginyl-phenylalanyl-alanineacetate (MRFA) and Ultramark 1621. External mass calibration was performed at the beginning of the sample set by infusing the positive calibration mix at 5 µL min^{-1} using the heated electrospray (HESI) ion source and the automated mass-calibration feature within the Q EXACTIVE Tune page was initiated.

18.6 DART-MS/MS Setup 2: API 4000 QTRAP

The same model DART SVP ion source was coupled to an API 4000 QTRAP (AB Sciex, Foster City, CA, USA) using a Vapur$^®$ interface (IonSense, Inc., Saugus, MA, USA). The QTRAP was operated as a quadrupole for selected reaction monitoring (SRM) scanning and the compound dependent SRM parameters needed to be tuned for before DART-MRM analyses could be conducted.

18.7 SRM Compound Tuning

The AB Sciex TurboV ion source was used for the compound tuning and a 5 µg mL^{-1} methanolic solution of the carbendazim was infused at 5 µL min^{-1}. The automated tuning feature within Analyst software (version 1.4.1) was used to determine the most-abundant transition to be m/z 190.1 → 160.1 with the following MS instrument parameters: declustering potential (DP) 56, collision energy (CE) 27, collision exit potential (CXP) 26 and a dwell time of 150 ms. These optimized ESI settings were used as the DART-MS/MS instrument conditions.

18.8 Data Processing

Automated peak-area integration for both the Q-Exactive and API 4000 MRM data sets was realized using a customized program for DART analysis, a version of the Gubbs Mass Spec Utilities (GMSU) software program (Gubbs, Inc., Alpharetta, Georgia, USA) and is also known as QuickCalc software for high-throughput sample analysis users on Thermo MS platforms. The raw data from both the Thermo and AB Sciex data sets were opened within the GMSU interface postacquisition and peak areas would automatically be calculated for the triple quadrupole data based on the MRM transition. For the tMS^2 data sets a Thermo Processing Method (.pmd file) was created first within Xcalibur (Thermo Fisher Scientific, San Jose, CA, USA) and then the Processing Method was linked to the original Xcalibur sequence files and the Xcalibur sequence files were opened in the GMSU user interface. A mass extraction window of ± 5 ppm was set for the product ion tMS^2 extracted ion chromatograms (XICs) within GMSU. After peak area integration in GMSU a peak-area report was generated and saved as an Excel spreadsheet where calibration curves were generated.

18.9 Quantitative Screening of Carbendazim in Orange Juices

The alert for monitoring carbendazim in orange juice with in the US hit in early January 2012, realizing the need for fast, robust and sensitive analytical techniques that can be easily implemented.[28] The US FDA action limit at 10 ppb for carbendazim in orange juice was set 20 times below the EU MRL of 200 ppb. The applicability of the targeted DART MS/MS approaches was evaluated keeping in mind these target levels, a number of oranges juices from the US, the EU and Indian were quantitatively screened on both the Q EXACTIVE and the API 4000 platforms. The triple quadrupole is the benchmark instrument when it comes to quantitative measurements over large linear dynamic ranges, however with the improvements developing in high-resolution accurate mass instrumentation being coupled with precursor mass selection there are new capabilities to be explored especially in food analysis.[29] Orange juices originating from the US market were screened by the DART-MS/MS method for carbendazim and used as the orange juice blank for generating the matrix-matched calibration curve. The calibration curves ranged over 3.5 orders of magnitude from 1–5000 ppb and the juice was again sampled from the stainless steel mesh in this transmission-mode DART approach coupled to high-resolution MS/MS and low-resolution MS/MS.

The linear dynamic range for the Q EXACTIVE tMS^2 detection was 2–2000 ppb with limit of detection at 1 ppb and for the API 4000 MRM method was 10–5000 ppb with limit of detection at 2 ppb and both curves having an R^2 value of 0.999. The results of quantitatively screening a number of orange juices purchased from the US, European and Indian markets are shown in Figure 18.5. It should be noted that the juices quantitatively screened for carbendazim on the two MS platforms were not identical, with the exception of the juices originating from the German market. The carbendazim quantified by both targeted MS/MS approaches in the German orange juice was nearly identical at 27 ppb by high resolution tMS^2 and 29 ppb by traditional triple quad MS/MS, nearly 7.5 times lower than the EU MRL. All of the non-US marketed juices all contained some level of carbendazim ranging from less than 10 ppb up to just under 30 ppb and all of the US marketed juices

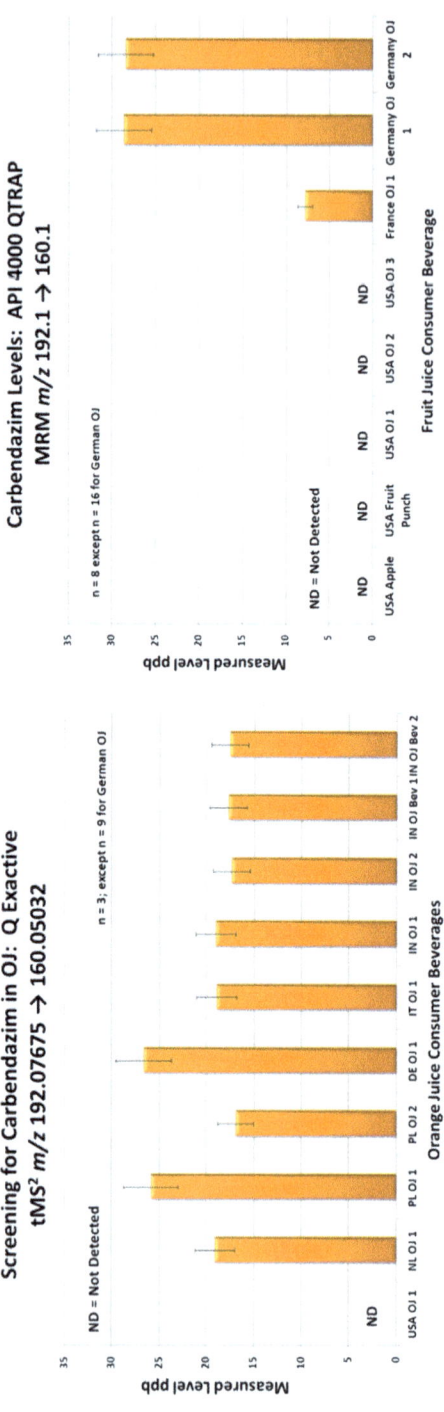

Figure 18.5 Results of implementing tMS² and MRM DART experiments for screening and quantifying carbendazim directly out of orange juices.

(purchased late January to early February 2012 after the US FDA carbendazim alert had occurred) screened by both targeted DART MS/MS methods did not test positive for carbendazim.

Figure 18.6 shows the raw XIC with a mass extraction window of ± 5 ppm for spiked concentrations of carbendazim in orange juice detected at 5, 10 and 20 ppb, which is at and around the US FDA's action limit using the high-resolution direct DART-tMS^2 method yielding good RDSs considering no internal standard correction and very good S/N values. The 10 ppb level was the limit of quantitation for the triple quad DART-MS/MS approach and what hindered this approach from detecting any lower than this with confidence was the much lower S/N at the same relative concentrations. At 2 ppb the LOD for the high-resolution tMS^2 approach yielded an average S/N 11 over three measurements with an average of 24 scans across each DART peak. A representative mass spectrum for the tMS^2 approach at the 10 ppb level is shown in Figure 18.7 and displays mass accuracy of 0.9 ppm for the monoisotopic peak. The use of an ultra high resolution MS platform facilitated separation of the orange juice matrix interference peak observed at m/z 161.05955, detected only 37 ppm away from the A+1 isotopic peak for carbendazim at m/z 161.05366. A resolution setting of 70 000 proved sufficient to resolve the two peaks when sampling using this direct DART-MS/MS method.[30] In summary, both the high-resolution tMS^2 and the low

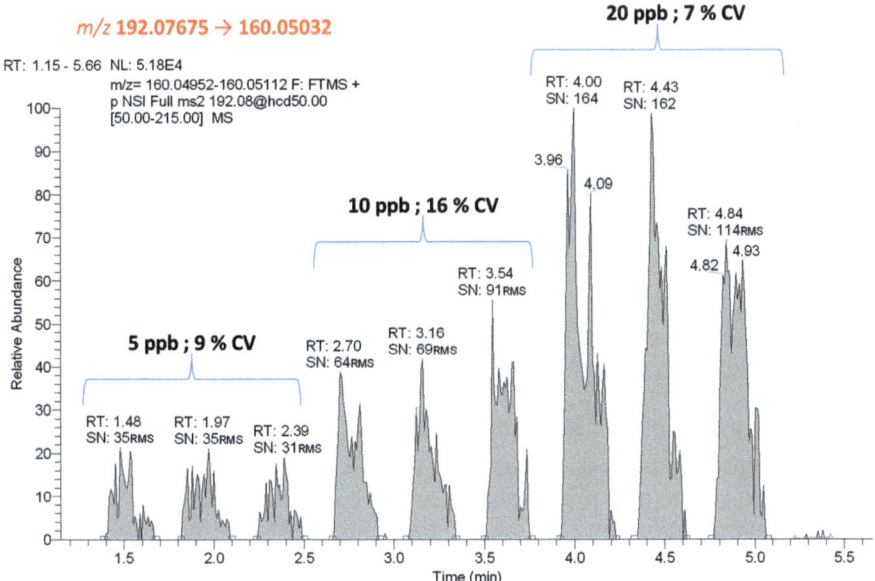

Figure 18.6 US FDA action limit for carbendazim in orange juice: 10 ppb spike and 2× range above and below displaying DART-tMS^2 XIC for major fragment ion at m/z 160.05032 using ± 5 ppm mass extraction window (140 000 FWHM). RSDs for $n=3$ using peak areas with no internal standard correction.

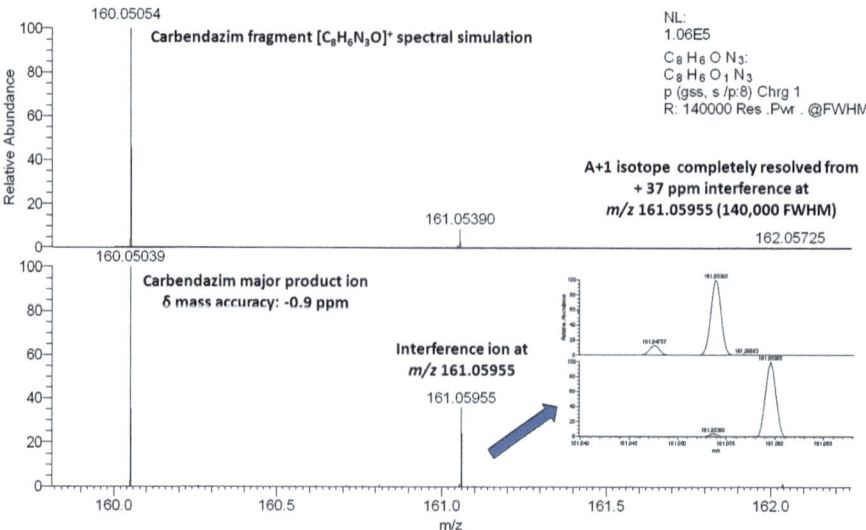

Figure 18.7   DART-tMS^2 spectrum for carbendazim in orange juice at 10 ppb level measured with −0.9 ppm mass accuracy for monoisotopic peak and A + 1 isotope completely resolved from interference peak within 37 ppm from the orange juice.

resolution triple quad DART methods quantified over 3 and 2.5 orders of magnitude, respectively. The LOQ for the high resolution approach was 2 ppb and for the triple quad method it was at 10 ppb.

The results present the successful first pass at quantification of pesticides directly out of fruit juice beverages and show the added benefit of coupling DART ambient ionization with high-resolution mass spectrometry in addition to specificity gained by operating under MS/MS conditions.

18.10   Conclusions

The implementation of a DART source with fixed flow and pressure, as well as good thermal control enabled improvements in sampling reliability during analysis. Development of a series of Experiment modules enabled rapid analysis of samples by presenting them for ionization in a more controlled manner. The utility of these modules for improving the quality of data and reproducibility of results has been presented using results from the successful analysis of liquids, tablets, and even complex samples presented by demonstrating quantification of pesticides directly out of fruit-juice beverages.

Acknowledgements

The authors would like to kindly thank New Objective, Inc., specifically Dr Gary Valaskovic and Amanda Berg for access to and help in setting up the

MS/MS experiments on the API 4000 QTRAP and Helena Svobodova for assistance in taking the high-resolution magnified photos of the spotted mesh; Dr Catharina Crone for insightful discussions on the Q EXACTIVE and Maciej Bromirski for access to the Q EXACTIVE and for gathering the orange-juice samples in Europe and India; Dr. Larry Elvebak from Gubbs, Inc. for customizing a version of the GMSU software for the automated peak integration for DART high-throughput data.

References

1. R. B. Cody, J. A. Laramee and H. D. Durst, Versatile new ion source for the analysis of materials in open air under ambient conditions, *Anal. Chem.*, 2005, 77, 2297–2302.
2. Z. Takats, J. M. Wiseman, B. Gologan and R. G. Cooks, Mass spectrometry sampling under ambient conditions with desorption electrospray ionization, *Science*, 2004, **306**, 471–473.
3. H. Kim and Y. P. Jang, Direct Analysis of Curcumin in Turmeric by DART-MS, *Phytochem. Anal.*, 2009, **20**, 372–377.
4. W. C. Samms, Y. J. Jiang, M. D. Dixon, S. S. Houck and A. Mozayani, Analysis of Alprazolam by DART-TOF Mass Spectrometry in Counterfeit and Routine Drug Identification Cases, *J. Forensic Sci.*, 2011, **56**, 993–998.
5. L. Vaclavik, T. Cajka, V. Hrbek and J. Hajslova, Ambient Mass Spectrometry Employing Direct Analysis in Real Time (DART) Ion Source for Olive Oil Quality and Authenticity Assessment, *Anal. Chim. Acta*, 2009, **645**, 56–63.
6. S. E. Edison, L. A. Lin, B. M. Gamble, J. Wong and K. Zhang, Surface swabbing technique for the rapid screening for pesticides using ambient pressure desorption ionization with high-resolution mass spectrometry, *Rapid Commun. Mass Spectrom.*, 2011, **25**, 127–139.
7. S. E. Edison, L. A. Lin and L. Parrales, Practical considerations for the rapid screening for pesticides using ambient pressure desorption ionisation with high-resolution mass spectrometry, *Food Add. Contam*, 2011, **28**, 1393–1404.
8. E. Jagerdeo and M. Abdel-Rehim, Screening of Cocaine and Its Metabolites in Human Urine Samples by Direct Analysis in Real-Time Source Coupled to Time-of-Flight Mass Spectrometry After Online Preconcentration Utilizing Microextraction by Packed Sorbent, *J. Am. Soc. Mass Spectrom.*, 2009, **20**, 891–899.
9. J. Schurek, L. Vaclavik, H. Hooijerink, O. Lacina, J. Poustka, M. Sharman, M. Caldow, M. W. F. Nielen and J. Hajslova, Control of Strobilurin Fungicides in Wheat Using Direct Analysis in Real Time Accurate Time-of-Flight and Desorption Electrospray Ionization Linear Ion Trap Mass Spectrometry, *Anal. Chem.*, 2008, **80**, 9567–9575.
10. Y. Zhao, M. Lam, D. Wu and R. Mak, Quantification of small molecules in plasma with direct analysis in real time tandem mass spectrometry,

without sample preparation and liquid chromatographic separation, *Rapid Commun. Mass Spectrom.*, 2008, **22**, 3217–3224.

11. S. Yu, E. Crawford, J. Tice, B. Musselman and J. T. Wu, Bioanalysis without Sample Cleanup or Chromatography: The Evaluation and Initial Implementation of Direct Analysis in Real Time Ionization Mass Spectrometry for the Quantification of Drugs in Biological Matrixes, *Anal. Chem.*, 2009, **81**, 193–202.

12. S. Yu, E. Crawford, J. Tice, B. Musselman and J. T. Wu, Improving the Reproducibility and Practicability of DART for the Direct Quantification of Drugs in Biological Fluids. Presented at: 56th ASMS Conference on Mass Spectrometry and Allied Topics. Denver, CO, USA, 1–5 June 2008.

13. E. Crawford, J. Gordon, J. T. Wu, B. Musselman, R. Liu and S. Yu, Direct analysis in real time coupled with dried spot sampling for bioanalysis in a drug-discovery setting, *Bioanalysis*, 2011, **3**, 1217–1226.

14. T. Cajka, L. Vaclavik, K. Riddellova and J. Hajslova, GC–TOF-MS and DART–TOF-MS: Challenges in the Analysis of Soft Drinks, *LC · GC Eur.*, 2008, **21**, 250–256.

15. T. Cajka, K. Riddellova, M. Tomaniova and J. Hajslova, Recognition of beer brand based on multivariate analysis of volatile fingerprint, *J. Chromatogr. A*, 2010, **1217**, 4195–4203.

16. M. Zachariasova, T. Cajka, M. Godula, A. Malachova, Z. Veprikova and J. Hajslova, Analysis of multiple mycotoxins in beer employing (ultra)-high resolution mass spectrometry, *Rapid Commun. Mass Spectrom.*, 2010, **24**, 3357–3367.

17. T. Cajka, K. Riddellova, M. Tomaniova and J. Hajslova, Ambient mass spectrometry employing a DART ion source for metabolomic fingerprinting/profiling: a powerful tool for beer origin recognition, *Metabolomics*, 2011, 7, 500–508.

18. H. J. Kim, E. H. Jee, K. S. Ahn, H. S. Choi and Y. P. Jang, Identification of Marker Compounds in Herbal Drugs on TLC with DART-MS, *Arch. Pharmacal Res.*, 2010, **33**(9), 1355–1359.

19. G. A. Harris and F. M. Fernandez, Simulations and Experimental Investigation of Atmospheric Transport in an Ambient Metastable-Induced Chemical Ionization Source, *Anal. Chem.*, 2009, **81**, 322–329.

20. E. Chernetsova, A. Revelsky and G. Morlock, Some new features of Direct Analysis in Real Time mass spectrometry utilizing the desorption at an angle option, *Rapid Commun. Mass Spectrom.*, 2011, **25**, 2275–2282.

21. B. Musselman, E. Crawford, C. Stacey and R. Cody, Strategies for Rapid Determination of Impurities using DART-MS from Surfaces. Presented at: *2009 Pittcon Conference*. Chicago, IL, USA, 8–13 March 2009.

22. J. J. Perez, G. A. Harris, J. E. Chipuk, J. S. Brodbelt, M. D. Green, C. Y. Hampton and F. M. Fernández, Transmission-mode direct analysis in real time and desorption electrospray ionization mass spectrometry of insecticide-treated bednets for malaria control, *Analyst*, 2010, **135**, 712–719.

23. A. H. Grange, An Autosampler and Field Sample Carrier for Maximizing Throughput Using an Open-Air Source for MS, *Am. Lab.*, 2008, **40**(16), 11–13.
24. E. Crawford and B. Musselman, Evaluating a direct swabbing method for screening pesticides on fruit and vegetable surfaces using direct analysis in real time (DART) coupled to an Exactive benchtop orbitrap mass spectrometer, *Anal. Bioanal. Chem.*, 2012, **403**, 2807–2812.
25. C. Ferrer, M. J. Martines-Bueno, A. Lozano and A. R. Fernandez-Alba, Pesticide residue analysis of fruit juices by LC-MS/MS direct injection. One year pilot survey, *Talanta*, 2011, **83**, 1552–1561.
26. J. Wang, W. Chow, D. Leung and J. Chang, Application of Ultrahigh-Performance Liquid Chromatography and Electrospray Ionization Quadrupole Orbitrap High-Resolution Mass Spectrometry for Determination of 166 Pesticides in Fruits and Vegetables, *J. Agric. Food Chem.*, 2012, **60**(49), 12088–13104.
27. E. Crawford, J. Tice, M. Festa and B. Musselman, High Throughput Open Air Desorption Ionization for Bioanalysis and Beyond. Presented at: 59th ASMS Conference on Mass Spectrometry and Allied Topics. Denver, CO, USA, 5–9 June 2011.
28. Risk Assessment for Safety of Orange Juice Containing Fungicide Carbendazim. Available from: http://www.epa.gov/pesticides/factsheets/chemicals/carbendazim-fs.htm (accessed February 15, 2012).
29. A. Kaufmann, The current role of high-resolution mass spectrometry in food analysis, *Anal. Bioanal. Chem.*, 2012, **403**, 1233–1249.
30. M. Kellmann, H. Muenster, P. Zomer and H. Mol, Full scan MS in comprehensive qualitative and quantitative residue analysis in food and feed matrices: How much resolving power is required? *J. Amer. Soc. Mass Spectrom.*, 2009, **20**, 1464–1476.

Laser-Ablation Electrospray Ionization Mass Spectrometry (LAESI®-MS): Ambient Ionization Technology for 2D and 3D Molecular Imaging

TRUST T. RAZUNGUZWA,* HOLLY D. HENDERSON,
BRENT R. RESCHKE, CALLEE M. WALSH AND
MATTHEW J. POWELL

Protea Biosciences Group Inc., 955 Hartman Run Road, Morgantown,
WV 26505, USA
*Email: trust.razunguzwa@proteabio.com

19.1 Introduction

19.1.1 Overview of Mass Spectrometry Imaging (MSI)

Mass-spectrometry imaging (MSI) is a very powerful technique for visualizing the spatial distribution of molecules in biological samples, such as tissue sections, whole-body sections, and cell colonies or populations. For this reason, MSI has seen recent advancements using ionization techniques such as matrix-assisted laser desorption ionization mass spectrometry (MALDI-MS),[1] secondary ion mass spectrometry (SIMS),[2–7] nanostructure-initiator mass spectrometry (NIMS),[8–13] and a variety of ambient ionization

New Developments in Mass Spectrometry No. 2
Ambient Ionization Mass Spectrometry
Edited by Marek Domin and Robert Cody
© The Royal Society of Chemistry 2015
Published by the Royal Society of Chemistry, www.rsc.org

techniques,[14–16] for sample introduction into a wide variety of mass spectrometers. Using the high resolving power, high mass accuracy, ion mobility, and MS/MS capabilities of mass spectrometers, a wide variety of small molecules, metabolites, lipids, peptides, and proteins can be identified and characterized in tissue-imaging analyses. The imaging component of MSI provides location information for the compounds of interest – whether they may be potential biomarkers or drug compounds. The ultimate result of an MSI experiment is that chemical and spatial information of previously known or unknown compounds is obtained by collecting mass spectra from many locations on a tissue section (acquired using the same instrumental settings), and this data is used to generate an "image" or ion map of the intensity distribution of the compound's mass signatures across every analysis location. In addition, MSI minimizes the amount of sample preparation required for analysis of target compounds, compared to LC-MS/MS methods where tissues are typically homogenized, and the compounds physically or chemically extracted for analysis, which results in a loss of spatial information. For these reasons, there have been significant developments of MSI techniques, with a number of publications on the use of MALDI imaging for pharmacokinetic[17–19] and histopathology studies.[20,21] Various ambient ionization mass-spectrometry techniques have also been applied to tissue imaging including laser-ablation electrospray ionization (LAESI),[22,23] electrospray-assisted laser desorption ionization (ELDI),[24] atmospheric-pressure matrix-assisted laser desorption mass spectrometry (AP-MALDI), desorption electrospray ionization (DESI),[25] and liquid extraction surface analysis (LESA).[26–28] This chapter will focus on the application of LAESI technology using the LAESI DP-1000 Direct Ionization System for tissue imaging, and it will begin by discussing the different ambient ionization techniques used for tissue imaging, of which LAESI technology is part.

19.1.2   Ambient Ionization Techniques for MSI

Ambient ionization mass spectrometry encompasses a group of techniques where sample ionization occurs at atmospheric pressure, (classification of these methods that are applied to tissue imaging is summarized in the flowchart in Figure 19.1). In addition to the ambient ionization techniques shown in Figure 19.1, mass-spectrometry imaging has been demonstrated using infrared matrix-assisted laser desorption electrospray ionization mass spectrometry (IR-MALDESI-MS),[29] and infrared laser-ablation metastable-induced chemical ionization mass spectrometry (IR-LAMICI-MS).[30] The configuration for IR-MALDESI-MS,[31] where water is defined as the matrix, is very similar to LAESI-MS.[32] Ionization for biological mass spectrometry, has traditionally been achieved by electrospray ionization (ESI), atmospheric-pressure chemical ionization (APCI), or matrix-assisted laser-desorption mass spectrometry (MALDI). The recently developed ambient ionization techniques, which include LAESI-MS, typically involve very minimal sample preparation, and in some cases, no sample preparation at all, prior to

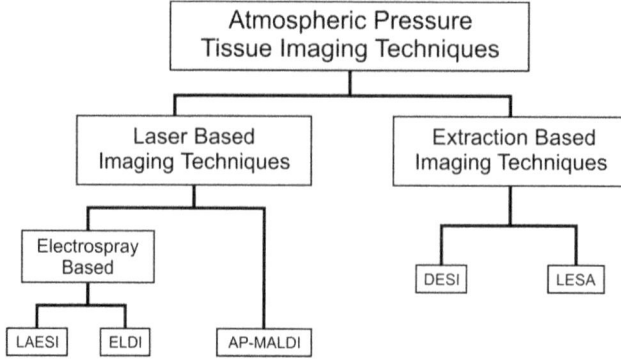

Figure 19.1 Classification of ambient ionization techniques discussed in this review.

sample analysis. The elimination or minimization of sample preparation for these sources is made possible by decoupling sample delivery (or molecule extraction into the gas phase) from sample ionization. In the case of LAESI, the sample is delivered *via* a laser-ablation event, and the analyte is subsequently ionized by coalescence with an electrospray plume, where acid-base chemistries interact in the ESI droplets. Consequently, ambient ionization techniques have been extremely well suited and applied to the direct analysis of tissue sections. Early efforts of MSI heavily involved the use of MALDI for tissue analysis, and MALDI imaging has since been used successfully for analyzing fairly abundant small molecules, membrane lipids and proteins in tissue sections, at spatial resolutions from ∼20 to 200 μm.[4,33–43] Typically, MALDI ionization is used together with time-of-flight (TOF) mass analyzers, because the speed of these analyzers match very well with discrete "packets" of ions generated from each laser pulse during MALDI. With a TOF instrument, a complete mass spectrum can be acquired within the nanosecond timescale of each laser pulse. Lately however, MALDI sources have been configured to work with other mass analyzers, such as the Fourier transform ion cyclotron resonance (FT-ICR) and linear ion traps. Conversely, ambient ionization sources, such as LAESI, are designed to integrate with a variety of mass analyzers from different mass-spectrometer suppliers, allowing these sources to be used with a wider variety of mass analyzers, particularly those having the highest resolving powers, for identification of compounds using accurate mass.

There are some challenges with the use of MALDI for tissue imaging, some of which are overcome by the use of ambient ionization techniques. One of the major drawbacks with MALDI is the requirement for the addition of an external MALDI matrix to the tissue sections prior to imaging. This matrix deposition process adds several steps in the sample-preparation process, which not only presents the potential of contaminating the tissue section, but, more importantly, can change the distribution of endogenous analytes of interest. The MALDI matrix can also limit the spatial resolution achievable if the droplet size of the matrix solvent during application is larger than the laser spot size. Researchers are therefore developing the use of sublimation for applying homogenous layers of matrices that have a very small crystal size.[44–47] Such layers can also be deposited using costly microsprayer MALDI matrix deposition devices. Ambient ionization techniques such as LAESI, DESI, and LESA do not require matrix deposition. The tissue section can be analyzed directly after cryosectioning. For LAESI, the tissue is analyzed frozen; therefore there is no need to thaw stored sections at the time of analysis, which preserves the spatial distribution of analytes for accurate ion maps. Another drawback for MALDI is use of vacuum during MALDI imaging, which can contribute to sample degradation, and to the sublimation of matrix in the source during data acquisition, thereby limiting the length of time a particular tissue section can be analyzed. Vacuum compatibility requirements can also limit the types of samples that are analyzed, especially for *in vivo* applications that are presented in this chapter. Lastly, depending on the MALDI matrix used, the added matrix can interfere with analysis of small molecules, as there can be numerous matrix adducts in the low *m/z* region of the mass spectrum. The elimination of MALDI matrix for tissue imaging circumvents these problems emanating from their use.

19.1.3   Laser-Ablation Electrospray Ionization (LAESI)

The configuration of the LAESI source is shown in Figure 19.2. In LAESI, a 2.94 μm mid-IR laser is used to ablate a sample with subsequent ionization using ESI mechanisms. This laser wavelength coincides with the O–H bond stretch present in water, which makes it very suitable for ablation of water-containing samples, such as animal and plant tissues. Since the water "matrix" is present within the sample, there is no need to actively coat the sample with any external matrix – as is required in UV-based laser methods (*e.g.* MALDI), where a UV-absorbing compound has to be applied to the sample.

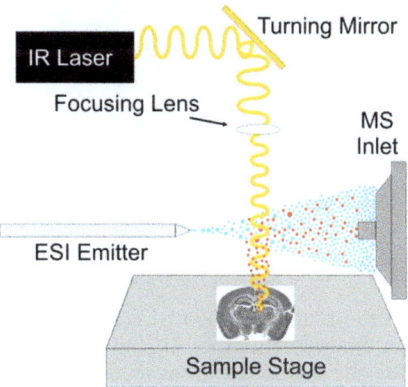

Figure 19.2   Schematic diagram of laser ablation electrospray ionization.

There are three phases to the process of mid-IR laser ablation: surface evaporation, followed by phase explosion and then fluid ejection in the form of an ablation plume that travels over 10 mm above the surface of the sample, and is comprised of small water droplets and sample particles.[48] The ablation plume contains a population of neutral molecules from the sample. Ionization occurs *via* coalescence of the sample molecules with the orthogonal electrospray (ESI) plume above the sample, and the sample ions pass into a mass spectrometer for detection.

19.2 LAESI DP-1000 Direct Ionization System Description

19.2.1 Overview: Protea LAESI DP-1000 Direct Ionization System

The Protea LAESI DP-1000 Direct Ionization System is the first commercial instrument using LAESI technology for sample delivery into mass spectrometers, and has been brought to market by Protea Biosciences Group, Inc. As shown in Figure 19.3, it is an integrated LAESI source containing an IR laser with a spot size of 200 μm and with an internal attenuator for adjusting output energy from 0 to 1.0 mJ for laser ablation. A conventional optic train directs and focuses the laser beam onto a thermoelectrically regulated 3D translational stage. The sample stage and electrospray emitter are enclosed

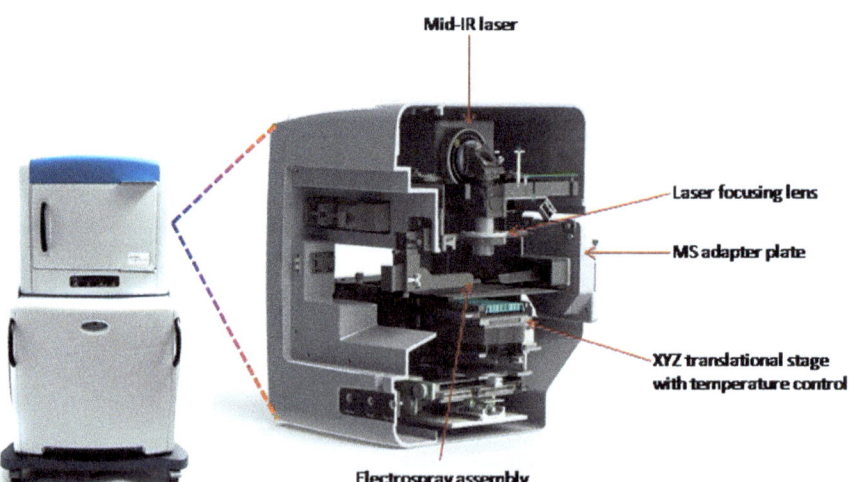

Figure 19.3   Diagram of the LAESI DP-1000 Direct Ionization System. An expanded view of the analysis chamber is shown on the right. Hardware components for the LAESI DP-1000 system include the electrospray assembly, midinfrared laser (2940 nm), laser focusing lens, MS adapter plate, and the temperature-regulated *XYZ* translational stage.

in an isolation shell to remove environmental effects. An on-board syringe pump, high-voltage power supply, and custom ESI emitter are integrated to provide a reproducible electrospray signal. This instrument platform is compatible with several major mass-spectrometer instrument vendors to provide a robust LAESI source on the front end. The 2.94 µm wavelength employed by the LAESI DP-1000 system is specifically tuned to the maximum absorption coefficient of water, making the LAESI DP-1000 system well suited for the analysis of water-containing samples. Additionally, this wavelength is not well absorbed by other biomolecules, such as lipids or proteins; thus, there is no fragmentation, ionization, or other disruption to the biomolecules themselves during the ablation event.

The LAESI DP-1000 system provides the control necessary to obtain valuable MS imaging data. Temperature control with a Peltier stage allows samples to be kept frozen at −10 °C during analysis, which preserves the water content necessary for ablation with the mid-IR laser, and prevents potential analyte migration. The temperature control system can also be used to heat samples to a temperature of 60 °C, allowing real-time monitoring of thermal reactions. On-board regulation of bath and sheath gases controls the humidity within the analysis chamber, preventing condensation and ice formation on the sample. Stability of the voltage supplied to the electrospray emitter is critical, and is accomplished using a precise high-voltage power supply monitored by high-voltage feedback circuits.

19.2.2   LAESI DP-1000 Software

The LAESI DP-1000 uses two software platforms, LAESI® Desktop Software (LDS) and ProteaPlot™, to provide a means of efficient data collection and analysis. As shown in Figure 19.4(A), LDS controls all aspects of the LAESI DP-1000 system for data collection, including ESI voltage, ESI flow rate, temperature of the Peltier stage, laser configuration, and stage movement. As well as this, LDS interfaces with major mass-spectrometer vendor software platforms for coordination of data collection and positional information. The LAESI DP-1000 system has two cameras that are controlled by LDS, an inline camera as well as a wide-angle camera. The wide-angle camera can be used to image the sample in real time, or pre- and post-analysis for examination of analysis locations and location definition and alignment, as shown in Figure 19.4(B). LDS can operate both in interactive mode, where instrument set-up and method optimization can be easily performed, and in project mode, where data-analysis parameters can be saved to be used for multiple samples or for future projects. ProteaPlot™ also interfaces with major mass-spectrometer vendor software platforms, and it is used in a postacquisition manner for general data processing, 2D and 3D ion mapping, image and data export, background subtraction, and spectral comparison (Figure 19.5). Both software platforms, LAESI Desktop Software (LDS) and ProteaPlot feature surface 2D and 3D analysis options, providing efficient paths for imaging workflows.

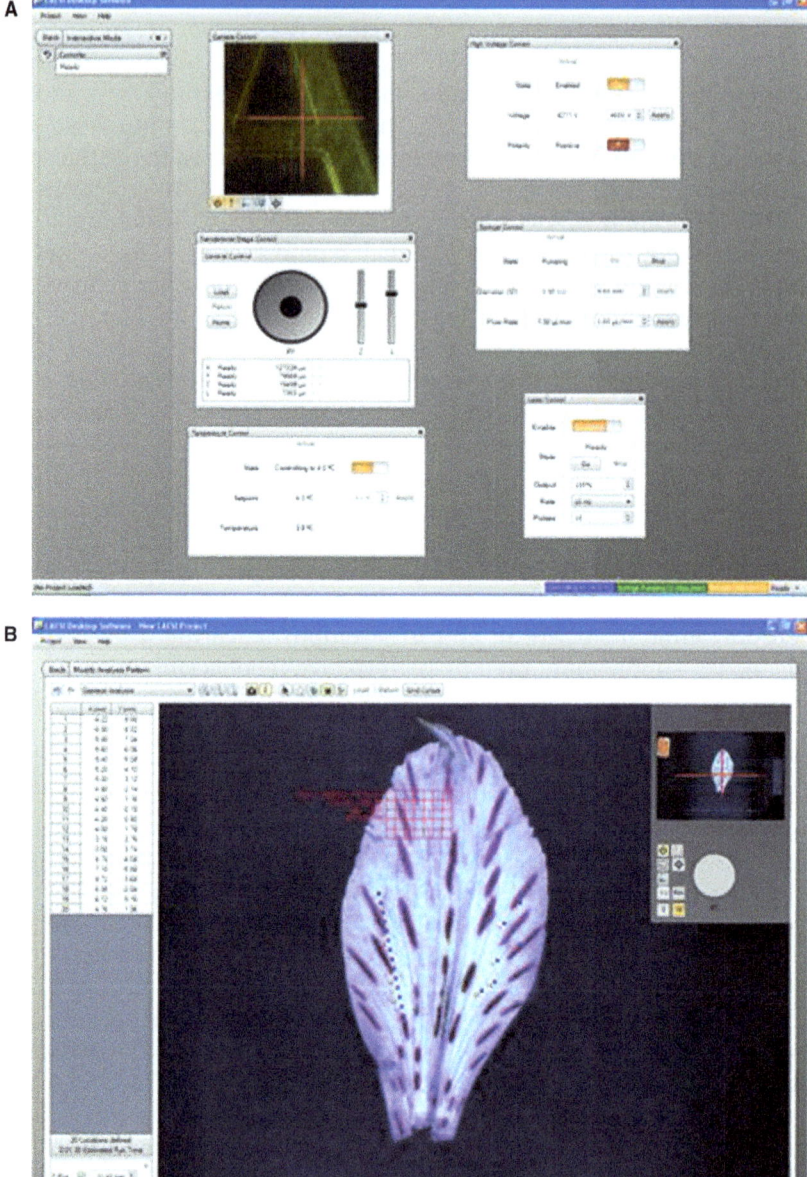

Figure 19.4 Graphical user interface for LDS software for data acquisition. Shown in (A) is the screen for setting up instrumental parameters, and shown in (B) is the screen for selection of area of interest to target.

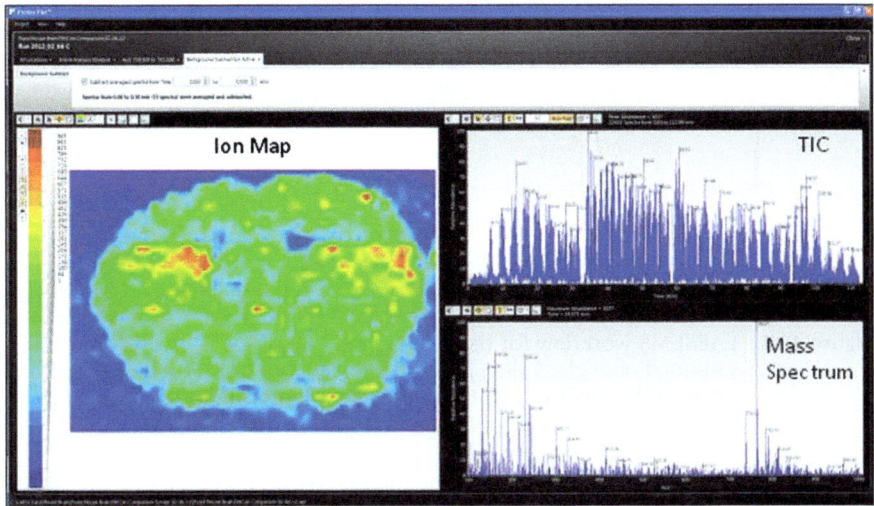

Figure 19.5 Graphical user interface for ProteaPlot postprocessing software, displaying a total ion chronogram (TIC) for the entire imaging run, an example mass spectrum, and an ion map from a mouse-brain coronal section tissue imaging experiment.

19.3 Molecular Imaging using the LAESI DP-1000 System

19.3.1 LAESI-MS Imaging Workflow

The LAESI mass-spectrometry imaging workflow is depicted in Figure 19.6 using animal-tissue imaging as an example. It offers unprecedented advantages with its ability to analyze a sample as it exists, and obtain mass spectrometric data. For LAESI-MS imaging, there is little to no sample preparation, as the technique has the ability to analyze frozen tissue samples directly without thawing them, using the tissue's natural water content as a matrix for absorption of the mid-IR laser energy. Compared to MALDI-MS imaging, there are no additional steps required to prepare the sample, such as matrix application or desiccation for analysis under vacuum, thereby reducing the time required to develop tissue imaging data after cryosectioning. Biological tissues can be taken directly from cryosectioning to the sample stage for analysis, biofluids can be analyzed directly from a plate, bacterial colonies can be analyzed on the agar where they have grown, and botanical samples can be collected and ablated. The remainder of the chapter describes results from imaging experiments of animal tissues, contact lenses, a live bacterial cell colony on an agar plate and a fungal colony on a potato agar plate.

19.3.2 Imaging Metabolites and Lipids in Mouse Brain Tissue

The LAESI DP-1000 system has been demonstrated to successfully map lipids and other small molecules in tissues. The downstream importance of

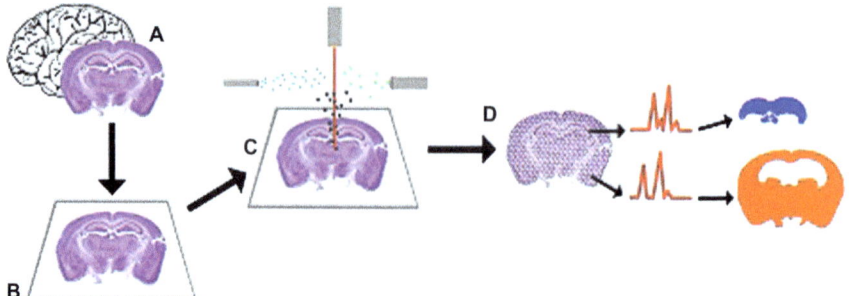

Figure 19.6   LAESI-MS workflow for tissue imaging. The tissue is cryosectioned (A), mounted on a glass slide (B), and subjected to LAESI analysis (C). Data analysis (D) involves extraction of spectral information from each pixel to construct ion maps of the tissue.

this function is that biomolecules can be localized and colocalized, shedding light on the relationship between their locations in an organism and their biological functions. The LAESI DP-1000 system can be integrated with gas-phase separation techniques, such as ion-mobility spectrometry (IMS), which allows the ionization of biological samples from their native state and discrimination of molecular species based on differences in the size, shape and charge state. After data acquisition, 2D drift time–m/z plots can be inspected for regions of interest characteristic of different molecular classes. In an example LAESI-ion-mobility experiment, lipids are targeted, which typically are observed at later drift times in the 2D plots due to the elongated structure of many lipid molecules, (see Figures 19.7 and 19.8). By identifying these regions of interest and exporting only the information from those regions, background signal can be significantly reduced. After exporting, the data is lockmass corrected and imported into ProteaPlot to generate ion maps of the analyzed frozen tissues. The ion maps are also overlaid upon an optical photograph of the sample taken prior to analysis to show localization of the different lipid species in the tissue. Using the IMS and MS data, many lipid species are putatively identified based on accurate mass, including fatty acids, fatty acyls, glycerolipids, di- and triglycerides, glycerophospholipids, phosphatidylcholines, phosphatidylethnaolamines, phosphatidylserines, sterols, and phosphoglycerols. Complex fields such as lipidomics can benefit greatly from the combination of direct ionization sources, like LAESI, and multidimensional gas-phase separations, like IMS-MS, to decrease the overall analysis time and increase the information gleaned from the analyzed samples.

19.3.3   Imaging Drug Compounds on Mouse-body Sections

An important step in drug development is the understanding of a drug's distribution through the body and its penetration into different organs. To acquire this information autoradiography has been used as the cornerstone

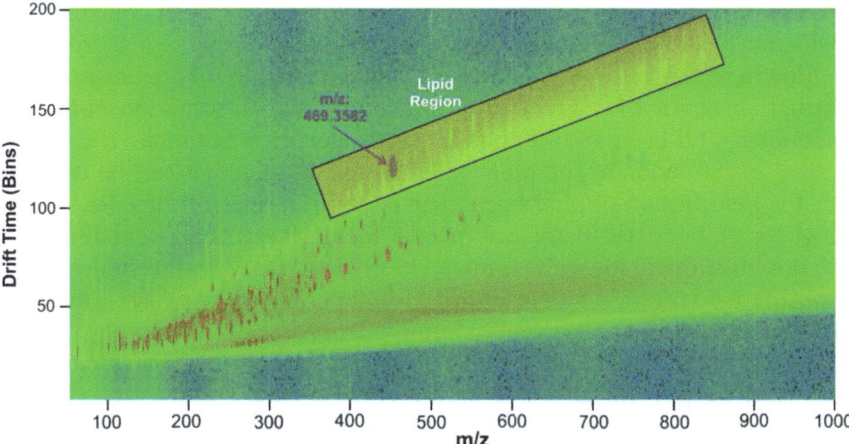

Figure 19.7   The positive-ion mode 2D *m/z*–drift time graph exported from Drift-Scope. The region highlighted in orange shows many of the lipid species detected in the analysis. Below the lipid region many of the background ions can be seen. Notice that upon inspection of the MS-IMS plot, separation of the lipids from the background is clearly discernible.

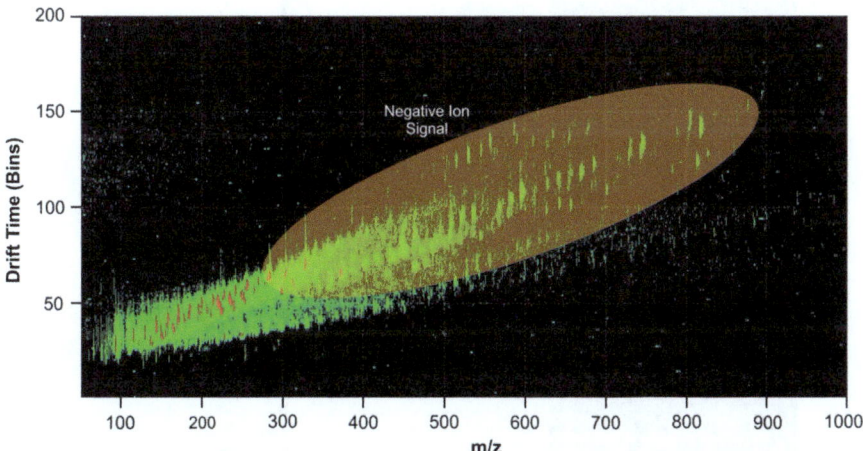

Figure 19.8   The negative-ion mode 2D *m/z*–drift time graph exported from Drift-Scope. The region highlighted in orange shows many of the lipid species detected in the analysis. Again, as in Figure 19.7, the background ions can be easily distinguished from the signal generated by the LAESI analysis. Based on the highlighted region of the 2D plot there are multiple ions that could be mapped across different drift time ranges. In addition, upon closer inspection of the 500–600 *m/z* range there appears to be isomeric separation of multiple species along the drift-time axis.

technique for such studies. More recently, mass-spectrometry imaging has provided a label-free alternative to autoradiography for studying drug distribution in whole-body mouse sections. This technique can provide not only information about the parent drug, but additional information about its metabolites, and even endogenous biomolecules in the animal. The LAESI DP-1000 system adds to the MSI toolkit of techniques that can be used to track the distribution of pharmaceutical drugs in tissue sections. Presented in Figures 19.9 and 19.10 are examples of raclopride and fexofenadine that were dosed in mice. Male CD-1 mice (25 g) were dosed intravenously with raclopride at 10 mg kg^{-1} and orally with fexofenadine at 50 mg kg^{-1}. After an animal was sacrificed and frozen in liquid nitrogen, the animal's whole body was adhered to a specimen disc with an ice slush made from deionized water, and 40 μm sections were cryosectioned in a cryostat at −20 °C. Imaging of the whole-body mouse sections was performed on the LAESI DP-1000 system connected to a Thermo LTQ Velos™ mass spectrometer. The LAESI DP-1000 settings included an electrospray voltage of 4 kV at 1 μL min^{-1} electrospray flow rate (0.1% acetic acid in 50% methanol), and 2 psi bath gas pressure to control humidity in the LAESI isolation shell during imaging. Whole-body mouse section slides were placed on a precooled (−10 °C) LAESI DP-1000 stage, and analyzed at 340 μm pitch (center–center distance) for several hours. A total of 10 laser pulses at 10 Hz were used at

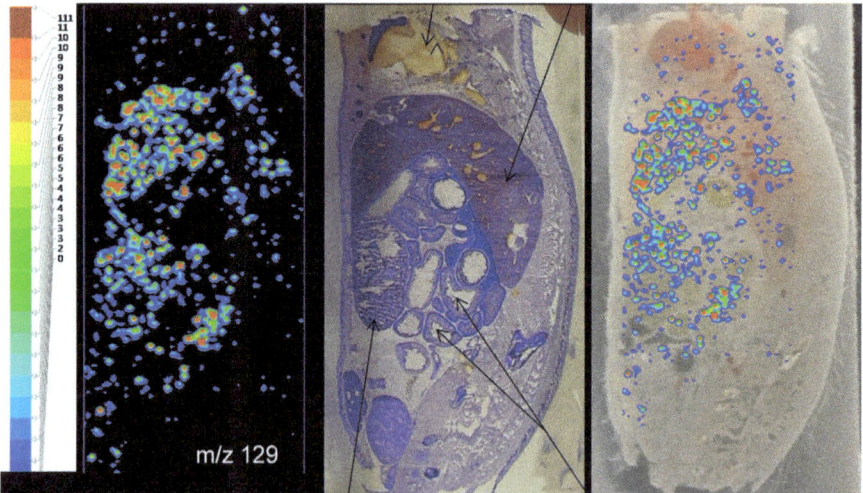

Figure 19.9   Ion map of raclopride (fragment ion $m/z = 129$), in a mouse-body section after 30 min after intravenous dosage (left). Stained optical image of tissue section details the organ structure, and optical image with overlaid ion map highlights the distribution profile. Drug is clearly visible in the liver and intestines. The presence in gastric organs supports the suggestion that raclopride is transported to the stomach after intravenous introduction, and excreted *via* the gastric system, which was previously shown by MALDI imaging in mice.[3]

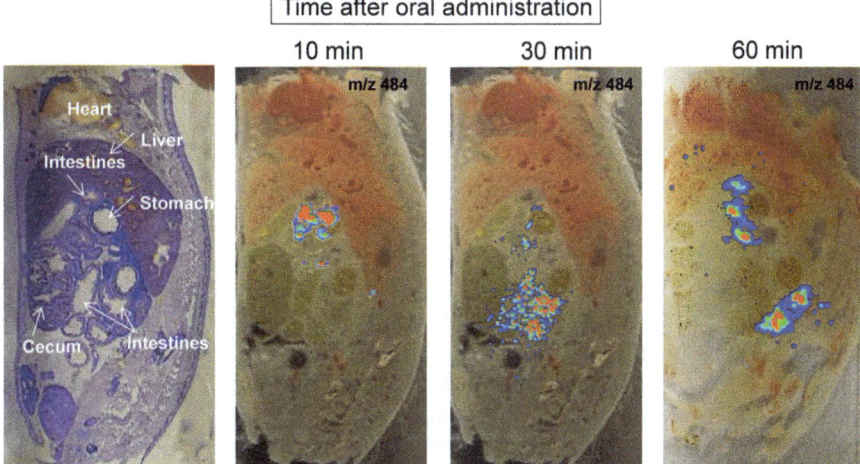

Figure 19.10   Ion maps representing the time course dosage study of fexofenadine (fragment ion *m/z* 484). These data demonstrate the drug through the stomach, intestines, and liver, and is consistent with its mode of excretion *via* the gastrointestinal tract with oral dosage. The maximum intensity at each location is represented in the ion maps.

each location. The mass spectrometer was operated in MS/MS mode, fragmenting *m/z* 348.3 for raclopride and *m/z* 502.7 for fexofenadine. Fragment peaks at *m/z* 129 (raclopride) and *m/z* 484 (fexofenadine), were chosen for construction of ion distribution maps. All acquired data was processed using ProteaPlot imaging software. The resulting molecular images are shown in Figures 19.9 and 19.10. Figure 19.9 shows an ion map (left), stained optical image (center), and an overlay of the ion map on the optical image (right), at the 30 min timepoint after dosage. Raclopride is clearly visible in the liver and gastric organ, suggesting that after intravenous dosage, raclopride is transported to the gastric system, an observation that has been made with MALDI imaging, LC-MS/MS and autoradigraphy.[19] In Figure 19.10, the distribution of fexofenadine drug is monitored at 10, 30, and 60 min after oral dosage. The passage of the drug through the gastric system is clearly seen in the ion maps. At 60 min the drug is detected in liver, showing the onset of accumulation of the drug in the liver at this timepoint. These experiments demonstrate the utility of LAESI-MS for pharmacokinetic studies.

19.3.4   Depth Profiling and 2D Imaging of Contact Lenses

Mechanisms of contact-lens spoilage are of interest to many researchers in the contact-lens industry due to the potential significance of lens discomfort to patients and providers alike. Biomolecules, including lipids and proteins, have been observed to accumulate on lenses during wear. The human eye contains tear film, which comprises three main components: aqueous (produced by the lacrimal glands), lipids (or meibum, produced by the meibomium glands), and mucins (proteins).[49,50] Because LAESI-MSI is performed at ambient pressure, contact lenses can be analyzed in the native state without sample dehydration or any other sample manipulation that could contaminate the lens or otherwise affect the quality of analysis. Furthermore, the natural water content of the hydrogel makes it particularly well-suited for analysis by the 2.94 µm mid-IR laser.

Numerous studies have characterized lipids and proteins common to human tear fluid. A comprehensive proteomic analysis of worn lenses performed by Zhao *et al.* identified sixty-eight proteins using LC-MS-MS.[51] The most frequently detected proteins were lysozyme and lipocalin. Research performed by Brown *et al.* presents characterization of polar and nonpolar lipids extracted from a worn lens.[52] Additionally, studies have been performed that identify many small molecules present from lens manufacture, such as phthalates and other plasticizers and polymers.[53] Information about the distribution of the inherent molecules may also provide information about biomolecule accumulation or lack thereof. LAESI-MSI can be a valuable

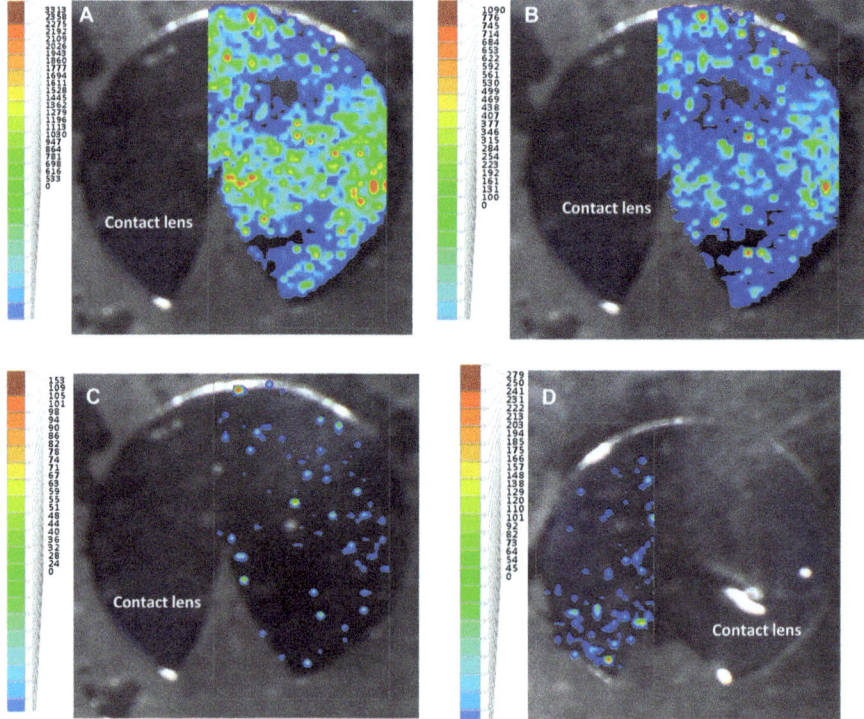

Figure 19.11   2D ion maps of contact lens showing distribution of two PDMS fragments from the contact lens material (A and B), distribution of lysozyme's charged   state *m/z* 1590.7 (C), phosphatidylcholine lipid *m/z* 760.8 (D).

analytical tool to help understand the extent and location of biomolecular accumulations. In addition, the structural characteristics that provide resistance to the accumulation of discomforting materials can also be studied. This information can then be correlated to the comfort of the lens wearer in terms of breathability and general comfort. Current and future studies can aid in the development of ways to minimize contact lens surface fouling by providing insight into the composition and spatial distribution of lipids and proteins on the lens, and therefore, the mechanisms of lens spoilage.

In a case study using the LAESI DP-1000 system for contact-lens analysis, Acuvue® Advance™ contact lenses were dosed with protein (lysozyme in PBS) and lipid (phosphatidycholine lipid) for 300 blinking cycles using an *in vitro* blinking cell. LAESI-MSI of the contact lenses was performed using a Protea LAESI DP-1000 coupled with a Thermo Scientific LTQ Velos mass spectrometer. Proteins and lipids adhered to the surface, and through the depth of the lens were detected and imaged. For 2D mapping, the lens was ablated with 20 laser pulses at 10 Hz at each location, at a spatial resolution of 360 μm. The mass spectrometer was operated in positive-ion mode using an electrospray solution of 0.1% acetic acid solution in 50% methanol.

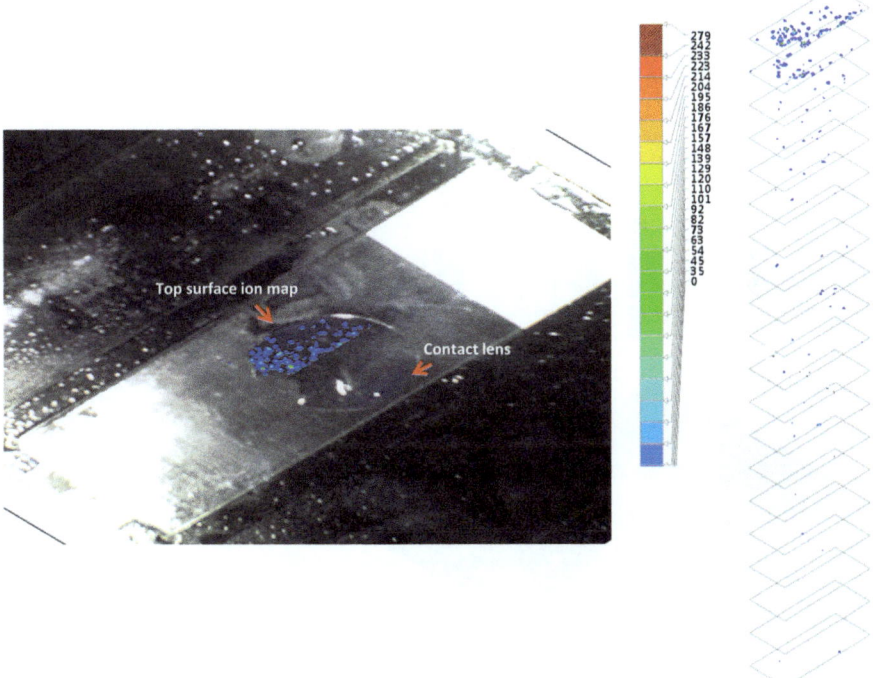

Figure 19.12 Ion maps for phosphatidylcholine lipid (*m/z* 760.7) for 20 laser pulses going through the entire thickness of the contact lens. The depth profile shows a reduced level of penetration of the lipid into the contact lens, compared to the lysozyme protein sample.

The electrospray solution was introduced at 1 μL min^{-1} with an electrospray voltage of 4 kV. Each contact lens analyzed was first rinsed in deionized water, placed on a glass slide and an incision was made to ensure that the contact lens laid flat on the microscope slide. After loading in the instrument, the sample stage was kept at −10 °C to prevent the lens from drying-out during analysis. The mass spectrometer was set to acquire data over a mass range 100–2000 *m/z*. After data acquisition, the data files were imported in ProteaPlot software into generate ion maps for select molecules. The resulting ion maps for PDMS polymer fragments (*m/z* 445, 673), lysozyme (*m/z* 1590.7) and phosphatidylcholine lipid (*m/z* 760.8) are shown in Figure 19.11, demonstrating the ability of LAESI-MS to detect biomolecules and polymer fragments on the surface of contact lenses.

For depth profiling, a pulse test was performed to determine the number of pulses required to fully perforate the contact lens. A count of 20 pulses was determined to be sufficient when employing a 1 Hz pulse rate to

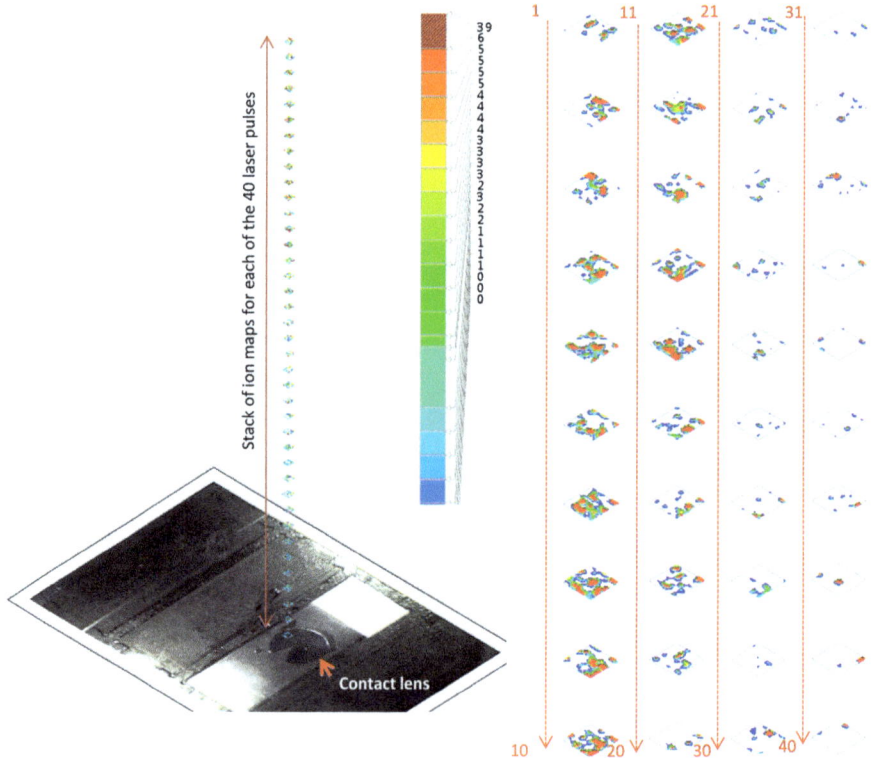

Figure 19.13   Ion maps for of lysozyme's charged state (*m/z* 1590.7). Numbers in red represent the laser pulse. A portion of the contact lens was analyzed for depth profiling as shown on the optical image of the contact lens on the left. The data shows the presence of lysozyme in the entire thickness of the contact lens, with the middle part of the lens showing greater presence of the protein.

completely ablate all the material in the 100 μm thickness of the contact lens. For depth profiling, 40 laser pulses were used at a frequency of 1 Hz at a spatial resolution of 200 μm. The resulting ion maps for the lipid and protein depth profiles are shown in Figures 19.12 and 19.13, respectively. These images show greater depth penetration of protein within the contact lens than the lipid that is primarily detected on and near the surface. The ability to detect biomolecules and components of the lens through its thickness is useful for biomaterial characterization, QC testing and biofouling studies.

19.3.5   *In vivo* Imaging of Cell Colonies

Direct analysis of living cell colonies has also been demonstrated using LAESI-MS. In experiments using the LAESI DP-1000 system, bacterial and fungal colonies growing on agar plates can be directly analysed with no sample preparation. Figure 19.14 shows example data from the direct analysis of *Aspergillus fumigatus* fungal colony on potato agar. The representative spectrum shows significant mass-spectral complexity from the cells

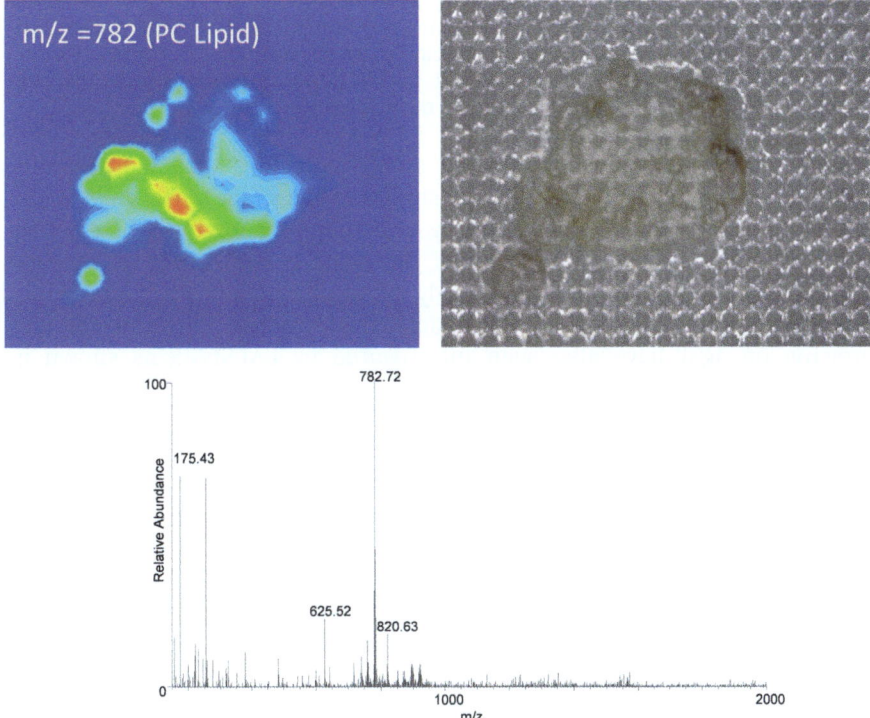

Figure 19.14   LAESI-MS representative spectrum from the direct analysis of fungal colony (*Aspergillus fumigatus*) acquired using the Protea LAESI DP-1000 system connected to a Thermo LTQ Velos mass spectrometer. From the spectrum the phosphatidylcholine lipid at $m/z = 782.7$ is mapped and the ion map closely resembles the shape of the colony.

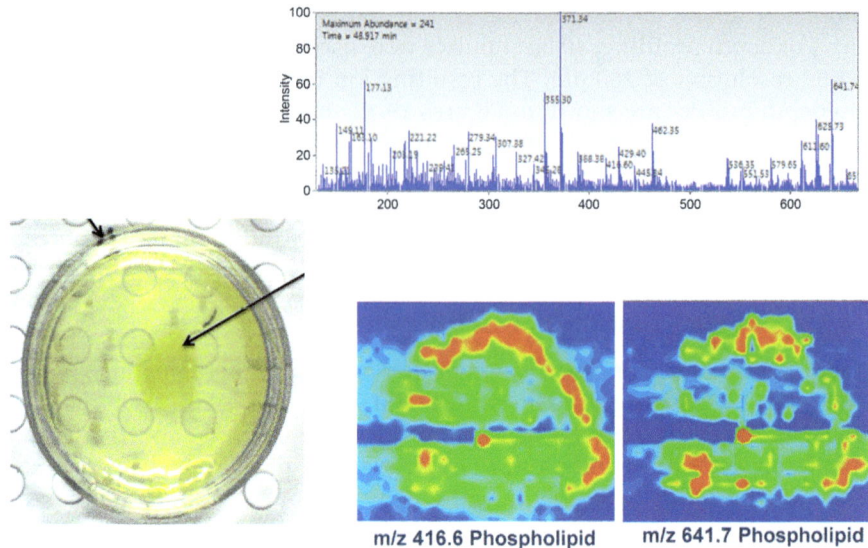

Figure 19.15   LAESI-MS representative spectrum from the direct analysis of an *E. coli* bacterial colony acquired using the Protea LAESI DP-1000 system connected to a Thermo LTQ Velos mass spectrometer. Ion maps for two phospholipids at $m/z = 416.6$ and 641.7 are mapped and the ion map closely resembles the shape of the colony.

representative of various lipids, and small molecules with a phosphatidylcholine lipid ($m/z = 782.7$) ion map similar to the shape of the colony. The ability to perform molecular profiling and mapping for live cell colonies allows for time-related monitoring of the same colony, study of the biodynamics of living populations, study of antibiotic/biofilm therapeutics stimulus response relations, and toxicological studies. Bacterial colonies growing on agar have also been interrogated by LAESI-MS as shown in Figure 19.15.

19.4   Future Perspectives

Current and future development efforts for LAESI technology are targeted towards reducing the spot size of the IR laser to improve spatial resolution. Current research efforts for MSI are undertaking cell-by-cell imaging and several research groups have demonstrated single-cell analysis using MALDI MSI[54–58] and secondary ion mass spectrometry (SIMS).[59–62] The main limitation to performing these experiments for laser-based methods is achieving the required spatial resolution, with challenges in configuring optics that achieve a small enough laser spot size. Furthermore, as the spot size decreases, there are greater challenges with obtaining the required sensitivities, as smaller amounts of analytes are available for detection. For LAESI-MS, single-cell imaging has already been demonstrated by the Vertes group using cell-by-cell imaging of plant tissue metabolites.[63,64]

Acknowledgements

We thank Pamela Williams and Gregory Boyce for sourcing and providing samples used in this work.

References

1. http://www.maldi-msi.org/.
2. A. V. Walker, *Anal. Chem.*, 2008, **80**, 8865–8870.
3. G. B. Eijkel, B. Kuekrer Kaletas, I. M. van der Wiel, J. M. Kros, T. M. Luider and R. M. A. Heeren, *Surf. Interface Anal.*, 2009, **41**, 675–685.
4. F. Benabdellah, A. Seyer, L. Quinton, D. Touboul, A. Brunelle and O. Laprevote, *Anal. Bioanal. Chem.*, **396**, 151–162.
5. B. K. Kaletas, I. M. van der Wiel, J. Stauber, C. Guezel, J. M. Kros, T. M. Luider and R. M. A. Heeren, *Proteomics*, 2009, **9**, 2622–2633.
6. E. B. Monroe, S. P. Annangudi, N. G. Hatcher, H. B. Gutstein, S. S. Rubakhin and J. V. Sweedler, *Proteomics*, 2008, **8**, 3746–3754.
7. D. Touboul, O. Laprevote and A. Brunelle, *Curr. Opin. Chem. Biol.*, **15**, 725–732.
8. O. Yanes, H.-K. Woo, R. Northen Trent, R. Oppenheimer Stacey, L. Shriver, J. Apon, N. Estrada Mayra, J. Potchoiba Michael, R. Steenwyk, M. Manchester and G. Siuzdak, *Anal. Chem.*, 2009, **81**, 2969–2975.
9. G. J. Patti, L. P. Shriver, C. A. Wassif, H. K. Woo, W. Uritboonthai, J. Apon, M. Manchester, F. D. Porter and G. Siuzdak, *Neuroscience*, **170**, 858–864.
10. D. Y. Lee, V. Platt, B. Bowen, K. Louie, C. A. Canaria, C. T. McMurray and T. Northen, *Integr. Biol.*, **4**, 693–699.
11. R. Calavia, F. E. Annanouch, X. Correig, O. Yanes J. Proteomics, Ahead of Print.
12. W. Reindl, P. Bowen Benjamin, A. Balamotis Michael, E. Green Jeffrey and R. Northen Trent, *Integrative biology : quantitative biosciences from nano to macro*, **3**, 460–467.
13. Z.-P. Yao, *Mass Spectrom. Rev.*, **31**, 437–447.
14. G. A. Harris, L. Nyadong and F. M. Fernandez, *Analyst*, 2008, **133**, 1297–1301.
15. A. Venter, M. Nefliu and R. G. Cooks, *TrAC, Trends Anal. Chem.*, 2008, **27**, 284–290.
16. M.-Z. Huang, C.-H. Yuan, S.-C. Cheng, Y.-T. Cho and J. Shiea, *Annu. Rev. Anal. Chem.*, 2010, **3**, 43–65.
17. G. Marko-Varga, T. E. Fehniger, M. Rezeli, B. Doeme, T. Laurell and A. Vegvari, *J. Proteomics*, **74**, 982–992.
18. S. R. Shanta, Y. Kim, Y. H. Kim and K. P. Kim, *Biomol. Ther.*, **19**, 149–154.
19. N. Takai, Y. Tanaka, K. Inazawa and H. Saji, *Rapid Commun. Mass Spectrom.*, 2012, **26**, 1549–1556.
20. S. Castellino, M. R. Groseclose and D. Wagner, *Bioanalysis*, **3**, 2427–2441.

21. F. Deutskens, J. Yang and M. Caprioli Richard, *J. Mass Spectrom.*, **46**, 568–571.
22. P. Nemes and A. Vertes, in *Mass Spectrometry Imaging*, ed. S. S. Rubakhin and J. V. Sweedler, Springer Science + Business Media, 2010, pp. 159–171.
23. P. Nemes, A. S. Woods and A. Vertes, *Anal. Chem.*, 2010, **82**, 982–988.
24. M.-Z. Huang, S.-S. Jhang, C.-N. Cheng, S.-C. Cheng and J. Shiea, *Analyst*, 2010, **135**, 759–766.
25. L. S. Eberlin, X. Liu, C. R. Ferreira, S. Santagata, N. Y. R. Agar and R. G. Cooks, *Anal. Chem.*, **83**, 8366–8371.
26. Q. Blatherwick Eleanor, J. Van Berkel Gary, K. Pickup, K. Johansson Maria, M.-E. Beaudoin, O. Cole Roderic, M. Day Jennifer, S. Iverson, D. Wilson Ian, H. Scrivens James and J. Weston Daniel, *Xenobiotica*, **41**, 720–734.
27. D. Eikel, M. Vavrek, S. Smith, C. Bason, S. Yeh, W. A. Korfmacher and J. D. Henion, *Rapid Commun. Mass Spectrom.*, **25**, 3587–3596.
28. S. Schadt, S. Kallbach, R. Almeida and J. Sandel, *Drug Metab. Dispos.*, **40**, 419–425.
29. G. Robichaud, J. A. Barry, K. P. Garrard and D. C. Muddiman, *J. Am. Soc. Mass Spectrom.*, 2013, **24**, 92–100.
30. A. S. Galhena, G. A. Harris, L. Nyadong, K. K. Murray and F. M. Fernandez, *Anal. Chem.*, **82**, 2178–2181.
31. A. Barry Jeremy and C. Muddiman David, *Rapid Commun. Mass Spectrom.*, 2011, **25**, 3527–3536.
32. P. Nemes and A. Vertes, *Anal. Chem.*, 2007, **79**, 8098–8106.
33. B. Balluff, C. Schoene, H. Hoefler and A. Walch, *Histochem. Cell Biol.*, **136**, 227–244.
34. A. Z. Berry Karin, A. Hankin Joseph, M. Barkley Robert, M. Spraggins Jeffrey, M. Caprioli Richard and C. Murphy Robert, *Chem. Rev.*, **111**, 6491–6512.
35. D. Bonnel, R. Legouffe, N. Willand, A. Baulard, G. Hamm, B. Deprez and J. Stauber, *Bioanalysis*, **3**, 1399–1406.
36. C. L. Carter, C. W. McLeod and J. Bunch, *J. Am. Soc. Mass Spectrom.*, **22**, 1991–1998.
37. R. Casadonte and R. M. Caprioli, *Nat. Protoc.*, **6**, 1695–1709.
38. L. H. Cazares, D. A. Troyer, B. Wang, R. R. Drake and O. John Semmes, *Anal. Bioanal. Chem.*, **401**, 17–27.
39. P. Chaurand, S. Cornett Dale, M. Angel Peggi and M. Caprioli Richard, *Mol. Cell. Proteomics*, **10**, O110 004259.
40. T. Gemoll, U. J. Roblick and J. K. Habermann, *Mol. Med. Rep.*, **4**, 1045–1051.
41. H. Cazares Lisa, A. Troyer Dean, B. Wang, R. Drake Richard and O. J. Semmes, *Anal. Bioanal. Chem.*, **401**, 17–27.
42. M. Lagarrigue, M. Becker, R. Lavigne, S.-O. Deininger, A. Walch, F. Aubry, D. Suckau and C. Pineau, *Mol. Cell. Proteomics*, **10**, M110 005991.

43. E. H. Seeley and R. M. Caprioli, *Trends Biotechnol.*, **29**, 136–143.
44. A. Thomas, L. Charbonneau Jade, E. Fournaise and P. Chaurand, *Anal. Chem.*, **84**, 2048–2054.
45. W. Bouschen, O. Schulz, D. Eikel and B. Spengler, *Rapid Commun. Mass Spectrom.*, **24**, 355–364.
46. R. C. Murphy, J. A. Hankin, R. M. Barkley and K. A. Zemski Berry, *Biochim. Biophys. Acta, Mol. Cell Biol. Lipids*, **1811**, 970–975.
47. J. Yang and R. M. Caprioli, *Anal. Chem.*, 2011, **83**, 5728–5734.
48. I. Apitz and A. Vogel, *Appl. Phys. A*, 2005, **81**, 329–338.
49. J. Chen, B. Green-Church Kari and K. Nichols Kelly, *Invest. Ophthalmol. Visual Sci.*, **51**, 6220–6231.
50. E. Shine Ward and P. McCulley James, *Curr. Eye Res.*, 2003, **26**, 89–94.
51. Z. Zhao, X. Wei, Y. Aliwarga, N. A. Carnt, Q. Garrett and M. D. P. Willcox, *Mol. Vision*, 2008, **14**, 2016–2024.
52. S. H. J. Brown, L. H. Huxtable, M. D. P. Willcox, S. J. Blanksby and T. W. Mitchell, *Analyst*, **138**, 1316–1320.
53. C. Maldonado-Codina, B. Morgan Philip, N. Efron and J.-C. Canry, *Optometry and vision science*, 2004, **81**, 455–460.
54. Y. Schober, S. Guenther, B. Spengler, A. Rompp, *Anal. Chem.*, Ahead of Print.
55. S. S. Rubakhin, W. T. Greenough and J. V. Sweedler, *Anal. Chem.*, 2003, **75**, 5374–5380.
56. K. J. Boggio, E. Obasuyi, K. Sugino, S. B. Nelson, N. Y. R. Agar and J. N. Agar, *Expert Rev. Proteomics*, **8**, 591–604.
57. F. Deutskens, J. Yang and R. M. Caprioli, *J. Mass Spectrom.*, **46**, 568–571.
58. T. Masujima, *Anal. Sci.*, 2009, **25**, 953–960.
59. A. F. M. Altelaar, S. L. Luxembourg, L. A. McDonnell, S. R. Piersma and R. M. A. Heeren, *Nat. Protoc.*, 2007, **2**, 1185–1196.
60. A. F. M. Altelaar and S. R. Piersma, *Methods Mol. Biol.*, **656**, 197–208.
61. H. F. Arlinghaus, C. Kriegeskotte, M. Fartmann, A. Wittig, W. Sauerwein and D. Lipinsky, *Appl. Surf. Sci.*, 2006, **252**, 6941–6948.
62. S. Fletcher John, S. Rabbani, A. Henderson, P. Blenkinsopp, P. Thompson Steve, P. Lockyer Nicholas and C. Vickerman John, *Anal. Chem.*, 2008, **80**, 9058–9064.
63. B. Shrestha.
64. B. Shrestha, J. M. Patt and A. Vertes, *Anal. Chem.*, **83**, 2947–2955.

CHAPTER 20

Liquid Extraction Surface Analysis Mass Spectrometry (LESA MS): Combining Liquid Extraction, Surface Profiling and Ambient Ionization Mass Spectrometry in One Novel Analysis Technique

DANIEL EIKEL* AND JACK D. HENION

Advion Inc., 10 Brown Rd, Ithaca, NY 14850, USA
*Email: deikel@advion.com

20.1   Introduction

In this chapter we give an overview of liquid extraction surface analysis mass spectrometry (LESA-MS), a novel analysis technique that combines liquid extraction from a surface of interest and ambient nanoelectrospray ionization combined with mass spectrometry to analyze compounds of interest. LESA MS was first described by Kertesz and Van Berkel[1] in 2009 and subsequently made commercially available by Advion Inc. by way of its TriVersa-NanoMate™ robotic nanoelectrospray ionization source (www.advion.com, Figure 20.1). LESA was initially intended as a complementary

New Developments in Mass Spectrometry No. 2
Ambient Ionization Mass Spectrometry
Edited by Marek Domin and Robert Cody
© The Royal Society of Chemistry 2015
Published by the Royal Society of Chemistry, www.rsc.org

Figure 20.1 The Advion TriVersa-NanoMate is currently the only commercially available system that can perform LESA infusion MS analysis.

analysis technique to MALDI imaging in pharmaceutical drug distribution and development; however, soon after the commercial availability of this technique, a broader use became apparent with applications ranging from biofilms on contact lenses,[2] antibiotics expressed by bacteria cultured in agar,[3] dried blood spot analysis,[4,5] surface properties of aged plastics[6] and aerosols from compactor material[7] – to mention only a few. Here, we will discuss selected applications and provide an outlook of LESA developments as they are currently unfolding, knowing full well that such a new technology will develop unexpectedly and in application areas not previously envisioned.

20.2 What Exactly is LESA?

LESA experiments usually require a robotic system to pick up a sample tip, aspirate a defined volume of extraction solvent from a reservoir (1–2 μL) and move the tip above the desired location on the target area of interest. The TriVersa-NanoMate robot, for example, can do so with 50 μm accuracy on a target surface and a distance to the surface usually within 200 μm. It then dispenses the extraction solvent onto the target, forming a microliquid junction, and a static microextraction process commences. Extraction time on the surface, repeated dispense/aspirate steps as well as wait times between the extraction cycles are all programmable and influence the extraction (Figure 20.2).

Following the extraction on the sample surface, the extract is delivered to a single nanoelectrospray microfabricated emitter located within an array of 400 emitter nozzles of a deep reactive ion etching (DRIE) microfabricated silicon chip, a voltage is applied and ions are generated by the

Step 1 – Solvent Selection
Extraction solvent is chosen and a disposable tip picks up the micro liter volume desired from a cooled reservoir

Step 2 – Analyte Extraction
Robot places extraction solvent on the surface, a liquid junction forms and aspirate/dispense cycles are initiated

Step 3 –Ionization
Robot delivers extracted analytes to a 400 nozzle ESI chip and a nano electrospray is generated

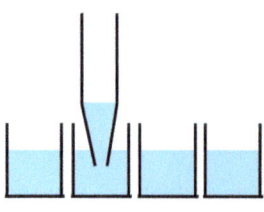  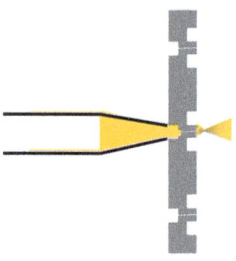

Figure 20.2 Schematic representation of the liquid-extraction surface analysis (LESA) workflow. Step 1 – Extraction solvent is robotically aspirated from the reservoir of choice. Step 2 – Solvent is placed on the surface, a liquid junction forms and the static extraction process starts. Step 3 – The extract is delivered to the nESI chip, a voltage applied and electrospray ionization commences with analysis *via* mass spectrometry. Reprinted and modified from ref. 27 with permission from John Wiley & Sons, Ltd.

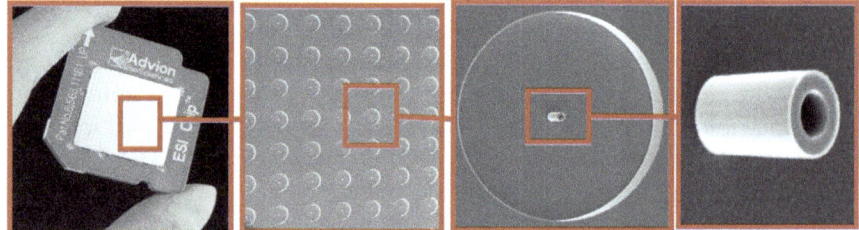

Figure 20.3 The enabling part of the Advion TriVersa-NanoMate technology is the microliter handling robot in combination with the nanoelectrospray chip, which consumes only about 400 nL solvent per minute spray time (a typical LESA experiment will result in *ca.* 5 min spray time) – sufficient for complex MS analysis from the surface of interest. The DRIE etched chip provides an array of 400 nozzles with dimensions of 5.5 μm ID, 28 μm OD and an emitter height from the surface of 55 μm.

nanoelectrospray process (Figure 20.3) that can subsequently be analyzed by a mass spectrometer.

The core of the analysis technique is the extraction of analytes into a very small volume of solvent in order to maintain a high analyte concentration (electrospray mass spectrometry is considered a concentration-dependent analysis technique) and the chip-based emitter system that allows an extended spray time based on the low volumes handled by the robot. A typical LESA experiment, utilizing 2 μL of spray solvent, can generate an MS data acquisition time of 4–5 min. The generated ions can be analyzed by a wide variety of MS platforms (so long as they are compatible with the TriVersa-NanoMate) running experiments such as IDA, MS^n, ultrahigh resolution, selected reaction monitoring (SRM) or any other combination of scan modes and fragmentations (collision-induced dissociation CID, electron transfer dissociation ETD, *etc.*) the respective MS provides. This allows for significantly improved data quality from the analytical surface of interest compared to other surface analysis approaches.

The formation of the microliquid junction on the sample surface also defines the spatial resolution of the LESA experiment and is dependent on the interplay of surface tensions from the processing tip, the extraction solvent and the surface itself. Using a typical solvent mixture of methanol/water 80/20, a spatial resolution of, *e.g.*, 0.8–1.0 mm can be achieved when sampling from a cryosectioned mouse brain (Figure 20.4).

Another unique aspect of LESA is the variety of extraction solvents usable (methanol, THF, iso-propanol, chloroform, ethylacetate, *etc.*) in order to optimize extraction and recovery of the analyte in question, the matrix of the surface, the hydrophobicity and porosity of the surface and the mode of ionization desired (positive and/or negative-ion mode). The restricting factor is the generation of a stable electrospray and the requirement of the spatial resolution of the experiment. Modifier addition such as formic acid or ammonium hydroxide in 0.1 vol% can make a significant difference to the ionization efficiency of the analyte and the overall sensitivity of the assay.

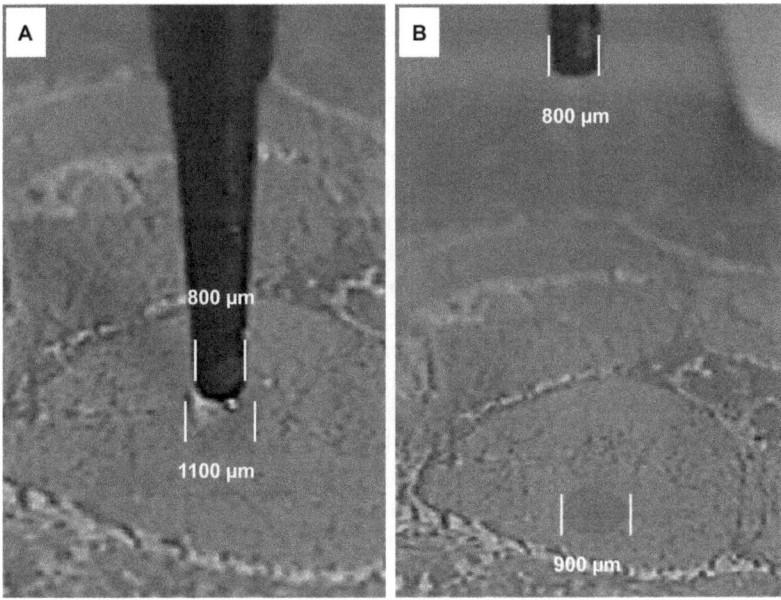

Figure 20.4   Example of a microliquid junction formed on a cryosection of brain during the LESA process. The area sampled is about 1.0 mm using an extraction solvent of 80/20 Methanol/Water 0.1 vol% formic acid. (A) The solvent drop shows a size of about 1.1 mm measured relative to the known tapered end of the pipette tip of 0.8 mm. (B) After the LESA process an area of about 0.9 mm appears wetted.
Reprinted from ref. 18 with permission from John Wiley & Sons, Ltd.

20.3   LESA Applied in the Field of Drug Distribution and Development

Mass-spectrometry imaging[8] (MSI) has become a major focus in studying the distribution of new chemical entities in the whole body of laboratory animals. Although quantitative whole-body autoradiography (QWBA) has traditionally been used for this application and offers both high spatial resolution (*ca.* 30 um) as well as quantitative data, it is limited to compounds and metabolites with a radiolabel, which are sometimes difficult and/or expensive to obtain. Any analysis technique utilizing a mass spectrometer as a detector could potentially overcome some of these limitations and analyze a range of compounds in the very same experiment with both qualitative and quantitative results. MSI techniques (ambient or otherwise) include analytical approaches such as matrix-assisted laser desorption ionization mass spectrometry (MALDI MS),[9–11] desorption electrospray ionization mass spectrometry (DESI-MS),[12] secondary ion mass spectrometry imaging (SIMS-imaging),[13] laser-ablation electrospray ionization (LAESI),[14] electrospray-assisted laser desorption/ionization MS[15] and others not explicitly mentioned here. The LESA-MS approach is one of the newer mass-spectrometry imaging techniques, with distinct characteristics compared to QWBA or other MSI methods in that no additional sample preparation is required, a soft ionization process is used and high-quality data based on the various scan modes or fragmentations the MS is capable of.

A good first example of this area of application is the spatial distribution of fluticasone, a topical corticosteroid used in the treatment of asthma, in lung tissue. ESI MS/MS fragmentation analysis shows a rich fragmentation pattern suitable for qualitative identification and quantification in both positive as well as negative nanoelectrospray MS/MS ion mode. However, the negative-ion mode shows a *ca.* 30-fold increase in signal and a *ca.* 6-fold improvement in signal-to-noise (data not shown). Here, fluticasone was dosed *ex vivo* into a mechanically vented and perfused guinea pig lung model to show its distribution pattern within the lung.[16] The lung was sectioned and LESA-SRM-MS profiling of the complete lung sections (Figure 20.5) showed one area of highest signal intensity in the longitudinal section A and a double peak signal in the vertical section B. This is consistent with a drug distribution throughout the trachea and into the bronchial part of the lung and is the expected result for an antiasthma drug reaching its target tissue by inhalation. Such a model system could be used to both screen for drugs or to optimize applicators for inhalants to, *e.g.*, optimize for drug localization, particle size and distribution within the lung.

Others have shown a skin penetration study of glucocorticoid receptor agonists *via* LESA.[17] The comparison of depth penetration of three test compounds into pig's ears was used to explain prior unusual findings with regard to a positive skin-blanching test for all three compounds and conflicting *in vitro* data regarding their agonist potency test results.

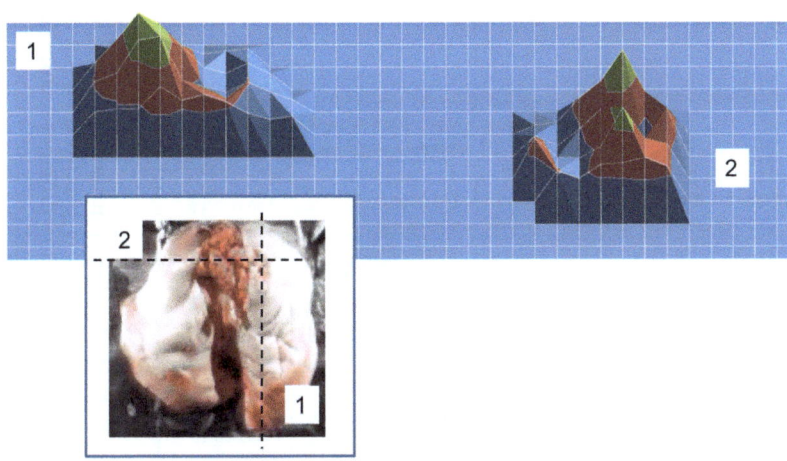

Figure 20.5 LESA-MS-SRM in negative-ion mode nESI profiling of fluticasone (SRM *m/z* 499.3 > *m/z* 413.2) in lung tissue after aerosol dosage into a mechanically vented and perfused guinea pig lung model. Distribution pattern shows a single area with intense analyte response along the longitudinal section A and two areas with strong analyte signal along the vertical section B. This is consistent with a drug distribution through the trachea and into the bronchial part of the lung only, the desired outcome for this topical corticosteroid drug used in asthma treatment.
Reprinted from ref. 36 with permission from John Wiley & Sons, Ltd.

As discussed above, LESA currently provides limited spatial resolution. However, in the case of whole-body drug distribution of drugs in dosed rodents, the respective tissues of interest are comparably large and a sampling from a limited number of locations on each organ and from three section depths is sufficient to gain a rapid understanding of the drug's distribution. A typical example is the distribution of terfenadine and its metabolite after 50 mg kg^{-1} PO terfenadine dosage to a mouse[18] (Figure 20.6). The drug is extensively metabolized and both drug and metabolite are distributed throughout the body with a distribution pattern supporting the assumption of hepatic clearance as well as a smaller renal component to the clearance. Compared to LC-MS/MS experiments it was estimated that LESA-MS was able to detect terfenadine and fexofenadine in the low pmol g^{-1} concentration in various tissues; however, this sensitivity will be compound and tissue dependent. Other publications in this area cover a range of drugs and metabolites from both phase I as well as phase II; *e.g.* GSH-adducts from propanolol,[1] phase I metabolites of figopitant,[19] hydroxylation and glucuronidation of diclofenac[20,21] and phase I metabolites from chloroquine.[22]

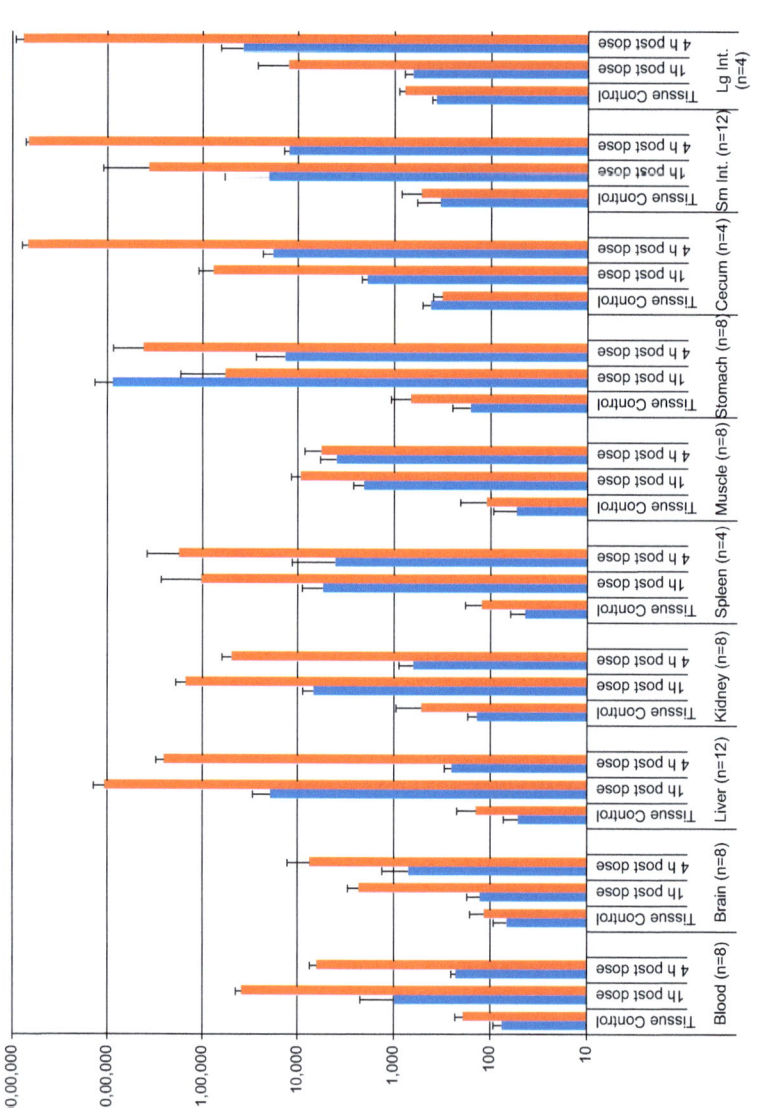

Figure 20.6 LESA-MS-SRM analysis of terfenadine (black columns, SRM *m/z* 472.3 > *m/z* 436.4) and fexofenadine (grey columns, SRM *m/z* 502.3 > *m/z* 171.1) in the positive-ion mode at different tissues/organs of mouse whole-body cryosections and at different times after P.O. dose of 50 mg kg⁻¹ terfenadine. Average cps signal of 4 to 12 locations from up to 3 different section depths of each organ and the respective standard deviation is shown. Drug-distribution pattern is in excellent agreement with known pharmacokinetic properties of the drug and its metabolite. Reprinted from ref. 18 with permission from John Wiley & Sons, Ltd.

20.4 LESA in other Biologically Oriented Applications

In the original LESA publication from Kertesz and Van Berkel[1] it had already been shown that LESA could also be utilized to analyze sitamaquine *via* punch outs from dried blood spot cards. However, in order to directly sample from the DBS card, a silicone spray treatment was required to facilitate the formation of the liquid junction.[23] A similar approach was used to quantify hydrochlorothiazide from dried blood spots later on.[24] LESA analysis from DBS cards for the analysis of hemoglobin variants from blood was recently demonstrated without the need for a silicone pretreatment.[5] Here, the authors showed the successful identification of a variety of hemoglobin variants, potentially allowing for a quick and simple screening tool for sickle-cell and thalassemia diseases.

In an attempt to analyze new antibiotics from bacterial agar cultures a solvent mixture containing ethyl acetate/acetone (10 µL aspirated to the tip, 65 : 35, v/v) containing 0.1% formic acid was used in the LESA extraction process.[3] This somewhat unusual solvent mixture allowed the formation of a liquid junction on the agar and at the same time allowed the generation of ions in the electrospray process. The MS data generated allowed for interrogation of antibiotics molecule production, bacterial signaling and function and could be used for screening purposes from cultured bacteria in the future.

LESA has also been used for shotgun lipidomic approaches for the characterization of artherosclerotic plaques.[25] Here, the LESA technique was compared to a standard liquid–liquid tissue extraction approach for shotgun lipidomics and the authors found equivalent data quality between the two approaches and could present comprehensive lipid profiling of artherosclerotic plaques and surrounding tissue. Lipid analysis *via* LESA was further shown to be useful for profiling of different brain regions of a rat. A comparison between LESA, MALDI imaging and a third approach utilizing laser microdissection followed by liquid extraction appears to favor the LESA approach in experimental simplicity and data quality obtained from brain tissue.[26]

20.5 LESA Applied to Challenging Surfaces and Applications

Although application areas are not clearly separated from each other, the type of surface analyzed usually provides reasonable distinction criteria. In a recent publication the analysis of biofilms forming on contact lenses was shown[2] – an interesting mix between the shotgun analysis of lipids (which formed the biofilm) and the underlying polymer (which formed the contact lens). This surface was challenging to analyze due its concave or convex shape; however, solvent conditions and a mounting scheme were found to successfully execute the LESA experiments.

Another interesting publication is the LESA analysis to identify pesticides from fruits and vegetables.[27] Here, the type of surface ranged from hard to smooth (*e.g.* apple pieces, kiwi skin, lettuce and dried fruit pulp) as well as shaped in usually convex form (*e.g.* grape halves or orange peels). The overall method sensitivity was sufficient to detect a variety of pesticides in the 100 ppb range in fruit pulp and 20-fold level below the EPA tolerance level directly from the fruit skin surface.

Direct analysis of pharmaceutical tablets is also possible using LESA-MS.[28] Here the challenge was not so much the surface shape but rather the porosity of the tablet on its surface, thus making the formation of the micro-liquid-junction challenging. The authors found LESA to be favorable compared to DART applied to the same samples, and point out the no carryover strategy using one tip–one sample–one nozzle as well as the method simplicity as strong points. However, sample porosity is certainly a limiting factor of the LESA analysis approach, as shown for toxicology applications testing the surface of feed pellets.[29] Feed had to be ground up and compressed into a wafer or the feed pellets had to be pretreated with silicone (see above also for dried blood spots) to enable the formation of a liquid junction and LESA analysis of, *e.g.*, 10 ppm lasalocid in alpaca feed. A small wood piece from dog stomach vomitus on the other hand was successfully analyzed without any prior treatment and Roquefortin C was readily confirmed as one of the two suspected toxins causing acute tremor in dog (Figure 20.7).

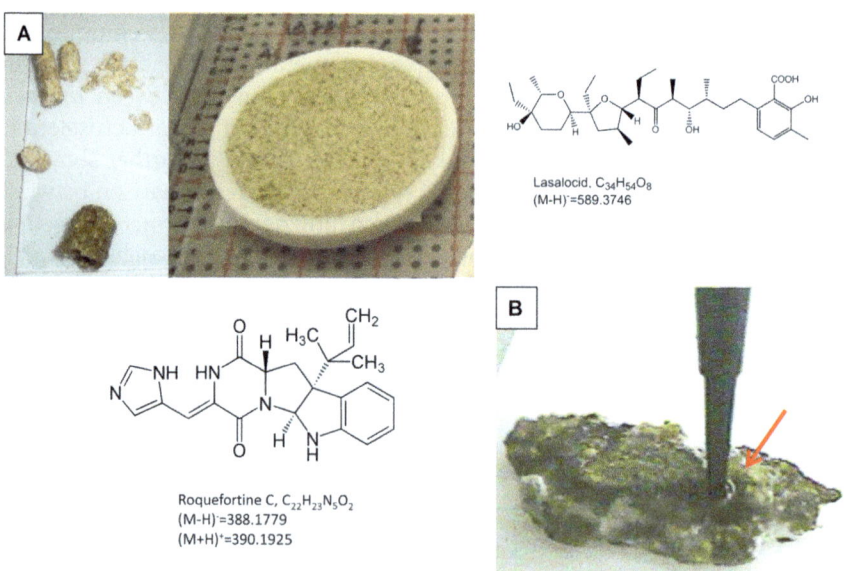

Lasalocid, $C_{34}H_{54}O_8$
$(M-H)^-=589.3746$

Roquefortine C, $C_{22}H_{23}N_5O_2$
$(M-H)^-=388.1779$
$(M+H)^+=390.1925$

Figure 20.7   LESA-MS applied to porous surfaces of toxicological interest. (A) feed pellets required either a silicone pretreatment or grinding and pressing into a feed waver. From pretreated samples, Lasalocid was detected in Alpaca feed at 10 ppm. (B) Roquefortin C was confirmed *via* LESA-HRMS from dog vomitus of a small piece of wood.
Reprinted from ref. 36 with permission from John Wiley & Sons, Ltd.

Another recent publication showed the use of LESA to automatically extract analytes from SPE cards in a 96-well format.[30] Both qualitative and quantitative analysis of herbicides, peptides and small-molecule drugs allowing parallel off-line sample preparation was demonstrated. Along similar lines, another research group recently analyzed collected aerosols on a rotating drum compactor applying the LESA technology.[7] The translation from the time to the spatial domain using the rotating drum and the following LESA analysis allowed a higher temporal sampling resolution with an indepth mass profiling of aerosols collected during the experiment time than available before.

A direct LESA analysis of the surface of sun-aged plastic helped to identify the underlying chemical process in the observed discoloration of exposed plastic sheets.[6] Here, the MS^n type of experiment was the enabling factor to determine which plastic component was degrading over time and contributing to its aging. Other studies are currently focusing on pigments and leachable chemicals from packaging materials, medical devices or plastic containers. We are not aware of scientific peer-reviewed publications on these particular topics as of this writing.

20.6 Outlook and Future Development of LESA

Aside from the broad range of new applications for LESA covering a variety of surface types and scientific questions from pharmacology, toxicology, material science or biology, there are other developments that will likely influence how this new ambient analysis tool will be utilized in the future. As mentioned above, the spatial resolution of LESA is currently limited to about 1.0 mm and is generally dependent on the surface tension interplay between the three physical objects involved (surface, solvent and sampling tip).

In a recent publication it was shown that using a large ID fused-silica capillary instead of a plastic tip one can achieve a resolution of *ca.* 200 µm on target[31] using the existing TirVersa-NanoMate robotic system. This would bring LESA spatial resolution closer to inner organ structures and within typical resolutions run by MALDI, DESI or QWBA.

In this particular LESA setup, the extract was also injected onto a nano-LC column and a chromatographic separation was coupled with LESA, allowing for separation of different lipids as well as separation of lipids from matrix.[31] As expected, the LESA-LC-MS approach can increase the sensitivity of LESA since it reduces matrix-ion suppression and concentrates the analyte into a narrow chromatographic peak. A similar approach was also used with a set-up consisting of a LEAP autosampler as the LESA sampling device. Here, the sampling size on target was *ca.* 1 mm and the extracted solvent was injected into an HPLC-MS system where propranolol and its metabolites were detected from the thin tissue section.[32]

Both smaller sampling size as well as the LC addition comes at the cost of increased analysis time given the extra time needed for chromatographic separation *versus* direct infusion and the handling of an increased number of locations per sample area. The latter is the classical conundrum of all mass-spectrometry imaging (MSI) methods striving for higher spatial resolution. One could envision a future LESA system though that would allow different modes of operation using lower-resolution drug profiling with an emphasis on whole-body sections and speed of analysis as well as a second analysis mode with higher spatial resolution combined with LC for added sensitivity and chromatographic separation on only selected locations on the target.

The second interesting development describes a LESA extraction step followed by an enzymatic or chemical modification of the analytes in solution prior to MS analysis. In such a work flow, *e.g.* a DBS card was extracted *via* LESA and the proteins contained were digested by trypsin prior to LC-MS/MS analysis of the resulting peptides.[4] The authors conclude that in this workflow the DBS card can provide a suitable sampling strategy for proteomic type analysis approaches. A very similar LESA extraction approach from FFPE tissues was also published.[33]

Both of these discussed applications and further developments of LESA point to the advantage of LESA in extracting the analyte of interest from the surface under ambient conditions into a liquid phase – which can subsequently be further manipulated as needed. A unique capability compared to other ambient ionization techniques where the ionization/extraction process is directly linked with each other (such as MALDI or DESI).

20.7   Conclusions

Liquid-extraction surface sampling (LESA) is a relatively new ambient surface analysis and ionization technique and as such has attracted considerable interest since its first presentation.[1] Despite its novelty, a commercial realization is available and LESA-MS is used in a variety of applications ranging from drug profiling in development and discovery to investigation of a variety of technical or biological surfaces as shown with selected examples above.

Within the toolbox of MSI approaches, LESA-MS has unique characteristics and nicely complements the analytical capabilities otherwise available to researchers as reviewed recently.[34–36] Its ease-of-use and fit-for-purpose approach make it readily adoptable by many labs and the significant increase of both contributions to conferences as well as publications in peer-reviewed journals shows the interest of the scientific community.

One of the strong points of LESA certainly is the fact that analyte extraction and ionization are separated from each other and can be optimized by choosing different solvent compositions and modifiers. In combination with the soft ionization and robustness against matrix effects that nanoelectrospray provides,[37] LESA appears suitable for a significant portion of the compound space including drugs, metabolites and toxins. The decoupling of extraction and ambient ionization process also allows for interesting future developments where the analyte is enzymatically or chemically modified after the extraction but prior to MS analysis and/or analytes are separated using a chromatographic approach prior to MS analysis.

The TriVersa-NanoMate LESA instrumentation is compatible with most commercially available MS systems and its nanoelectrospray chip is the enabling technology for LESA since nanoelectrospray not only reduces ion matrix suppression effects,[37] but also generates an extended electrospray time from a very small volume of solvent, which can be used for extensive MS experiments allowing indepth analysis of the surface of interest.

The limitations of LESA-MS lay in the *ca.* 1 mm spatial resolution, thus limiting the ability to investigate smaller regions within biological structures of interest – a reason why LESA should be considered to be a profiling approach. This spatial resolution can be improved to *ca.* 200 μm resolution. However, as with other MSI approaches, this raises the question of analysis time, data file size generated and sensitivity when sampling from smaller areas of a surface.

Another crucial point for LESA is the static extraction process from, *e.g.*, thin tissue sections. This process is currently not well characterized or understood, making the optimization an empirical process. And although sensitivity comparisons to QWBA, MALDI imaging, DESI or DART show favorable results for LESA-MS analysis, a better understanding of the extraction process is still needed to improve extraction efficiency and drive sensitivity. Also, LESA depends on the formation of the liquid junction itself, which requires a surface that is sufficiently hydrophobic for the solvent used for extraction.

It is interesting to note that there is an increasing number of publications in nonclassical LESA application areas (if one can describe any use of a 5 year young analytical technique as classic) with interesting scientific questions addressed in material science, biology and chemistry.

In summary, we believe LESA MS is an exciting novel approach in the field of ambient ionization mass spectrometry that complements other tools available to scientists in diverse fields of research.

References

1. V. Kertesz and G. J. Van Berkel, *J. Mass Spectrom.*, 2010, **45**, 252.
2. S. H. Brown, L. H. Huxtable, M. D. Willcox, S. J. Blanksby and T. W. Mitchell, *Analyst*, 2013, **138**, 1316.
3. M. Kai, I. González, O. Genilloud, S. B. Singh and A. Svatoš, *Rapid Commun. Mass Spectrom.*, 2012, **26**, 2477.
4. N. J. Martin, J. Bunch and H. J. Cooper, *J. Am. Soc. Mass Spectrom.*, 2013, **24**, 1242.
5. R. L. Edwards, A. J. Creese, M. Baumert, P. Griffiths, J. Bunch and H. J. Cooper, *Anal. Chem.*, 2011, **6**, 2265.
6. M. R. Paine, P. J. Barker, S. A. Maclauglin, T. W. Mitchell and S. J. Blanksby, *J. Am. Soc. Mass Spectrom.*, 2012, **26**, 412.

7. S. J. Fuller, Y. Zhao, S. S. Cliff, A. S. Wexler and M. Kalberer, *Anal. Chem.*, 2012, **84**, 9858.
8. S. S. Rubakhin and J. V. Sweedler, *Methods in Molecular Biology*, ed. J. M. Walker, Springer, New York, 1st edn, 2010, p. 487.
9. L. A. McDonnell and R. M. Heeren, *Mass Spectrom. Rev.*, 2007, **26**, 606.
10. P. Chaurand, S. A. Schwartz, M. L. Reyzer and R. M. Caprioli, *Toxicol. Pathol.*, 2005, **33**, 92.
11. P. Chaurand, S. A. Schwartz and R. M. Caprioli, *Curr. Opin. Chem. Biol.*, 2002, **6**, 676.
12. J. M. Wiseman, D. R. Ifa, Y. Zhu, C. B. Kissinger, N. E. Manicke, P. T. Kissinger and R. G. Cooks, *PNAS*, 2008, **105**, 18120.
13. M. S. Burns, *J. Microsc.*, 1982, **127**, 237.
14. P. Nemes and A. Vertes, *Anal. Chem.*, 2007, **21**, 8098.
15. J. Shiea, M. Z. Huang, H. J. Hsu, C. Y. Lee, C. H. Yuan, I. Beech and J. Sunner, *Rapid Commun. Mass Spectrom.*, 2005, **19**, 3701.
16. D. Eikel, C. Alpha, G. S. Rule, S. J. Prosser, J. D. Henion, Liquid Extraction Surface Analysis (LESA) combined with nano-electrospray MS as a novel tool in early ADME studies of drug candidates. 58th Conference of the American Society for Mass Spectrometry 2010 – oral contribution.
17. P. Marshall, V. Toteu-Djomte, P. Bareille, H. Perry, G. Brown, M. Baumert and K. Biggadike, *Anal Chem*, 2010, **18**, 7787.
18. D. Eikel, M. Vavrek, W. A. Korfmacher and J. D. Henion, *Rapid Commun. Mass Spectrom.*, 2011, **25**, 3587.
19. S. Schadt, S. Kallbach, R. Almeida and J. Sandel, *Drug Metab. Dispos.*, 2012, **40**, 419.
20. D. Cox, T. Covey, J. C. Y. Leblanc, B. Schnieder, G. J. VanBerkel, V. Kertesz, P. Moench, J. Flarakos, Analysis of Tissue Samples by Liquid Extraction Surface Analysis (LESA) Coupled to Differential Ion Mobility (DMS) High Resolution MS/MS. 59th Conference of the American Society for Mass Spectrometry 2011 – poster contribution.
21. S. J. Prosser, D. Eikel, S. T. Linehan, K. Murphy, D. Heller, P. J. Rudewicz, J. D. Henion, Liquid Extraction Surface Analysis Mass Spectrometry (LESA MS): Drug Distribution and Metabolism of Diclofenac in Mouse. 59th Conference of the American Society for Mass Spectrometry 2011 – oral contribution.
22. W. B. Parson, S. L. Koeniger, R. W. Johnson, J. Erickson, Y. Tian, C. Stedman, A. Schwartz, E. Tarcsa, R. Cole and G. J. Van Berkel, *J. Mass Spectrom.*, 2012, **47**, 1420.
23. J. J. Stankovich, M. J. Walworth, V. Kertesz, R. King, G. J. VanBerkel, Liquid Microjunction Surface Sampling Probe Analysis of Dried Blood Spots using an Automated Chip-Based Nano-ESI Infusion Device. 58th conference of the American Society of Mass Spectrometry 2010 – oral contribution.
24. J. Henion, D. Eikel, G. Rule, J. Vega, S. Prosser, J. Jones, Liquid Extraction Surface Analysis (LESA) of Dried Blood Spot Cards via Chip-Based Nanoelectrospray for Drug and Drug Metabolite Monitoring

Studies. 58th conference of the American Society of Mass Spectrometry 2010 – oral contribution.

25. C. Stegemann, I. Drozdov, J. Shalhoub, J. Humphries, C. Ladroue, A. Didangelos, M. Baumert, M. Allen, A. H. Davies, C. Monaco, A. Smith, Q. Xu and M. Mayr, *Circ.: Cardiovasc. Genet.*, 2011, **3**, 232.
26. R. Taguchi, K. Ikeda, Y. Tajima, Several different approaches for the analysis in localization profile of lipid molecular species by mass spectrometry. 59th conference of the American Society of Mass Spectrometry 2011 – poster contribution.
27. D. Eikel and J. D. Henion, *Rapid Commun. Mass Spectrom.*, 2011, **16**, 2345.
28. A. D. Ray, D. J. Weston, E. Q. Blatherwick, J. R. Snelling, J. H. Scrivens, Investigating Liquid Extraction Surface Analysis (LESA) for Rapid Analysis of Pharmaceutical Tablets. Annual meeting of the British Mass Spectrometry Society 2010 – poster contribution.
29. J. D. Henion, D. Eikel, Y. Li, J. Ebel, Automated Surface Sampling of Veterinary Toxicological Samples with Analysis by Nano Electrospray Mass Spectrometry. 59th conference of the American Society of Mass Spectrometry 2011 – poster contribution.
30. M. J. Walworth, M. S. Elnaggar, J. J. Stankovich, C. Witkowski, J. L. Norris and G. J. Van Berkel, *Rapid Commun. Mass Spectrom.*, 2011, **17**, 2389.
31. R. Almeida, Liquid extraction surface analysis (LESA) combined with nano-electrospray MS for direct sampling of planar tissues. 1st AB SCIEX European Conference on MS/MS 2011, Noordwijkerhout, The Netherlands – oral contribution.
32. V. Kertesz and G. J. Van Berkel, *Bioanalysis*, 2013, **5**, 819.
33. M. Wisztorski, B. Fatou, J. Franck, A. Desmons, I. Farré, E. Leblanc, I. Fournier and M. Salzet, *Proteomics Clin Appl*, 2013, **7**, 234.
34. E. Q. Blatherwick, G. J. Van Berkel, K. Pickup, M. K. Johansson, M. E. Beaudoin, R. O. Cole, J. M. Day, S. Iverson, I. D. Wilson, J. H. Scrivens and D. J. Weston, *Xenobiotica*, 2011, **8**, 720.
35. V. Kertesz and G. J. van Berkel, 2013 (TBD).
36. D. Eikel and J. D. Henion, in *Mass Spectrometry for Drug Discovery and Drug Development*, ed. W. Korfmacher, Wiley, 1st edn, 2013, ch. 8, pp. 221–238.
37. M. Karas, U. Bahr and T. Duelcks, *Fresenius' J. Anal. Chem.*, 2000, **6**, 669–676.

Subject Index

Page numbers in **bold** indicate a comprehensive treatment of the topic.
Page numbers in *italics* refer to figures and tables.